Mechanical Reliability Improvement

MECHANICAL ENGINEERING
A Series of Textbooks and Reference Books

Founding Editor

L. L. Faulkner

*Columbus Division, Battelle Memorial Institute
and Department of Mechanical Engineering
The Ohio State University
Columbus, Ohio*

Additional Volumes in Preparation

Mechanical Reliability Improvement

Probability and Statistics for Experimental Testing

R. E. Little

The University of Michigan–Dearborn
Dearborn, Michigan, U.S.A.

Assisted by
D. M. Kosikowski

MARCEL DEKKER, INC.   NEW YORK · BASEL

ISBN: 0-8247-0812-1

This book is printed on acid-free paper.

Headquarters
Marcel Dekker, Inc.
270 Madison Avenue, New York, NY 10016
tel: 212-696-9000; fax: 212-685-4540

Eastern Hemisphere Distribution
Marcel Dekker AG
Hutgasse 4, Postfach 812, CH-4001 Basel, Switzerland
tel: 41-61-260-6300; fax: 41-61-260-6333

World Wide Web
http://www.dekker.com

The publisher offers discounts on this book when ordered in bulk quantities. For more information, write to Special Sales/Professional Marketing at the headquarters address above.

Current printing (last digit):
10 9 8 7 6 5 4 3 2 1

PRINTED IN THE UNITED STATES OF AMERICA

To my grandchildren:
Isabella Maria, Sophia Victoria,
James Robert John, and Lucas Roy

Preface

Mechanical reliability analysis is no longer limited to a small collection of classical statistical analyses. The speed of the present generation of micro-computers makes it possible to program and evaluate alternative computer-intensive analyses for each mechanical reliability application of specific interest. Thus computer-intensive analyses are now an indispensable part of improving mechanical reliability.

This is a self-contained mechanical reliability reference/text book. It covers the probability and statistics background required to plan, conduct, and analyze mechanical reliability experiment test programs. Unfortunately this background is not adequately conveyed by a traditional probability and statistics course for engineers because it (1) does not provide adequate information regarding test planning and the associated details of test con-duct, (2) does not employ vector and matrix concepts in stating conceptual statistical models, (3) does not exploit direct analogies between engineering mechanics concepts and probability and statistics concepts, (4) does not exploit the use of microcomputers to perform computer-intensive simula-tion-based, randomization-based, and enumeration-based statistical ana-lyses, and (5) is woefully inept relative to practical mechanical reliability models. This book attempts to overcome each of these fundamental defi-ciencies.

Typesetting costs have traditionally forced authors to use overly succinct nomenclature and notation when presenting probability and statistics
concepts. But 30 years of teaching experience clearly indicates that overly
succinct notation exacts an extremely heavy price in terms of perspective
and understanding. Accordingly, acronyms are employed throughout this
book to convey explicitly the technical presumptions that the traditional
notations are intended to convey implicitly. Although it may take some
time to become comfortable with these acronyms, their use highlights the
technical presumptions that underlie each reliability analysis, thereby providing valuable perspective regarding its applicability and practicality.

Test planning details and orthogonal conceptual statistical models are
presented in Chapters 1 and 2 for completely randomized design test programs with equal replication, and for unreplicated randomized complete
block design and split-plot design experiment test programs. The respective
conceptual statistical models are stated in volume vector notation to demonstrate relevant orthogonality relationships. This presentation provides intuition regarding the construction of the associated orthogonal augmented
contrast arrays. Use of orthogonal augmented contrast arrays in statistical
analysis markedly enhances understanding the mechanics of partitioning
statistically relevant sums of squares and the enumeration of the associated
degrees of freedom.

The enumeration-based and simulation-based microcomputer programs presented in Chapters 3–6 establish and illustrate the probability
and statistics concepts of fundamental interest in mechanical reliability.
Several elementary statistical tests of hypotheses are presented and illustrated. The relationship of these tests of hypotheses to their associated
statistical confidence intervals is explained. Computer-intensive statistical
tests of hypotheses that serve as viable alternatives to classical statistical
tests of hypotheses are also presented. In turn, linear regression analysis is
presented in Chapter 7 using both column vector and matrix notation.
Emphasis is placed on testing the adequacy of the presumed conceptual
regression model and on allocation of test specimens to the particular independent variable values that have statistical advantage.

Chapters 1–7 establish the test planning and probability and statistics
background to understand the mechanical reliability analyses that are presented, discussed, and then illustrated using example microcomputer programs in Chapter 8. Mechanical reliability cannot rationally be separated
from mechanical metallurgy. The appropriate reliability improvement
experiment test program depends on the relevant mode(s) of failure, the
available test equipment, the test method and its engineering objective, as
well as on various practical and economic considerations. Thus, to excel, a
reliability engineer must have the ability to program and evaluate mechan-

ical reliability analyses that are consistent with the actual details of the experiment test program conduct. In particular, it is important that (1) statistically effective test specimen allocation strategies be employed in conducting each individual test, (2) the statistical adequacy of the presumed failure model be critically examined, and (3) the accuracy and precision of the resulting statistical estimates be evaluated and properly interpreted.

R. E. Little

Introduction

The first step in mechanical design for a new product is to synthesize (configure) the product and its components such that it performs the desired function. Design synthesis is enhanced by first recognizing functional analogies among existing designs that are known to perform well in service and then suggesting several alternative designs based on these functional analogies. In turn, when well-defined objective criteria have been employed to compare these alternative designs to establish the design that has the greatest overall advantage, the proposed design can reasonably be viewed as being both feasible and practical. The next step in mechanical design for a new product is to attempt to assure that the proposed design will exhibit adequate reliability in service operation. Tentative assurance of adequate reliability for the new product requires a combination of (1) pseudo-quantitative design analyses that involve analytical bogies such as design allowables and/or factors of safety and (2) laboratory tests involving experimental bogies based on (reasonably) extreme load and environment histories. However, it is imperative to understand that adequate reliability for the new product can be demonstrated only by its actual (future) performance in service. Nevertheless, a combination of pseudo-quantitative design analysis and laboratory testing can generally be employed either to maintain or to improve the reliability of an existing product.

When the mechanical design objective is to maintain the service-proven reliability of a re-designed product, the re-design must meet the analytical and experimental bogies that were met by the present design. However, when the mechanical design objective is to improve the reliability of a re-designed product, the re-design must excel these analytical and experimental bogies. Moreover, the improved laboratory test performance for the re-design must be demonstrated statistically before it is rational to presume that the reliability of the re-design will excel the reliability of the present design. This statistical demonstration is clearly much more credible when (1) the reliability improvement experiment test program is conducted using load and environment histories that are as nominally identical to the actual service load and environment histories as practical and, in particular, (2) all of the respective laboratory test failures are identical in location, mode of failure, and fracture appearance to the failures that presumably will occur in service.

This text is primarily concerned with the statistical analyses of life and strength data generated by reliability improvement experiment test programs. Accordingly, experiment test program planning and probability concepts are presented and discussed before presenting and illustrating various statistical analyses and their mechanical reliability applications.

Nomenclature and Acronyms and Microcomputer Program Index

Nomenclature and Acronyms

anc	aggregated number of cycles (includes both *fnc* and *snc* datum values in a life experiment test program)
ANOVA	(statistical) analysis of variance (see Chapter 6)
APRCRDV's	all possible replicate conceptual random datum values (pertains only to the block and the treatment or treatment combination of specific interest)
APRCREE's	all possible replicate conceptual random experimental errors (pertains to all blocks and all treatments or treatment combinations)
APRCRHNDDV's	all possible replicate conceptual random homoscedastic normally distributed datum values (pertains only to the block and the treatment or treatment combination of specific interest)
APRCRHNDRDV's	all possible replicate conceptual random homoscedastic normally distributed regression datum values (pertains only to the *ivv* of specific interest)
APRCRHNDSDDV's	all possible replicate conceptual random homoscedastic normally distributed stopping distance datum values (pertains only to the *isv* of specific interest)

APRCRHNDEE's	all possible replicate conceptual random homoscedastic normally distributed experimental errors (pertains to all blocks and all treatments or treatment combinations)
APRCRHNDREE's	all possible replicate conceptual random homoscedastic normally distributed regression experimental errors (pertains to all *ivv*'s)
c	a constant or a generic coefficient
cbe	conceptual block effect
cbec	conceptual block effect contrast
cbesc	conceptual block effect scalar coefficient
cbmptie	conceptual block, main-plot treatment interaction effect
cbmptiec	conceptual block, main-plot treatment interaction effect contrast
cbmptiesc	conceptual block, main-plot treatment interaction effect scalar coefficient
cbmptsptie	conceptual block, main-plot treatment, split-plot treatment interaction effect
cbmptsptiec	conceptual block, main-plot treatment, split-plot treatment interaction effect contrast
cbmptsptiesc	conceptual block, main-plot treatment, split-plot treatment interaction effect scalar coefficient
cbsptie	conceptual block, split-plot treatment interaction effect
cbsptiec	conceptual block, split-plot treatment interaction effect contrast
cbsptiesc	conceptual block, split-plot treatment interaction effect scalar coefficient
cbtie	conceptual block, treatment interaction effect
cbtiec	conceptual block, treatment interaction effect contrast
cbtiesc	conceptual block, treatment interaction effect scalar coefficient
ccc	conceptual correlation coefficient
CDF	cumulative distribution function, typically denoted $F(-)$
cdpj	conceptual distribution parameter for a two-parameter CDF whose probability paper is constructed using a *logarithmic* abscissa metric, $j = 1,2$
clp	conceptual location parameter for a (two-parameter) CDF whose probability paper is constructed using a *linear* abscissa metric

$clp0$	conceptual location parameter pertaining to a CDF with more than two conceptual parameters, or conceptual location parameter pertaining to $(ivv)^0$ in simple linear regression
$clp1$	conceptual location parameter pertaining to $(ivv)^i$ in simple linear regression
$clpj$	conceptual location parameter pertaining to $(ivjv\text{'s})^i$ in multiple linear regression, or conceptual location parameter pertaining to $(ivj\text{'s})^i$ in multiple linear polynomial regression
$clpjk$	conceptual location parameter pertaining to $(ivjv\text{'s})^i$ $(ivkv\text{'s})^k$ in multiple linearpolynomial regression
cm	conceptual mean (of a collection of $ctKm$'s)
$cmlp$	*fictitious* conceptual minimum life parameter for a three-parameter distribution
$cmpte$	conceptual main-plot treatment effect
$cmptec$	conceptual main-plot treatment effect contrast
$cmptesc$	conceptual main-plot treatment effect scalar coefficient
$cmptm$	conceptual main-plot treatment mean
$cmptsptie$	conceptual main-plot treatment, split-plot treatment interaction effect
$cmptsptiec$	conceptual main-plot treatment, split-plot treatment interaction effect contrast
$cmptsptiesc$	conceptual main-plot treatment, split-plot treatment interaction effect scalar coefficient
covar	covariance (of paired random variables or statistics)
cp	conceptual parameter (viz., a parameter in a conceptual statistical model)
CRD	Completely Randomized Design experiment test program
$CRDV$'s	conceptual random datum values
$CRHDV$'s	conceptual random homoscedastic datum values
$CRHEE$'s	conceptual random homoscedastic experimental errors
$CRHNDDV$'s	conceptual random homoscedastic normally distributed datum values
$CRHNDEE$'s	conceptual random homoscedastic normally distributed experimental errors
$CRHNDMPTEEE$'s	conceptual random homoscedastic normally distributed main-plot treatment effect experimental errors
$CRHNDMPTSPTIEEE$'s	conceptual random homoscedastic normally distributed main-plot treatment, split-plot treatment interaction effect experimental errors

CRHNDREE's	conceptual random homoscedastic normally distributed regression experimental errors, viz., the deviations of the respective *CRHNDRDV*'s from their associated [mean$_i$(*APRCRHNDRDV*'s)]'s established by the conceptual simple linear regression statistical model
CRHNDSDDV's	conceptual random homoscedastic normally distributed stopping distance datum values (Supplemental Topic 8.F)
CRHNDSDEE's	conceptual random homoscedastic normally distributed stopping distance experimental errors (Supplemental Topic 8.F)
CRHNDSPTEEE's	conceptual random homoscedastic normally distributed split-plot treatment effect experimental errors
CRHNDSubPlotEE's	conceptual random homoscedastic normally distributed sub-plot experimental errors, viz., the *CRHNDEE*'s formed by aggregating the *CRHNDSPTEEE*'s and *CHNDMPTSPTIEEE*'s in an unreplicated split-plot experiment test program
CRSIDV's	conceptual random statistically identical datum values
CRSIEE's	conceptual random statistically identical experimental errors
CSD	conceptual stopping distance (a random variable)
csdm	conceptual statistical distribution mean, viz., the actual value for the mean of the conceptual statistical distribution that is comprised of *APRCRDV*'s in a quantitative (CRD) experiment test program
csmm	conceptual statistical model mean, viz., the actual value for the mean of the conceptual statistical distribution that is comprised of *APRCRDV*'s in a comparative experiment test program
csmmsc	conceptual statistical model mean scalar coefficient
csp	conceptual scale parameter for a (two-parameter) CDF whose probability paper is constructed using a *linear* abscissa metric
cspj	generic conceptual scale pertaining to a CDF with more than one scale parameter, $j = 0, 1, 2 \cdots$
cspte	conceptual split-plot treatment effect
csptec	conceptual split-plot treatment effect contrast
csptesc	conceptual split-plot treatment effect scalar coefficient
csptm	conceptual split-plot treatment mean
cte	conceptual treatment effect
ctec	conceptual treatment effect contrast

ctesc	conceptual treatment effect scalar coefficient
ctKm	technically verbalized as the actual value for the mean of the conceptual sampling distribution comprised of *APRCRHND(Treatment K)DV*'s
ctm	conceptual treatment mean
d	generic duration to failure – a parameter
d*	a specific value of the generic duration to failure
di	duration interval (used in simulation-based microcomputer programs to estimate subsystem reliability)
divv	*different* independent variable values used in a linear regression experiment test program
em	elastic modulus
est(-)	technically verbalized as the estimate of the actual value for the (-)
est[mean(-)]	technically verbalized as the estimate of the actual value for the mean of the conceptual statistical or sampling distribution that consists of all possible replicate realization values for random variable or statistic (-)
est{mean[*APR(-)DV*'s]}	technically verbalized as the estimate of the actual value for the mean of the conceptual statistical distribution that consists of *APR(-)DV*'s (pertains to the block and the treatment or treatment combination, or to the *ivv* of specific interest)
est[var(-)]	technically verbalized as the estimate of the actual value for the variance of the conceptual statistical or sampling distribution that consists of all possible replicate realization values for the random variable or statistic (-)
est{var[*APR(-)EE*'s]}	technically verbalized as the estimate of the actual value for the variance of the conceptual statistical distribution that consists of *APR(-)EE*'s (pertains to all blocks and all treatments or treatment combinations, or to all *ivv*'s of specific interest)
f(-)	generic probability density function (PDF), technically written as *f*(- |*cdp*'s) in which |*cdp*'s is verbalized as given numerical values for the respective *cdp*'s
F(-)	generic cumulative distribution function (CDF), technically written as *F*(- |*cdp*'s) in which |*cdp*'s is verbalized as given numerical values for the respective *cdp*'s
f_a	alternating force amplitude

fnc	number of (alternating stress) cycles to (fatigue) failure – a parametric value
fnc*	a specific number of (alternating stress) cycles to (fatigue) failure
fnc(*pf*)	number of (alternating stress) cycles to (fatigue) failure pertaining to a parametric value for the probability of failure
fnc(*pf**)	number of (alternating stress) cycles for (fatigue) failure pertaining to a specific value for the probability of failure, where *pf* is stated in per cent, e.g., *fnc*(50) is the median number of (alternating stress) cycles to (fatigue) failure
g	generic function (functional relationship)
h	generic function (functional relationship)
H_a	alternative hypothesis
H_n	null hypothesis
HRF	hazard rate function (also called the instantaneous failure rate function IFRF)
i	generic index
IFRF	instantaneous failure rate function (also called the hazard rate function HRF)
isv	initial speed value (the independent variable in a stopping distance experiment test program)
ivv	independent variable value, the abscissa metric in a linear regression experiment test program
ivv*	specific value for the independent variable in simple linear regression *ivv*
ivv_i	the ith *ivv* used in conducting the linear regression experiment test program
j	generic index
k	generic index
kps	Kendall's positive score test statistic value
kr	index for n_{rkdivv} in linear regression, viz., the number of replicate tests conducted at the kth different independent variable value *divv*
ktau	Kendall's *tau* test statistic value
lsd	least significant difference (the test statistic in Fisher's protected t test)
m	generic index
mean(-)	technically verbalized as the actual value for the mean of the conceptual statistical or sampling distribution that consists of all possible realization values for the random variable or statistic (-)

mean[$APR(\text{-})DV$'s]	technically verbalized as the actual value for the mean of the conceptual statistical distribution that consists of $APR(\text{-})DV$'s (pertains to the block and the treatment or treatment combination, or to the ivv of specific interest)
mpd	minimum practical difference
mpr	minimum practical ratio
(MS)	mean square $= (SS)/n_{sdf}$, where n_{sdf} is the $(SS)_{sdf}$, viz., the number of statistical degrees of freedom pertaining to the associated sum(s) of squares(SS)
n	generic index
n_a	number of independent observations (datum values) averaged
n_b	number of blocks in an experiment test program
n_{bt}	number of binomial trials
n_{cdp}	number of conceptual (statistical) distribution parameters
n_{clp}	number of conceptual location parameters in a statistical model
n_{cp}	number of conceptual parameters in a statistical model
n_{delo}	number of distinct equally-likely outcomes
n_{digit}	number of digits (in each pseudorandom integer number)
n_{divv}	number of *different* independent variable values in simple linear regression
n_{dsdf}	number of denominator statistical degrees of freedom for Snedecor's central F conceptual sampling distribution and associated test statistic
n_{dv}	number of datum values
n_{dyv}	number of different discrete y values that random variable Y can take on
n_f	number of flips
n_{fo}	number of favorable outcomes
n_h	number of heads
n_{if}	number of items that failed prior to enduring test duration d^* in n_{st} independent strength tests
n_{is}	number of items that survived in n_{rt} independent reliability tests
n_{it}	number of items tested
n_l	number of (treatment) levels
n_{lt}	number of life tests in a given life (reliable life) experiment test program
n_{mpt}	number of main-plot treatments

n_{nsdf}	number of numerator statistical degrees of freedom for Snedecor's central F conceptual sampling distribution and associated test statistic
n_{oosi}	number of outcomes of specific interest
n_{pc}	number of paired comparisons
n_{pdv}	number of *paired* datum values
n_{ps}	number of positive signs
n_r	number of replicates (replicate datum values, replicate measurement values)
n_{rbelo}	number of randomization-based equally-likely experiment test program outcomes
n_{rdv}	number of regression datum values in a regression experiment test program
n_{rkdivv}	number of replicates at the kth different independent variable value *divv* in simple linear regression, where k is the index for the *divv* and kr is the index for the n_{rkdivv}
n_{rmv}	number of replicate measurement values
n_{rprv}	number of replicate *paired* realization values (datum values, measurement values)
n_{rt}	number of independent reliability tests conducted in a reliability experiment test program
n_{rvos}	number of random variables or statistics
n_s	number of independent datum values summed
n_{sbelo}	number of simulation-based equally-likely experiment test program outcomes
n_{sdf}	number of statistical degrees of freedom
n_{spt}	number of split-plot treatments
n_{st}	number of strength tests conducted in a strength experiment test program
n_t	number of treatments in an experiment test program
n_{tc}	number of treatment combinations in an experiment test program
n_{wdv}	number of weighted datum values
n_{wrdv}	number of weighted regression datum values in a simple linear weighted regression experiment test program
p	probability
PDF	probability density function, typically denoted $f(\cdot)$
pf	probability of failure before duration d – a parametric value or the invariant probability of failure before a predetermined duration d^* in each independent reliability test
pf^*	specific value of pf, viz., a selected value of the CDF percentile of specific interest in reliability analysis

pfo	probability of a favorable outcome
p_{oosi}	probability of an outcome of specific interest
p(pp)	plotting position (*pp*) stated in terms of the nonlinear *p* ordinate metric on probability paper
ps	probability of surviving *at least* duration d – a parametric value or the invariant probability of surviving for *at least* a predetermined duration d^* in each independent reliability test
RCBD	randomized complete block design experiment test program
rdv_i's	respective linear regression experiment test program datum values, where each rdv_i is associated with its underlying ivv_i
rnc	run-out number of (alternating stress) cycles
rnc^*	preselected run-out number of cycles
s(50)	actual value for the metric pertaining to the median of the presumed conceptual strength (resistance) statistical distribution
s_a	alternating stress amplitude
sc	scalar coefficient
scp	statistical confidence probability
sddv's	stopping distance datum values
smpvmd	standardized minimum practical value of the maximum difference among the respective *ctKm*'s
snc	suspension number of cycles, SNC* number of cycles imposed before the *Type I* suspension of the given life (reliable life) test
sp	statistical power
spsr	sum of the positive signed ranks (the test statistic in a signed-ranks test)
(SS)	sum(s) of squares
sw	statistical weight (also relative statistical weight in Supplemental Topic 7.B)
var(-)	technically verbalized as the actual value for the variance of the conceptual statistical or sampling distribution that consists of all possible realization values for the random variable or statistic (-)
var[*APR(-)EE*'s]	technically verbalized as the actual value for the variance of the conceptual statistical distribution that consists of *APR(-)EE*'s (pertains to all blocks and all treatments or treatment combinations, or to all *ivv*'s of specific interest)
wdv	weighted datum values

$WRDV_i$'s	conceptual weighted simple linear regression datum values (overly succinct notation for conceptual random *heteroscedastic* normally distributed weighted simple linear regression datum values); weighted regression datum values carry the subscript i to connote that the associated (concomitant) ivv_i
$wrdv_i$'s	weighted simple linear regression experiment test program datum values, viz., the realizations of the corresponding $WRDV_i$'s
$y(pp)$	plotting position (pp) stated in terms of the linear y ordinate metric on probability paper

NOTE: *Each of these microcomputer programs writes its output into a microcomputer file with the same name.*

Acronym	Program function	Example input data file (if required) (a)	Page (b)
ABLNOR	Computes a slightly biased A-basis statistical tolerance limit given 6 to 32 uncensored replicate datum values randomly selected from a conceptual two-parameter $\log_e$-normal life (endurance) distribution	Prompt	(507)
ABLNSTL	Computes the classical exact A-basis statistical tolerance limit given 6 to 32 uncensored replicate datum values randomly selected from a conceptual two-parameter $\log_e$-normal life (endurance) distribution	*STLDATA*	466
ABNOR	Computes a slightly biased A-basis statistical tolerance limit given 6 to 32 uncensored replicate datum values randomly selected from a conceptual (two-parameter) normal life (endurance) distribution	Prompt	(507)
ABNSTL	Computes the classical exact A-basis statistical tolerance limit given 6 to 32 uncensored replicate datum values randomly selected from a conceptual (two-parameter) normal life (endurance) distribution	*STLDATA*	461
ABW	Computes a slightly biased A-basis statistical tolerance limit given 6 to 16 uncensored replicate datum values randomly selected from a conceptual two-parameter Weibull life (endurance) distribution	*STLDATA*	(507)
AGESTCV	Aggregates the elements of adjacent column vectors in the estimated complete analytical model	See text example	54
ANOVA	Performs a classical ANOVA using Snedecor's central F test statistic	See text examples	249
AANOVADTA	Generates normally distributed pseudorandom data underlying the four Chapter 6 RCBD experiment test program examples that are intended to provide insight regarding ANOVA fundamentals	Prompt	279

(a) Parentheses indicate that the required microcomputer example input data file is not printed in the text.
(b) Parentheses indicate that the associated microcomputer program example output file is not printed in the text.

Acronym	Program function	Example input data file (if required) (a)	Page (b)
ANOVANT	Tests the null hypothesis of normality for the est($CRHNDEE$'s) of specific interest in ANOVA using a generalized version of the modified Michael's $MDSPP$ test statistic	*AANOVDTA*	278
ATCMLRM	Performs a statistical test of the adequacy of the conceptual multiple linear regression model	See text example	353
ATCSLRM	Performs a statistical test of the adequacy of the conceptual simple linear regression model (and also tests the null hypothesis that the actual value for the $clp1$ is equal to zero)	(*EXSLRDTA*)	338
AVE1	Simulates the statistical behavior of the arithmetic average of n_a independent uniformly distributed pseudorandom numbers–version 1	Prompt	111
AVE2	Simulates the statistical behavior of the arithmetic average of n_a independent uniformly distributed pseudorandom numbers–version 2	Prompt	113
AVE3A	Examines the simulation errors pertaining to the sum of n_s normal pseudorandom numbers—employs Knuth polar method with Wichmann–Hill generator. (Microcomputer program *AVE3A2* is an extension of microcomputer program *AVE3A* with 1,000,000 simulations)	Prompt	162
AVE3B	Examines the normal approximation errors pertaining to the sum of n_s uniform pseudorandom numbers—employs Wichmann–Hill generator. (Microcomputer program *AVE3B2* is an extension of microcomputer program *AVE3B* with 1,000,000 simulations)	Prompt	164
AVE3C	Examines the normal approximation errors pertaining to the sum of n_s exponential pseudorandom numbers—employs Wichmann–Hill generator. (Microcomputer program *AVE3C2* is an extension of microcomputer program *AVE3C* with 1,000,000 simulations)	Prompt	166

Acronym	Program function	Example input data file (if required) (a)	Page (b)
AVE3D	Examines the simulation errors pertaining to the sum of n_s normal pseudorandom numbers—employs IBM SSP algorithm with Wichmann–Hill generator. (Microcomputer program *AVE3D2* is an extension of microcomputer program *AVE3D* with 1,000,000 simulations)	Prompt	(162)
BARTLETT	Performs Bartlett's likelihood ratio test for homoscedasticity (that technically pertains to the respective est(*CRHNDEE*'s) generated in classical ANOVA for an equally replicated CRD experiment test program, but is applied in an *ad hoc* manner to the nonrepeated est(*CRHNDEE*'s) generated in classical ANOVA's pertaining to either an unreplicated RCBD or SPD experiment test program)	*RBBHTDTA*	(280)
BBLNOR	Computes a slightly biased *B*-basis statistical tolerance limit given 6 to 32 uncensored replicate datum values randomly selected from a conceptual two–parameter $\log_e$–normal life (endurance) distribution	Prompt	(507)
BBLNSTL	Computes the classical exact *B*-basis statistical tolerance limit given 4 to 32 uncensored replicate datum values randomly selected from a conceptual two–parameter $\log_e$–normal life (endurance) distribution	*STLDATA*	466
BBNOR	Computes a slightly biased *B*-basis statistical tolerance limit given 6 to 32 uncensored replicate datum values randomly selected from a conceptual (two–parameter) normal life (endurance) distribution	Prompt	(507)
BBNSTL	Computes the classical exact *B*-basis statistical tolerance limit given 4 to 32 uncensored replicate datum values randomly selected from a conceptual (two–parameter) normal life (endurance) distribution	*STLDATA*	462
BBW	Computes a slightly biased *B*-basis statistical tolerance limit given 6 to 16 uncensored replicate datum values randomly selected from a conceptual two–parameter Weibull life (endurance) distribution	Prompt	(507)

Acronym	Program function	Example input data file (if required) (a)	Page (b)
FCOIN2	Simulates flipping a fair coin—version 2	Prompt	108
FEBMPDT	Performs Fisher's enumeration-based test for paired-comparison datum values, given any value for the minimum practical difference *mpd* of specific interest	(*FEMPDDTA*)	143
FEBT	Performs Fisher's enumeration-based test for paired-comparison datum values	(*EXPCDTA1*)	94
FP	Computes the numerical value of Snedecor's central F test statistic that corresponds to the probability *p* value of specific interest	Prompt	(209)
FRBT	Performs Fisher's randomization-based test for paired-comparison datum values	(*EXRBPCD1*)	100
HISTPRO1	Generates histogram data for the observed proportions in 10 equal-width intervals for 1000 uniformly distributed pseudorandom numbers, zero to one	Prompt	115
HISTPRO2	Generates histogram data for the observed proportions in 10 equal-width intervals for 100,000 uniformly distributed pseudorandom numbers, zero to one	Prompt	116
IBPSCI	Computes an intuitive 100(*scp*)% (two-sided) statistical confidence interval that allegedly includes the actual value for the fixed binomial probability that a given binomial trial (paired-comparison, reliability test) will generate a favorable outcome	Prompt	415
ISLRCLNS	Computes the so-called inverse simple linear regression statistical confidence limits, given that the actual value for the *clp1* is negative	(*IRNSDATA*)	344
ISLRCLPS	Computes the so-called inverse simple linear regression statistical confidence limits, given that the actual value for the *clp1* is positive	(*IRPSDATA*)	343
ISLRTLNS	Computes the so-called inverse simple linear regression statistical tolerance limits, given that the actual value for the *clp1* is negative	(*IRNSDATA*)	345
ISLRTLPS	Computes the so-called inverse simple linear regression statistical tolerance limits, given that the actual value for the *clp1* is positive	(*IRPSDATA*)	345
LEV	Generates pseudorandom datum values from a conceptual (two–parameter) largest-extreme-value distribution	Prompt	(200)

Acronym	Program function	Example input data file (if required) (a)	Page (b)
LSEV4A	Performs a maximum likelihood analysis given a conceptual (two–parameter) $\log_e$ smallest-extreme-value life (endurance) distribution and computes classical $100(scp)\%$ (one-sided) asymptotic statistical confidence limits that allegedly bound the actual value for the *fnc(pf)* of specific interest–parameterization 4	*WBLDATA*	427
LSEV1B	Performs a maximum likelihood analysis given a conceptual (two–parameter) $\log_e$ smallest-extreme-value life (endurance) distribution and computes classical and LR lower $100(scp)\%$ (one-sided) asymptotic statistical confidence bands that allegedly bound the actual conceptual CDF–parameterization 1	*WBLDATA*	546
LSEV2B	Performs a maximum likelihood analysis given a conceptual (two–parameter) $\log_e$ smallest-extreme-value life (endurance) distribution and computes classical and LR lower $100(scp)\%$ (one-sided) asymptotic statistical confidence bands that allegedly bound the actual conceptual CDF–parameterization 2	*WBLDATA*	547
LSEV3B	Performs a maximum likelihood analysis given a conceptual (two–parameter) $\log_e$ smallest-extreme-value life (endurance) distribution and computes classical and LR lower $100(scp)\%$ (one-sided) asymptotic statistical confidence bands that allegedly bound the actual conceptual CDF–parameterization 3	*WBLDATA*	548
LSEV4B	Performs a maximum likelihood analysis given a conceptual (two–parameter) $\log_e$ smallest-extreme-value life (endurance) distribution and computes classical and LR lower $100(scp)\%$ (one-sided) asymptotic statistical confidence bands that allegedly bound the actual conceptual CDF–parameterization 4	*WBLDATA*	549
LSEV1C	Performs a maximum likelihood analysis given a conceptual (two–parameter) $\log_e$ smallest-extreme-value life (endurance) distribution and computes classical and LR lower $100(scp)\%$ (one-sided) asymptotic statistical confidence limits that allegedly bound the actual value for the *fnc(pf)* of specific interest– parameterization 1	*WBLDATA*	565

Acronym	Program function	Example input data file (if required) (a)	Page (b)
LSEV2C	Performs a maximum likelihood analysis given a conceptual (two–parameter) $\log_e$ smallest-extreme-value life (endurance) distribution and computes classical and LR lower 100(*scp*)% (one-sided) asymptotic statistical confidence limits that allegedly bound the actual value for the *fnc(pf)* of specific interest– parameterization 2	*WBLDATA*	565
LSEV3C	Performs a maximum likelihood analysis given a conceptual (two–parameter) $\log_e$ smallest-extreme-value life (endurance) distribution and computes classical and LR lower 100(*scp*)% (one-sided) asymptotic statistical confidence limits that allegedly bound the actual value for the *fnc(pf)* of specific interest– parameterization 3	*WBLDATA*	566
LSEV4C	Performs a maximum likelihood analysis given a conceptual (two–parameter) $\log_e$ smallest-extreme-value life (endurance) distribution and computes classical and LR lower 100(*scp*)% (one-sided) asymptotic statistical confidence limits that allegedly bound the actual value for the *fnc(pf)* of specific interest– parameterization 4	*WBLDATA*	566
L1A	Performs a maximum likelihood analysis given a conceptual (one–parameter) logistic strength (resistance) distribution and computes classical and LR lower 100(*scp*)% (one-sided) asymptotic statistical confidence limits that allegedly bound the actual value for *s*(50)	*UADDATA*	439
L2ALCL	Performs a maximum likelihood analysis given a conceptual (two–parameter) logistic strength (resistance) distribution and computes a classical lower 100(*scp*)% (one-sided) asymptotic statistical confidence limit that allegedly bounds the actual value for the *s(pf)* of specific interest	*(ASDATA)*	497

Acronym	Program function	Example input data file (if required) (a)	Page (b)
L2AS50	Performs a maximum likelihood analysis given a conceptual (two–parameter) logistic strength (resistance) distribution and computes a classical lower 100(scp)% (one-sided) asymptotic statistical confidence limit that allegedly bounds the actual value for $s(50)$	(ASDATA)	495
L2B	Performs a maximum likelihood analysis given a conceptual (two–parameter) logistic strength (resistance) distribution and computes classical and LR lower 100(scp)% (one-sided) asymptotic statistical confidence bands that allegedly bound the actual CDF	(ASDATA)	562
L2C	Performs a maximum likelihood analysis given a conceptual (two–parameter) logistic strength (resistance) distribution and computes classical and LR lower 100(scp)% (one-sided) asymptotic statistical confidence limits that allegedly bound the actual value for $s(pf)$	(ASDATA)	569
MDFBBSTL	Computes modified distribution-free (nonparametric) B-basis statistical tolerance limits	MDFDATA	475
MINREL	Computes a one-sided lower statistical confidence limit that allegedly bounds the actual value for the reliability, viz., the invariant binomial probability that a test item will survive the reliability test of specific interest	Prompt	140
NOR	Generates pseudorandom datum values from a conceptual (two–parameter) normal distribution	Prompt	197
NORTEST	Tests the null hypothesis of normality for replicate (presumed replicate) datum values using the modified MDSPP test statistic	ANORDATA	198
NTCMLRM	Tests the null hypothesis of normality for the est(CRHNDREEs) pertaining to the multiple linear regression experiment test program that was actually conducted	(NTMLRDTA)	(350)
NTCSLRM	Tests the null hypothesis of normality for the est(CRHNDREEs) pertaining to the simple linear regression experiment test program that was actually conducted	(NTSLRDTA)	(350)

Acronym	Program function	Example input data file (if required) (a)	Page (b)
N1A	Performs a maximum likelihood analysis given a conceptual (one–parameter) normal strength (resistance) distribution and computes classical and LR lower 100(*scp*)% (one-sided) asymptotic statistical confidence limits that allegedly bound the actual value for *s*(50)	*UADDATA*	438
N2ALCL	Performs a maximum likelihood analysis given a conceptual (two–parameter) normal strength (resistance) distribution and computes a classical lower 100(*scp*)% (one-sided) asymptotic statistical confidence limit that allegedly bounds the actual value for the *s*(*pf*) of specific interest	(*ASDATA*)	496
N2AS50	Performs a maximum likelihood analysis given a conceptual (two–parameter) normal strength (resistance) distribution and computes a classical lower 100(*scp*)% (one-sided) asymptotic statistical confidence limit that allegedly bounds the actual value for *s*(50)	(*ASDATA*)	494
N2B	Performs a maximum likelihood analysis given a conceptual (two–parameter) normal strength (resistance) distribution and computes classical and LR lower 100(*scp*)% (one-sided) asymptotic statistical confidence bands that allegedly bound the actual CDF	(*ASDATA*)	561
N2C	Performs a maximum likelihood analysis given a conceptual (two–parameter) normal strength (resistance) distribution and computes classical and LR lower 100(*scp*)% (one-sided) asymptotic statistical confidence limits that allegedly bound the actual value for *s*(*pf*)	(*ASDATA*)	569
OTPNLCLS	Computes the optimal stimulus level for the next test item, given a conceptual (two–parameter) normal strength distribution, viz., it computes the *s* value that maximizes the value of the asymptotic lower (one-sided) confidence limit that allegedly bounds the actual value for *s*(*pf*)–version S	*OSADATA*	500

Acronym	Program function	Example input data file (if required) (a)	Page (b)
SAFNCM11	Performs a maximum likelihood analysis for a linear s_a–$\log_e(fnc)$ model with a homoscedastic fatigue strength and computes the pragmatic bias-corrected estimate of the median of the presumed conceptual (two–parameter) normal fatigue strength distribution at $fnc = fnc^*$	(SAFNCDBC)	532
SAFNCM12	Performs a maximum likelihood analysis for a linear s_a–$\log_e(fnc)$ model with a homoscedastic fatigue strength and computes a pragmatic bias-corrected lower $100(scp)\%$ (one-sided) statistical confidence limit that allegedly bounds the median of the presumed conceptual (two–parameter) normal fatigue strength distribution at $fnc = fnc^*$	(SAFNCDBC)	533
SAFNCM13	Performs a maximum likelihood analysis for a linear s_a–$\log_e(fnc)$ model with a homoscedastic fatigue strength and computes a pragmatic bias-corrected lower $100(scp)\%$ (one-sided) statistical confidence limit that allegedly bounds the metric value for the pth percentile of the presumed conceptual (two–parameter) normal fatigue strength distribution at $fnc = fnc^*$	(SAFNCDBC)	533
SAFNCM31	Performs a maximum likelihood analysis for a quadratic s_a–$\log_e(fnc)$ model with a homoscedastic fatigue strength and computes the pragmatic bias-corrected estimate of the median of the presumed conceptual (two–parameter) normal fatigue strength distribution at $fnc = fnc^*$ (inclusive analysis). (See SAFNCM34 for the corresponding exclusive analysis)	(SAFNCDBC)	534
SAFNCM32	Performs a maximum likelihood analysis for a quadratic s_a–$\log_e(fnc)$ model with a homoscedastic fatigue strength and computes a pragmatic bias-corrected lower $100(scp)\%$ (one-sided) statistical confidence limit that allegedly bounds the median of the presumed conceptual (two–parameter) normal fatigue strength distribution at $fnc = fnc^*$ (inclusive analysis). (See SAFNCM35 for the corresponding exclusive analysis)	(SAFNCDBC)	536

Acronym	Program function	Example input data file (if required) (a)	Page (b)
SAFNCM33	Performs a maximum likelihood analysis for a quadratic s_a–$\log_e(fnc)$ model with a homoscedastic fatigue strength and computes a pragmatic bias-corrected lower $100(scp)\%$ (one-sided) statistical confidence limit that allegedly bounds the metric value for the pth percentile of the presumed conceptual (two–parameter) normal fatigue strength distribution at $fnc = fnc^*$ (inclusive analysis). (See SAFNCM36 for the corresponding exclusive analysis)	(SAFNCDBC)	536
SAFNCM34	Performs a maximum likelihood analysis for a quadratic s_a–$\log_e(fnc)$ model with a homoscedastic fatigue strength and computes the pragmatic bias-corrected estimate of the median of the presumed conceptual (two–parameter) normal fatigue strength distribution at $fnc = fnc^*$ (exclusive model)	(SAFNCDBC)	537
SAFNCM35	Performs a maximum likelihood analysis for a quadratic s_a–$\log_e(fnc)$ model with a homoscedastic fatigue strength and computes a pragmatic bias-corrected lower $100(scp)\%$ (one-sided) statistical confidence limit that allegedly bounds the median of the presumed conceptual (two–parameter) normal fatigue strength distribution at $fnc = fnc^*$ (exclusive model)	(SAFNCDBC)	538
SAFNCM36	Performs a maximum likelihood analysis for a quadratic s_a–$\log_e(fnc)$ model with a homoscedastic fatigue strength and computes a pragmatic bias-corrected lower $100(scp)\%$ (one-sided) statistical confidence limit that allegedly bounds the metric value for the pth percentile of the presumed conceptual (two–parameter) normal fatigue strength distribution at $fnc = fnc^*$ (exclusive model)	(SAFNCDBC)	539
SEED	Generates 50 new sets of three, three-digit odd seed numbers for subsequent use in the Wichmann–Hill pseudorandom number generator	(START)	5
SEV	Generates pseudorandom datum values from a conceptual (two–parameter) smallest-extreme-value distribution	Prompt	(200)

Acronym	Program function	Example input data file (if required) (a)	Page (b)
SEV1A	Performs a maximum likelihood analysis given a conceptual (one–parameter) smallest-extreme-value strength (resistance) distribution and computes classical and LR lower 100(*scp*)% (one-sided) asymptotic statistical confidence limits that allegedly bound the actual value for *s*(50)	*UADDATA*	439
SEV2ALCL	Performs a maximum likelihood analysis given a conceptual (two–parameter) smallest-extreme-value strength (resistance) distribution and computes a classical lower 100(*scp*)% (one-sided) asymptotic statistical confidence limit that allegedly bounds the actual value for the *s(pf)* of specific interest	*(ASDATA)*	498
SEV2AS50	Performs a maximum likelihood analysis given a conceptual (two–parameter) smallest-extreme-value strength (resistance) distribution and computes a classical lower 100(*scp*)% (one-sided) asymptotic statistical confidence limit that allegedly bounds the actual value for *s*(50)	*(ASDATA)*	495
SEV2B	Performs a maximum likelihood analysis given a conceptual (two–parameter) smallest-extreme-value strength (resistance) distribution and computes classical and LR lower 100(*scp*)% (one-sided) asymptotic statistical confidence bands that allegedly bound the actual CDF	*(ASDATA)*	563
SEV2C	Performs a maximum likelihood analysis given a conceptual (two–parameter) smallest-extreme-value strength (resistance) distribution and computes classical and LR lower 100(*scp*)% (one-sided) asymptotic statistical confidence limits that allegedly bound the actual value for *s(pf)*	*(ASDATA)*	570
SIMNOR	Generates pseudorandom data from a conceptual (two–parameter) normal distribution (see Figures 5.4 and 5.5)	Prompt	191
SLNABSTL	Simulates the variability of *A*-basis statistical tolerance limits for replicate datum values that are presumed to have been randomly selected from a conceptual two parameter log$_e$–normal life (endurance) distribution	Prompt	(467)

Acronym	Program function	Example input data file (if required) (a)	Page (b)
SLNBBSTL	Simulates the variability of B-basis statistical tolerance limits for replicate datum values that are presumed to have been randomly selected from a conceptual two parameter $\log_e$−normal life (endurance) distribution	Prompt	(467)
SNABSTL	Simulates the variability of A-basis statistical tolerance limits for replicate datum values that are presumed to have been selected from a conceptual (two−parameter) normal distribution.	Prompt	(463)
SNBBSTL	Simulates the variability of B-basis statistical tolerance limits for replicate datum values that are presumed to have been randomly selected from a conceptual (two−parameter) normal distribution	Prompt	(463)
SSLOSSCL	Simulates the proportion of 100(scp)% lower (one-sided) statistical confidence limit assertions that are actually correct when a quantitative CRD experiment test program is replicated 1000 times	Prompt	145
SSTSSCI1	Simulates the proportion of classical (shortest) 100(scp)% (two-sided) statistical confidence intervals that correctly include the actual value for the mean of a conceptual (two−parameter) normal distribution when the associated quantitative CRD experiment test program is replicated 1000 times	Prompt	213
SSTSSCI2	Computes 12 replicate classical (shortest) 100(scp)% (two-sided) statistical confidence intervals that allegedly (individually) include the actual value for the mean of a conceptual (two−parameter) normal distribution	Prompt	215
SSTSSCI3	Generates the empirical sampling distribution for the statistic [the ratio of the half-width of the classical (shortest) 100(scp)% (two-sided) statistical confidence interval that allegedly includes the actual value of the mean of a conceptual (two−parameter) normal distribution to its associated midpoint]	Prompt	217

Acronym	Program function	Example input data file (if required) (a)	Page (b)
SSTSSCI4	Computes 12 "replicate" classical (shortest) $100(scp)\%$ (two-sided) statistical confidence intervals that allegedly (individually) include the actual value for the mean of a conceptual (two–parameter) normal distribution	Prompt	219
SSTSSCI5	Generates the pragmatic sampling distribution for the statistic [the ratio of the half-width of the classical (shortest) $100(scp)\%$ (two-sided) statistical confidence interval that allegedly includes the actual value of the mean of a conceptual (two–parameter) normal distribution to its associated midpoint]	Prompt	(221)
SWABSTL	Simulates the variability of *A*-basis statistical tolerance limits for replicate datum values that are presumed to have been randomly selected from a conceptual two–parameter Weibull life (endurance) distribution.	Prompt	(467)
SWBBSTL	Simulates the variability of *A*-basis statistical tolerance limits for replicate datum values that are presumed to have been randomly selected from a conceptual two–parameter Weibull life (endurance) distribution.	Prompt	(467)
TP	Computes the numerical value of Student's central t test statistic that corresponds to the probability p value of specific interest	Prompt	(209)
UNI	Generates pseudorandom datum values from a conceptual (two–parameter) uniform distribution	Prompt	(200)
UNIFORM	Generates pseudorandom numbers that are uniformly distributed over the interval from zero to one	Prompt	103
UWLOSSCB	Computes a straight-line lower $100(scp)\%$ (one-sided) statistical confidence band in simple linear regression that is parallel to est[mean($APRCRHNDDV$'s)] given ivv] and pertains simultaneously to all ivv in the interval from ivv_{low} to ivv_{high}	*UWLCBDTA*	348

Contents

6.   Statistical Analysis of Variance (ANOVA)   231

7.   Linear Regression Analysis   308

l   **Contents**

Mechanical Reliability Improvement

1

Experiment Test Program Planning and Statistical Analysis Fundamentals

1.1. INTRODUCTION

The ideal test procedure is the classical scientific method in which each successive step in the experiment test program is logical and effective as a direct result of prior planning. We should always emulate this classical scientific method as much as possible. To accomplish this goal, we must first establish a *hypothesis–experiment–analysis* sequence for our experiment test program, and then focus on improving the effectiveness of each respective sequence segment.

The effectiveness of each *hypothesis–experiment–analysis* sequence segment can be enhanced by improving and increasing its input (Figure 1.1). Hypotheses are improved using information obtained from experience, preliminary testing, and the technical literature. Statistical planning of the conduct and details of experiment test programs markedly increases the efficiency of testing the hypotheses of specific interest. In turn, competent statistical analysis generates less subjective and more dependable conclusions regarding these hypotheses.

1.2. TEST OBJECTIVE

Every well-planned experiment test program has a well-defined objective. This test objective must subsequently be restated in terms of the correspond-

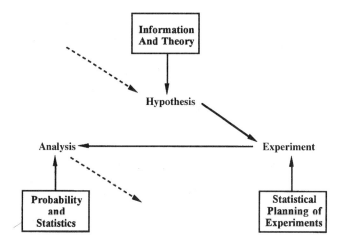

Figure 1.1 Technical inputs to each experiment test program segment.

ing null and alternative hypotheses to permit statistical analysis. There must always be a clear understanding of (*a*) how the test objective and the associated null and alternative test hypotheses are related, and (*b*) how the organizational structure of the experiment test program and its corresponding statistical model are related. The test objective must be evident in every aspect of the *hypothesis–experiment–analysis* methodology. Moreover, it should be the basis for deciding exactly what will be done in the experiment test program and specifically when and how it will be done, *before* beginning the experiment test program.

Suppose a product, denoted *A*, is unreliable in service. Then, the relevant test objective is to "assure" that the proposed redesign, denoted *B*, will be more reliable than *A*. The null hypothesis is $B = A$ statistically for this experiment test program, whereas the alternative hypothesis is that $B > A$ statistically. (The technical meaning of $B = A$ statistically and $B > A$ statistically is explained in Section 3.5) The null hypothesis is not intended to be physically realistic. Rather, its purpose is to provide a rational basis for computing all hypothetical probabilities of potential interest in the subsequent statistical analysis. In contrast, the alternative hypothesis must be physically realistic, because it must be consistent with the test objective. The alternative hypothesis determines which of the computed hypothetical probabilities are actually used in the subsequent statistical analysis.

When the test objective is to measure some material behavior parameter, say a tensile yield strength, the null and alternative hypotheses are generally submerged into a statistical confidence interval statement. A statistical confidence interval is a probability interval that contains all of those hypothetical values that could have been assigned to a null hypothesis without leading to its rejection in the associated statistical analyses.

1.3. EXPERIMENT TEST PROGRAM PLANNING FUNDAMENTALS

All well-planned experiment test programs incorporate the following fundamentals:

1. Randomization
2. Replication
3. Planned grouping

Several other important concepts and procedures that improve the effectiveness and efficiency of well-planned experiment test programs are discussed in Section 1.4.

1.3.1. Randomization and Random Samples

It is seldom practical to make measurements on every nominally identical experimental unit (e.g., laboratory test specimen, component, or device) of specific interest. Rather, we generally must select a subset of these experimental units, called a statistical sample, and then make measurements on this sample. The sampling process is rational only if the sample is *statistically representative* of the population of experimental units from which it was selected. It will be demonstrated later in this text that when the sample is randomly selected from a population, the sample will exhibit a statistical behavior that is related to the statistical behavior of that population—and therefore the statistical behavior of the population can be inferred from the statistical behavior of this random sample.

> Acceptable procedures for obtaining a random sample are such that each nominally identical member of the population of specific interest is equally likely to be selected at each stage of the selection procedure.

1.3.1.1. Random Selection Example

Suppose we have two nominally identical 8 ft long, 3/4 in. diameter AISI 1020 cold-drawn steel rods and we wish to machine eight tension test specimens, each 6 in. long. To obtain the desired random sample, we first cut these two rods into 32 nominally identical test specimen blanks numbered consecutively from 1 to 32 (starting at either end of either rod), and then we randomly select eight of these 32 specimen blanks (as outlined below).

The required random selection is conveniently accomplished in this example by running microcomputer program *RANDOM1*, which selects n_{elpri} equally-likely pseudorandom integers from the integers 1 through n_i, such that each of these n_i integers is equally likely to occur at each location in a time-order sequence (arrangement) of the n_{elpri} integers. Microcomputer program *RANDOM1* requires a new set of three, three-digit odd seed numbers as input data each time that it is run. These seed numbers initiate its pseudorandom number generator (Wichmann and Hill, 1981). To run microcomputer program *RANDOM1*, merely type RANDOM1 and follow the prompts.

```
C> RANDOM1
```

Input the number of equally-likely pseudorandom integers n_{elpri} of specific interest

```
8
```

Input the number of integers, 1 through n_i, from which to select these n_{elpri} equally likely pseudorandom integers

```
32
```

Input a new set of three, three-digit odd seed numbers (obtained by running microcomputer program *SEED*)

```
587 367 887
```

Time Order	Pseudorandom Numbers
1	12
2	22
3	26
4	19
5	24
6	17
7	11
8	18

However, microcomputer program *SEED* must first be run to generate each new set of three, three-digit odd seed numbers that are subsequently used as input to microcomputer program *RANDOM1*. Each successive set of three, three-digit seed numbers generated by microcomputer program *SEED* is such that the numbers 1 through 9 are equally likely to occur at each of the first two digits and the numbers 1, 3, 5, 7, and 9 are equally likely to occur at the third digit. To run microcomputer program *SEED*, use a convenient editor to enter the last nine digits of your driver's license number into microcomputer file *START* using the format, three digits, space, three digits, space, three digits. If any of these three-digit integers is even, make them odd by flipping a coin and adding 1 for a head and subtracting 1 for a tail. Then, type SEED as many times as desired to generate successive sets of 50 three, three-digit odd seed numbers as desired. When these seed numbers are nearly exhausted, merely type SEED again, as many times as desired. (Microcomputer program *SEED* continually updates microcomputer file *START* with a new set of three, three-digit odd seed numbers.)

Microcomputer program *RANDOM1* pertains to random sampling from the population of integers, 1 through n_i, such that no integer is repeated. This random sampling is *without replacement*, viz., the size of this population decreases by one after each successive random selection. In contrast, microcomputer program *RANDOM2* pertains to random sampling *with replacement* of the randomly selected integer back into the invariant population of integers. Note, however, for microcomputer program *RANDOM2*, the invariant population of integers is either 0 through 9 for

```
c> TYPE START

745 237 391

C> SEED

Your 50 new sets of three, three-digit odd seed numbers are

        587 367 887
        225 949 431
        757 171 333
        823 199 245
        131 449 153
            et cetera
```

C> RANDOM2   w / replacement

Input the number of equally-likely pseudorandom integers n_{elpri} of specific interest

12

Input the number of digits, n_{digits}, of specific interest for each of these n_{elpri} equally likely pseudorandom integers

2

Input a new set of three, three-digit odd seed numbers (obtained by running microcomputer program *SEED*)

225 949 431

Time Order	Pseudorandom Numbers
1	7
2	49
3	94
4	20
5	90
6	7
7	18
8	0
9	38
10	60
11	95
12	5

---

one-digit integers, or 00 through 99 for two-digit integers, or 000 through 999 for three-digit integers, etc.

Microcomputer program *RANDOM2* is used in this text to generate pseudorandom numbers in microcomputer simulation examples that demonstrate the probability behaviors of specific interest in statistical analysis. (This program generates pseudorandom numbers that, in theory, are such that the integers 0 through 9 are equally likely to occur at each digit of the pseudorandom number, regardless of its number of digits and which integers are selected at its other digits.)

1.3.2. Replication

Replicate measurements (observations), by definition, are independent measurements (observations) made on nominally identical experimental units under nominally identical conditions. The greater the number of replicate measurements (observations) made, the more precise the resulting statistical estimates, viz., the smaller their intrinsic estimation components. In turn, the smaller these intrinsic estimation components, the greater the probability of correctly rejecting the null hypothesis in favor of the relevant alternative hypothesis when this alternative hypothesis is correct.

Consider the four respective measurement values for treatments A and B depicted by time-order-of-testing in Figure 1.2. (Treatment is the technical term for the variable being studied.) If we know by experience that these measurement values always exhibit negligible variability (differences), e.g., as illustrated in (a), then we can rationally conclude that $B > A$ statistically without conducting a statistical analysis. On the other hand, if we observe that the respective replicate measurement values for A and B exhibit marked

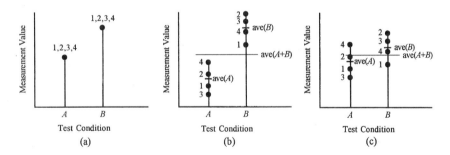

Figure 1.2 Schematic illustrating three extreme situations: (a) the respective replicate measurement values for treatments A and B exhibit negligible variability; (b) the variability (differences) *between* the arithmetic averages of the respective replicate measurement values for treatments A and B is relatively large compared to the variability (differences) *within* (among) these replicate measurement values; (c) the variability (differences) *between* the arithmetic averages of the respective replicate measurement values for treatments A and B is relatively small compared to the variability (differences) *within* (among) these replicate measurement values. (Only four replicate measurement values for treatments A and B are depicted by time-order-of-testing for simplicity of presentation. The minimum amount of replication that is statistically adequate is discussed in Chapter 6.)

variability (differences), e.g., as illustrated in (b) and (c), then a statistical analysis is required to conclude that $B > A$ statistically.

In the classical statistical analysis called analysis of variance (Chapter 6) the variability among the respective replicate measurement values for treatments A and B is partitioned into two statistically independent components: the *between* variability and the *within* variability. In turn, the magnitude of the ratio of these two variability components can be used to test the null hypothesis that $B = A$ statistically versus the alternative hypothesis that $B > A$ statistically. Note that in (b) the variability (differences) *between* the arithmetic averages of the treatments A and B replicate measurement values is relatively large compared to the variability (differences) *within* (among) these replicate measurement values, whereas in (c) the variability (differences) *between* the arithmetic averages of the treatments A and B replicate measurement values is relatively small compared to the variability (differences) *within* (among) these replicate measurement values. Because the ratio of the *between* variability to the *within* variability is much larger for (b) than for (c), the null hypothesis that $B = A$ statistically is much more likely to be rejected in favor of the alternative hypothesis that $B > A$ statistically for (b) than for (c). Or, from another perspective, the primary distinction between (b) and (c) is that, when the alternative hypothesis that $B > A$ statistically is correct, many more replicate measurement values are required for (c) than for (b) to have the same probability of correctly rejecting the null hypothesis that $A = B$ statistically in favor of this alternative hypothesis.

Exercise Set 1

These exercises are intended to provide insight regarding *within* and *between* variabilities, technically termed mean squares, by using analogous mechanics concepts, viz., the mass moment of inertia concept and the associated parallel-axis theorem, to compute their underlying sums of squares.

Consider the four replicate measurement values for treatment A and treatment B depicted in Figures 1.2(b) and (c). Presume that each of these measurement values has a unit (dimensionless) mass. Then the *within* mean square, *within*(MS); can be computed as the sum of the mass moments of inertia of the treatments A and B measurement values about their respective centroids (arithmetic averages), divided by $n_t \cdot (n_{rmv} - 1)$, where n_t is the number of treatments, and n_{rmv} is the number of replicate measurement values. The *between* mean square, *between*(MS); can be computed as the mass moment of inertia of the treatments A and B measurement values, coalesced at their respective centroids (arithmetic averages), about the centroid (arithmetic average) of all measurement values, divided by $(n_t - 1)$. The respective divisors are the associated number of statistical degrees of freedom (defined later). We rationally reject the null hypothesis that $B = A$ statistically in favor of the alternative hypothesis that $B > A$ statistically when the ratio of the *between* mean square to the *within* mean square is so large that it casts serious doubt on the statistical credibility of the null hypothesis.

1. Run microcomputer program *RANDOM2* twice with $n_{elpri} = 4$ and $n_{digit} = 1$ to generate two sets of four pseudorandom integers. Then add 185.5 to the four pseudorandom integers in the first set to construct the treatment A measurement values. In turn, add 195.5 to the four pseudorandom integers in the second set to construct the treatment B measurement values. Next, compute the arithmetic average of the treatment A measurement values, of the treatment B measurement values, and of both treatments A and B measurement values.

2. Plot these measurement values and the three arithmetic averages as illustrated in Figures 1.2(b) and (c). Then use this plot and the parallel-axis theorem to compute the *within* and *between* sums of squares. In turn, compute the *within* and *between* mean squares and the ratio of the *between*(MS) to the *within*(MS). If this ratio is sufficiently large in a statistical sense, viz., greater than 10 for this exercise set, reject the null hypothesis that $B = A$ statistically in favor of the alternative hypothesis that $B > A$ statistically.

3. Repeat 1 and 2 with the same pseudorandom integers, but add 190.0 to the four pseudorandom integers in the first data set to generate the treatment A measurement values and add 191.0 to the four pseudorandom integers in the second data set to generate the treatment B measurement values.

1.3.3. Planned Grouping (Constrained Randomization, Blocking)

Planned grouping has two basic applications: (a) to help assure that the sample experimental units are more representative of the actual population of experimental units of specific interest, and (b) to help mitigate spurious effects caused by nuisance variables, viz., unavoidable differences among experimental units and among test conditions. Planned grouping in application (a) involves constrained randomization, whereas planned grouping in application (b) involves blocking.

1.3.3.1. Constrained Randomization Example

Consider Figures 1.3(a)–(d). Recall that complete randomization is appropriate only for a population of nominally identical experimental units, e.g., the 32 test specimen blanks in our randomization example. This population of nominally identical experimental units is depicted in Figure 1.3(a). A representative sample from this population is depicted in Figure 1.3(c). It is obtained by running microcomputer program *RANDOM1* with $n_{elpri} = 8$ and $n_i = 32$. However, it is seldom prudent to presume that any two 8 ft. long 3/4 in. diameter AISI 1020 cold-drawn steel rods are nominally identical. The population of experimental units from two rods that are not presumed to be nominally identical is depicted in Figure 1.3(b). Constrained randomization is required to generate a representative sample from this population. For example, we obtain a *proportionate* random sample by running microcomputer program *RANDOM1* twice, once for rod 1 and once for rod 2, each time with $n_{elpri} = 4$ and $n_i = 16$. This constrained randomization is such that the same proportion, 1/4, of the experimental units in each rod *must be* randomly selected.

Note that the proportionate random sample in (d) is equally representative of the population in (a). Thus, proportionate random sampling is appropriate whenever there are potential (possible) differences either between or within the sources of the alleged nominally identical experimental units that comprise the population of specific interest. For example, a random sample from the population in (a) can be obtained by first partitioning each rod into four groups (blocks) of four *adjacent* experimental units, and then *randomly* selecting one experimental unit from each of the eight resulting groups (blocks).

1.3.3.2. Blocking Example

Suppose our test objective is to compare two plastics, *A* and *B*, with regard to their respective estimated median axial-load fatigue strengths at 10 million cycles. Typical test specifications call for conditioning the fatigue test specimens in a 50% relative humidity, 23°C environment for at least 48 hr and then maintaining this standard laboratory environment throughout the duration of the experiment test program. (Note that a standard laboratory environment is mandatory in quantitative experiment test programs when the respective test outcomes are sensitive to a change in the test environment.) Suppose, however, that our laboratory has no humidity control so that the relative humidity changes from season to season. We will now demonstrate that blocking (planned grouping) can be used in a comparative experiment test program to help make our comparison of *A* and *B* as fair as

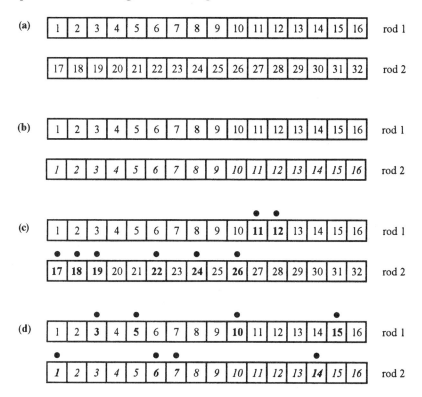

Figure 1.3   Random samples for our randomization example with two 8 ft long, 3/4 in. diameter AISI 1020 cold-drawn steel rods, each cut into 16 test specimen blanks that are presumed to be nominally identical. (a) The population of specific interest consists of two rods that are presumed to be nominally identical. (b) the population of specific interest consists of two rods are not presumed to be nominally identical. (c) A random sample of size 8 from (a) obtained by running microcomputer program *RANDOM*1 with seed numbers 587 367 887, as in the text randomization example. (d) A *proportionate* random sample of size 8 from (b) is such that the *same proportion* of test specimen blanks in each rod *must be* randomly selected for inclusion in the random sample. Accordingly, for a sample size equal to 8, microcomputer program *RANDOM*1 is run *twice*, once to establish the four test specimen blanks randomly selected from rod 1 and once to establish the four test specimen blanks randomly selected from rod 2.

possible, despite our inability to maintain the required laboratory relative humidity at 50%.

To illustrate the concept of blocking, consider the effect of a small steady change in laboratory relative humidity on the outcome of a comparative experiment test program with 16 axial-load fatigue tests that are conducted sequentially using a single fatigue test machine. Each fatigue specimen is tested to failure or until it has endured 10 million stress cycles without failing (92 hr, 35 min, and 33 sec at 30 Hz), at which time the test is suspended (termed *Type I* censoring). Our goal is to make the experiment test program comparison of A and B as unbiased (fair) as possible. The null hypothesis in this example is that $B = A$ statistically. The alternative hypothesis is that $B > A$ statistically, but the proper alternative hypothesis is not an issue now. Rather, we wish to plan (structure) our blocking such that, presuming that the null hypothesis is correct, both A and B have an equal chance of excelling in each comparison.

Now consider four alternative experiment test programs, each with eight A's and eight B's. Figure 1.4(a) depicts an experiment test program in which all eight A's are tested before testing all eight B's. Presuming that a small increase in relative humidity slightly decreases the fatigue strength of both A and B by the same incremental amount, then the comparison of A and B generated by this experiment test program is clearly biased in favor of A. In contrast, Figure 1.4(b) depicts an experiment test program in which A and B are tested in *time* blocks of size two. However, because A precedes B in each time block, the comparison of A and B generated by this experiment test program is still slightly biased in favor of A. Next, suppose that we use pseudorandom numbers to determine whether A or B is tested first in each time block, Figure 1.4(c). Note that, depending on the specific randomization outcome, it is possible that Figure 1.4(c) is identical to Figure 1.4(b). Thus, we must employ constrained randomization to assure an unbiased (fair) comparison of A and B, viz., we must require that A is tested first in exactly four time blocks and in turn that these four time blocks be determined using pseudorandom numbers, Figure 1.4(d.1). Then, the time order of testing for all eight time blocks should be determined by using pseudorandom numbers, Figure 1.4(d.2).

This elementary example is intended to illustrate that the comparison of A and B is unbiased (fair) even if the individual tests do not conform exactly to certain test specifications. Moreover, the combination of blocking and *subsequent* randomizations can provide unbiased (fair) comparisons even when working with populations consisting of different batches of material, slowly changing test conditions, markedly different test conditions, etc. Blocking is intended to balance (counteract) the major sources of comparison bias, either actual or potential. Subsequent randomization is intended to balance (mitigate) more subtle sources of comparison bias. Several examples of combined blocking and subsequent randomization are presented in Chapter 2.

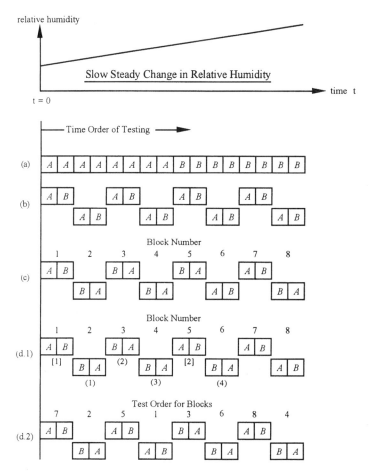

Figure 1.4 Alternative time orders of testing in the presence of a small steady change in the test environment. (a) Test all eight A specimens, then test all eight B specimens. Although this is the most common experiment test program it results in a comparison of A and B that is clearly biased in favor of A when a small increase in relative humidity causes a slight decrease in the fatigue strength. (b) Employ time blocks of size two with A preceding B in each time block. The comparison of A and B is now only slightly biased in favor of A. (c) Use pseudorandom numbers to determine whether A precedes B or B precedes A in each time block. The comparison of A and B generated by this experiment test program is slightly biased in favor of A only when more time blocks begin with A than with B, or vice versa. Accordingly, we require that the same number of time blocks begin with A as with B. Then, we randomly select which four time blocks begin with A (or B) as in (d.1). In turn, we (should) randomly select the time order of testing for these time blocks as in (d.2). Note that the resulting comparison of A and B is unbiased (fair) even when change in the test environment is negligible. *Thus the use of time blocks in experiment test program planning is always good practice.*

Caveat: Suppose that the fatigue strength of a plastic, say nylon, is much more sensitive to a change in humidity than another plastic, say polypropylene. In this situation, blocks are said to *interact* with the materials being studied and the comparison generated by the experiment test program in Figure 1.4(d) is not unbiased (fair) because the respective block-effect incremental change amounts are not the same for both materials.

Exercise Set 2

These exercises are intended to enhance your understanding of the time blocking example summarized in Figures 1.4(d.1) and (d.2).

1. Assume a slowly changing test environment as shown in Figure 1.4. Suppose we have two fatigue machines, one old and one new. Devise an experiment test program with eight time blocks of size two that provides an unbiased comparison of *A* and *B*. Completely specify the associated time order of testing on each machine. Would it make a difference if both machines were nominally identical?

2. (a) Rework the text blocking example summarized in Figures 1.4(d.1) and (d.2) twice using structured time blocks of size four, first, *ABBA* and *BAAB* (or vice versa), and then *ABAB* and *BABA*(or vice versa). (b) Is one of these two experiment test programs objectively (or subjectively) preferable to the other or to the text example summarized in Figures 1.4(d.1) and (d.2)?

3. Suppose that we have a single fatigue machine and each individual fatigue test takes (up to) 4 days. In turn, suppose that the test environment can rationally be considered to change slowly for successive segments of (a) 32 days, (b) 16 days, and (c) 8 days. Devise experiment test programs with time blocks of size two, if practical, that provide unbiased comparisons of *A* and *B*. Completely specify the associated time orders of testing for each practical experiment test program.

4. Presume that the effect of the nuisance variable in the ith block, whatever its nature, changes the response of both *A* and *B* by (exactly) the same incremental amount. Use the algebraic expression $[B + (\text{block effect incremental amount})_i] - [A + (\text{block effect incremental amount})_i] = (B - A)$ to demonstrate that the actual magnitude of the ith block effect incremental amount is irrelevant to an unbiased (fair) comparison of *A* and *B*. (Note

that the actual magnitude of the ith block-effect incremental amount can differ from block to block.)

1.4. EXPERIMENT TEST PROGRAM PLANNING TIPS

1.4.1. Preliminary Testing

The most important requisite in statistical planning of an experiment test program is preliminary information regarding the physical nature of the variables (treatments) to be studied. This information is most trustworthy when it is based on preliminary tests employing the same equipment, technicians, and test methods that will be used in the statistically planned experiment test program. Moreover, planning decisions made relative to the size and cost of the proposed experiment test program are more objective when based on preliminary testing. Thus, in new and unique mechanical reliability applications, it may be reasonable to expend as much as one-half of the test resources on preliminary testing.

1.4.2. Orthogonality

Unless there is a specific reason for doing otherwise, the experiment test program should be structured such that when its conceptual statistical model is expressed in column vector format, its respective column vectors are *mutually orthogonal* (Chapter 2). Statistical analyses for orthogonal conceptual statistical models have simple geometric interpretations and intuitive mechanics analogies (Section 1.5).

1.4.3. Equal Replication

Unless there is a specific reason for doing otherwise, all comparative experiment test programs should include equal replication, viz., the same number of replicates for each treatment (treatment level, treatment combination) being compared. Chapter 2 introduces the concept of statistical degrees of freedom for the mutually orthogonal column vectors comprising the estimated conceptual statistical model. Statistical degrees of freedom provide an intuitive index to the amount of replication. (The greater the amount of replication, the greater the precision of the resulting statistical estimates.)

1.4.4. Time Blocks

Time blocks should always be used in conducting an experiment test program, either to mitigate the (actual or potential) spurious effects of time trends, or as a precautionary measure against an inadvertent interruption of the experiment test program (e.g., by equipment breakdown). However, unless a statistical analysis indicates that a time-trend effect has occurred or unless some inadvertent experiment test program interruption has occurred, it is seldom necessary to include these time blocks in the conceptual statistical model ultimately presumed in statistical analysis.

1.4.5. Batch-to-Batch Effects

Never presume that batch-to-batch effects (differences) are negligible when conducting an experiment test program whose objective is to make a quantitative assessment, e.g., the yield strength in a conventional laboratory tension test for a given material and its processing. Unless batch-to-batch effects have been clearly demonstrated to be negligible, always include at least two relatively diverse batches of experimental units in the experiment test program and then statistically estimate the batch-to-batch effect. The physical interpretation of the associated quantitative assessment is clearly dubious whenever the magnitude of the estimated batch-to-batch effect is not negligible. (See Supplemental Topic 6.C.)

1.5. STATISTICAL ANALYSIS FUNDAMENTALS

Two fundamental abstractions establish the foundation of classical statistical analyses. First, the experiment test program of specific interest, and each of the sets of nominally identical experimental units and/or test conditions that comprise this experiment test program, can be continually replicated indefinitely to generate the conceptual collection of all possible experiment test program outcomes. Moreover, the outcome for the experiment test program that is (was) actually conducted is (was) randomly selected from this collection of all possible equally-likely experiment test program outcomes (see Table 1.1). Second, its conceptual random datum values ($CRDV_i$'s) are (were) randomly selected from their corresponding conceptual statistical distributions that respectively consist of all possible replicate conceptual random datum values ($APRCRDV_i$'s).

Although in concept the experiment test program that is (was) actually conducted can be continually replicated indefinitely, the number of its equally likely outcomes can be either finite or infinite. The measurement metric is discrete for a conceptual statistical distribution that consists of a finite number of equally likely $CRDV_i$'s. It is continuous for a conceptual statistical distribution that consists of an infinite number of equally-likely $CRDV_i$'s. Figure 1.5 depicts all possible equally-likely outcomes for (a) the digits 0 through 9 when each digit is equally likely to be selected in a random selection process, and for (b) the sum of the number of dots that actually

Table 1.1 Illustration of the Infinite Conceptual Collection of Replicate Experiment Test Programs and Associated Infinite Collection of Equally-likely Experiment Test Program Outcomes

Conceptual experiment test program[a]	Associated *equally-likely* outcome
1	$[CRDV_1, CRDV_2, CRDV_3, \cdots CRDV_{n_{dv}}]_1$
2	$[CRDV_1, CRDV_2, CRDV_3, \cdots CRDV_{n_{dv}}]_2$
3	$[CRDV_1, CRDV_2, CRDV_3, \cdots CRDV_{n_{dv}}]_3$
$\vdots$	$\vdots$
m	$[CRDV_1, CRDV_2, CRDV_3, \cdots CRDV_{n_{dv}}]_m$
$\vdots$	$\vdots$
∞	$[CRDV_1, CRDV_2, CRDV_3, \cdots CRDV_{n_{dv}}]_\infty$

[a]The first fundamental statistical concept is the experiment test program that is (was) actually conducted is (was) randomly selected from this conceptual collection.

appear on a pair of dice when the numbers of dots that actually appear on each die, 1 through 6, are independent and equally likely to appear after each random toss.

1.5.1. Depicting Finite Conceptual Statistical Distributions by Point Masses

It is intuitively convenient to depict a conceptual statistical distribution that consists of a finite number of equally likely $CRDV_i$'s by a corresponding finite number of identical point masses, as in Figure 1.5. The centroid of these identical point masses is termed the *mean* of the conceptual statistical distribution. It is also termed the *expected value* of a randomly selected $CRDV_i$. The units of a mean or an expected value are the units for the measurement metric. The mass moment of inertia of these identical point masses about their centroid (mean, expected value), divided by the sum of these masses, is termed the *variance* of this finite conceptual statistical distribution. Its units are the square of the units for the measurement metric. In turn, because mass in statistical jargon is termed weight, each equally likely $CRDV_i$ has the same value for its statistical weight (sw), viz.,

$$sw \text{ (by definition)} = 1/[\text{var}(APRCRDV_i\text{'s}]$$

Moreover, because the variance of a finite conceptual statistical distribution is, by definition, the mass moment of inertia about the centroid, divided by the sum of the masses, its value is independent of the value for the centroid. Accordingly, each of the finite conceptual statistical distributions depicted in Figure 1.5 can be translated either to the right or to the left along their measurement metrics, and the respective variances of these conceptual sta-

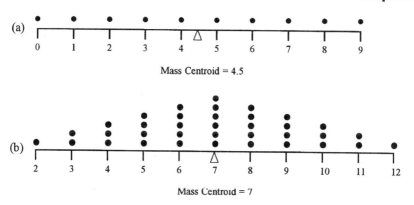

Figure 1.5 Two example conceptual statistical distributions that consist of a finite number of equally-likely $CRDV_i$'s. (The graphical depiction of a conceptual statistical distribution that consists of an infinite number of equally-likely $CRDV_i$'s is presented and discussed in Chapter 3.)

tistical distributions do not change. Thus, given $CRDV_i$'s randomly selected from two (or more) different conceptual distributions that differ only by the respective values for their means, the associated statistical weights all have identical values.

1.5.2. Statistically Identical and Homoscedastic Conceptual Random Datum Values

When an experiment test program consists only of $CRDV_i$'s pertaining to a single set of nominally identical test conditions and experimental units, these $CRDV_i$'s have identical values for their statistical weights (by definition) and thus are statistically identical (by definition). We explicitly denote statistically identical $CRDV_i$'s as $CRSIDV_i$'s. The intuitive least-squares estimator of the actual value for the mean of the conceptual statistical distribution that consists of $APRCRSIDV_i$'s is the arithmetic average of the experiment test program $CRSIDV_i$'s. On the other hand, when the experiment test program datum values pertain to two or more sets of nominally identical test conditions and/or experimental units, an arithmetic average is statistically credible only if all of the experiment test program $CRDV_i$'s are presumed to have identical values for their statistical weights (Supplemental Topic 6.B). The presumption that two or more sets of $CRDV_i$'s have identical values for their statistical weights so markedly simplifies classical statistical analyses that this presumption is almost universally employed. When all of the respective sets of experiment test program $CRDV_i$'s have identical values for their respective statistical weights, these $CRDV_i$'s are termed *homoscedastic.* We explicitly denote homoscedastic $CRDV_i$'s as $CRHDV_i$'s. Accordingly, the intuitive least-squares estimator of the actual value for the mean of the conceptual statistical model that consists of two or more homoscedastic conceptual statistical distributions is the arithmetic average of all of the respective $CRHDV_i$'s. Clearly, $CRHDV_i$'s are obtained by randomly selecting $CRDV_i$'s from two or more conceptual statistical distributions that differ only by the values for their means, as shown in Figure 1.6.

1.5.3. Arithmetic Averages and Orthogonal Estimated Conceptual Statistical Models

The orthogonality relationships of specific interest in classical statistical analyses pertain to the estimated conceptual statistical model. These orthogonality relationships are the direct consequence of the use of arithmetic averages in least-squares estimation. Suppose that the experiment test program of specific interest consists only of replicate datum values pertaining to a given set of nominally identical experimental units and test conditions. Then, the experiment test program $CRSIDV_i$'s can be completely explained by a conceptual statistical model with only two terms: (a) the mean of the conceptual statistical distribution that consists of $APRCRSIDV_i$'s, and (b) the associated conceptual statistically identical experimental errors ($CRSIEE_i$'s). The least-squares estimated statistical model also has two terms: (*a*) the least-squares estimate of the actual value for the mean of the conceptual statistical distribution that consists of $APRCRSIDV_i$'s, viz., the arithmetic average of the experiment test program datum values, and (*b*) the associated est($CRSIEE_i$'s). The orthogonality of the least-squares estimated statistical model is obvious when it is numerically

Figure 1.6 Two finite conceptual statistical distributions whose $CRSIDV_i$'s have identical statistical weights and are therefore homoscedastic. The conceptual distribution that consists of $APRCSIDV_i$'s for B is identical to conceptual distribution that consists of $APRCSIDV_i$'s for A, except that the former is translated a positive distance along its discrete measurement metric. (See also Figure 3.29.)

depicted in a column vector format. Suppose that our hypothetical datum values are 0.50, 0.49, and 0.51. Then the least-squares estimated statistical model, stated in column vector notation, is

$$\underset{\text{hypothetical datum value}_i\text{'s}}{\begin{vmatrix} 0.50 \\ 0.49 \\ 0.51 \end{vmatrix}} = \underset{\text{est[mean}(APRCRSIDV_i\text{'s})]}{\begin{vmatrix} 0.50 \\ 0.50 \\ 0.50 \end{vmatrix}} + \underset{\text{est}(CRSIEE_i\text{'s})}{\begin{vmatrix} +0.00 \\ -0.01 \\ +0.01 \end{vmatrix}}$$

Note that the dot product of the |est[mean($APRCRSIDV_i$'s)]| column vector and the |est($CRSIEE_i$'s)| column vector is equal to zero. Accordingly, these two column vectors are orthogonal. Note also that $(0.50)^2 + (0.49)^2 + (0.51)^2 = 0.7502 = (0.50)^2 + (0.50)^2 + (0.50)^2 + (0.00)^2 + (-0.01)^2 + (0.01)^2$.

1.5.4. Intrinsic Statistical Estimation Errors

Statistical estimates are computed by substituting the realization values for the experiment test program $CRDV_i$'s into the corresponding statistical estimators (estimation expressions). However, since these realization values (datum values) differ from replicate experiment test program to replicate program, statistical estimates also differ from replicate program to replicate experiment test program. On the other hand, the actual value for the quantity being estimated is invariant. Thus, under continual replication of the experiment test program, each statistical estimator, whatever its nature, generates replicate realizations of an intrinsic statistical estimation error that is confounded with the actual value for the quantity being estimated.

The following numerical example is intended to enhance your intuition regarding intrinsic statistical estimation errors. Suppose that the experiment test program of specific interest consists of four replicate datum values pertaining to a single set of nominally identical experimental units and test conditions. Suppose also that these replicate datum values are randomly selected from the conceptual statistical distribution that consists of the equally-likely integers 00 through 99. Then, the conceptual statistical model, written in our hybrid column vector notation, is

$$|CRSIDV_i\text{'s}| = |csdm| + |CRSIEE_i\text{'s}| = |49.5| + |CRSIEE_i\text{'s}|$$

in which the $csdm$ is the conceptual statistical distribution mean, viz., 49.5. The corresponding least-squares estimated statistical model is

$$|\text{replicate datum value}_i\text{'s}| = |\text{est}(csdm)| + |\text{est}(CRSIEE_i\text{'s})|$$

in which est($csdm$) is equal to the arithmetic average of the four replicate datum values. Next, suppose we run microcomputer program $RANDOM2$

with $n_{elpri} = 4$, $n_{digit} = 2$, and seed numbers 115 283 967 to generate the four replicate datum values 68, 96, 42, and 9. Accordingly, for the conceptual statistical model:

$$
CRSIDV_i\text{'s} \quad\quad csdm \quad\quad CRSIEE_i\text{'s}
$$

$$
\begin{vmatrix} 68 \\ 96 \\ 42 \\ 9 \end{vmatrix} = \begin{vmatrix} 49.5 \\ 49.5 \\ 49.5 \\ 49.5 \end{vmatrix} + \begin{vmatrix} +18.5 \\ +46.5 \\ -7.5 \\ -40.5 \end{vmatrix}
$$

$$\text{sum} \neq 0.0$$

whereas for the estimated conceptual statistical model:

$$
\text{replicate datum value}_i\text{'s} \quad\quad \text{est}(csdm) \quad\quad \text{est}(CRSIEE_i\text{'s})
$$

$$
\begin{vmatrix} 68 \\ 96 \\ 42 \\ 9 \end{vmatrix} = \begin{vmatrix} 53.75 \\ 53.75 \\ 53.75 \\ 53.75 \end{vmatrix} + \begin{vmatrix} +14.25 \\ +42.25 \\ -11.75 \\ -44.75 \end{vmatrix}
$$

$$\text{sum} = 0.0$$

Obviously, est($csdm$) is not exact. Rather, est($csdm$) is equal to the sum of the actual value of the $csdm$, 49.5, plus the realization of its intrinsic statistical estimation error component, which, in this numerical example, is $53.75 - 49.5 = 4.25$. However, each time that our example experiment test program is replicated (as required in Exercise Set 3) a different realization value for this statistical estimation error component will (very likely) be computed. Note that we do not know *a priori* either its magnitude or sign. Note also that the associated est($CRSIEE_i$'s) are not exact. Rather each est($CRSIEE_i$) is equal to the actual value for its corresponding $CRSIEE_i$ plus the realization of its intrinsic statistical estimation error component, which, in this numerical example, is equal to -4.25.

This elementary example is intended to support two fundamental statistical concepts. First, every statistical estimate, whatever its associated statistical estimator (estimation expression), is the sum of the actual value for the quantity being estimated plus an intrinsic statistical estimation error component. Second, the collection of all possible intrinsic statistical estimation error components that would occur under continual replication of the experiment test program establishes the conceptual statistical distribution (technically termed a conceptual sampling distribution) for this intrinsic statistical estimation error component. Clearly, the more we understand about the analytical and numerical nature of this conceptual statistical distribution (conceptual sampling distribution), the more dependable the inferences (conclusions) that we can draw from the appropriate statistical analysis for a randomly selected outcome of a continually replicated experiment test program.

Exercise Set 3

These exercises are intended to support the notions that (a) every statistical estimate includes an intrinsic statistical estimation error component, and (b) the collection of all possible intrinsic statistical estimation error components that would occur under continual replication of the experiment test program establishes the conceptual sampling distribution for this intrinsic statistical estimation error component.

1. Replicate the text example experiment test program (at least) five times by running microcomputer program *RANDOM2* with $n_{elpri} = 4$, $n_{digit} = 2$, each time using a different set of three, three-digit odd seed numbers. Write the associated conceptual and estimated statistical models. Then, numerically explain the respective [est($csdm$)]'s and their associated [est($CRSIEE_i$'s)]'s as the sum of their actual values plus realizations of the corresponding intrinsic statistical estimation error components. Do these realizations differ from replicate experiment test program outcome to replicate experiment test program outcome? Discuss.

2. Suppose that microcomputer program *RANDOM2* is repeatedly run with $n_{elpri} = 4$, $n_{digit} = 2$ until all equally-likely experiment test program outcomes (distinct sets of pseudorandom datum values) have been enumerated. (a) Is the conceptual sampling distribution that consists of the collection of all possible realizations of the intrinsic statistical estimation error component more akin to Figure 1.5(a) or (b)? (b) Is the mean of its conceptual sampling distribution equal to zero? If so, then (i) the expected value of a randomly selected realization of the intrinsic statistical estimation error component is equal to zero, and (ii) the least-squares statistical estimator is unbiased. (c) Is this conceptual sampling distribution symmetrical about its mean? Discuss.

3. Is it practical to suppose that microcomputer program *RANDOM2* is repeatedly run with $n_{elpri} = 4$, $n_{digit} = 2$ until all equally-likely experiment test program outcomes have been enumerated? (Hint: How many equally-likely experiment test program outcomes can be enumerated?) Suppose instead that a microcomputer program is written to generate n_{rep} equally-likely experiment test program outcomes, where n_{rep} is very large, say, 10,000 or 100,000. (a) Could the resulting empirical sampling

distribution that consists of these n_{rep} realizations of the intrinsic statistical estimation error component be used to approximate the conceptual sampling distribution that consists of all possible realizations of the intrinsic statistical estimation error component? If so, (b) would the mass centroid of this empirical sampling distribution be exactly equal to the actual value for the *csdm*, or would it also have an intrinsic statistical estimation error component?

Exercise Set 4

These exercises are intended to review the mechanics concept of mass moment of inertia and the associated parallel-axis theorem.

1. Recall that the respective statistical weights (*sw*'s) pertaining to a collection of n_{dv} $CRSIDV_i$'s have identical values. Next, suppose that the mass moment of inertia of this collection of n_{dv} $CRSIDV_i$'s about the point c is expressed as

    mass moment of inertia about point c

    $$= \sum_{i=1}^{n_{dv}} sw \cdot (CRSIDV_i - c)^2$$

    Then, set the derivative of the mass moment of inertia with respect to c equal to zero to demonstrate that the mass moment of inertia takes on its minimum values when

    $$c = \sum_{i=1}^{n_{dv}} \frac{CRSIDV_i}{n_{dv}}$$

    viz., when c is equal to the *arithmetic average* of the respective $CRSIDV_i$'s.

2. Suppose that the mass moment of inertia of a collection of n_{dv} $CRSIDV_i$'s about point c is re-expressed as

    mass moment of inertia about point c

    $$= \sum_{i=1}^{n_{dv}} sw \cdot \{[CRSIDV_i - \text{ave}(CRSIDV_i\text{'s})]$$
    $$+ [\text{ave}(CRSIDV_i\text{'s} - c]\}^2$$

    in which ave($CRSIDV_i$'s) is the arithmetic average of the respective n_{dv} $CRSIDV_i$'s. (a) Derive the parallel-axis theorem by (i) expanding the squared term in this revised expression, (ii) stating the physical (mechanics) interpretations for the first and last terms, and (iii) explaining why the middle (the cross-product) term is equal to zero. In turn, (b) rewrite this parallel-axis theorem expression to pertain to the special situation where the sw for (mass of) each of the respective $CRSIDV_i$'s is dimensionless and is equal to one.

3. Given the respective $CRSIDV_i$'s re-expressed in our hybrid column vector notation as

$$|CRSIDV_i\text{'s} - c| = |CRSIDV_i\text{'s} - \text{ave}(CRSIDV_i\text{'s})| + |\text{ave}(CRSIDV_i\text{'s}) - c|$$

in which ave($CRSIDV_i$'s) is the arithmetic average of the respective n_{dv} $CRSIDV_i$'s, (a) demonstrate that the $|CRSIDV_i\text{'s} - \text{ave}(CRSIDV_i\text{'s})|$ column vector and the $|\text{ave}(CRSIDV_i\text{'s}) - c|$ column vector are orthogonal by explaining why their dot product is equal to zero. Next, (b) express the sum of squares of the elements in the $|CRSIDV_i\text{'s} - c|$ column vector as the sum of the respective sum of squares of the elements in the $|CRSIDV_i\text{'s} - \text{ave}(CRSIDV_i\text{'s})|$ and the $|\text{ave}(CRSIDV_i\text{'s}) - c|$ column vector. Then, (c) demonstrate that this expression is *identical* to the expression in Exercise 2(b).

Remark: The identity in (c) is exploited in Chapter 7 by using the mass moment of inertia expression pertaining to specific collections of $CRSIDV_i$'s with unit (dimensionless) masses to establish expressions for the *between* and *within* sums of squares in simple linear regression.

1.6. CLOSURE

Treatments with different random assignments to experimental units (or vice versa) will have different numbers of replicates and different statistical analyses. There is a monumental difference between a single measurement conducted on each of 10 different experimental units and 10 repeated measurements conducted on a specific experimental unit, even though there are 10 measurement datum values to be statistically analyzed in each case. This difference leads to the following axiom:

There is no valid statistical analysis without understanding exactly how the experiment test program was actually conducted.

Accordingly, Chapter 2 is concerned with the organizational structure of, and the test conduct details for, statistically planned experiment test programs with orthogonal conceptual statistical models. In turn, Chapter 3 is concerned with the statistical analysis of these experiment test programs. The sequentially arranged terminology presented in Supplemental Topic 1.A is intended to provide the background that is fundamental to the presentations and discussions in Chapters 2 and 3.

1.A. SUPPLEMENTAL TOPIC: PLANNED EXPERIMENT TEST PROGRAM STATISTICAL TERMINOLOGY

The following statistical terminology is intended to enhance understanding of the discussions found in Chapters 2–5.

Conceptual Statistical Distribution . . . . . . Given any specific set of test conditions within a given experiment test program, the associated conceptual statistical distribution consists of all possible replicate conceptual random datum values that conceptually would be obtained if this experiment test program were continually replicated indefinitely.

Conceptual Sampling Distribution . . . . Given any specific experiment test program, a conceptual sampling distribution consists of all possible replicate realization values for a statistical estimator (or for a statistic) that conceptually would be obtained if this experiment test program were continually replicated indefinitely.

Conceptual Statistical Model . . . . . . The analytical expression that consists of the presumed deterministic physical model for the experiment test program datum values plus a conceptual experimental error term.

Conceptual Experimental Error . . . The term that must be added to the presumed deterministic physical model to explain the variability of experiment test program datum values. The mean of the conceptual statistical distribution that consists of all possible replicate realization values for the conceptual experimental errors is equal to zero (because the deterministic portion of the conceptual statistical model is presumed to be correct).

Mean . . . . For conceptual statistical (sampling) distributions with a discrete measurement metric, the statistical analog to the mass centroid in mechanics. However, for conceptual statistical (sampling) distributions with a continuous measurement metric, it is the statistical analog to an area centroid in mechanics. The units for the mean are the units for the measurement metric that is employed in the experiment test program. Although the mean is an intuitive location parameter for all conceptual statistical (sampling) distribution, its statistical use as a location parameter is typically limited to symmetrical conceptual statistical (sampling) distributions.

Variance . . . . For conceptual statistical (sampling) distributions with a discrete measurement metric, the statistical analog to the mass moment of inertia about the mass centroid, divided by the sum of the masses. However, for conceptual statistical (sampling) distributions with a continuous measurement metric, it is the statistical analog to an area moment of inertia about the area centroid, divided by the (total) area. The units for the variance are the square of the units for the measurement metric that is employed in the experiment test program.

Standard Deviation . . . . The square root of the variance. Its units are same as the units for the mean, viz., the units for the measurement metric that is employed in the experiment test program. Although the standard deviation is an intuitive scale parameter for all conceptual statistical (sampling) distributions, its statistical use as a scale parameter is typically limited to conceptual (two-parameter) normal (statistical and sampling) distributions (Chapter 5).

Conceptual Parameter	A parameter in a conceptual statistical distribution or in a conceptual statistical model. In our hybrid column vector notation, its magnitude is established by its scalar coefficient.
Statistical Estimator	The estimation expression (algorithm) that is used to compute the statistical estimate of the actual value for a conceptual parameter (or its scalar coefficient). Least-squares and maximum likelihood statistical estimators are employed in this text.
Statistical Estimate	The realization value obtained by appropriately substituting the experiment test program datum values into the corresponding statistical estimator (estimation expression, estimation algorithm). Statistical estimates always include the realization of an intrinsic statistical estimation error component confounded with the actual value for the quantity being estimated.
Unbiased Statistical . . . . *Estimator*	A statistical estimator whose statistical bias is equal to zero, viz., whose difference between actual value being estimated and the mean of its conceptual sampling distribution is equal to zero. The expected value for the intrinsic statistical estimation error component is equal to zero for unbiased statistical estimators, viz., the mean (centroid) of the conceptual sampling distribution that consists of all possible replicate values for the intrinsic statistical estimation error is equal to zero.
Statistical Bias	The difference between the actual value for the quantity being estimated and the mean of the conceptual sampling distribution that consists all possible replicate values for the associated statistical estimate. The expected value for the intrinsic statistical estimation error component is equal to the statistical bias for biased statistical estimators, viz., the mean (centroid) of the conceptual sampling distribution that con-

sists of all possible replicate values for the intrinsic statistical estimation error component is equal to the statistical bias.

Statistic . . . . The expression (algorithm) that, given the respective experiment test program datum values, is used to establish the realization value of specific interest. This realization value (also termed a statistic) is presumed to have been randomly selected from its associated conceptual sampling distribution. (Statistical estimators and statistical estimates are statistics.)

Test Statistic . . . . A statistic that is used in statistical analysis to test a given null hypothesis versus its physically relevant alternative hypothesis.

Treatments . . . . Treatments are the test variables (test conditions) being studied in an experiment test program. Treatments are distinct, e.g., either zinc-plated or shot-peened, whereas treatment levels are quantitative, e.g., zinc-plated, either 1, 2, or 3 mils thick. Treatments (treatment levels, treatment combinations) are randomly assigned to nominally identical experimental units.

Blocks . . . . Blocks are groups of experimental units and/or test conditions that are deliberately selected in an experiment test program to be as nominally identical as practical. Blocks are employed in the experiment test program to mitigate the spurious effects of batch-to-batch effects and differences among test conditions on the respective estimated values for the treatment effects of specific interest.

2

Planned Experiment Test Programs with Orthogonal Conceptual Statistical Models

2.1. INTRODUCTION

The basic feature of a planned experiment test program is the overt relationship between its conceptual statistical model and its organizational structure. Column vector notation is used in the planned experiment test program examples that follow to highlight the orthogonality relationships underlying their conceptual statistical models.

2.2. COMPLETELY RANDOMIZED DESIGN EXPERIMENT TEST PROGRAMS

2.2.1. Quantitative (Single Treatment) Experiment Test Programs

The simplest statistically planned experiment is a completely randomized design (CRD) experiment test program with a single treatment. It is called a quantitative experiment test program. The usual test objective is to estimate a mechanical behavior value. If replicate measurements of this mechanical behavior exhibit obvious statistical variability, then a statistical model is appropriate. This statistical model essentially asserts that the mechanical behavior value is actually the mean of the conceptual statistical distribution

29

that consists of all possible replicate mechanical behavior measurement values.

Recall that replicate measurements made under nominally identical conditions on nominally identical test specimens (experimental units) are presumed to generate (mutually independent) conceptual random statistically identical datum values, denoted $CRSIDV_i$'s. Thus, the conceptual statistical model for a quantitative CRD experiment test program with n_r replicate measurement values is stated in our hybrid column vector notation as

$$|CRSIDV_i\text{'s}| = |csdm| + |CRSIEE_i\text{'s}| = csdm \cdot | + 1\text{'s}| + |CRSIEE_i\text{'s}|$$

in which the $csdm$ denotes the conceptual statistical distribution mean, viz., the centroid of the conceptual statistical distribution that consists of all possible realization values for the $CRSIDV_i$'s under continual replication of the experiment test program, and the $CRSIEE_i$'s are deviations of the associated $CRSIDV_i$'s from the $csdm$. Note that the expected value of each of the respective $CRSIEE_i$'s is equal to zero (by definition). Note also that if the conceptual statistical distribution consisting of $APRCRSIDV$'s is symmetrical about its mean, then both negative and positive $CRSIEE_i$'s are equally likely.

> *Remark*: Our notation employs capital letters to connote conceptual random datum values and conceptual random errors and lower case letters to connote the actual (numerical) values for conceptual model parameters and experiment test program datum values. This notational distinction, however, is not critical in this chapter.

The associated estimated conceptual statistical column vector model for this quantitative CRD experiment test program is written as

|experiment test program datum value$_i$'s|

$$= |\text{est}(csdm)| + |\text{est}(CRSIEE_i\text{'s}|$$
$$= \text{est}(csdm) \cdot | + 1\text{'s}| + |\text{est}(CRSIEE_i\text{'s})|$$

in which, under continual replication of the experiment test program, the |experiment test program datum values| column vector has n_r statistical degrees of freedom (defined later), the |est($csdm$)| column vector has one statistical degree of freedom, and the |est($CRSIEE_i$'s)| column vector has $(n_r - 1)$ statistical degrees of freedom.

The conceptual statistical model also involves the explicit presumption that the respective $CRSIEE_i$'s are random and the implicit presumption that these $CRSIEE_i$'s are mutually independent. Thus, a critical step in statistical analysis for a quantitative CRD experiment test program is to examine the respective est($CRSIEE_i$'s) relative to a lack of randomness, e.g., the presence of an obvious time-order-of-testing trend.

2.2.1.1. Numerical Example of a Quantitative CRD Experiment Test Program

Suppose we have 16 nominally identical fatigue test specimens that (we allege) are representative of the population of all possible nominally identical test specimens of specific interest. However, we choose to conduct only eight fatigue tests in our quantitative CRD experiment test program. Depicting each of the eight required fatigue test specimens by a rectangle in Figure 2.1, we randomly select eight of the 16 nominally identical test specimens by running microcomputer program *RANDOM1*; with $n_{elpri} = 8$ and $n_i = 16$. The resulting randomization outcome is indicated by the test specimen number printed inside each of the rectangles that are intended to depict a fatigue test specimen. In turn, the experiment test program fatigue tests are conducted back-to-back in random time order by the same test technician using the same fatigue test machine. The random time order of testing is indicated by the subscript number in parentheses located at the lower right-hand corner of each fatigue test specimen rectangle. It is established by running microcomputer program *RANDOM1* with $n_{elpri} = 8$ and $n_i = 8$.

We now construct hypothetical data for this example CRD quantitative experiment test program such that an obvious time-order-of-testing effect exists. The conventional algebraic least-squares estimate of the actual value for the *csdm* is the arithmetic average of the respective experiment test program datum values. The resulting est($CRSIEE_i$'s) are computed as "residuals," viz., as the numerical deviations of the respective experiment test

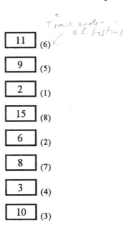

Figure 2.1 Schematic of the organizational structure of a quantitative completely randomized design (CRD) experiment test program. Each rectangle is intended to depict a fatigue test specimen. The respective randomization details of this CRD experiment test program are discussed in the text.

program datum values from their arithmetic average. These deviations sum to zero (by definition). Accordingly, the $|+1\text{'s}|$ column vector and the $|\text{est}(CRSIEE_i\text{'s})|$ column vector are orthogonal (because the dot product of the $|+1\text{'s}|$ column vector and the $|\text{est}(CRSIEE_i\text{'s})|$ column vector is equal to zero).

hypothetical datum value$_i$'s, arranged in time order of testing		est($csdm$)		est($CRSIEE_i$'s)
1		4.5		−3.5
2		4.5		−2.5
3	=	4.5	+	−1.5
4		4.5		−0.5
5		4.5		+0.5
6		4.5		+1.5
7		4.5		+2.5
8		4.5		+3.5

sum = 0.0

These est($CRSIEE_i$'s) are plotted in Figure 4.4(a). Visual inspection of this plot indicates an obvious time-order-of-testing trend. Consequently we have a rational basis to doubt the credibility of the presumed conceptual statistical model. (Time-order-of-testing trends are generally caused by improper test conduct and/or the absence of proper randomization.)

Suppose, however, that the hypothetical datum values for this quantitative CRD experiment test program example actually pertain to the randomly selected time order of testing (from smallest to largest) for the fatigue test specimens depicted schematically in Figure 2.1, viz.,

hypothetical datum value$_i$'s, re-arranged in the random time order of testing in Figure 2.1		est($csdm$)		est($CRSIEE_i$'s)
6		4.5		+1.5
5		4.5		+0.5
1	=	4.5	+	−3.5
8		4.5		+3.5
2		4.5		−2.5
7		4.5		+2.5
4		4.5		−0.5
3		4.5		−1.5

sum = 0.0

These est($CRSIEE_i$'s) are plotted in Figure 4.4(b). Supplemental Topic 4.A. As expected, visual inspection of this plot indicates no obvious time-order-of-testing trend. Accordingly, we do not have a rational basis to doubt the credibility of the presumed conceptual statistical model.

In turn, suppose that these randomly rearranged hypothetical datum values are again rearranged to conform with the location order (from smallest to largest) of the randomly selected test specimen blanks within the rod that was cut to form the 16 blanks that comprise the population of specific interest. Then,

hypothetical datum value$_i$'s, re-rearranged to conform to the random specimen blank order in Figure 2.1		est($csdm$)		est($CRSIEE_i$'s)
1		4.5		−3.5
4		4.5		−0.5
2	=	4.5	+	−2.5
7		4.5		+2.5
5		4.5		+0.5
3		4.5		−1.5
6		4.5		+1.5
8		4.5		+3.5

$$\text{sum} = 0.0$$

These est($CRSIEE_i$'s) are plotted in Figure 4.4(c). Again, as expected, visual inspection of this plot indicates no obvious time-order-of-testing trend. However, the suggestion of a possible test specimen blank location effect is sufficiently strong to require that a formal statistical test be conducted. For the present, the fundamental issue is that we are obliged to examine the respective est($CRSIEE_i$'s) relative to each presumption underlying the conceptual statistical model (whenever possible).

2.2.1.2. Discussion

Test specimens and their materials are always produced and processed in batches. Thus, when conducting a quantitative experiment test program, it is always prudent to employ test specimens (experimental units) that are selected from two or more diverse sources and thus potentially exhibit marked batch-to-batch effects. Then it is statistically rational to estimate the actual value for the $csdm$ if and only if the null hypothesis that there are no batch-to-batch effects is not rejected in subsequent statistical analysis.

It is never prudent to presume that the experimental unit batch-to-batch effect is negligible without competent experimental verification. Similarly, it is never prudent to presume that test machine and test technician effects are negligible without competent experimental verification.

Exercise Set 1

These exercises are intended to introduce an array of mutually orthogonal column vectors that establish the estimated conceptual statistical model when certain of these column vectors are appropriately aggregated.

Consider the following estimated complete analytical model that consists of eight mutually orthogonal column vectors. This estimated complete analytical model completely explains the first text example |hypothetical datum value$_i$'s| column vector, viz., the algebraic sum of the ith element in each of the j column vectors is equal to the ith hypothetical datum value. The |est($CRSIEE_i$'s)| column vector in the estimated statistical model is constructed by aggregating the respective elements of the last seven column vectors. (The criteria for selecting the column vectors to be aggregated are discussed later.)

hypothetical		seven mutually orthogonal column vectors that establish						
datum value$_i$'s	est($csdm$)	the est($CRSIEE_i$'s) column vector when aggregated						
1	4.5	−0.5	−0.5	−0.5	−0.5	−0.5	−0.5	−0.5
2	4.5	+0.5	−0.5	−0.5	−0.5	−0.5	−0.5	−0.5
3	4.5	0.0	+1.0	−0.5	−0.5	−0.5	−0.5	−0.5
4	4.5	0.0	0.0	+1.5	−0.5	−0.5	−0.5	−0.5
5	4.5	0.0	0.0	0.0	+2.0	−0.5	−0.5	−0.5
6	4.5	0.0	0.0	0.0	0.0	+2.5	−0.5	−0.5
7	4.5	0.0	0.0	0.0	0.0	0.0	+3.0	−0.5
8	4.5	0.0	0.0	0.0	0.0	0.0	0.0	+3.5
sum =	0.0	0.0	0.0	0.0	0.0	0.0	0.0	0.0

(where each column after "datum value$_i$'s" is joined by $=$ and $+$ signs respectively)

1. (a) Verify that the eight column vectors that comprise the estimated complete analytical model are mutually orthogonal. Next, (b) aggregate the last seven of these column vectors to obtain the first text example |est($CRSIEE_i$'s)| column vector. Then, (c) demonstrate that the sum of the respective sums of squares of the elements of these seven mutually orthogonal column vectors equals the sum of squares of the elements of this |est($CRSIEE_i$'s)| column vector. Finally, (d) demonstrate that the sum of squares of the elements of this |est($CRSIEE_i$'s)| column vector plus the sum of squares of the elements of the associated |est($csdm$)| column vector is equal to the sum of squares of the elements of the |hypothetical datum value$_i$'s| column vector.

2. We will demonstrate later that (i) the direction of each of the column vectors that comprise the estimated complete analytical model is fixed, and (ii) the length of each of these column vectors

is established by the observed experiment test program datum values. If so, (a) how many statistical degrees of freedom does each of these column vectors have under continual replication of the experiment test program? In turn, (b) how many statistical degrees of freedom does a column vector have that is established by aggregating seven mutually orthogonal column vectors in the estimated complete analytical model?

3. Consider again the estimated conceptual statistical model for each of the text numerical examples. (a) How many statistical degrees of freedom does the |hypothetical datum value$_i$'s)| column vector have? (b) How many statistical degrees of freedom does the |est($csdm$)| column vector have? (c) How many statistical degrees of freedom does the |est($CRSIEE_i$'s)| column vector have? Is the sum of (b) and (c) equal to (a)? Comment appropriately.

2.2.2. Comparative (Multiple Treatment) Experiment Test Programs

The simplest comparative CRD experiment test program involves either (a) two treatments, or (b) one treatment with two levels. In turn, the simplest conceptual statistical model asserts that the two conceptual statistical distributions that respectively consist of $APRCRSIDV_i$'s for these two treatments (treatment levels) are identical except that the actual values for their respective means (may) differ. Then, as discussed in Section 1.4, all of the respective experiment test program datum values have identical values for their statistical weights and thus are homoscedastic. Accordingly, the actual value for the mean of each conceptual statistical distribution of specific interest can be rationally estimated by the arithmetic average of the appropriate experiment test program datum values.

It is convenient in stating the conceptual statistical model to define a conceptual treatment effect (cte) as the actual value for the mean of the conceptual statistical distribution that consists of $APRCRHDV_i$'s for that treatment (treatment level) minus the actual value for the mean of the conceptual statistical distribution that consists of $APRCRHDV_i$'s for all treatments (viz., for the entire experiment test program). The latter mean is termed the conceptual statistical model mean and is denoted $csmm$. In more general perspective, a conceptual statistical effect is defined as the deviation of the actual value for its associated conceptual mean from the actual value for the $csmm$. Then, when the corresponding estimates are computed using arithmetic averages, (a) these estimates always sum to zero, (b) the associated estimated statistical model consists of mutually orthogonal column vectors, and (c) the associated estimated scalar coefficients are independent under continual replication of the experiment test program.

It is good statistical practice to keep the number of treatments (treatment levels) of specific interest as small as practical and to increase the replication of these treatments (treatment levels) as much as possible. On the other hand, suppose that (a) n_t treatments (treatment levels) are to be compared, (b) each treatment (treatment level) has n_r replicates, and (c) each individual replicate test is conducted on a nominally identical experimental unit. The resulting equally replicated CRD experiment test program requires $(n_r) \cdot (n_t)$ nominally identical experimental units and nominally invariant test conditions throughout the entire experiment test program. Accordingly, unless our experimental units are routinely processed in homogeneous batches at least as large as $(n_r) \cdot (n_t)$, and unless the test conditions can be viewed as being invariant for practical purposes throughout the entire experiment test program, an equally replicated CRD experiment test program is inappropriate. Even then we must presume that the spurious effects of all experiment test program nuisance variables effects are negligible.

The conceptual statistical model for a comparative CRD experiment test program with two or more (equally replicated) treatments (treatment levels) is written in our hybrid column vector notation as

$$|CRHDV_i\text{'s}| = |csmm| + |cte_i\text{'s}| + |CRHEE_i\text{'s}|$$

in which the *csmm* is the conceptual statistical model mean and the *cte*'s are the respective conceptual treatment effects. The associated estimated statistical model for this CRD experiment test program is

|experiment test program datum value$_i$'s|

$$= |\text{est}(csmm)| + |\text{est}(cte_i\text{'s}| + |\text{est}(CRHEE_i\text{'s})|$$

in which the |experiment test program datum value$_i$'s| column vector has $(n_r) \cdot (n_t)$ statistical degrees of freedom, the |est($csmm$)| column vector has one statistical degree of freedom, the |est(cte_i's)| column vector has $(n_t - 1)$ statistical degrees of freedom, and the |est($CRHEE_i$'s)| column vector has $(n_t - 1) \cdot (n_t)$ statistical degrees of freedom.

The conventional algebraic least-squares estimate of the actual value for the *csmm* is the arithmetic average of all experiment test program datum values. (It is of little interest in a comparative experiment test program.) In turn, the conventional algebraic least-squares estimates of the actual values for respective *cte$_i$*'s are the arithmetic averages of the experiment test program datum values pertaining to each respective treatment (treatment level) minus the arithmetic average of all experiment test program datum values.

Least-squares estimation assures that (a) the respective est(cte_i's) sum to zero, and (b) the |est(cte_i's)| column vector is orthogonal to the |est($csmm$)| column vector. Finally, the resulting est($CRHEE_i$'s) are computed numerically as "residuals," viz., as values that satisfy the scalar equations associated with the respective experiment test program datum value$_i$'s. These est($CRHEE_i$'s) sum to zero collectively, *and* for each respective conceptual treatment (treatment level), thereby assuring that the |est($CRHEE_i$'s)| column vector is orthogonal to both the |est($csmm$)| column vector and the |est(cte_i's)| column vector. The orthogonality relationships generated by least-squares estimators assure that the *within* and *between* sums of squares that underlie the classical statistical analysis of variance (Chapter 6) are statistically independent.

2.2.2.1. Numerical Example for a Comparative CRD Experiment Test Program with (Only) Two Treatments (Treatment Levels)

Suppose we wish to compare the fatigue lives of laboratory specimens with circumferentially and longitudinally polished surfaces. Let circumferentially polished be denoted treatment A and longitudinally polished be denoted treatment B. Then, the null hypothesis is that the two conceptual statistical distributions that respectively consist of $APRCRHDV_i$'s for treatments A and B are identical, whereas the alternative hypothesis is that these two conceptual statistical distributions differ only in the actual values for their respective means.

Suppose also, that to test this null hypothesis versus its given alternative hypothesis, (a) we prepare 16 nominally identical fatigue test specimen blanks, each depicted by a rectangle in Figure 2.2, and (b) we randomly assign treatments A and B to these test specimen blanks by running microcomputer program $RANDOM1$ with $n_{elpri} = 16$ and $n_i = 16$. The outcome of this randomization procedure is indicated by the test specimen blank number printed inside each of the respective rectangles. In turn, the respective laboratory fatigue tests must be conducted in random time order using the same test machine, test technician, and test procedure. Otherwise, the spurious effects of different test machines, different test technicians, or different test procedures will be confounded with the est(cte's)—unless these spurious effects are appropriately balanced (as in a randomized complete block experiment test program, Section 2.3). The random time order of fatigue testing is established by running microcomputer program $RANDOM1$ with $n_{elpri} = 16$ and $n_i = 16$, but using a different set of three three-digit odd seed numbers. This random time order is indicated by the

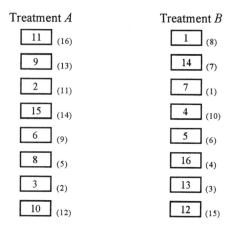

Treatment A Treatment B

| 11 | (16) | | 1 | (8) |

| 9 | (13) | | 14 | (7) |

| 2 | (11) | | 7 | (1) |

| 15 | (14) | | 4 | (10) |

| 6 | (9) | | 5 | (6) |

| 8 | (5) | | 16 | (4) |

| 3 | (2) | | 13 | (3) |

| 10 | (12) | | 12 | (15) |

Figure 2.2 Schematic of the organizational structure of a comparative CRD experiment test program involving two treatments, viz., treatments A and B, each with eight replicates.

subscript number in parentheses located at the lower right-hand corner of each test specimen rectangle.

However, we recommend that all mechanical reliability experiment test programs be conducted with time blocks for the time order of testing (even though these time blocks are ignored in the statistical analysis for a CRD experiment test program). Consider Figure 2.3, where the time order

Treatment A Treatment B

| 11 | (2) | | 1 | (7) |

| 9 | (5) | | 14 | (6) |

| 2 | (1) | | 7 | (2) |

| 15 | (8) | | 4 | (4) |

| 6 | (6) | | 5 | (5) |

| 8 | (7) | | 16 | (8) |

| 3 | (4) | | 13 | (3) |

| 10 | (3) | | 12 | (1) |

Figure 2.3 Recommended time order of testing for the comparative CRD experiment test program whose organizational structure is depicted in Figure 2.2. See the text discussion for further details.

of testing is now randomized for each treatment. Accordingly, we recommend that treatment A specimen 2 and treatment B specimen 12 are tested back-to-back in random time order, before treatment A specimen 11 and treatment B specimen 7 are tested back-to-back in random time order, and so forth, until treatment A specimen 15 and treatment B specimen 16 are tested back-to-back in random time order. Recall, moreover, that one-half of these random time orders should start with a treatment A specimen and the other one-half should start with a treatment B specimen.

We now construct hypothetical datum values for this CRD experiment test program such that a direction of polishing effect (as well as a time-order-of-testing trend) clearly exists, viz.,

hypothetical datum value$_i$'s, arranged in time order of testing		est($csmm$)		est(cte_i's)		est($CRHEE_i$'s)	
1		8.5		−4.0		−3.5	
2		8.5		−4.0		−2.5	
3		8.5		−4.0		−1.5	
4		8.5		−4.0		−0.5	
5		8.5		−4.0		+0.5	
6		8.5		−4.0		+1.5	
7		8.5		−4.0		+2.5	
8		8.5		−4.0		+3.5	
=			+		+	—	sum = 0.0
9		8.5		+4.0		−3.5	
10		8.5		+4.0		−2.5	
11		8.5		+4.0		−1.5	
12		8.5		+4.0		−0.5	
13		8.5		+4.0		+0.5	
14		8.5		+4.0		+1.5	
15		8.5		+4.0		+2.5	
16		8.5		+4.0		+3.5	
						—	sum = 0.0
				sum = 0.0	sum = 0.0		

First, recall that est($csmm$) is the arithmetic average of all experiment test program datum values, viz., 8.5 in this numerical example. Then, since the arithmetic average of the treatment A datum values is equal to 4.5, the est(cte_i's) for treatment A are equal to $(4.5 - 8.5) = -4.0$. In turn, since the arithmetic average of the treatment B datum values is equal to 12.5, the est(cte_i's) for treatment B are equal to $(12.5 - 8.5) = +4.0$. Next, note that these est(cte_i's) sum to zero. Accordingly, the |est(cte_i's)| column vector is orthogonal to the |est($csmm$)| column vector. Finally, the resulting est($CRHEE_i$'s) are computed numerically as "residuals," viz., as values that satisfy the scalar equations associated with each respective experiment test program datum value$_i$. These est($CRHEE_i$'s) also sum to zero collectively and for each treatment. Thus, the respective column vectors in this estimated conceptual statistical model are mutually orthogonal.

> *Remark*: Although it may appear that the magnitudes of the est(cte_i's) are much too large in this (hypothetical) example to be explained as the sole consequence of the random variability of the $CRHEE_i$'s, proper statistical inference is burdened by the lack of credibility regarding the conceptual statistical model presumption that the respective $CRHEE_i$'s are random.

Next, we rearrange our hypothetical datum values 1 through 16 to conform to the random order of test specimen blank assignment to treatments A and B in Figure 2.2. Then, we compute the conventional algebraic least-squares estimates of the actual values for the *csmm*, the *cte*$_i$'s, and the $CRHEE_i$'s in our conceptual statistical model, viz.,

hypothetical datum value$_i$'s, rearranged in random test specimen blank assignment order in Figure 2.2		est(*csmm*)		est(*cte$_i$*'s)		est(*CRHEE$_i$*'s)	
11		8.5		−0.5		−3.0	
9		8.5		−0.5		−1.0	
2		8.5		−0.5		−6.0	
15		8.5		−0.5		+7.0	
6		8.5		−0.5		−2.0	
8		8.5		−0.5		+0.0	
3		8.5		−0.5		−5.0	
10		8.5		−0.5		+2.0	
	=		+		+	—	sum = 0.0
1		8.5		+0.5		−8.0	
14		8.5		+0.5		+5.0	
7		8.5		+0.5		−2.0	
4		8.5		+0.5		−5.0	
5		8.5		+0.5		−4.0	
16		8.5		+0.5		+7.0	
13		8.5		+0.5		+4.0	
12		8.5		+0.5		+3.0	
						— sum = 0.0	
		sum = 0.0		sum = 0.0			

Next, recall that every statistical estimate is equal to the actual value for the quantity being estimated plus an intrinsic statistical estimation error. However, the actual values of the *cte*$_i$'s is equal to zero for our second numerical example, because the expected values for the arithmetic averages of the eight datum values 1 through 16 randomly assigned to treatments A and B are both equal to the actual value for the $csmm$, viz., 8.5. Accordingly, the realization values for the respective intrinsic statistical estimation error components of the est(cte_i's) are exactly equal to the respective numerical values for the est(cte_i's).

We now demonstrate that the realization values for the intrinsic statistical estimation error components of the est(cte_i's) depend on the specific details of the random reassignment of our hypothetical experiment test program datum values to treatments A and B. Accordingly, we now reassign the hypothetical experiment test program datum values to treatments A and B, this time to conform to the respective random time orders of testing given in Figure 2.2. Then, we recompute the conventional algebraic least-squares estimates of the actual values for the $csmm$, the cte_i's, and the $CRHEE_i$'s in our conceptual statistical model, viz.,

hypothetical datum value$_i$'s, rearranged in random stime order of testing in Figure 2.2		est($csmm$)		est(cte_i's)		est($CRHEE_i$'s)	
16		8.5		+1.75		+5.75	
13		8.5		+1.75		+2.75	
11		8.5		+1.75		+0.75	
14		8.5		+1.75		+3.75	
9		8.5		+1.75		−1.25	
5		8.5		+1.75		+5.25	
2		8.5		+1.75		−8.25	
12		8.5		+1.75		+1.75	
=			+		+		sum = 0.0
8		8.5		−1.75		+1.25	
7		8.5		−1.75		+0.25	
1		8.5		−1.75		−5.75	
10		8.5		−1.75		+3.75	
6		8.5		−1.75		−0.75	
4		8.5		−1.75		−2.75	
3		8.5		−1.75		−3.75	
15		8.5		−1.75		+8.25	

sum = 0.0

sum = 0.0   sum = 0.0

Clearly, the intrinsic statistical estimation error components of the respective est(cte_i's) differ from our second to our third numerical examples. In fact, if the random reassignment of hypothetical experiment test program datum values were continually repeated, we would generate the conceptual statistical (sampling) distribution that consists of all possible realizations of the intrinsic statistical estimation error.

> *Remark*: This example comparative CRD experiment test program has a broader practical application when it is expanded to test the null hypothesis that no batch-to-batch effects exist for the fatigue specimen material of specific interest. Suppose that eight fatigue specimen blanks are selected from each of two different batches of the material of specific interest and that, in each batch, four are circumferentially polished and four are longitudinally polished. Then, the null hypothesis that no batch-to-batch effect exists can be tested statistically (Chapter 6). (The null hypothesis that this batch-to-batch effect does not interact with the direction of polishing effect can also be tested statistically.)

2.2.2.2. Discussion

A comparative CRD experiment test program can be extended to as many treatments (treatment levels) as desired, but its practicality in mechanical reliability diminishes markedly as the number of required nominally identical experimental units increases. These treatments (treatment levels) can be either *qualitative*, e.g., surface preparations such as shot-peening, zinc plating, or induction hardening, or *quantitative* levels, e.g., surface preparations, such as shot-peening, to several different Almen intensities, zinc plating to attain several different thicknesses, or induction hardening to several different case depths.

When the treatments (treatment levels) involve experimental unit (test specimen) processing procedures such as shot-peening, zinc plating, or induction hardening, the processing equipment controls must be reset to zero and then reset for each successive experimental unit (test specimen) in an independent attempt to meet the required treatment (treatment level) specification target value. Otherwise, the implicit presumption that the respective replicate $CRHDV_i$'s are mutually independent is not credible.

> *Remember*: However convenient and expedient it may seem, it is not statistically proper to have identical settings of the processing equipment controls for each replicate experimental unit receiving the same treatment (treatment level). Rather, it is the statistically proper experimental unit preparation and test conduct randomization details that make the respective replicate $CRHDV_i$'s as mutually independent as practical.

2.2.3. Treatment Combinations in Factorial Arrangements

It is often necessary in comparative experiment test programs to assign two or more treatments *in combination* to an experimental unit, applied either simultaneously or sequentially without specific regard to order of application (completely randomized design and randomized complete block design experiment test programs) or sequentially with specific regard to order of application (split-plot design experiment test programs). To begin, we presume that our treatments can be applied simultaneously or sequentially without regard to order of application. We also presume that each of the n_t treatments, at each of its n_{tl} treatment levels, are applied in combination with all other treatments, at each of their n_{tl} treatment levels. This treatment combination configuration is called a *factorial arrangement*. When the factorial arrangement is employed in a CRD experiment test program, the latter is called a factorial design experiment test program.

An unreplicated $(2)^2$ factorial design experiment test program appears in Figure 2.4. Its $(n_{tl})^{n_t} = (2)^2 = 4$ treatment combinations are randomly assigned to four nominally identical experimental units (test specimens). The associated conceptual statistical column vector model has four mutually orthogonal column vectors, viz.,

$$|CRHDV_i\text{'s}| = |csmm| + |ct1e_i\text{'s}| + |ct2e_i\text{'s}| + |ct1t2ie_i\text{'s}|$$

in which $ct1e$ and $ct2e$ denote conceptual treatment effect one and conceptual treatment effect two, also called main effects, and $ct1t2ie$ denotes their interaction, viz., the conceptual *t1, t2* (two-factor) interaction effect. The physical basis for this two-factor interaction (synergistic) effect is almost

		Treatment Two	
		low level	*high level*
Treatment One	*low level*	experimental unit 3(1)	experimental unit 2(2)
	high level	experimental unit 1(4)	experimental unit 4(3)

Figure 2.4 Organizational structure for an unreplicated CRD $(2)^2$ factorial design experiment test program.

always much more difficult to understand than the physical basis for the associated conceptual treatment (main) effects. In fact, from a pragmatic perspective, the interaction effect term represents the failure of a simple additive (main effects) conceptual statistical model to explain the observed experiment test program datum values in a statistically adequate manner.

The treatment levels in a factorial arrangement may be either quantitative, e.g., temperatures equal to 25°C and 100°C, or qualitative, e.g., shot-peened and not shot-peened. The respective treatments (main effects) must be such that their time order of application is irrelevant.

2.2.4. Orthogonality and Yate's Enumeration Algorithm

We now use Yate's enumeration algorithm to state the main effect contrasts for $(2)^2$ and $(2)^4$ factorial arrays, Figure 2.5, where the low level of each

Specimen Number	$|ct1ec_i\text{'s}|$	$|ct2ec_i\text{'s}|$
1	−1	−1
2	+1	−1
3	−1	+1
4	+1	+1

(a) $(2)^2 = 4$

Specimen Number	$|ct1ec_i\text{'s}|$	$|ct2ec_i\text{'s}|$	$|ct3ec_i\text{'s}|$	$|ct4ec_i\text{'s}|$
1	−1	−1	−1	−1
2	+1	−1	−1	−1
3	−1	+1	−1	−1
4	+1	+1	−1	−1
5	−1	−1	+1	−1
6	+1	−1	+1	−1
7	−1	+1	+1	−1
8	+1	+1	+1	−1
9	−1	−1	−1	+1
10	+1	−1	−1	+1
11	−1	+1	−1	+1
12	+1	+1	−1	+1
13	−1	−1	+1	+1
14	+1	−1	+1	+1
15	−1	+1	+1	+1
16	+1	+1	+1	+1

(b) $(2)^4 = 16$

Figure 2.5 Main effects contrast array for an unreplicated $(2)^{nt}$ factorial design experiment test program, where $n_t = 2$ for (a) and $n_t = 4$ for (b). The elements of the main effects contrast array for a factorial design are established using Yate's enumeration algorithm in which the two levels for treatment one are denoted (coded) −1, +1, −1, +1, ..., the two levels for treatment two are denoted (coded) −1, −1, +1, +1, ..., etc. Note that (i) the elements of each contrast column vector sum to zero (by definition), and (ii) the respective contrast column vectors are mutually orthogonal.

treatment corresponds to a minus one and the high level corresponds to a plus one. Note that (a) the elements of the respective contrast column vectors sum to zero (by definition), and (b) the respective contrast column vectors are mutually orthogonal.

The associated complete analytical model consists of $(2)^{nt}$ terms that completely explain the experiment test program datum values without employing an experimental error term. For example, the complete analytical model for an unreplicated $(2)^2$ factorial design experiment test program is written in our hybrid column vector notation as

$$|CRHDV_i\text{'s}| = (csmmsc) \cdot |+1\text{'s}| + (ct1esc) \cdot |ct1ec_i\text{'s}| + (ct2esc) \cdot |ct2ec_i\text{'s}|$$
$$+ (ct1t2iesc) \cdot |ct1t2iec_i\text{'s}|$$

in which $csmmsc$ denotes the conceptual statistical model mean scalar coefficient, $ct1esc$ the conceptual treatment-one effect scalar coefficient, $ct2esc$ the conceptual treatment-two effect scalar coefficient, and $ct1t2iesc$ the conceptual treatment-one, treatment-two interaction effect scalar coefficient.

Similarly, the complete analytical vector model for an unreplicated $(2)^4$ factorial design experiment test program can be written in expanded column vector format as

$$|CRHDV_i\text{'s}| = (csmmsc) \cdot |+1\text{'s}| + (ct1esc) \cdot |ct1ec_i\text{'s}| + (ct2esc) \cdot |ct2ec_i\text{'s}|$$
$$+ (ct3esc) \cdot |ct3ec_i\text{'s}| + (ct4esc) \cdot |ct4ec_i\text{'s}|$$
$$+ (ct1t2iesc) \cdot |ct1t2iec_i\text{'s}| + (ct1t3iesc) \cdot |ct1t3iec_i\text{'s}|$$
$$+ (ct1t4iesc) \cdot |ct1t4iec_i\text{'s}| + (ct2t3iesc) \cdot |ct2t3iec_i\text{'s}|$$
$$+ (ct2t4iesc) \cdot |ct2t4iec_i\text{'s}| + (ct3t4iesc) \cdot |ct3t4iec_i\text{'s}|$$
$$+ (ct1t2t3iesc) \cdot |ct1t2t3iec_i\text{'s}| + (ct1t2t4iesc) \cdot |ct1t2t4iec_i\text{'s}|$$
$$+ (ct1t3t4iesc) \cdot |ct1t3t4iec_i\text{'s}| + (ct2t3t4iesc) \cdot |ct2t3t4iec_i\text{'s}|$$
$$+ (ct1t2t3t4iesc) \cdot |ct1t2t3t4iec_i\text{'s}|$$

The correspondence between the complete analytical model and its associated orthogonal augmented contrast array is evident only when the main effects contrast array (Figure 2.5) is augmented. As illustrated in Figure 2.6, the first column vector in the orthogonal augmented contrast array is the $|+1\text{'s}|$ identity column vector. It is orthogonal to each of the contrast column vectors because the elements of each respective contrast column vector sum to zero (by definition). The respective elements of each of the interaction effect contrast column vectors are *computed* by multiplying the corresponding elements of the associated main effect contrast column vectors. For example, the respective elements of the two-factor $ct1t3ie$ contrast column vector in Figure 2.6 are

(a)	Specimen Number	$\|+1\text{'s}\|$	$\|ct1ec_i\text{'s}\|$	$\|ct2ec_i\text{'s}\|$	$\|ct1t2iec_i\text{'s}\|$
Orthogonal Augmented	1	+1	-1	-1	+1
Unreplicated $(2)^2$	2	+1	+1	-1	-1
Factorial Design	3	+1	-1	+1	-1
Contrast Array	4	+1	+1	+1	+1

(b)	Specimen Number	$\|+1\text{'s}\|$	$\|ct1ec_i\text{'s}\|$	$\|ct2ec_i\text{'s}\|$	$\|ct3ec_i\text{'s}\|$	$\|ct1t2iec_i\text{'s}\|$	$\|ct1t3iec_i\text{'s}\|$	$\|ct2t3iec_i\text{'s}\|$	$\|ct1t2t3iec_i\text{'s}\|$
	1	+1	-1	-1	-1	+1	+1	+1	-1
Orthogonal	2	+1	+1	-1	-1	-1	-1	+1	+1
Augmented	3	+1	-1	+1	-1	-1	+1	-1	+1
Unreplicated $(2)^3$	4	+1	+1	+1	-1	+1	-1	-1	-1
Factorial Design	5	+1	-1	-1	+1	+1	-1	-1	+1
Contrast	6	+1	+1	-1	+1	-1	+1	-1	-1
Array	7	+1	-1	+1	+1	-1	-1	+1	-1
	8	+1	+1	+1	+1	+1	+1	+1	+1

Figure 2.6 Orthogonal augmented contrast arrays for unreplicated $(2)^2$ and $(2)^3$ factorial design experiment test programs. The respective elements of each interaction contrast column vector is *computed* by multiplying the corresponding elements of the associated main effect contrast column vectors.

$$ct1t3iec = \begin{vmatrix} -1 \cdot -1 = +1 \\ +1 \cdot -1 = -1 \\ -1 \cdot -1 = +1 \\ +1 \cdot -1 = -1 \\ -1 \cdot +1 = -1 \\ +1 \cdot +1 = +1 \\ -1 \cdot +1 = -1 \\ +1 \cdot +1 = +1 \end{vmatrix}$$

Exercise Set 2

These exercises are intended to demonstrate (a) the mutual orthogonality of all of the column vectors in the orthogonal augmented contrast array pertaining to an unreplicated $(2)^{n_i}$ factorial design experiment test program, and (b) the correspondence between the individual terms of the complete analytical model and the associated column vectors in the respective orthogonal augmented contrast arrays.

1. Verify that the column vectors in each of the orthogonal augmented contrast arrays in Figure 2.6 are respectively mutually orthogonal by evaluating the appropriate respective dot products.

2. (a) Construct the orthogonal augmented contrast array for an unreplicated $(2)^4$ factorial design experiment test program. (b)

How many dot products must be equal to zero to assure that all of the column vectors in this orthogonal augmented contrast array are in fact mutually orthogonal?

3. (a) Construct the orthogonal augmented contrast array for an unreplicated $(2)^5$ factorial design experiment test program. Then, (b) demonstrate that there are various alternative ways to compute the elements of a three-factor and higher-order interaction effect contrast column vector.

2.2.5. Column-Vector-Based Least-Squares Estimation

In the column-vector-based least-squares estimation procedure (as opposed to the conventional algebraic least-squares estimation procedure based on arithmetic averages), each scalar coefficient in the complete analytical model is estimated as follows: (a) evaluate the dot product of the experiment test program datum values column vector and the corresponding column vector in the orthogonal augmented contrast array and then (b) divide this dot product value by the sum of squares of the integer value elements of the corresponding column vector. Accordingly, the column-vector-based least-squares estimate of the jth scalar coefficient in the complete analytical model is

$$
\text{est}(sc_j) = \frac{\displaystyle\sum_{i=1}^{n_{dv}} c_{j,i}(\text{experiment test program datum value})_i}{\displaystyle\sum_{i=1}^{n_{dv}} c_{j,i}^2}
$$

in which $c_{j,i}$ is the integer value of the ith element in the jth column vector of the orthogonal augmented contrast array. For example, suppose the test specimen numbers in Figure 2.6(b) are hypothetical experiment test program datum values. Then, est($ct1t2iesc$) is computed as follows:

$$
\text{est}(ct1t2iesc) = \frac{(+1 \cdot 1) + (-1 \cdot 2) + (-1 \cdot 3) + (+1 \cdot 4) + (+1 \cdot 5) + (-1 \cdot 6) + (-1 \cdot 7) + (+1 \cdot 8)}{(+1^2) + (-1^2) + (-1^2) + (+1^2) + (+1^2) + (-1^2) + (-1^2) + (+1^2)} = \frac{0}{8} = 0.0
$$

In turn, each of the elements in the $|ct1t2iec_i$'s$|$ column vector is multiplied by est($ct1t2iesc$) to obtain the corresponding elements in the $|\text{est}(ct1t2ie_i$'s)$|$ column vector. The following array is obtained when this three-step procedure is used to compute the respective elements of each of the column vectors in the estimated complete analytical model:

hypothetical datum value$_i$'s	est (csmm)	est (ct1e$_i$'s)	est (ct2e$_i$'s)	est (ct3e$_i$'s)	est (ct1t2ie$_i$'s)	est (ct1t3ie$_i$'s)	est (ct2t3ie$_i$'s)	est (ct1t2t3ie$_i$'s)
1	4.5	−0.5	−1.0	−2.0	0.0	0.0	0.0	0.0
2	4.5	+0.5	−1.0	−2.0	0.0	0.0	0.0	0.0
3	4.5	−0.5	+1.0	−2.0	0.0	0.0	0.0	0.0
4	4.5	+0.5	+1.0	−2.0	0.0	0.0	0.0	0.0
5	4.5	−0.5	−1.0	+2.0	0.0	0.0	0.0	0.0
6	4.5	+0.5	−1.0	+2.0	0.0	0.0	0.0	0.0
7	4.5	−0.5	+1.0	+2.0	0.0	0.0	0.0	0.0
8	4.5	+0.5	+1.0	+2.0	0.0	0.0	0.0	0.0

The rows are combined as: $= \; + \; + \; + \; + \; + \; + \; +$

Exercise Set 3

These exercises are intended to generate an understanding of the column-vector-based least-squares estimation procedure by requiring hand calculations for the respective estimated scalar coefficients and the respective elements of each of the column vectors in the estimated complete analytical model.

1. Given the hypothetical datum values in Exercise Set 1, use the following orthogonal augmented array to compute the respective elements of the |est(*csdm*)| column vector and the aggregated |est(*CRSIEE$_i$*'s)| column vector.

+1's	|crsieec(1)$_i$'s|	|crsieec(2)$_i$'s|	|crsieec(3)$_i$'s|	|crsieec(4)$_i$'s|	|crsieec(5)$_i$'s|	|crsieec(6)$_i$'s|	|crsieec(7)$_i$'s|
+1	−1	−1	−1	−1	−1	−1	−1
+1	+1	−1	−1	−1	−1	−1	−1
+1	0	+2	−1	−1	−1	−1	−1
+1	0	0	+3	−1	−1	−1	−1
+1	0	0	0	+4	−1	−1	−1
+1	0	0	0	0	+5	−1	−1
+1	0	0	0	0	0	+6	−1
+1	0	0	0	0	0	0	+7

2. Use the orthogonal array whose *transpose* appears below to compute the aggregated |est(*CRHEE$_i$*'s)| column vectors for any one of the three examples pertaining to Figure 2.2. First compute the scalar coefficients associated with the 14 |crheec(j)$_i$'s| column vectors, then compute the respective elements of the 14 corresponding column vectors in the estimated complete analytical model, and in turn aggregate the respective elements of these fourteen column vectors into the elements of the |est(*CRHEE$_i$*'s)| column vector.

+1	+1	+1	+1	+1	+1	+1	+1	+1	+1	+1	+1	+1	+1	+1	+1	$\lvert +1\text{'s}\rvert$
−1	−1	−1	−1	−1	−1	−1	−1	+1	+1	+1	+1	+1	+1	+1	+1	$\lvert ctec_i\text{'s}\rvert$
−1	+1	0	0	0	0	0	0	0	0	0	0	0	0	0	0	$\lvert crheec(1)_i\text{'s}\rvert$
−1	−1	+2	0	0	0	0	0	0	0	0	0	0	0	0	0	$\lvert crheec(2)_i\text{'s}\rvert$
−1	−1	−1	+3	0	0	0	0	0	0	0	0	0	0	0	0	$\lvert crheec(3)_i\text{'s}\rvert$
−1	−1	−1	−1	+4	0	0	0	0	0	0	0	0	0	0	0	$\lvert crheec(4)_i\text{'s}\rvert$
−1	−1	−1	−1	−1	+5	0	0	0	0	0	0	0	0	0	0	$\lvert crheec(5)_i\text{'s}\rvert$
−1	−1	−1	−1	−1	−1	+6	0	0	0	0	0	0	0	0	0	$\lvert crheec(6)_i\text{'s}\rvert$
−1	−1	−1	−1	−1	−1	−1	+7	0	0	0	0	0	0	0	0	$\lvert crheec(7)_i\text{'s}\rvert$
0	0	0	0	0	0	0	0	−1	+1	0	0	0	0	0	0	$\lvert crheec(8)_i\text{'s}\rvert$
0	0	0	0	0	0	0	0	−1	−1	+2	0	0	0	0	0	$\lvert crheec(9)_i\text{'s}\rvert$
0	0	0	0	0	0	0	0	−1	−1	−1	+3	0	0	0	0	$\lvert crheec(10)_i\text{'s}\rvert$
0	0	0	0	0	0	0	0	−1	−1	−1	−1	+4	0	0	0	$\lvert crheec(11)_i\text{'s}\rvert$
0	0	0	0	0	0	0	0	−1	−1	−1	−1	−1	+5	0	0	$\lvert crheec(12)_i\text{'s}\rvert$
0	0	0	0	0	0	0	0	−1	−1	−1	−1	−1	−1	+6	0	$\lvert crheec(13)_i\text{'s}\rvert$
0	0	0	0	0	0	0	0	−1	−1	−1	−1	−1	−1	−1	+7	$\lvert crheec(14)_i\text{'s}\rvert$

3. Run microcomputer program $RANDOM1$ with $n_{dv} = n_{elpri} = 8$ to generate hypothetical datum values for an unreplicated $(2)^3$ factorial design experiment test program (Figure 2.6). First, (a) compute the respective elements of each of the eight column vectors in the estimated complete analytical model. Then, (b) check whether the sum of the ith element in each of these eight column vectors is equal to the ith hypothetical datum value. In turn, (c) check whether the sum of the sum of squares pertaining to the respective elements of these eight column vectors is equal to the sum of squares for the eight hypothetical datum values.

2.2.6. Microcomputer Programs

Microcomputer program $CALESTCV$ calculates the respective n_{dv} elements in each of the n_{dv} orthogonal column vectors in the estimated complete analytical model. It requires input information from two microcomputer files: $DATA$, which contains the respective n_{dv} elements of the |experiment test program datum value$_i$'s| column vector, and $ARRAY$, which contains the transpose of the associated n_{dv} by n_{dv} orthogonal augmented contrast array for the given experiment test program. Microcomputer program $CALESTCV$ is supplemented by microcomputer programs $CKSUMSQS$ and $AGESTCV$, both of which require input information from microcomputer files $DATA$ and $ARRAY$. Microcomputer program $CKSUMSQS$ sums the sums of squares of the elements of the respective n_{dv} orthogonal column vectors in the estimated complete analytical model and compares this sum to the sum of squares of the n_{dv} experiment test program datum values. These two sums are equal for all orthogonal arrays. (Thus, microcomputer program *CKSUMSQS* should be run before running either microcomputer program *CALESTCV* or *AGESTCV*.)

```
COPY C2EXDATA DATA

  1 file(s) copied

C>TYPE DATA

16   Number of datum values, n_dv, followed by the experiment test
     program datum values that are properly ordered relative to the
     corresponding n_dv by n_dv experiment test program orthogonal
     augmented contrast array
1
2
3
4
5
6
7
8
9
10
11
12
13
14
15
16
```

(These 16 hypothetical datum values pertain to the comparative CRD experiment test program example whose organizational structure appears in Figure 2.2.)

```
COPY C2EXARRY ARRAY

 1 file(s) copied

C>TYPE ARRAY
```

+1	+1	+1	+1	+1	+1	+1	+1	+1	+1	+1	+1	+1	+1	+1	+1	1	$\lvert+1\text{'s}\rvert$
−1	−1	−1	−1	−1	−1	−1	−1	+1	+1	+1	+1	+1	+1	+1	+1	2	$\lvert ctec_i\text{'s}\rvert$
−1	+1	0	0	0	0	0	0	0	0	0	0	0	0	0	0	3	$\lvert crheec(1)_i\text{'s}\rvert$
−1	−1	+2	0	0	0	0	0	0	0	0	0	0	0	0	0	4	$\lvert crheec(2)_i\text{'s}\rvert$
−1	−1	−1	+3	0	0	0	0	0	0	0	0	0	0	0	0	5	$\lvert crheec(3)_i\text{'s}\rvert$
−1	−1	−1	−1	+4	0	0	0	0	0	0	0	0	0	0	0	6	$\lvert crheec(4)_i\text{'s}\rvert$
−1	−1	−1	−1	−1	+5	0	0	0	0	0	0	0	0	0	0	7	$\lvert crheec(5)_i\text{'s}\rvert$
−1	−1	−1	−1	−1	−1	+6	0	0	0	0	0	0	0	0	0	8	$\lvert crheec(6)_i\text{'s}\rvert$
−1	−1	−1	−1	−1	−1	−1	+7	0	0	0	0	0	0	0	0	9	$\lvert crheec(7)_i\text{'s}\rvert$
0	0	0	0	0	0	0	0	−1	+1	0	0	0	0	0	0	10	$\lvert crheec(8)_i\text{'s}\rvert$
0	0	0	0	0	0	0	0	−1	−1	+2	0	0	0	0	0	11	$\lvert crheec(9)_i\text{'s}\rvert$
0	0	0	0	0	0	0	0	−1	−1	−1	+3	0	0	0	0	12	$\lvert crheec(10)_i\text{'s}\rvert$
0	0	0	0	0	0	0	0	−1	−1	−1	−1	+4	0	0	0	13	$\lvert crheec(11)_i\text{'s}\rvert$
0	0	0	0	0	0	0	0	−1	−1	−1	−1	−1	+5	0	0	14	$\lvert crheec(12)_i\text{'s}\rvert$
0	0	0	0	0	0	0	0	−1	−1	−1	−1	−1	−1	+6	0	15	$\lvert crheec(13)_i\text{'s}\rvert$
0	0	0	0	0	0	0	0	−1	−1	−1	−1	−1	−1	−1	+7	16	$\lvert crheec(14)_i\text{'s}\rvert$

(This array is the transpose of the 16 by 16 orthogonal augmented contrast array that pertains to the comparative CRD experiment test program example whose organizational structure appears in Figure 2.2.)

```
C> CKSUMSQS
```

This program presumes that the transpose of the appropriate experiment test program orthogonal augmented contrast array appears in microcomputer file *ARRAY* and that the corresponding properly ordered experiment test program datum values appear in microcomputer file *DATA*.

This program computes (i) the sum of squares of the experiment test program datum values, and (ii) the sum of the sums of squares of the respective elements of the n_{dv} column vectors in the estimated complete analytical model. The proposed experiment test program augmented contrast array is not orthogonal if these two sums of squares are not exactly equal

$\qquad$ (i) .149600000000000D + 04 $\quad$ (ii) .149600000000000D + 04

```
C> CALESTCV
```

This program presumes that the transpose of the appropriate experiment test program orthogonal augmented contrast array appears in microcomputer file *ARRAY* and that the corresponding properly ordered experiment test program datum values appear in microcomputer file *DATA*.

This program computes each of the n_{dv} elements in each of the n_{dv} column vectors in the estimated complete analytical model

Column Vector ($j=$)	Element ($i=$)	Element Value
1	1	$.8500000000D + 01$
1	2	$.8500000000D + 01$
1	3	$.8500000000D + 01$
1	4	$.8500000000D + 01$
1	5	$.8500000000D + 01$
1	6	$.8500000000D + 01$
1	7	$.8500000000D + 01$
1	8	$.8500000000D + 01$
1	9	$.8500000000D + 01$
1	10	$.8500000000D + 01$
1	11	$.8500000000D + 01$
1	12	$.8500000000D + 01$
1	13	$.8500000000D + 01$
1	14	$.8500000000D + 01$
1	15	$.8500000000D + 01$
1	16	$.8500000000D + 01$

Column Vector ($j=$)	Element ($i=$)	Element Value
2	1	$-.4000000000D + 01$
2	2	$-.4000000000D + 01$
2	3	$-.4000000000D + 01$
2	4	$-.4000000000D + 01$
2	5	$-.4000000000D + 01$
2	6	$-.4000000000D + 01$
2	7	$-.4000000000D + 01$

2	8	$-.4000000000D + 01$
2	9	$.4000000000D + 01$
2	10	$.4000000000D + 01$
2	11	$.4000000000D + 01$
2	12	$.4000000000D + 01$
2	13	$.4000000000D + 01$
2	14	$.4000000000D + 01$
2	15	$.4000000000D + 01$
2	16	$.4000000000D + 01$

Column Vector ($j =$)	Element ($i =$)	Element Value
3	1	$-.5000000000D + 00$
3	2	$.5000000000D + 00$
3	3	$.0000000000D + 00$
3	4	$.0000000000D + 00$
3	5	$.0000000000D + 00$
3	6	$.0000000000D + 00$
3	7	$.0000000000D + 00$
3	8	$.0000000000D + 00$
3	9	$.0000000000D + 00$
3	10	$.0000000000D + 00$
3	11	$.0000000000D + 00$
3	12	$.0000000000D + 00$
3	13	$.0000000000D + 00$
3	14	$.0000000000D + 00$
3	15	$.0000000000D + 00$
3	16	$.0000000000D + 00$

Column Vector ($j =$)	Element ($i =$)	Element Value
4	1	$-.5000000000D + 00$
4	2	$-.5000000000D + 00$
4	3	$.1000000000D + 01$
4	4	$.0000000000D + 00$
4	5	$.0000000000D + 00$
4	6	$.0000000000D + 00$
4	7	$.0000000000D + 00$

et cetera

C> AGESTCV

This program presumes that the transpose of the appropriate experiment test program orthogonal augmented contrast ARRAY appears in microcomputer file *ARRAY* and that the corresponding properly ordered experiment test program datum values appear in microcomputer file *DATA*.

This program aggregates *adjacent* estimated column vectors, e.g., to aggregate estimated column vectors $j = 3$ through $j = 16$, type 3 space 16. (Note that aggregating estimated column vectors 1 through n_{dv} will verify the input experiment test program datum values.)

3 16

Element ($i=$)	Element Value
1	$-.3500000000D + 01$
2	$-.2500000000D + 01$
3	$-.1500000000D + 01$
4	$-.5000000000D + 01$
5	$+.5000000000D + 01$
6	$+.1500000000D + 01$
7	$+.2500000000D + 01$
8	$+.3500000000D + 01$
9	$-.3500000000D + 01$
10	$-.2500000000D + 01$
11	$-.1500000000D + 01$
12	$-.5000000000D + 01$
13	$+.5000000000D + 01$
14	$+.1500000000D + 01$
15	$+.2500000000D + 01$
16	$+.3500000000D + 01$

Microcomputer program *AGESTCV* aggregates only *adjacent* estimated column vectors (that are of specific interest in subsequent statistical analysis). This restriction must be considered, for example, in the analysis of variance (Chapter 6) when constructing the associated orthogonal augmented contrast array.

Exercise Set 4

These exercises are intended to (a) acquaint you with the microcomputer programs *CALESTCV*, *CKSUMSQS*, and *AGESTCV*, and (b) broaden your perspective regarding the measurement metric versus the measurement metric of relevance in the conceptual statistical model. It is expected that you will conclude that the magnitudes of all terms in the estimated complete analytical model depend on the measurement metric employed in statistical analysis and that, in particular, it may be possible to find a physically relevant metric that makes the estimated values of the interaction effects negligible for practical purposes. It should be obvious that the hypothetical datum values for the unreplicated $(2)^3$ factorial design experiment test program in Figure 2.6(b) were deliberately selected so that each of the four estimated interaction effects are equal to zero. Accordingly, the associated conceptual statistical model is easy to understand physically, viz., it is geometrically interpreted as a plane in $t1$, $t2$, $t3$ space. The fundamental issue is that we should always employ a measurement metric that establishes a comprehensible conceptual statistical model. Accordingly, although linear measurement metrics for distance and time are straightforward and intuitive, the physics of the given phenomenon (whether we understand it or nor) may dictate employing nonlinear measurement metrics, say distance squared and logarithmic time. Thus, it is always appropriate in statistical analysis to explore alternative measurement metrics.

1. (a) Given the transpose of the array in Figure 2.6(b) and the hypothetical datum values for the text unreplicated $(2)^3$ factorial design experiment test program numerical example, run microcomputer programs *CKSUMSQS* and *CALESTCV* to compute the respective eight elements for each of the eight column vectors in the estimated complete analytical model. Then, (b), repeat (a) using the natural logarithms of the hypothetical example datum values in (a) as your hypothetical datum values. Do you recommend adopting this new measurement metric?

2. Given the array in Example 2 of Exercise Set 3 and the hypothetical datum values pertaining to any of the three text numerical examples associated with Figure 2.2, run microcomputer programs *CKSUMSQS* and *AGESTCV* to verify the respective elements of the est($CRHEE_i$'s).

2.2.7. Pseudoreplication in Unreplicated Factorial Design Experiment Test Programs

Recall that every statistical estimate consists of the actual value of the quantity being estimated plus an intrinsic statistical estimation error.

However, when the actual value for the quantity being estimated is equal to zero, its associated intrinsic statistical estimation error must rationally be regarded as an *experimental* error. Accordingly, when it is presumed that there is no physical basis for certain terms in the estimated complete analytical model, the elements of the associated column vectors are reinterpreted as estimated experimental errors and are then aggregated to form the elements of the resulting $|\text{est}(CRHEE_i\text{'s})|$ column vector. This $|\text{est}(CRHEE_i\text{'s})|$ column vector is *statistically equivalent* to an analogous $|\text{est}(CRHEE_i\text{'s})|$ column vector that pertains to actual replication—provided that the resulting conceptual statistical model is correct. Moreover, since each of the column vectors in the estimated complete analytical model has one statistical degree of freedom, when k of these column vectors are aggregated to form a column vector in the estimated conceptual statistical model, e.g., the $|\text{est}(CRHEE_i\text{'s})|$ column vector, this aggregated column vector has k statistical degrees of freedom.

Consider, for example, the unreplicated $(2)^4$ factorial design experiment test program whose estimated complete analytical column vector model is

$$
\begin{aligned}
|\text{experiment test program datum value}_i\text{'s}| = \text{est}(csmmsc) \cdot |+1\text{'s}| \\
+ \text{est}(ct1esc) \cdot |ct1ec_i\text{'s}| + \text{est}(ct2esc) \cdot |ct2ec_i\text{'s}| \\
+ \text{est}(ct3esc) \cdot |ct3ec_i\text{'s}| + \text{est}(ct4esc) \cdot |ct4ec_i\text{'s}| \\
+ \text{est}(ct1t2iesc) \cdot |ct1t2iec_i\text{'s}| + \text{est}(ct1t3iesc) \cdot |ct1t3iec_i\text{'s}| \\
+ \text{est}(ct1t4iesc) \cdot |ct1t4iec_i\text{'s}| + \text{est}(ct2t3iesc) \cdot |ct2t3iec_i\text{'s}| \\
+ \text{est}(ct2t4iesc) \cdot |ct2t4iec_i\text{'s}| + \text{est}(ct3t4iesc) \cdot |ct3t4iec_i\text{'s}| \\
+ \text{est}(ct1t2t3iesc) \cdot |ct1t2t3iec_i\text{'s}| + \text{est}(ct1t2t4iesc) \cdot |ct1t2t4iec\text{'s}| \\
+ \text{est}(ct1t3t4iesc) \cdot |ct1t3t4iec\text{'s}| + \text{est}(ct2t3t4iesc) \cdot |ct2t3t4iec\text{'s}| \\
+ \text{est}(ct1t2t3t4iesc) \cdot |ct1t2t3t4iec\text{'s}|
\end{aligned}
$$

Because even two-factor interaction effects are often difficult to understand physically, it is generally rationalized that there is seldom a rational physical basis for a three-factor interaction effect and almost never a rational physical basis for a higher-order interaction effect. Accordingly, all estimated three-factor and higher-order interaction column vectors are typically aggregated to form a preliminary $|\text{est}(CRHEE_i\text{'s})|$ column vector. In turn, this preliminary $|\text{est}(CRHEE_i\text{'s})|$ column vector forms a basis for deciding whether it statistically rational to aggregate one or more of the estimated main-effect and two-factor interaction effect column vectors into the $|\text{est}(CRHEE_i\text{'s})|$ column vector that is subsequently used in statistical analysis. This aggregation methodology is statistically effective because the larger the number of statistical degrees of freedom for the aggregated $|\text{est}(CRHEE_i\text{'s})|$ column vector, the greater the precision of the remaining est($ctesc$'s) and the greater the statistical power of the experiment test program (Chapter 6). It is important to understand, however, that the aggregated $|\text{est}(CRHEE_i\text{'s})|$ column vector is statistically proper *only* when the resulting conceptual statistical model is indeed correct. Box, et al. (1978) present a procedure for deciding which column vectors in the estimated complete analytical model can rationally be aggregated to form a statistically credible est($CRHEE_i$'s) column vector.

Suppose that, after appropriate analysis, our example estimated complete analytical model is reinterpreted in terms of the following estimated conceptual statistical model:

$$|\text{experiment test program datum value}_i\text{'s}| = \text{est}(csmmsc) \cdot |+1\text{'s}|$$
$$+ \text{est}(ct1esc) \cdot |ct1ec_i\text{'s}| + \text{est}(ct2esc) \cdot |ct2ec_i\text{'s}|$$
$$+ \text{est}(ct3esc) \cdot |ct3ec_i\text{'s}] + \text{est}(ct1t3iesc) \cdot |ct1t3iec_i\text{'s}|$$
$$+ \text{est}(ct2t3iesc) \cdot |ct2t3iec_i\text{'s}| + |\text{est}(CRHEE_i\text{'s})|$$

in which the elements of 10 estimated column vectors have been aggregated to form the elements of the aggregated $|\text{est}(CRHEE_i\text{'s})|$ column vector (which thus has 10 statistical degrees of freedom, all resulting from pseudoreplication). The column vectors remaining in the estimated conceptual statistical model have presumably been retained only because it is quite unlikely that the magnitudes of their respective estimated scalar coefficients could be the sole consequence of substituting (random) experiment test program datum values into the least-squares estimation expressions. If so, then it is statistically rational to assert that these estimated column vectors physically explain the respective magnitudes of the experiment test program datum values.

2.3. UNREPLICATED RANDOMIZED COMPLETE BLOCK DESIGN EXPERIMENT TEST PROGRAMS

Comparative experiment test programs that have practical mechanical reliability application generally involve blocking to mitigate the spurious effects of nuisance experiment test program variables. Figure 2.7 depicts an unreplicated randomized complete block design (RCBD) experiment test program where, by definition, there is a single experiment test program datum value for each of the n_t treatments (treatment levels, treatment combinations) within each block. The experimental units and the test conditions

	Treatment			
Block 1 $_{(2)}$	$A_{(1,2)}$	$B_{(4,4)}$	$C_{(2,3)}$	$D_{(3,1)}$
Block 2 $_{(1)}$	$A_{(2,1)}$	$B_{(1,3)}$	$C_{(4,2)}$	$D_{(3,4)}$
Block 3 $_{(3)}$	$A_{(3,3)}$	$B_{(2,4)}$	$C_{(1,2)}$	$D_{(4,1)}$

Figure 2.7 Example of the organizational structure of an unreplicated randomized complete block design (RCBD) experiment test program. The randomization details for this unreplicated RCBD experiment test program with four treatments in each of its three blocks are discussed in the text.

comprising each respective block must be as nominally identical as practical, but the respective blocks may differ markedly. The random assignment of treatments (treatment levels, treatment combinations) to the nominally identical experimental units in each block is indicated by the first number in the treatment subscript parentheses. The random time order of testing for blocks is indicated by the number in the block subscript parentheses. (All testing of experimental units in a block must be completed before beginning testing in another block.) The random time order of testing in each respective block is indicated by the second number in the treatment subscript parentheses.

The complete analytical model for an unreplicated RCBD experiment test program includes conceptual block effects, cbe's, and conceptual block, treatment interaction effect terms, $cbtie$'s. It is succinctly stated in hybrid column vector notation as

$$|CRHDV_i\text{'s}| = |csmm| + |cbe_i\text{'s}| + |cte_i\text{'s}| + |cbtie_i\text{'s}|$$

The associated conceptual statistical model has no replication or pseudo-replication *unless* it is explicitly presumed that all blocking variables are deliberately selected so that no $cbtie$ has a physical basis. Then, based on this explicit presumption, the $cbtie_i$'s column vector can be reinterpreted as a $CRHEE_i$'s column vector and the conceptual statistical model is rewritten as

$$|CRHDV_i\text{'s}| = |csmm| + |cbe_i\text{'s}| + |cte_i\text{'s}| + |CRHEE_i\text{'s}|$$

in which the physical interpretation of the $csmm$ is even more nebulous than in a CRD experiment test program.

The corresponding estimated column vector statistical model for an unreplicated RCBD experiment test program is

|experiment test program datum value$_i$'s|

$$= |\text{est}(csmm)| + |\text{est}(cbe_i\text{'s}| + |\text{est}(cte_i\text{'s})| + |\text{est}(CRHEE_i\text{'s})|$$

where the |experiment test program datum value$_i$'s| column vector has $(n_b)(n_t)$ statistical degrees of freedom, the |est($csmm$)| column vector has one statistical degree of freedom, the |est(cbe_i's)| column vector has $(n_b - 1)$ statistical degrees of freedom, the |est(cte_i's)| column vector has $(n_t - 1)$ statistical degrees of freedom, and the |est($CRHEE_i$'s)| column vector has $(n_b - 1) \cdot (n_t - 1)$ statistical degrees of freedom.

Remember that the test conduct details for an unreplicated RCBD experiment test program must be properly randomized. First, the respective treatments (treatment levels, treatment combinations) must be randomly assigned to the nominally identical experimental units in each block. Then, the respective treatments (treatment levels, treatment combinations) in a randomly selected block are tested back-to-back in random time order (by the same test technician using the same test procedure) before beginning the tests in another (randomly selected) block. Microcomputer program *RANDOM1* should be run to establish the pseudorandom numbers underlying each of these randomization details.

2.3.1. Paired-Comparison Experiment Test Programs

An unreplicated RCBD experiment test program with only two treatments (treatment levels) is called a *paired-comparison* experiment test program. Suppose we wish to compare two different fatigue test machines in our laboratory relative to the null hypothesis that the difference in calibration between these two machines is indeed equal to zero. However, we only have a few short segments of left-over 3/4 in. diameter rods of various materials available in our stock room. This left-over material is quite adequate to form the individual blocks in our paired-comparison experiment test program (and may even be preferable). We merely require that the two fatigue test specimens that form each block be cut from adjacent blanks with each rod segment, regardless of its material. We further require that both fatigue test specimens within each block be machined back-to-back in random time order by the same lathe hand on the same lathe and tested back-to-back in random time order by the same test technician using the same test machine and test procedure, etc. In short, our goal is to make all details of the fatigue test specimen preparation and testing as uniform as practical *within* each block, even though the respective blocks may differ markedly.

2.3.1.1. Numerical Example for a Paired-Comparison Experiment Test Program with Four Blocks (Paired-Comparisons)

Figure 2.8 depicts a paired-comparison experiment test program with four blocks (paired-comparisons). The hypothetical datum values pertaining to the respective block/treatment combinations are given in the corresponding respective block/treatment locations.

The corresponding estimated statistical column vector model appears below. Recall that the elements of each of the column vectors in the estimated statistical model are conventionally computed using algebraic least-squares estimators that are based on arithmetic averages. Accordingly, est($csmm$) is conventionally computed as the arithmetic average of all experiment test program datum values. In turn, the elements of the |est(cte_i's)| column vector are conventionally computed as the arithmetic average of the experiment test program datum values pertaining to each respective treatment (treatment level, treatment combination) minus the arithmetic average of all experiment test program datum values. Similarly, the elements of the |est(cbe_i's)| column vector are conventionally computed as the experiment test program datum values pertaining to the respective blocks minus the arithmetic average of all experiment test program datum values. Finally, the resulting est($cbtie_i$'s) are conventionally computed as "residuals," values that satisfy the scalar equations associated

	Treatment A	Treatment B	Block Arithmetic Average
Block 1 $_{(2)}$	$1.0_{(1,2)}$	$2.0_{(2,1)}$	1.5
Block 2 $_{(3)}$	$3.0_{(2,2)}$	$4.0_{(1,1)}$	3.5
Block 3 $_{(4)}$	$5.0_{(1,1)}$	$6.0_{(2,2)}$	5.5
Block 4 $_{(1)}$	$7.0_{(1,2)}$	$8.0_{(2,1)}$	7.5
Treatment Arithmetic Average	4.0	5.0	(4.5)

Figure 2.8 Hypothetical datum values for an example paired-comparison experiment test program with four blocks (paired comparisons). Randomization details for this paired-comparison experiment test program are indicated by the numbers in the subscript parentheses.

with the respective experiment test program datum value$_i$'s. These est($cbtie_i$'s) must rationally be re-interpreted as est($CRHEE$'s) in the estimated statistical model when it is presumed that no $cbtie$ has a credible physical basis.

hypothetical datum value$_i$'s	est($csmm$)	est(cbe_i's)	est(cte_i's)	est($CRHEE_i$'s)
$\begin{vmatrix} 1.0 \\ 3.0 \\ 5.0 \\ 7.0 \\ 2.0 \\ 4.0 \\ 6.0 \\ 8.0 \end{vmatrix}$ =	$\begin{vmatrix} 4.5 \\ 4.5 \\ 4.5 \\ 4.5 \\ 4.5 \\ 4.5 \\ 4.5 \\ 4.5 \end{vmatrix}$ +	$\begin{vmatrix} -3.0 \\ -1.0 \\ +1.0 \\ +3.0 \\ -3.0 \\ -1.0 \\ +1.0 \\ +3.0 \end{vmatrix}$ +	$\begin{vmatrix} -0.5 \\ -0.5 \\ -0.5 \\ -0.5 \\ +0.5 \\ +0.5 \\ +0.5 \\ +0.5 \end{vmatrix}$ +	$\begin{vmatrix} 0.0 \\ 0.0 \\ 0.0 \\ 0.0 \\ 0.0 \\ 0.0 \\ 0.0 \\ 0.0 \end{vmatrix}$

Note that the hypothetical datum values in this unreplicated RCBD experiment test program example were deliberately constructed such that the resulting est($CRHEE_i$'s), viz., the resulting est($cbtiesc_j$'s), were all equal to zero.

The estimated statistical model in this paired-comparison example can also be computed by constructing the associated orthogonal augmented contrast array and then running microcomputer program *AGESTCV*, as required in Exercise Set 5 below.

Exercise Set 5

These exercises pertain to the text paired-comparison numerical example and are intended to demonstrate that the contrasts in the respective orthogonal augmented contrast arrays that will be used to compute the column vectors of the estimated statistical model are not unique, but nevertheless generate the same aggregated results. Note that the elements of the three $|cbtiec_j$'s$|$ column vectors are established by multiplication of the respective elements of the three $|cbec_j$'s$|$ column vectors and the $|ctec_j$'s$|$ column vector.

1.   Consider the following orthogonal augmented contrast array. Run microcomputer program *AGESTCV* in conjunction with this orthogonal augmented contrast array and the text hypothetical datum values column vector to verify the text paired-comparison numerical example.

$\lvert +1\text{'s}\rvert$	$\lvert cbec_i\text{'s}\rvert$			$\lvert ctec_i\text{'s}\rvert$	$\lvert cbtiec_i\text{'s}\rvert$		
+1	−1	−1	+1	−1	+1	+1	−1
+1	+1	−1	−1	−1	−1	+1	+1
+1	−1	+1	−1	−1	+1	−1	+1
+1	+1	+1	+1	−1	−1	−1	−1
+1	−1	−1	+1	+1	−1	−1	+1
+1	+1	−1	−1	+1	+1	−1	−1
+1	−1	+1	−1	+1	−1	+1	−1
+1	+1	+1	+1	+1	+1	+1	+1

2. Repeat Example 1 for the following orthogonal augmented contrast array.

$\lvert +1\text{'s}\rvert$	$\lvert cbec_i\text{'s}\rvert$			$\lvert ctec_i\text{'s}\rvert$	$\lvert cbtiec_i\text{'s}\rvert$		
+1	−1	−1	−1	−1	+1	+1	+1
+1	+1	−1	−1	−1	−1	+1	+1
+1	0	+2	−1	−1	0	−2	+1
+1	0	0	+3	−1	0	0	−3
+1	−1	−1	−1	+1	−1	−1	−1
+1	+1	−1	−1	+1	+1	−1	−1
+1	0	+2	−1	+1	0	+2	−1
+1	0	0	+3	+1	0	0	+3

3. Repeat Example 1 for the following orthogonal augmented contrast array.

$\lvert +1\text{'s}\rvert$	$\lvert cbec_i\text{'s}\rvert$			$\lvert ctec_i\text{'s}\rvert$	$\lvert cbtiec_i\text{'s}\rvert$		
+1	−3	+1	−1	−1	+3	−1	+1
+1	−1	−1	+3	−1	+1	+1	−3
+1	+1	−1	−3	−1	−1	+1	+3
+1	+3	+1	+1	−1	−3	−1	−1
+1	−3	+1	−1	+1	−3	+1	−1
+1	−1	−1	+3	+1	−1	−1	+3
+1	+1	−1	−3	+1	+1	−1	−3
+1	+3	+1	+1	+1	+3	+1	+1

4. Consider the orthogonal augmented contrast array in Example 1. Rearrange the *rows* of this orthogonal augmented contrast array and the associated elements of the text hypothetical datum values column vector so that the elements of the $ctec_i$'s column vector alternate from −1 to +1 (as would occur if Yate's enumeration algorithm had been used to establish these $ctec_i$'s). Then, run

microcomputer program *AGESTCV* to verify the text paired-comparison numerical example.

5. (a) Repeat Example 4 for the orthogonal augmented contrast array in Example 2. (b) Repeat Example 4 for the orthogonal augmented contrast array in Example 3.

6. Rework Example 1, this time using the natural logarithms of the hypothetical experiment test program datum values. Are the estimated interaction effect scalar coefficients still equal to zero?

2.3.2. Other Experiment Test Programs with Blocking

In certain applications it may be impractical or impossible to obtain (create) blocks of sufficient size to assign each treatment of specific interest to a test specimen (experimental unit) within each block. Randomized incomplete block design (RIBD) experiment test programs (Natrella, 1963) can be used in these applications. In other applications, two or more blocking variables must simultaneously be employed in a structured arrangement to assure a fair (unbiased) comparison of the respective treatments. For example, Latin-square experiment test programs (Natrella, 1963) have two independent blocking variables in a structured arrangement that balances (mitigates) the spurious effects of the respective combinations of these nuisance variables in the same manner that a RCBD experiment test program balances (mitigates) the spurious effects of a single collection (aggregation) of independent nuisance variables.

Suppose we wish to conduct fatigue tests on longitudinal test specimens cut from a 1/8 in. thick sheet of cold-rolled AISI 1018 steel. Although it is seldom considered, cold-rolled steel sheet stock can exhibit a noticeable widthwise crowning and a concomitant variation in cold-work, hardness, static strength, and fatigue strength. In addition, lengthwise variations in hardness, static strength, and fatigue strength sometimes occur. Accordingly, Figure 2.9 depicts the worst-case variation in fatigue strength for longitudinal test specimen blanks. When the associated *structured* width location and length location blocking variables are presumed to be independent (viz., do not interact), a Latin-square experiment test program is appropriate for balancing the (potentially) spurious effects of these two blocking variables (Figure 2.10).

2.4. UNREPLICATED SPLIT-PLOT DESIGN EXPERIMENT TEST PROGRAMS

When practicalities require that the individual treatments be applied sequentially in a specific order to obtain the desired treatment combinations, then

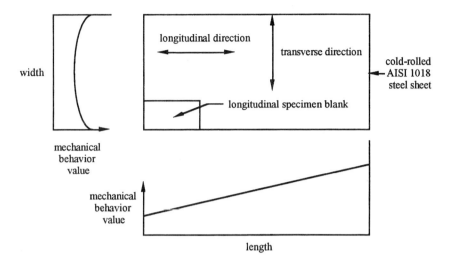

Figure 2.9 Worst-case variation in hardness, static strength, and fatigue strength for longitudinal test specimen blanks cut from a cold-rolled AISI 1018 steel sheet.

split-plot (nested) design experiment test programs are appropriate. For example, consider a $(2)^2$ factorial arrangement with the following 4 treatment combinations:

1. no treatment
2. zinc plated (only)
3. induction hardened (only)
4. *first* induction hardened and *then* zinc plated

D	C	B	A
C	D	A	B
B	A	D	C
A	B	C	D

cold-rolled
← AISI 1018
steel sheet

Figure 2.10 Organizational structure of a 4-by-4 Latin-square experiment test program with two independent *structured* blocking variables (width location and length location) and four treatments (*A*, *B*, *C*, and *D*). See Natrella (1963) for the appropriate randomization details. (Note that if the actual fatigue test specimen blanks are sufficiently small, each of these 16 longitudinal blanks can be partitioned further into 4-by-4 Latin-square arrangements.)

We now illustrate an unreplicated split-plot design (SPD) experiment test program that includes a $(2)^2$ factorial arrangement of these four treatment combinations in each of four blocks. Suppose we select an 8 ft long 5/8 in. diameter AISI 1045 hot-rolled rod to make the 16 corrosion-fatigue specimen blanks that are required for our example unreplicated SPD experiment test program. By partitioning this rod into four blocks of four adjacent specimen blanks, Figure 2.11(a), we can statistically insure, at least in part, against the problem of a test equipment failure during this lengthy test program with its severe test environment. Next, we subdivide each block into two nominally identical main-plots that subsequently receive the main-plot treatment, viz., induction hardening (because it must be done before zinc plating). In turn, we subdivide each of the main-plots into two nominally identical split-plots which subsequently receive the zinc-plating split-plot treatment. As indicated in Figure 2.11(d), this procedure generates eight experimental units for main-plots and 16 experimental units for split-plots.

Next, consider the process of randomly assigning treatments to experimental units for our example unreplicated SPD experiment test program in Figure 2.11(d). This randomization process has three stages: (a) we randomly select a block (and complete that block before beginning another randomly selected block), (b) we randomly assign the main-plot treatments

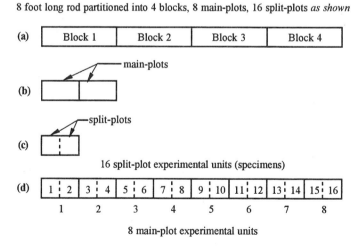

Figure 2.11 Development of an unreplicated split-plot design (SPD) experiment test program with a $(2)^2$ factorial arrangement of *hierarchical* treatment combinations in each of four blocks.

to the main-plot experimental units within the given block, and (c) we randomly assign the split-plot treatments to the split-plot experimental units within each respective main-plot within the given block. The essential feature of a SPD experiment test program is that the assignment of the split-plot treatments to split-plot experimental units is not randomized throughout each block, as are main-plot treatments, but only throughout each main-plot. As a direct consequence of this hierarchical constraint on their random assignment of split-plot treatments to split-plot experimental units, the conceptual main-plot experimental errors are different from the conceptual split-plot experimental errors. Accordingly, the conceptual statistical model for an unreplicated RCBD experiment test program is not valid for an unreplicated SPD experiment test program.

The random assignment of the main-plot treatments to the main-plot experimental units within blocks is illustrated in Figure 2.12. The two adjacent split-plot experimental units that comprise each induction-hardened main-plot experimental unit must be induction hardened as uniformly as possible. Accordingly, within each block, these two split-plot experimental units must be induction-hardened back-to-back in random time order *without* changing the control settings of the induction-hardening equipment. In contrast, the control settings must be zeroed and reset between each main-plot, in an *independent* attempt to meet the induction-hardening specification target value for each respective main-plot experimental unit. (This main-plot treatment application procedure is required to make the conceptual main-plot treatment effect experimental errors as independent as practical.)

In turn, as illustrated in Figure 2.13, split-plot treatments are randomly assigned to the split-plot experimental units *within* main-plots *within* blocks. The eight split-plot experimental units (corrosion-fatigue test specimens) that require zinc plating must be plated such that there is an inde-

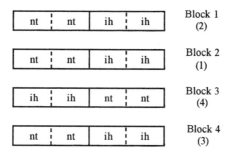

Figure 2.12 Randomization details for the main-plot treatment assignment to main-plot experimental units (ih = induction-hardening, nt = no treatment).

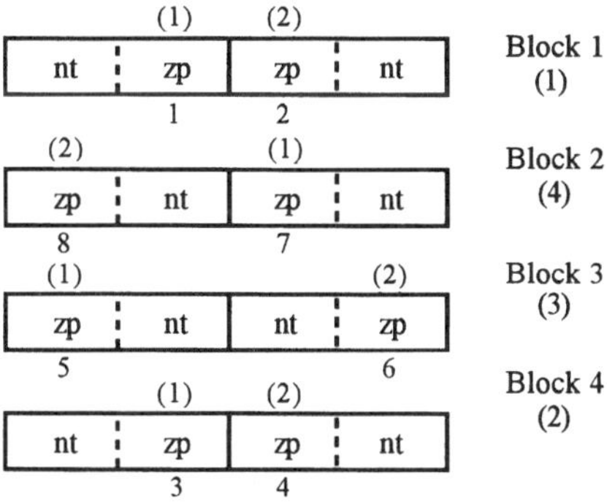

Figure 2.13 Randomization details for the split-plot treatment assignment to split-plot experimental units (zp = zinc plated, nt = no treatment).

pendent attempt to meet the zinc-plating specification target value for each respective split-plot experimental unit. However, if these eight experimental units (corrosion-fatigue test specimens) are sent to a zinc-plating vendor, these test specimens would almost surely be plated in a single batch—and the statistical credibility of the conceptual split-plot treatment effect experimental errors would be jeopardized. We can avoid improper processing of split-plot (and main-plot) experimental units by understanding when certain processing problems exist and how to avoid these processing problems (if possible).

Finally, we illustrate the random time order of testing for the 16 corrosion-fatigue test specimens in our example unreplicated SPD experiment test program. First, all tests on the experimental units comprising a block must be completed before starting tests on the experimental units comprising another block. The random time order of selecting blocks is indicated by the numbers in parentheses located below the block number in Figure 2.14. Next, all tests on experimental units comprising each main-plot must be completed before starting tests on the experimental units comprising another main-plot. The random time order of selecting main-plot experimental units for testing is indicated by numbers in parentheses located directly above the main-plot experimental units depicted in Figure 2.14.

Finally, the random time order of selecting split-plot experimental units for testing is indicated by the numbers in parentheses inside the respective split-plot experimental units depicted in Figure 2.14. The resulting time order of testing, presuming a single test machine and a single test technician, is indicated by the numbers 1 through 16 located below the respective split-plot experimental units depicted in Figure 2.14.

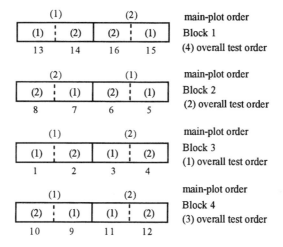

Figure 2.14 Randomization details for time order of testing of the 16 experimental units (test specimens) comprising our example unreplicated split-plot experiment test program.

2.4.1. Unreplicated Split-Plot Design Conceptual Statistical Model

The complete analytical model for this example unreplicated SPD experiment test program is written in our hybrid column vector notation as

$$|CRHDV_i\text{'s}| = |csmm| + |cbe_i\text{'s}| + |cmpte_i\text{'s}| + |cbmptie_i\text{'s}|$$
$$+ |cspte_i\text{'s}| + |cbsptie_i\text{'s}|$$
$$+ |cmptsptie_i\text{'s}| + |cbmptsptie_i\text{'s}|$$

When the n_b blocks are deliberately selected so that no block, treatment interaction has a physical basis, then the respective aggregated block, treatment interaction effect column vectors can be reinterpreted as the associated conceptual random homoscedastic experimental error column vectors, viz.,

$$|CRHDV_i\text{'s}| = |csmm| + |cbe_i\text{'s}| + |cmpte_i\text{'s}| + |CRHMPTEEE_i\text{'s}|$$
$$+ |cspte_i\text{'s}| + |CRHSPTEEE_i\text{'s}|$$
$$+ |cmptsptie_i\text{'s}| + |CRHMPTSPTIEEE_i\text{'s}|$$

in which $CRHMPTEEE_i$'s denotes the conceptual random homoscedastic main-plot treatment effect experimental errors, $CRHSPTEEE_i$'s denotes the conceptual random homoscedastic split-plot treatment effect experimental errors, and $CRHMPTSPTIEEE_i$'s denotes the conceptual random homoscedastic main-plot treatment, split-plot treatment interaction effect experimental errors. In turn, the corresponding estimated statistical column vector model is

$$|\text{experiment test program datum value}_i\text{'s}| = |\text{est}(csmm)| + |\text{est}(cbe_i\text{'s})|$$
$$+ |\text{est}(cmpte_i\text{'s})| + |\text{est}(CRHMPTEEE_i\text{'s})|$$
$$+ |\text{est}(cspte_i\text{'s})| + |\text{est}(CRHSPTEEE_i\text{'s})|$$
$$+ |\text{est}(cmptsptie_i\text{'s})| + |\text{est}(CRHMPTSPTIEEE_i\text{'s})|$$

in which the $|\text{experiment test program datum value}_i\text{'s}|$ column has $(n_b) \cdot (n_{mpt}) \cdot (n_{spt})$ statistical degrees of freedom, the est($csmm$) column vector has one statistical degree of freedom, the est(cbe_i's) column vector has $(n_b - 1)$ statistical degrees of freedom, the est($cmpte_i$'s) column vector has $(n_{mpt} - 1)$, the est($CRHMPTEEE_i$'s) column vector has $(n_b - 1) \cdot (n_{mpt} - 1)$, the est($cspte_i$'s) column vector has $(n_{spt} - 1)$, the est($CRHSPTEEE_i$'s) vector column has $(n_b - 1) \cdot (n_{spt} - 1)$, the est($mptsptie_i$'s) column vector has $(n_{mpt} - 1) \cdot (n_{spt} - 1)$, and the est($CRHMPTSPTIEEE_i$'s) column vector has $(n_b - 1) \cdot (n_{mpt} - 1) \cdot (n_{spt} - 1)$, where the number of main-plot treatments is denoted n_{mpt} and the number of split-plot treatments is denoted n_{spt}.

> *Remark*: The respective $csptesc_j$'s are generally estimated more precisely than the respective $cmptesc_j$'s in split-plot experiment test programs. Accordingly, split-plot designs are usually regarded as being more effective when the test objective places more emphasis on precise estimation of the actual values for the respective $csptesc_j$'s than on precise estimation of the actual values for the respective $cmptesc_j$'s. In contrast, all $ctesc_j$'s are estimated with equal precision in an unreplicated RCBD experiment test program.

The conventional algebraic least-squares estimates based on arithmetic averages are tedious to compute for an unreplicated SPD. Accordingly, we employ the column-vector-based least-squares estimation procedure in microcomputer program *AGESTCV* to compute the scalar coefficients in the estimated complete analytical model. The orthogonal augmented contrast array that was used in conjunction with microcomputer program *AGESTCV* to compute the estimated values of the scalar coefficients in the complete analytical model for our example unreplicated SPD experiment test program appears in Figure 2.15. Two alternative orthogonal augmented contrast arrays that can be used in conjunction with microcomputer program *AGESTCV* to verify these estimated values appear in Exercises 1 and 2 of Exercise Set 6. Obviously, these orthogonal augmented contrast arrays array are not unique.

$|+1\text{'s}|$	$|cbec_i\text{'s}|$			$|cmptec_i\text{'s}|$	$|cbmptiec_i\text{'s}|$			$|csptec_i\text{'s}|$	$|cbsptiec_i\text{'s}|$			$|cmptsptiec_i\text{'s}|$	$|cbmptsptiec_i\text{'s}|$		
+1	−1	−1	−1	−1	+1	+1	+1	−1	+1	+1	+1	+1	−1	−1	−1
+1	−1	−1	−1	−1	+1	+1	+1	+1	−1	−1	−1	−1	+1	+1	+1
+1	−1	−1	−1	+1	−1	−1	−1	−1	+1	+1	+1	−1	+1	+1	+1
+1	−1	−1	−1	+1	−1	−1	−1	+1	−1	−1	−1	+1	−1	−1	−1
+1	+1	−1	−1	−1	−1	+1	+1	−1	−1	+1	+1	+1	+1	−1	−1
+1	+1	−1	−1	−1	−1	+1	+1	+1	+1	−1	−1	−1	−1	+1	+1
+1	+1	−1	−1	+1	+1	−1	−1	−1	−1	+1	+1	−1	−1	+1	+1
+1	+1	−1	−1	+1	+1	−1	−1	+1	+1	−1	−1	+1	+1	−1	−1
+1	0	+2	−1	−1	0	−2	+1	−1	0	−2	+1	+1	0	+2	−1
+1	0	+2	−1	−1	0	−2	+1	+1	0	+2	−1	−1	0	−2	+1
+1	0	+2	−1	+1	0	+2	−1	−1	0	−2	+1	−1	0	−2	+1
+1	0	+2	−1	+1	0	+2	−1	+1	0	+2	−1	+1	0	+2	−1
+1	0	0	+3	−1	0	0	−3	−1	0	0	−3	+1	0	0	+3
+1	0	0	+3	−1	0	0	−3	+1	0	0	+3	−1	0	0	−3
+1	0	0	+3	+1	0	0	+3	−1	0	0	−3	−1	0	0	−3
+1	0	0	+3	+1	0	0	+3	+1	0	0	+3	+1	0	0	+3

Figure 2.15 Nonunique orthogonal augmented contrast array that can be used to compute the elements in the column vectors of the estimated complete analytical model for the text example SPD experiment test program.

Each of these three orthogonal arrays can also be used to estimate the column vectors for an unreplicated RCBD design experiment test program with four blocks and four treatment combinations in a $(2)^2$ factorial arrangement, except that the column vectors must be rearranged to place all nine of the estimated interaction effect column vectors involving blocks in adjacent columns before running microcomputer program $AGESTCV$. These nine adjacent column vectors, when aggregated, form the $|\text{est}(CRHEE_i\text{'s})|$ column vector for this unreplicated RCBD experiment test program with nine statistical degrees of freedom. Moreover, each of these three arrays can also be used to estimate the column vectors for an equally replicated CRD experiment test program with four replicates of the four treatment combinations in a $(2)^2$ factorial arrangement. Then, all nine of the estimated interaction effect column vectors involving (nonexistent) blocks are aggregated with the three estimated block column vectors to form the $|\text{est}(CRHEE_i\text{'s})|$ column vector for this equally replicated CRD experiment test program with 12 statistical degrees of freedom.

> *Remark*: Because the same orthogonal augmented contrast array can be used for an equally replicated CRD, an unreplicated RCBD, or an unreplicated split-plot design experiment test program, each of its column vectors must be explicitly identified using the appropriate acronym.

2.4.1.1. Numerical Example for an Unreplicated Split-Plot Design Experiment Test Program with a $(2)^2$ Factorial Arrangement for the Main-Plot and Split-Plot Treatments

Suppose that the following hypothetical datum values pertain to our example unreplicated SPD experiment test program with its main-plot and split-plot treatments structured in a $(2)^2$ factorial arrangement:

	mpt low level		*mpt high level*	
	spt low level	*spt high level*	*spt low level*	*spt high level*
Block 1	1	2	3	4
Block 2	5	6	7	8
Block 3	9	10	11	12
Block 4	13	14	15	16

Running microcomputer program *AGESTCV* generates the following estimated column vectors pertaining to the associated estimated unreplicated SPD conceptual statistical model:

hypothetical
datum value$_i$'s $=$ est($csmm$) $+$ est(cbe_i's) $+$ est($cmpt_i$'s) $+$ est($cspte_i$'s)

i	est($csmm$)	est(cbe_i's)	est($cmpt_i$'s)	est($cspte_i$'s)
1	8.5	−6.0	−1.0	−5.0
2	8.5	−6.0	−1.0	+0.5
3	8.5	−6.0	+1.0	−0.5
4	8.5	−6.0	+1.0	+0.5
5	8.5	−2.0	−1.0	−0.5
6	8.5	−2.0	−1.0	+0.5
7	8.5	−2.0	+1.0	−0.5
8	8.5	−2.0	+1.0	+0.5
9	8.5	+2.0	−1.0	−0.5
10	8.5	+2.0	−1.0	+0.5
11	8.5	+2.0	+1.0	−0.5
12	8.5	+2.0	+1.0	+0.5
13	8.5	+6.0	−1.0	0.5
14	8.5	+6.0	−1.0	+0.5
15	8.5	+6.0	+1.0	−0.5
16	8.5	+6.0	+1.0	+0.5

These hypothetical datum values were deliberately constructed so that all estimated interaction effect scalar coefficients were equal to zero.

Exercise Set 6

These exercises pertain to our unreplicated SPD experiment test program numerical example and are intended to enhance your understanding of alternative orthogonal augmented contrast arrays that can be used to compute the respective column vectors in the estimated complete analytical model.

$|+1$'s$|$	$|cbec_i$'s$|$			$|cmptec_i$'s$|$	$|cbmptiec_i$'s$|$			$|csptec_i$'s$|$	$|cbsptiec_i$'s$|$			$|cmptsptiec_i$'s$|$	$|cbmptsptiec_i$'s$|$		
+1	−1	−1	−1	−1	+1	+1	+1	−1	+1	+1	+1	+1	−1	−1	−1
+1	+1	−1	−1	−1	−1	+1	+1	−1	−1	+1	+1	+1	+1	−1	−1
+1	0	+2	−1	−1	0	−2	+1	−1	0	−2	+1	+1	0	+2	−1
+1	0	0	+3	−1	0	0	−3	−1	0	0	−3	+1	0	0	+3
+1	−1	−1	−1	−1	+1	+1	+1	+1	−1	−1	−1	−1	+1	+1	+1
+1	+1	−1	−1	−1	−1	+1	+1	+1	+1	−1	−1	−1	−1	+1	+1
+1	0	+2	−1	−1	0	−2	+1	+1	0	+2	−1	−1	0	−2	+1
+1	0	0	+3	−1	0	0	−3	+1	0	0	+3	−1	0	0	−3
+1	−1	−1	−1	+1	−1	−1	−1	−1	+1	+1	+1	−1	+1	+1	+1
+1	+1	−1	−1	+1	+1	−1	−1	−1	−1	+1	+1	−1	−1	+1	+1
+1	0	+2	−1	+1	0	+2	−1	−1	0	−2	+1	−1	0	−2	+1
+1	0	0	+3	+1	0	0	+3	−1	0	0	−3	−1	0	0	−3
+1	−1	−1	−1	+1	−1	−1	−1	+1	−1	−1	−1	+1	−1	−1	−1
+1	+1	−1	−1	+1	+1	−1	−1	+1	+1	−1	−1	+1	+1	−1	−1
+1	0	+2	−1	+1	0	+2	−1	+1	0	+2	−1	+1	0	+2	−1
+1	0	0	+3	+1	0	0	+3	+1	0	0	+3	+1	0	0	+3

1. Use the array in Figure 2.15 to verify the results of our unreplicated SPD experiment test program numerical example by running microcomputer program *AGESTCV* appropriately.
2. Consider the previous orthogonal augmented contrast array on page 72 and reorder the corresponding hypothetical datum values column vector appropriately to verify the results of our unreplicated SPD experiment test program numerical example by running microcomputer program *AGESTCV*.
3. Use the following orthogonal augmented contrast array to verify the results of our unreplicated SPD experiment test program numerical example by running microcomputer program *AGESTCV*.

$\|+1\text{'s}\|$	$\|cbec_i\text{'s}\|$			$cmptec_i\text{'s}$	$\|cbmptiec_i\text{'s}\|$			$\|csptec_i\text{'s}\|$	$\|cbsptiec_i\text{'s}\|$			$\|cmptsptiec_i\text{'s}\|$	$\|cbmptsptiec_i\text{'s}\|$		
+1	−1	−1	+1	−1	+1	+1	−1	−1	+1	+1	−1	+1	−1	−1	+1
+1	−1	−1	+1	−1	+1	+1	−1	+1	−1	−1	+1	−1	+1	+1	−1
+1	−1	−1	+1	+1	−1	−1	+1	−1	+1	+1	−1	−1	+1	+1	−1
+1	−1	−1	+1	+1	−1	−1	+1	+1	−1	−1	+1	+1	−1	−1	+1
+1	+1	−1	−1	−1	−1	+1	+1	−1	−1	+1	+1	+1	+1	−1	−1
+1	+1	−1	−1	−1	−1	+1	+1	+1	+1	−1	−1	−1	−1	+1	+1
+1	+1	−1	−1	+1	+1	−1	−1	−1	−1	+1	+1	−1	−1	+1	+1
+1	+1	−1	−1	+1	+1	−1	−1	+1	+1	−1	−1	+1	+1	−1	−1
+1	−1	+1	−1	−1	+1	−1	+1	−1	+1	−1	+1	+1	−1	+1	−1
+1	−1	+1	−1	−1	+1	−1	+1	+1	−1	+1	−1	−1	+1	−1	+1
+1	−1	+1	−1	+1	−1	+1	−1	−1	+1	−1	+1	−1	+1	−1	+1
+1	−1	+1	−1	+1	−1	+1	−1	+1	−1	+1	−1	+1	−1	+1	−1
+1	+1	+1	+1	−1	−1	−1	−1	−1	−1	−1	+1	+1	+1	+1	+1
+1	+1	+1	+1	−1	−1	−1	−1	+1	+1	+1	−1	−1	−1	−1	−1
+1	+1	+1	+1	+1	+1	+1	+1	−1	−1	−1	−1	−1	−1	−1	−1
+1	+1	+1	+1	+1	+1	+1	+1	+1	+1	+1	+1	+1	+1	+1	+1

2.5. MECHANICAL RELIABILITY APPLICATIONS OF STATISTICALLY PLANNED EXPERIMENT TEST PROGRAMS

The purpose of this section is to provide perspective regarding the fundamental role of statistically planned experiment test programs in mechanical reliability applications. The probability and statistics concepts presented in subsequent chapters should be critically examined relative to their mechanical reliability interpretation and implementation.

Adequate reliability can only be established (demonstrated) by the actual service performance of the given product. (Recall that the proof of the pudding is in the eating.) Accordingly, all design methodologies attempt in some manner or form to extrapolate service-proven experience for past designs to the proposed design.

2.5.1. Quantitative Reliability Experiment Test Programs

The most common mechanical reliability experiment test program is the so-called quality assurance test in which the proposed design is required to pass some arbitrary quantitative performance bogey. The presumption is that if all similar and analogous past designs that passed this bogey test have subsequently performed reliably in service, then if the proposed design passes this bogey test, it will perform reliably in service. However, for this methodology to be trustworthy, it must also be such that the mechanical mode of failure in its bogey test is not only identical to the mechanical mode of failure in service operation, but the characteristics of the respective failure surfaces are very similar if not almost identical.

A quality assurance bogey is ideally established using service-proven performance data from several years of prior service experience with similar or analogous designs. In this context, it is clear that a quality assurance bogey experiment test program indirectly compares sequential mechanical designs. Obviously, the more direct the comparison between service conditions and the test bogey, the more trustworthy the quality assurance test pass/fail conclusion.

> *Remark*: A quality assurance test bogey effectively performs the same function that a factor of safety performs in design analysis. Both use service-proven performance as a rational basis for predicting adequate mechanical reliability for the proposed design.

Now suppose we continually replicate a quality assurance experiment test program for a particular design that is known to perform reliably in service. Clearly, unless the given test bogey is overly extreme, eventually one or more of these replicate quality assurance experiment test programs will generate datum values that fail to pass the test bogey and thus establish a quality assurance dilemma. The use of an overly extreme test bogey eliminates the need for statistical analysis in making the pass/fail decision—because a design that passes this test bogey will surely perform reliably in service. However, the use of an overly extreme test bogey can generate a substantial possibility that a design that fails to pass the test bogey would perform reliably in service if it is actually given the opportunity. On the other hand, the use of only a moderately extreme test bogey generates the possibility that a design that passes the test bogey will not perform reliably in service. This possibility is so undesirable that mechanical designers have almost universally opted in favor of the use of overly extreme test bogies, viz., in favor of overdesign.

The issue is whether we opt to overdesign, or whether we establish a test bogey with a statistics-based pass/fail criteria. The choice is not as simple as it may seem. Overdesign is inherently relatively immune to the batch-to-batch variability of the test experimental units. This is particularly important because it is seldom reasonable to assert that the test experimental units for the proposed design can be viewed as having been randomly selected from the conceptual collection (population) that consists of all possible future production test experimental units. It is even more unreasonable to ignore the batch-to-batch variability of these future production test experimental units in a statistics-based pass/fail criteria. Unfortunately, the batch-to-batch variability of future production test experimental units is the most important and least studied factor in establishing adequate service reliability (see Supplemental Topic 6.C). Accordingly, so-called quality assurance tests seldom involve a trustworthy statistics-based pass/fail criteria.

2.5.2. Comparative Experiment Test Programs

Comparative experiment test programs have the obvious advantage of side-by-side comparisons of the past and proposed designs (or past, present, and proposed designs). Their primary disadvantage is that more testing is required. However, the additional test time and cost is almost always worthwhile when the reliability of a design must be improved.

Consider again the paired-comparison experiment test program of Section 2.3. Let treatment A be a past or present design that has demonstrated an inadequate reliability in service. This reliability problem can be alleviated by requiring proposed design B to excel the past or present design A statistically by at least a certain increment or ratio. Paired-comparison experiment test programs are particularly effective in this mechanical reliability improvement application (Chapter 3). In contrast, suppose that the past or present design exhibits adequate reliability in service. Paired-comparison experiment test programs are not statistically appropriate in this situation (for reasons that will be apparent later).

2.6. CLOSURE

In Chapter 3 we introduce elementary probability and statistics concepts and illustrate their application in paired-comparison experiment test programs conducted to improve mechanical reliability. Then, in Chapters 4 and 5 we develop the additional statistics background required to conduct more sophisticated statistical analyses for equally replicated CRD and unreplicated RCBD and SPD experiment test programs.

2.A. SUPPLEMENTAL TOPIC: CHOOSING PHYSICALLY RELEVANT CONCEPTUAL TREATMENT EFFECT CONTRASTS

Typically, the test objective, the associated null and alternative hypotheses, and the experiment test program organizational structure dictate the choice of the $ctec_i$'s that are used to construct the orthogonal augmented contrast array of specific interest. However, when two or more sets of $ctec_i$'s suffice, the actual experiment test program outcome determines the physically relevant choice for the $ctec_i$'s.

> *Remark*: Any set of $ctec_i$'s suffices when the omnibus null hypothesis is that all of the actual values of the associated $ctesc_j$'s are equal to zero and the alternative hypothesis is that not all of the actual values of these $ctesc_i$'s are equal to zero.

Consider the following two example alternative orthogonal augmented contrast arrays (that include only two blocks for simplicity of presentation). Although the respective $ctec_i$'s may appear to be similar, each is properly employed to explain a markedly different experiment test program outcome.

Example 1 Orthogonal Augmented Contrast Array

experiment test program datum value$_i$'s	+1's	$ctec_i$'s		$cbec_i$'s	$ctbiec_i$'s	
Block 1, Treatment A datum value	+1	−1	−1	−1	+1	+1
Block 1, Treatment B datum value	+1	+1	−1	−1	−1	+1
Block 1, Treatment C datum value	+1	0	+2	−1	0	−2
Block 2, Treatment A datum value	+1	−1	−1	+1	−1	−1
Block 2, Treatment B datum value	+1	+1	−1	+1	+1	−1
Block 2, Treatment C datum value	+1	0	+2	+1	0	+2

Example 2 Orthogonal Augmented Contrast Array

experiment test program datum value$_i$'s	+1's	$ctec_i$'s		$cbec_i$'s	$ctbiec_i$'s	
Block 1, Treatment A datum value	+1	−1	+1	−1	+1	−1
Block 1, Treatment B datum value	+1	0	−2	−1	0	+2
Block 1, Treatment C datum value	+1	+1	+1	−1	−1	−1
Block 2, Treatment A datum value	+1	−1	+1	+1	−1	+1
Block 2, Treatment B datum value	+1	0	−2	+1	0	−2
Block 2, Treatment C datum value	+1	+1	+1	+1	+1	+1

Example 1

Little (1997) examined the effect of three different thicknesses on the long-life fatigue performance of an automotive composite for both axial loading and four-point bending by conducting unreplicated RCBD experiment test programs, each with six time blocks. Cursory examination of the experiment test program datum values indicated that the 3/16 and 1/4 in. thick axial-load and four-point-bending fatigue specimens exhibited similar lives and that the 1/8 in. thick axial-load and four-point-bending specimens exhibited longer lives than either the 3/16 or 1/4 in. thick specimens. Given these datum values, the $ctec_i$'s in the Example 1 Orthogonal Augmented Contrast Array were chosen for use in a statistical analysis in which treatment A was the 1/4 in. thick specimens, treatment B was the 3/16 in. thick specimens, and treatment C was the 1/8 in. thick specimens. Observe that contrast $ctec(1)$ compares the fatigue lives of the 3/16 in. thick specimens (+1's) to the fatigue lives of the 1/4 in. thick specimens (−1's). Stated more technically, contrast $ctec(1)$ compares the actual value for the $ctBm$ (+1) to the actual value for the $ctAm$ (−1), where the null hypothesis is that the actual values for the $ctAm$ and the $ctBm$ are equal and the alternative hypothesis is that the actual values for the $ctAm$ and the $ctBm$ are not equal. As expected, the null hypothesis could not be rejected for the given experiment test program datum values. Next, observe that contrast $ctec(2)$ compares the fatigue lives of the 1/8 in. thick specimens (+2's) to the arithmetic average of the fatigue lives of the 3/16 and 1/4 in. thick specimens (both −1's). Stated more technically, contrast $ctec(2)$ compares the actual value for the $ctCm$ (+2) to the actual value for the sum ($ctAm + ctBm$) (both −1), where the null hypothesis is that the actual value for the $ctCm$ is equal to the actual value for the sum ($ctAm + ctBm$) and the alternative hypothesis is that the actual value for the $ctCm$ is greater than the actual value for the sum ($ctAm + ctBm$). (This comparison might be more intuitive if contrast $ctec(2)$ were scaled down to $[-1/2, -1/2, +1]$. However, contrasts are traditionally expressed as the smallest set of integers that have the desired ratios.) As expected, the null hypothesis was rejected in favor of the alternative hypothesis.

Remark 1: The actual value for the sum ($ctAm + ctBm$) is more intuitively expressed as the actual value for the mean of the conceptual statistical distribution that is comprised of all possible replicate datum values for both treatment A and treatment B.

Remark 2: If the experiment test program datum values are appropriately rearranged, contrast *ctec(1)* can also be used to compare either the actual value for the *ctCm* to the actual value for the *ctAm*, or to compare the actual value for the *ctCm* to the actual value for the *ctBm*. However, the associated three statistical analyses will not be independent because each analysis will involve two of the same three sets of treatment datum values. (See Section 6.3.)

Example 2

Little and Thomas (1993) examined the effect of grade on the endurance limit of grades 2, 5, and 8 machines screws by conducting an unreplicated SPD experiment test program with its individual fatigue tests conducted using a modified up-and-down strategy (Chapter 8). The respective endurance limits were estimated by maximum likelihood analyses (Chapter 8) prior to conducting a statistical analysis of variance (Chapter 6). A plot of these three endurance limit estimates versus machine screw grade clearly indicated that the endurance limit increases with grade, but the statistical issue is whether a quadratic term is required in addition to a linear term to explain this increase. Accordingly, the *ctec$_i$*'s in the Example 2 Orthogonal Augmented Contrast Array were chosen for use in a statistical analysis in which main-plot treatment *A* was the grade 2 machine screws, main-plot treatment *B* was the grade 5 machine screws, and main-plot treatment *C* was the grade 8 machine screws. Given equally spaced main-plot treatment levels, these two *ctec$_i$*'s, technically termed *orthogonal polynomial contrasts*, can be used to compute a second-order polynomial (parabolic) expression that passes through the three estimated endurance limit points plotted versus machine screw grade. Contrast *ctec(1)* is used to estimate the actual value of the linear effect scalar coefficient in this second-order polynomial expression. The null hypothesis associated with *ctec(1)* is that the actual value for the linear effect scalar coefficient is equal to zero, whereas the alternative hypothesis is that the actual value for the linear effect scalar coefficient is greater than zero. As expected this null hypothesis was emphatically rejected. In turn, contrast *ctec(2)* is used to estimate the actual value of its quadratic effect scalar coefficient in this second-order polynomial expression. The null hypothesis associated with *ctec(2)* is that the actual value for the quadratic effect scalar coefficient is equal to zero, whereas the alternative hypothesis is that the actual value for the quadratic effect scalar coefficient is less than zero, viz., is negative. This null hypothesis was also rejected, but not as emphatically as the null hypothesis pertaining to contrast *ctec(1)*.

Remark: A very brief table of contrasts pertaining to orthogonal polynomial contrasts for a single quantitative treatment, equally replicated at each of k equally spaced treatment levels, $k = 3$ to 6, can be found in Table 8.1 of Little and Jebe (1975).

2.B. SUPPLEMENTAL TOPIC: FRACTIONAL FACTORIAL ARRANGEMENTS AND STATISTICAL CONFOUNDING

Sometimes only a fraction of a $(2)^{n_i}$ factorial array is employed in an experiment test program, say a one-half, a one-quarter, or even a one-eighth fraction. Fractional factorial design experiment test programs are particularly appropriate when test specimens (experimental units) are very expensive or are available in very limited quantities. For example, suppose that we wish to examine the effect of three treatments, but only four specimens (experiment units) are available for testing. If so, our estimated conceptual statistical model can have only four estimated scalar coefficients (at most). On the other hand, recall that the estimated complete analytical model for a $(2)^3$ factorial design experiment test program has eight estimated scalar coefficients. Thus, these eight estimated scalar coefficients must be parsed into four groups of the sum of two *statistically confounded* (inseparable) estimated scalar coefficients to establish the estimated conceptual statistical model.

Consider the standard orthogonal augmented $(2)^3$ factorial arrangement contrast array [originally presented in Figure 2.6(b)]:

Constructed   *Augmented*

	Specimen Number	$\lvert +1\text{'s}\rvert$	$\lvert ct1ec_i\text{'s}\rvert$	$\lvert ct2ec_i\text{'s}\rvert$	$\lvert ct3ec_i\text{'s}\rvert$	$\lvert ct1t2iec_i\text{'s}\rvert$	$\lvert ct1t3iec_i\text{'s}\rvert$	$\lvert ct2t3iec_i\text{'s}\rvert$	$\lvert ct1t2t3iec_i\text{'s}\rvert$
Orthogonal	1	+1	−1	−1	−1	+1	+1	+1	−1
Augmented	2	+1	+1	−1	−1	−1	−1	+1	+1
$(2)^3$	3	+1	−1	+1	−1	−1	+1	−1	+1
Factorial	4	+1	+1	+1	−1	+1	−1	−1	−1
Arrangement	5	+1	−1	−1	+1	+1	−1	−1	+1
Contrast	6	+1	+1	−1	+1	−1	+1	−1	−1
Array	7	+1	−1	+1	+1	−1	−1	+1	−1
	8	+1	+1	+1	+1	+1	+1	+1	+1

Note that we cannot arbitrarily select specimen numbers (rows) 1, 2, 3, and 4 for our one-half fraction $(2)^3$ factorial design experiment test program because the first four elements of the $\lvert ct3ec_i\text{'s}\rvert$ column vector do not sum to zero. Thus, we must select the specific specimen numbers (rows) that are consistent with our test objective. Since our conceptual statistical model will have four terms (at most), the most intuitive one-half fraction orthogonal augmented $(2)^3$ factorial design contrast array has its first four column vectors identical to the four column vectors that comprise the standard orthogonal augmented contrast array for a $(2)^2$ factorial design experiment test program, viz.,

	Specimen Number	$\lvert +1\text{'s}\rvert$	$\lvert ct1ec_i\text{'s}\rvert$	$\lvert ct2ec_i\text{'s}\rvert$	$\lvert ct1t2iec_i\text{'s}\rvert$
Orthogonal Augmented	1	+1	−1	−1	+1
$(2)^2$	2	+1	+1	−1	−1
Factorial Arrangement	3	+1	−1	+1	−1
Contrast Array	4	+1	+1	+1	+1

Now consider the one-half fraction orthogonal augmented $(2)^3$ factorial design contrast array that is comprised of specimen numbers (rows) 5, 2, 3, and 8 (in that order) taken from the original orthogonal augmented $(2)^3$ factorial arrangement contrast array. Note that the first four column vectors in this proposed one-half fraction array are indeed identical to the four column vectors that comprise the standard orthogonal augmented $(2)^2$ factorial design contrast array. The issue now is to establish the nature of the statistical confounding that is intrinsic in this proposed one-half fraction orthogonal augmented $(2)^3$ factorial arrangement contrast array.

Specimen Number	$\lvert +1\text{'s}\rvert$	$\lvert ct1ec_i\text{'s}\rvert$	$\lvert ct2ec_i\text{'s}\rvert$	$\lvert ct3ec_i\text{'s}\rvert$	$\lvert ct1t2iec_i\text{'s}\rvert$	$\lvert ct1t3iec_i\text{'s}\rvert$	$\lvert ct2t3iec_i\text{'s}\rvert$	$\lvert ct1t2t3iec_i\text{'s}\rvert$
5	+1	−1	−1	+1	+1	−1	−1	+1
2	+1	+1	−1	−1	−1	−1	+1	+1
3	+1	−1	+1	−1	−1	+1	−1	+1
8	+1	+1	+1	+1	+1	+1	+1	+1

The last four column vectors in this proposed one-half fraction orthogonal augmented contrast array establish its intrinsic statistical confounding. Note that the $\lvert ct1ec_i\text{'s}\rvert$ column vector is identical to the $\lvert ct2ct3iec_i\text{'s}\rvert$ column vector. Thus the associated estimated scalar coefficient will actually be the statistically confounded sum [est($ct1esc$) + est($ct2t3iesc$)]. Next, note that the $\lvert ct2ec_i\text{'s}\rvert$ column vector is identical to the $\lvert ct1ct3iec_i\text{'s}\rvert$ column vector. Thus, the associated estimated scalar coefficient will actually be the statistically confounded sum [est($ct2esc$) + est($ct1t3iesc$)]. Then, note that the $\lvert ct3ec_i\text{'s}\rvert$ column vector is identical to the $\lvert ct1ct2iec_i\text{'s}\rvert$ column vector. Thus, the associated estimated scalar coefficient will actually be the statistically confounded sum [est($ct3esc$) + est($ct1t2iesc$)]. Finally, note that the

|+ 1's| identity column vector is identical to the $|ct1t2t3iec_i\text{'s}|$ column vector. Thus, the associated estimated scalar coefficient will actually be the confounded sum [est($csmmsc$) + est($ct1t2t3iesc$)]. These four estimated scalar coefficients have practical physical interpretation *only when* it is arbitrarily presumed that the actual values for the respective interaction effects are equal to zero. If so, then the four estimated scalar coefficients will be, respectively, the est($csmmsc$), the est($ct1esc$), the est($ct2esc$), and the est($ct3esc$).

> *Remark*: The economy of this one-half fraction experiment test program is so great that there can be a strong tendency to rationalize the presumption that the actual value for the interaction effects are equal to zero when, in fact, this presumption is false.

This very brief presentation is intended to provide the perspective that fractional factorial experiment test programs can be advantageous when (a) the number of treatments (main effects) of specific interest is large, (b) the available test resources are very limited, and (c) the actual values for all (or certain) interaction effects can objectively be presumed to be negligible. These three constraints usually limit the use of fractional factorial experiment test programs to either screening studies (in which the test objective is to discern which main effects dominate the test outcome) and to ruggedness studies [in which several dominant test variables are slightly changed from their nominal specification (target) values to ascertain their relative effects on the resulting test outcomes]. Box, et al. (1978) is an authoritative reference for the planning, conduct, and statistical analysis of factorial and fractional factorial design experiment test programs.

Exercise Set 7

These exercises are intended to provide additional perspective regarding the statistical confounding intrinsic in our elementary example.

1. Verify that the statistical confoundings intrinsic in the ordered specimen numbers (rows), 1, 4, 7, and 6, of the standard orthogonal augmented $(2)^3$ factorial design contrast array are, respectively, the differences: [est($ct1esc$) − est($ct2t3iesc$)], [est($ct2esc$) − est($ct1t3iesc$)], [est($ct3esc$) − est($ct1t2iesc$)], and [est($csmmsc$) − est($ct1t2t3iesc$)].

2. Verify that the four statistical confoundings in Example 1 can be added to and subtracted from the four corresponding statistical confoundings elaborated above to obtain exactly the same

numerical estimates for the actual values of the eight scalar
coefficients that are obtained by computing the eight scalar
coefficients in the estimated complete analytical model associated with the standard $(2)^3$ factorial design experiment test
program.

2.B.1. Extension

Suppose that a standard $(2)^3$ factorial design experiment test program is
appropriate, but nominally identical test specimens (experimental units)
are available only in two batches of size four. Then, a batch effect contrast
column vector must be added to the eight column vectors that comprise the
standard orthogonal augmented $(2)^3$ factorial design contrast array.
However, its estimated (batch effect) scalar coefficient will be statistically
confounded with one of the other estimated scalar coefficients. Accordingly,
the basic issue is which of the possible statistical confoundings is most
consistent with the test objective. Consider the following proposed orthogonal augmented contrast array:

Proposed Orthogonal Augmented Contrast Array for a $(2)^3$ Factorial Design
Experiment Test Program

Specimen Number	$|bec_i\text{'s}|$	$|+1\text{'s}|$	$|ct1ec_i\text{'s}|$	$|ct2ec_i\text{'s}|$	$|ct3ec_i\text{'s}|$	$|ct1t2iec_i\text{'s}|$	$|ct1t3iec_i\text{'s}|$	$|ct2t3iec_i\text{'s}|$	$|ct1t2t3iec_i\text{'s}|$
5	+1	+1	−1	−1	+1	+1	−1	−1	+1
2	+1	+1	+1	+1	−1	−1	−1	+1	+1
3	+1	+1	−1	−1	−1	−1	+1	−1	+1
8	+1	+1	+1	+1	+1	+1	+1	+1	+1
1	−1	+1	−1	−1	−1	+1	+1	+1	−1
4	−1	+1	+1	+1	−1	+1	−1	−1	−1
7	−1	+1	−1	+1	+1	−1	−1	+1	−1
6	−1	+1	+1	−1	+1	−1	+1	−1	−1

This proposed array consists of the proposed one-half fraction orthogonal
augmented $(2)^3$ factorial design contrast array in our elementary statistical
confounding example above plus the remaining specimen numbers (rows)
in the standard orthogonal augmented $(2)^3$ factorial design contrast array
rearranged in the order 1, 4, 7, and 6. The estimated batch effect for this
proposed array is statistically confounded with the estimated three-factor
interaction effect. Accordingly, the estimated three-factor interaction effect
cannot subsequently be reinterpreted as the est($CRHEE_i$'s) in the estimated
conceptual statistical model (even if its actual value is equal to zero).
Nevertheless, it is rational to assert that a statistically credible estimate
of the actual value for the $ct1t2t3iesc$, if it were attainable, would likely be

quite small compared to the actual value for batch effect scalar coefficient. If so, the effect of statistical confounding on the estimated value of the actual batch effect would also be quite small. One way or the other, given the proposed statistical confounding, the actual values for all three main effect and all three of the two-factor interaction scalar coefficients can be estimated, as well as the actual value for the conceptual statistical model mean scalar coefficient. Then, if it is presumed that the actual value for one or more of the three two-factor interaction effects is equal to zero, the associated estimated interaction effect(s) can be reinterpreted as est($CRHEE_i$'s) in the estimated conceptual statistical model.

Remark: The estimate of the actual value for the batch effect (and its associated statistical confidence interval) can well be the most important information obtained from an experiment test program. Never pass up a practical opportunity to estimate the actual value for a batch effect.

3

Basic Probability and Statistics Concepts and Their Mechanical Reliability Applications

3.1. INTRODUCTION

The successive outcomes of continually replicated experiment test programs exhibit random variability. The value of specific interest that we associate with each respective outcome is usually termed a statistic (test statistic), but it is more broadly termed a random variable. We subsequently use a capital letter to denote a random variable and a lower-case letter to denote its realization value.

3.2. EXACT ENUMERATION-BASED PROBABILITY

Recall that the first fundamental abstraction in statistical analysis is that the experiment test program that is (was) actually conducted can be continually replicated indefinitely to generate the conceptual collection of all possible equally likely experiment test program outcomes. Moreover, the outcome for the experiment test program that is (was) actually conducted is (was) randomly selected from this collection of all possible equally likely experiment test program outcomes. Given this fundamental statistical abstraction, suppose that the experiment test program that is (was) actually conducted has only n_{elo} equally likely outcomes. The enumeration-based probability of obtaining an outcome of specific interest (p_{oosi}) when this experiment test

84

program is actually conducted is defined as the ratio of the number of experiment test program outcomes of specific interest (n_{oosi}) to the number of equally likely experiment test program outcomes (n_{elo}), viz.,

$$\text{exact enumeration-based probability } p_{oosi} = \frac{n_{oosi}}{n_{elo}}$$

We now present the classical fair-coin experiment for which the respective enumerations of n_{elo} and n_{oosi} are easily accomplished and intuitively understood.

3.2.1. Enumeration of All Equally Likely Outcomes for the Classical Fair-Coin Experiment Test Program

The classical fair-coin experiment test program consists of n_f independent coin flips, where (a) heads and tails are the only possible outcomes for each successive flip, and (b) heads and tails are equally likely to occur. Let n_{oosi} pertain to the observed number of heads, n_h, during n_f flips. We use Yate's enumeration algorithm in Table 3.1 to enumerate all of the respective equally likely outcomes of the classical fair-coin experiment test program by associating tails with -1 and heads with $+1$.

Two fundamental probability concepts are illustrated in this classical fair-coin experiment test program example. First, the probabilities of occurrence for independent outcomes can be multiplied to establish the probability that the corresponding collection of outcomes will occur (together), e.g., the probability of observing HH as the outcome of two independent flips is $(1/2) \cdot (1/2) = 1/4$. Second, the probabilities pertaining to a collection of mutually exclusive experiment test program outcomes can be summed to establish the probability that one of these outcomes will occur. (Note that the sum of probabilities for mutually exclusive and exhaustive outcomes is always equal to one.)

Exercise Set 1

These exercises are intended to introduce a factorial expression for establishing the number of equally likely classical fair-coin experiment test program outcomes of specific interest.

1. Suppose a fair-coin is (independently) flipped five times. (a) Enumerate the respective n_{elo} equally likely outcomes (sequences of heads and tails) by using Yate's enumeration algorithm and associating tails with -1 and heads with $+1$. Then, (b) plot your

Table 3.1 Enumeration of All Equally-Likely Outcomes for Classical Fair-Coin Experiment Test Program—and Corresponding Enumeration-Based Probability of Obtaining an Outcome of Specific Interest

number of flips = n_f	number of equally-likely experiment test program outcomes = n_{elo}	enumeration of all respective equally-likely experiment test program outcomes	number of equally-likely experiment test program outcomes such that $n_{oosi} = n_h$	enumeration-based probability that a randomly selected equally-likely experiment test program outcome will have its $N_H = n_h$	enumeration-based probability that a randomly selected equally-likely experiment test program outcome will have its $N_H \geq n_h$
1	2	T	0	1/2	2/2
		H	1	1/2	1/2
2	4	TT	0	1/4	4/4
		HT	(1)		
		TH	1	2/4	3/4
		HH	2	1/4	1/4
3	8	TTT	0	1/8	8/8
		HTT	(1)		
		THT	(1)		
		HHT	(2)		
		TTH	1	3/8	7/8
		HTH	(2)		
		THH	2	3/8	4/8
		HHH	3	1/8	1/8

results in the format illustrated in Figure 1.5(b), using the number of heads n_h as the abscissa metric.

2. The respective numbers of equally likely classical fair-coin experiment test program outcomes with n_h heads occurring in n_f flips can be computed using the following factorial expression for n_h given n_f, where n_h ranges from zero to n_f (and zero factorial is equal to one by definition):

$$\text{number of distinct replicate fair-coin experiment test programs with } n_h \text{ heads in } n_f \text{ flips} = \frac{n_f!}{(n_h)! \cdot (n_f - n_h)!}$$

(a) Use this factorial expression to verify your results in Example 1, viz., let $n_f = 5$ and let n_h successively take on the values 0, 1, 2, 3, 4, and 5. Then, (b) verify that the total number of equally likely classical fair-coin experiment test program outcomes sum to 2^5. In turn, (c) verify the enumeration-based probabilities computed in Example 1.

3.2.1.1. Discussion

The null hypothesis for the classical fair-coin experiment test program is $H = T$ statistically. This null hypothesis is statistically equivalent to the null hypothesis that $B = A$ statistically for a paired-comparison experiment test program. Note that, given the null hypothesis that $H = T$ statistically, the probability that a head will occur as the outcome of any given flip is exactly equal to 1/2. Correspondingly, given the null hypothesis that $B = A$ statistically, the probability that the $(b - a)$ difference for any given paired-comparison will be positive is exactly equal to 1/2. Thus, the fair-coin experiment test program outcome that a head will occur for any given flip is statistically equivalent to the paired-comparison experiment test program outcome that a positive $(b - a)$ difference will occur for any given paired-comparison. Accordingly, the outcome of any given flip of a fair-coin, heads or tails, is statistically equivalent to the sign of the observed $(b - a)$ difference, positive or negative, for any given paired comparison. This statistical equivalence underlies the three example statistical tests of hypothesis for a paired-comparison experiment test program that appear below. The first two of these three examples are primarily intended to illustrate the calculation of enumeration-based null hypothesis rejection probabilities.

3.2.2. Three Statistical Tests of the Null Hypothesis that $B = A$ Statistically Versus the Simple (One-Sided) Alternative Hypothesis that $B > A$ Statistically, Given the Outcome of a Paired-Comparison Experiment Test Program

The minimal probability background just presented is sufficient to perform three alternative statistical tests of hypotheses that have direct application in improving the mechanical reliability of a product. First, recall that a paired-comparison experiment test program with n_{pc} paired comparisons is actually an unreplicated randomized complete block design (RCBD) experiment test program with two treatments and n_b blocks. Let its two treatments of specific interest be denoted A and B, where A pertains to the present design and B pertains to the proposed design. Suppose, to alleviate a mechanical reliability problem with A, we require that B excels A statistically. Accordingly, the null hypothesis is that $B = A$ statistically, whereas the simple (one-sided) alternative hypothesis is that $B > A$ statistically. Next, presume that we have conducted a paired-comparison experiment test program with $n_{pc} = 8$ paired comparisons (with $n_b = 8$ blocks) and that we obtained the following outcomes:

	a	b	sign of the $(b - a)$ difference
Block 1	112	125	+
Block 2	156	173	+
Block 3	113	141	+
Block 4	197	219	+
Block 5	234	255	+
Block 6	166	177	+
Block 7	121	131	+
Block 8	143	159	+

3.2.3. The Classical Sign Test

The test statistic that is used in the classical sign test is the sum of positive signs for the collection of $(b - a)$ differences that constitute the outcome of a paired-comparison experiment test program. Accordingly, microcomputer program *EBST* (enumeration-based sign test) first calculates the data-based reference value of this test statistic for the collection of $(b - a)$ differences that constitute the outcome of the paired-comparison experiment test program that was actually conducted. Then, given the null hypothesis that $B =$

A statistically, it constructs all equally likely outcomes for this paired-comparison experiment test program by using Yate's enumeration algorithm to reassign positive and negative signs to its $(b - a)$ differences. In turn, it counts the number of these constructed outcomes that have its test statistic value equal to or greater than the data-based reference value. Finally, it calculates the enumeration-based probability that the randomly selected outcome of this paired-comparison experiment test program when continually replicated will have its test statistic value equal to or greater than the data-based reference value. If this probability is sufficiently small, the credibility of the null hypothesis underlying this enumeration-based probability calculation is dubious. If, in addition, the actual experiment test program outcome is consistent with the simple (one-sided) alternative hypothesis that $B > A$ statistically, then the null hypothesis must rationally be rejected. The value of the null hypothesis rejection probability should (must) be selected before the actual paired-comparison experiment test program outcome is known. The size of the null hypothesis rejection probability is subjective—but it is traditionally selected to be either 0.10, 0.05, or 0.01.

Two issues involved in testing the null hypothesis warrant further discussion. First, we can incorrectly reject the null hypothesis when it is correct. Returning to the classical fair-coin experiment test program to illustrate this notion, we assert that one possible, but unlikely, equally likely classical fair-coin experiment test program outcome is eight heads in eight flips. In fact, the probability of this classical fair-coin experiment test program outcome is exactly the same as for our example paired-comparison experiment test program outcome, viz., $p = 0.0039$ (1/256). Thus, given a classical fair-coin experiment test program outcome with eight heads in eight flips, we would rationally (and incorrectly) reject the null hypothesis that the actual values for the probabilities of heads and tails are each exactly equal to 1/2. On the other hand, given our example paired-comparison experiment test program outcome, it is impossible to know whether our rational test of hypothesis decision is correct or not. Accordingly, we must always be aware of the possibility of committing a *Type I* error, viz., of incorrectly rejecting a null hypothesis when it is correct. In addition, there is a second type of error that occurs in a statistical test of the null hypothesis, viz., the failure to reject the null hypothesis when the alternative hypothesis is correct. This second error, termed a *Type II* error, can be more detrimental in some mechanical reliability applications than a *Type I* error.

Remark: Remember that the null hypothesis does not have to be physically credible. Its purpose is to provide a rational basis for calculating the probability that is used to test the null hypothesis versus the physically relevant alternative hypothesis.

```
C> COPY EXPCDTA1 DATA

  1 file(s) copied

C>EBST
```

The data-based sum of positive signs for the collection of $(b - a)$ differences that constitute the outcome of the paired-comparison experiment test program that was actually conducted is equal to 8.0.

Given the null hypothesis that $B = A$ statistically, this microcomputer program constructed exactly 256 equally-likely outcomes for this paired-comparison experiment test program by using Yate's enumeration algorithm to reassign positive and negative signs to its $b - a)$ differences. The number of these outcomes that had its sum of positive signs for its $(b - a)$ differences equal to or greater than 8.0 is equal to 1. Thus, given the null hypothesis that $B = A$ statistically, the enumeration-based probability that a randomly selected outcome of this paired comparison experiment test program when continually replicated will have its sum of positive signs for its $(b - a)$ differences equal to or greater than 8.0 is equal to 0.0039. When this probability is sufficiently small, reject the null hypothesis in favor of the simple (one-sided) alternative hypothesis that $B > A$ statistically.

The second issue of concern in testing the null hypothesis that $B = A$ statistically pertains to the reason for tabulating cumulative probabilities in the format given in the last column of Table 3.1. We always tabulate the probabilities pertaining to experiment test program outcomes *equal to or greater than* the experiment test program outcome of specific interest because the probability of any given outcome, even the most probable one, approaches zero as n_{elo} increases without bound.

Table 3.2 presents the enumeration-based probabilities of specific interest for a classical sign test of the null hypothesis that $B = A$ statistically, given a paired-comparison experiment test program with $n_{pc} = 8$. Note that the last column in this tabulation is deliberately stated in a

Table 3.2 Enumeration-Based Probabilities of Specific Interest (Column 4) for a Classical Sign Test of the Null Hypothesis that $B = A$ Statistically, Given the Respective $(b - a)$ Differences for a Paired-Comparison Experiment Test Program with $n_{pc} = 8$

number of outcomes of specific interest = n_{oosi} = number of positive signs = n_{ps}	number of equally-likely paired-comparison experiment test program outcomes with n_{ps} positive signs for respective $(b - a)$ differences = n_{elo}	enumeration-based probability that $N_{PS} = n_{ps}$	enumeration-based probability that $N_{PS} \geq n_{ps}$
0	1	1/256	1.0000
1	8	8/256	0.9961
2	28	28/256	0.9648
3	56	56/256	0.8555
4	70	70/256	0.6367
5	56	56/256	0.3633
6	28	28/256	0.1445
7	8	8/256	0.0352
8	1	1/256	0.0039

format that agrees with our simple (one-sided) alternative hypothesis. Thus, we reject the null hypothesis in favor of this alternative hypothesis *only* when the experiment test program generates a sufficiently large number of positive signs.

3.2.4. Wilcoxon's Signed-Rank Test

Obviously the sign test sacrifices important information regarding the magnitude of the data-based $(b - a)$ differences for the paired-comparison experiment test program that was actually conducted. The test statistic that is used in Wilcoxon's signed-rank test is the sum of the positive signed-ranks for the collection of $(b - a)$ differences that constitute the outcome of a paired-comparison experiment test program.

Before presenting microcomputer program *EBSRT* (enumeration-based signed-rank test), we first illustrate how the signed-rank is established for each of the $(b - a)$ differences that collectively constitute the outcome of a paired-comparison experiment test program. Consider the following tabulation based on the example paired-comparison datum values that were used to illustrate the classical sign test:

	a	b	respective $(b - a)$ differences	rank of the [absolute value of each $(b - a)$ difference]	respective signed-ranks
Block 1	112	125	+13	3	+3
Block 2	156	173	+17	5	+5
Block 3	113	141	+28	8	+8
Block 4	197	219	+22	7	+7
Block 5	234	255	+21	6	+6
Block 6	166	177	+11	2	+2
Block 7	121	131	+10	1	+1
Block 8	143	159	+16	4	+4

sum of the positive signed-ranks = 36

Although it is not an issue now, we note that both (unsigned) rank ties and zero $(b - a)$ differences can occur in certain paired-comparison experiment test programs. Accordingly, when (unsigned) rank ties occur, microcomputer program *EBSRT* assigns the average rank to each (unsigned) tied rank, and when one or more zero $(b - a)$ differences occur, microcomputer program *EBSRT* assigns 1/2 of the associated unsigned ranks to the sum of the positive signed-ranks.

We now illustrate the computation of the enumeration-based probability that the sum of positive signed-ranks given all equally likely outcomes for a paired-comparison experiment test program, ignoring ties and zero $(b - a)$ differences in this illustrative example. Consider, for simplicity, a paired comparison experiment test program with only four paired comparisons (blocks). Given the null hypothesis that $B = A$ statistically, we require that (a) each rank is equally likely to appear in each block, and (b) positive and negative signs are equally likely for each rank. Accordingly, if we arbitrarily pick any order for the four ranks that is convenient, say, 1, 2, 3, and 4, $2^4 = 16$ equally likely sign sequences for the signed-ranks can be constructed using Yate's enumeration algorithm. In turn, the sums of positive signed-ranks, *spsr*, can be computed and summarized.

ranks 1, 2, 3, 4 with positive and negative signs assigned using Yates's enumeration algorithm				respective equally-likely outcomes and their sum of positive signed-ranks, *spsr*	enumeration-based probability that $SPSR = spsr$	enumeration-based probability that $SPSR \geq spsr$
−1	−2	−3	−4	0	1/16	16/16
+1	−2	−3	−4	1	1/16	15/16
−1	+2	−3	−4	2	1/16	14/16
+1	+2	−3	−4	(3)		
−1	−2	+3	−4	3	2/16	13/16
+1	−2	+3	−4	(4)		
−1	+2	+3	−4	(5)		
+1	+2	+3	−4	(6)		
−1	−2	−3	+4	4	2/16	11/16
+1	−2	−3	+4	5	2/16	9/16
−1	+2	−3	+4	6	2/16	7/16
+1	+2	−3	+4	(7)		
−1	−2	+3	+4	7	2/16	5/16
+1	−2	+3	+4	8	1/16	3/16
−1	+2	+3	+4	9	1/16	2/16
+1	+2	+3	+4	10	1/16	1/16

```
C> COPY EXPCDTA1 DATA

  1 file(s) copied

C> EBSRT
```

The data-based sum of positive signed-ranks for the collection of $(b - a)$ differences that constitute the outcome of the paired-comparison experiment test program that was actually conducted is equal to 36.0.

Given the null hypothesis that $B = A$ statistically, this microcomputer program constructed exactly 256 equally-likely outcomes for this paired-comparison experiment test program by using Yate's enumeration algorithm to reassign positive and negative signs to its $(b - a)$ differences. The number of these outcomes that had its sum of positive signed-ranks for its $(b - a)$ differences equal to or greater than 36.0 is equal to 1. Thus, given the null hypothesis that $B = A$ statistically, the enumeration-based probability that a randomly selected outcome of this paired comparison experiment test program when continually replicated will have its sum of positive signed-ranks for its $(b - a)$ differences equal to or greater than 36.0 is equal to 0.0039. When this probability is sufficiently small, reject the null hypothesis in favor of the simple (one-sided) alternative hypothesis that $B > A$ statistically.

Remark: The relationship between the sign test and the signed-rank test is clear when it is understood that microcomputer program *EBST* is the same as microcomputer program *EBSRT*, except that the "ranks" in the former are all equal to one.

3.2.5. Fisher's Enumeration-Based Test

Even Wilcoxon's signed-rank test sacrifices information regarding the actual magnitudes of the individual $(b - a)$ differences in a paired-comparison experiment test program. The test statistic that is used in Fisher's enumeration-based test is the sum of the actual values for the $(b - a)$ differences that constitute the outcome of a paired-comparison experiment test program. Microcomputer program *FEBT* (Fisher's enumeration-based test) is analogous to microcomputer programs *EBST* and *EBSRT*.

```
C> COPY EXPCDTA1 DATA

  1 file(s) copied

C> FEBT
```

The data-based sum of the actual values for the collection of $(b - a)$ differences that constitute the outcome of the paired-comparison experiment test program that was actually conducted is equal to 138.

Given the null hypothesis that $B = A$ statistically, this microcomputer program constructed exactly 256 equally-likely outcomes for this paired-comparison experiment test program by using Yate's enumeration algorithm to reassign positive and negative signs to its $(b - a)$ differences. The number of these outcomes that had its sum of the actual values for its $(b - a)$ differences equal to or greater than 138 is equal to 1. Thus, given the null hypothesis that $B = A$ statistically, the enumeration-based probability that a randomly selected outcome of this paired comparison experiment test program when continually replicated will have its sum of the actual values for its $(b - a)$ differences equal to or greater than 138 is equal to 0.0039. When this probability is sufficiently small, reject the null hypothesis in favor of the simple (one-sided) alternative hypothesis that $B > A$ statistically.

Remark: Because Fisher's enumeration-based test is based on the actual values for the individual paired-comparison $(b - a)$ differences, it requires an appropriate microcomputer program to compute the null hypothesis rejection probability. In contrast, because the classical sign test and Wilcoxon's signed-rank test do not depend on the actual values for the individual paired-comparison $(b - a)$ differences, these tests have traditionally been performed using hand calculations and tables.

3.2.5.1. Discussion

Microcomputer programs *EBST*, *EBSRT*, and *FEBT* are arbitrarily limited to paired-comparison experiment test programs with 16 or fewer blocks. (It is unlikely that a paired-comparison experiment test program of specific interest in mechanical test or mechanical reliability applications will have more than 16 blocks.) Despite this arbitrary size limitation, we can still use this enumeration-based methodology to test the null hypothesis that $B = A$ statistically for larger paired-comparison experiment test programs. Clearly, regardless of the size of a paired-comparison experiment test program, each of its paired comparisons provides a positive or a negative sign for its $(b - a)$ difference that is equally credible for purposes of testing the null hypothesis that $B = A$ statistically. Thus, we can construct a *surrogate* paired-comparison experiment test program by running microcomputer program *RANDOM1* to select 16 $(b - a)$ differences from the paired-comparison experiment test program that was actually conducted. Then, these 16 $(b - a)$ differences are input data to microcomputer programs *EBST*, *EBSRT*, and/or *FEBT*. This random selection process can be repeated several times to demonstrate that the respective outcomes of the resulting surrogate experiment test programs generate null hypothesis rejection probabilities that are numerically consistent.

Remark: The only statistical drawback to using a randomly selected subset of paired-comparison experiment test program $(b - a)$ differences is that, given an acceptable probability of committing a *Type I* error, the associated probability of committing a *Type II* error is increased.

Exercise Set 2

The purpose of these exercises is to familiarize you with the use of the microcomputer programs *EBST*, *EBSRT*, and *FEBT*.

1. Verify that microcomputer programs *EBST*, *EBSRT*, and *FEBT* compute the same value for the null hypothesis rejection probability when microcomputer file *EXPCDTA1* is employed as input data. Then, examine the respective $(b - a)$ differences in *EXPCDTA1* to explain this unusual result.
2. Run programs *EBST*, *EBSRT*, and *FEBT* with microcomputer file *EXPCDTA2* as input data and compare the respective computed null hypothesis rejection probabilities. Examine the respective $(b - a)$ differences in *EXPCDTA2* and explain why the three respective results differ.
3. Suppose that the *b* and *a* datum values were transformed *before* the respective $(b - a)$ differences are computed, e.g., by taking the logarithms of the *a* and *b* datum values. (a) Can a strictly monotonic transformation affect the sum of positive signs for the respective $(b - a)$ differences? (b) Can a strictly monotonic transformation affect the sum of the positive signed-ranks for the respective $(b - a)$ differences? (c) Can a strictly monotonic transformation affect the sum of the respective $(b - a)$ differences?
4. In defining reliability, the test stimulus severity and required test duration are specified. The only issue is whether or not an individual specimen survives the imposed test stimulus for the required test duration. Which of the three microcomputer programs, *EBST*, *EBSRT*, or *FEBT*, if any, is relevant to reliability estimation?

Exercise Set 3

The purpose of these exercises is to examine the effect of zero $(b - a)$ differences on the resulting values for the null hypothesis rejection probability.

1. Modify the datum values in microcomputer file *EXPCDTA1* by setting *a* equal to *b* in a randomly selected paired comparison. Then, run microcomputer programs *EBST*, *EBSRT*, and *FEBT* and compare the respective results with and without this zero $(b - a)$ difference. Repeat this process, each time generating another zero $(b - a)$ difference, until all $(b - a)$ differences in the modified data set are equal to zero. Discuss your results.
2. Given the datum values in the *EXPCDTA1* file, successively add more and more additional zero $(b - a)$ differences to the existing datum values to observe the effect of a substantial proportion of

zero $(b - a)$ differences on the resulting null hypothesis rejection probability. Discuss your results.

3.3. EMPIRICAL SIMULATION-BASED PROBABILITY

We can simulate flipping a fair coin using a computer-intensive methodology that employs pseudorandom numbers. Suppose that we continually generate (mutually independent) pseudorandom numbers. Suppose in addition, one-half of these pseudorandom numbers correspond to a head and the other half correspond to a tail. Then, when we generate a pseudorandom number, its realization is equivalent to the outcome of flipping a fair coin. Now suppose that we generate n_f pseudorandom numbers and count the number of the associated simulated coin flips that actually correspond to a head. It is demonstrated in Section 3.5 that the observed simulation-based proportion of heads asymptotically approaches the exact enumeration-based probability of a head as n_f increases without bound. Stated in more generic terms, the empirical simulation-based probability p_{oosi} asymptotically approaches the corresponding exact enumeration-based probability p_{oosi} as the number n_{sbelo} of simulation-based equally likely experiment test program outcomes increases without bound. This asymptotic behavior provides an intuitive justification for defining the empirical simulation-based probability p_{oosi} as follows:

$$\text{empirical simulation-based probability } p_{oosi} = \frac{n_{oosi}}{n_{sbelo}}$$

One of the primary goals of this chapter is to demonstrate that this empirical simulation-based probability value provides a reasonably accurate approximation to the exact enumeration-based probability value even when n_{sbelo} is limited to a size that is practical in microcomputer programs.

Simulation-based microcomputer programs have two primary applications: (a) to provide perspective regarding the accuracy of the empirical simulation-based probability value in situations where the exact enumeration-based probability value is known, and (b) to generate reasonably accurate empirical simulation-based probability values in situations where the exact enumeration-based probability value is unknown. Both of these applications will be illustrated later in this text.

In Supplemental Topic 8.D we explain how to generate pseudorandom numbers (simulated datum values) from any known (two-parameter) conceptual statistical distribution of specific interest. However, in applications where the conceptual statistical distribution is unknown, we cannot generate simulated datum values. Thus, simulation-based methodologies are impotent in these applications and must be replaced by randomization-based methodologies.

3.4. EMPIRICAL RANDOMIZATION-BASED PROBABILITY

Recall that under the null hypothesis that $B = A$ statistically, the $(b - a)$ differences generated by a paired-comparison experiment test program are equally likely to be either positive or negative. Accordingly, when a randomization-based methodology is employed to compute an empirical null hypothesis rejection probability value, the positive and negative signs that are reassigned to the $(b - a)$ differences under the null hypothesis are selected using uniform pseudorandom numbers such that positive and negative signs are equally likely. Thus, this randomization-based probability value calculation requires that we (a) generate n_{rbelo} randomization-based equally likely experiment test program outcomes, each with a randomly reassigned sequence of equally likely positive and negative signs for its collection of $(b - a)$ differences, (b) count the number n_{oosi} of these randomization-based equally likely experiment test program outcomes that have the outcome of specific interest, and then (c) compute the empirical randomization-based probability p_{oosi} as the ratio of n_{oosi} to n_{rbelo}, viz.,

$$\text{empirical randomization-based probability } p_{oosi} = \frac{n_{oosi}}{n_{rbelo}}$$

However, the following empirical randomization-based probability expression given by Noreen (1989) is recommended on the basis of its (slightly) improved statistical behavior:

$$\text{Noreen's empirical randomization-based probability } p_{oosi} = \frac{n_{oosi} + 1}{n_{rbelo} + 1}$$

In theory, each of these empirical randomization-based probability values asymptotically approach the exact enumeration-based probability value as n_{rbelo} increases without bound. However, when n_{rbelo} is sufficiently large that the empirical randomization-based probability value is reasonably accurate, then this value is interchangeable for practical purposes with the exact enumeration-based probability value.

Randomization-based statistical methodologies are almost universally applicable in mechanical reliability applications. These computer-intensive analyses are becoming much more diverse relative to their practical application.

3.4.1. Example of Randomization-Based Test of Hypothesis

We now present an example of a randomization-based test of hypothesis for a paired-comparison experiment test program. Recall that the test statistic outcome for Fisher's enumeration-based test is the sum of the actual values for the individual paired-comparison $(b - a)$ differences. Recall also that given the null hypothesis that $B = A$ statistically, positive and negative signs are equally likely for each of these $(b - a)$ differences. Thus, for Fisher's enumeration-based test, we construct all equally likely outcomes for the paired-comparison experiment test program that was actually conducted by using Yate's enumeration algorithm to reassign positive and negative signs to its $(b - a)$ differences. Analogously, for Fisher's randomization-based test, we construct n_{rbelo} equally likely outcomes for the paired-comparison experiment test program that was actually conducted by using pseudorandom numbers to reassign equally-likely positive and negative signs to these $(b - a)$ differences. This pseudorandom sign reassignment is akin to flipping a fair coin to establish whether the sign for each individual $(b - a)$ difference is positive or negative. It is accomplished in microcomputer program *FRBT* (Fisher's randomization-based test) using uniform pseudorandom numbers (Section 3.5). This random sign reassignment is repeated until n_{rbelo} randomization-based equally likely outcomes for the paired-comparison experiment test program of specific interest have been constructed. Then, n_{oosi} is the number of these outcomes that has its sum of the actual values for the $(b - a)$ differences equal to or greater than the sum for the paired-comparison experiment test program that was actually conducted. In turn, the empirical randomization-based null hypothesis rejection probability value is computed using Noreen's expression.

Remark: An example randomization-based test of the null hypothesis that $A = B$ statistically versus the composite (two-sided) alternative hypothesis that $A \neq B$ statistical is presented in Supplemental Topic 3.D for a CRD experiment test program with two treatments denoted A and B.

```
COPY EXRBPCD1 DATA

  1 file(s) copied

C> FRBT
```

The data-based sum of the actual values for the collection of $(b - a)$ differences that constitute the outcome of the paired-comparison experiment test program that was actually conducted is equal to 138.

Given the null hypothesis that $B = A$ statistically, this microcomputer program constructed 9999 equally-likely outcomes for this paired-comparison experiment test program by using uniform pseudorandom numbers to reassign equally-likely positive and negative signs to its $(b - a)$ differences. The number of these outcomes that had its sum of its actual values for its $(b - a)$ differences equal to or greater than 138 is equal to 43. Thus, given the null hypothesis that $B = A$ statistically, the randomization-based probability that a randomly selected outcome of this paired-comparison experiment test program when continually replicated will have its sum of the actual values for its $(b - a)$ differences equal to or greater than 138 is equal to 0.0044. When this probability is sufficiently small, reject the null hypothesis in favor of the simple (one-sided) alternative hypothesis that $B > A$ statistically.

Exercise Set 4

These exercises are intended to generate confidence in the accuracy of randomization-based probability values. The basic notion is that the empirical randomization-based probability value asymptotically approaches the exact enumeration-based probability value as n_{rbelo} increases without bound. Accordingly, the empirical randomization-based probability value is a reasonably accurate estimate of the exact enumeration-based probability value when n_{rbelo} is sufficiently large.

     1. (a) Run microcomputer program *FRBT* with microcomputer file *EXRBPCD1* as input data and with the number of replicate randomization-based paired-comparison experiment test programs

to 99, 999, 9999, and 99999. Then, (b) compare the respective empirical randomization-based null hypothesis rejection probability values to the corresponding exact enumeration-based null hypothesis rejection probability values obtained by running microcomputer program *FEBT* with the same input data, and comment on the apparent relationship between the accuracy of the empirical randomization-based probability value and the corresponding value for n_{rbelo}.

2. Repeat Example 1, but use microcomputer file *EXRBPCD2* as input data.

3.4.1.1. Discussion

The only difference between Fisher's enumeration-based test and Fisher's randomization-based test for a paired-comparison experiment test program lies in the way that the sign sequences for the respective unsigned $(b - a)$ differences are generated. The enumeration-based methodology requires that no sign sequence is repeated and that all possible sign sequences must be generated. In contrast, sign sequences can be repeated in the randomization-based methodology before all sign sequences are generated (as in random selection *with replacement*), and not all sign sequences need necessarily be generated. However, it should be intuitively clear by analogy with the empirical simulation-based estimate of p_{oosi} that, as n_{rbelo} increases without bound, the empirical randomization-based estimate of p_{oosi} statistically becomes more and more accurate. This fundamental simulation-based (randomization-based) probability behavior is demonstrated in the next section using pseudorandom numbers.

3.5. SIMULATION-BASED DEMONSTRATIONS OF FUNDAMENTAL PROBABILITY BEHAVIORS

We now present several simulation-based demonstrations of fundamental probability behaviors that form the foundation for a broad range of statistical analyses of specific interest in mechanical reliability.

3.5.1. The Uniform Pseudorandom Number Generator

For the simulation-based demonstrations of specific interest in this section, we wish to generate a series of conceptual numbers, each with n_{digit} digits, such that each of the $10^{n_{digit}}$ numbers in the interval $0.000 \ldots 0$ to

0.999 . . . 9 is equally likely to be randomly selected at each stage of the simulation process. These equally likely conceptual random numbers are said to be uniform. Fortunately, recursive algorithms can be employed in microcomputer programs to generate a sequence of pseudorandom numbers that (more or less) emulate the desired conceptual random numbers. These recursive algorithms have deficiencies that range from terrible to tolerable. Thus, each candidate pseudorandom number generator should have its credibility established by a wide variety of appropriate simulation-based demonstrations, viz., by having its output favorably compared to the behavior of a conceptual uniform random number generator.

Remark: We restrict our uniform pseudorandom numbers to the interval zero to one so that their realization values have an intuitive probability interpretation.

Pseudorandom number generators require seed numbers to initiate their generation process. Subsequent seed numbers are computed as an integral part of the pseudorandom number generation process. All sequences of pseudorandom numbers initiated by the same seed numbers are identical.

3.5.1.1. Simulation Example One: Eight-Digit Uniform Pseudorandom Numbers

Suppose we seek (mutually independent) eight-digit numbers that are equally likely to occur anywhere in the interval, 0.00000000 to 0.99999999. These numbers are conceptually generated by allowing each number (0,1,2,3,4,5,6,7,8,9) to have an equal opportunity of appearing at each digit location, independently of the numbers that appear at the other digit locations. The Wichmann and Hill (1981) pseudorandom number generator appears to generate numbers that, for practical purposes, are random, mutually independent, and equally likely.

   Microcomputer program *UNIFORM* is intended to demonstrate that local peculiarities will surely appear in any long sequence of pseudorandom numbers (or conceptual random numbers). Note that the first four digits are 1's for Realization Value 17, whereas there are four 5's consecutively for Realization Value 20. However, Realization Values 17 and 20 are just as random as any of the 48 other realization values. Obviously, random does not connote thoroughly mixed.

```
C> UNIFORM

Input a new set of three, three-digit odd seed numbers

315 527 841
```

50 Uniform 8-Digit Pseudorandom Numbers

Realization Value 1 = 0.48530650
Realization Value 2 = 0.26336943
Realization Value 3 = 0.93063729
Realization Value 4 = 0.05385052
Realization Value 5 = 0.95320082
Realization Value 6 = 0.78554568
Realization Value 7 = 0.04344359
Realization Value 8 = 0.86942954
Realization Value 9 = 0.89197557
Realization Value 10 = 0.17496878
Realization Value 11 = 0.36369488
Realization Value 12 = 0.47844548
Realization Value 13 = 0.31804622
Realization Value 14 = 0.89448952
Realization Value 15 = 0.77663169
Realization Value 16 = 0.80202186
Realization Value 17 = 0.11116467
Realization Value 18 = 0.58226093
Realization Value 19 = 0.51807672
Realization Value 20 = 0.46755559
Realization Value 21 = 0.09092041
Realization Value 22 = 0.07729161
Realization Value 23 = 0.57270917
Realization Value 24 = 0.33268900
Realization Value 25 = 0.62262042
Realization Value 26 = 0.00963998
Realization Value 27 = 0.75832743
Realization Value 28 = 0.38012507
Realization Value 29 = 0.31219037
Realization Value 30 = 0.28146538
Realization Value 31 = 0.99146934
Realization Value 32 = 0.23620845

Realization Value 33 = 0.65813461
Realization Value 34 = 0.31369959
Realization Value 35 = 0.75173686
Realization Value 36 = 0.64987329
Realization Value 37 = 0.03442018
Realization Value 38 = 0.83189163
Realization Value 39 = 0.80251659
Realization Value 40 = 0.78838620
Realization Value 41 = 0.52509928
Realization Value 42 = 0.60097733
Realization Value 43 = 0.08983323
Realization Valuc 44 = 0.53603702
Realization Value 45 = 0.34250420
Realization Value 46 = 0.18520123
Realization Value 47 = 0.40569386
Realization Value 48 = 0.58325702
Realization Value 49 = 0.48702976
Realization Value 50 = 0.88043365

Exercise Set 5

These exercises are intended to enhance your understanding of pseudorandom numbers and the enumeration-based definition of probability.

1. Run microcomputer program *UNIFORM* using a new set of three, three-digit odd seed numbers. Examine the output carefully, looking for repeated digits, simple patterns (e.g., 12345 or 12321), and repeated patterns (e.g., 123123) when both rows and columns are scrutinized. If you do not find something in your output that you consider to be "peculiar," repeat this exercise until you do.

2. Run program *UNIFORM* using a new set of three, three-digit odd seed numbers. Inspect the output relative to the number of times any of the digits 0 through 9 appears in the first column. (a) What is the maximum number of times that one of these digits actually appears? (b) What is the expected number of times that each of the digits 0 through 9 will appear? (c) If the actual maximum is less than eight, repeat this exercise until it is equal to or

greater than eight. State the number of times that this exercise was repeated and the associated sets of seed numbers.

3. Run program *UNIFORM* using a new set of three, three-digit odd seed numbers. Inspect the entire output carefully with regard to the number of times the digit 5 appears in any row. (a) What is the maximum number of times that the digit 5 actually appears? (b) What is the expected number of times that the digit 5 will appear? Is this expected value necessarily an integer? (c) If the observed maximum number of times that the digit 5 actually appeared is less than four, repeat this exercise until it is at least equal to four. Then state the number of times that this exercise was repeated and the associated sets of seed numbers.

4. Run microcomputer program *RANDOM2* using the same new set of three, three-digit odd seed numbers as was used originally in Example 1. Let $n_{elpri} = 50$ and $n_{digits} = 8$. Compare the respective outputs and comment appropriately.

3.5.1.2. Simulation Example Two: Flipping A Fair Coin Using Realization Values for Uniform Pseudorandom Numbers

Suppose we simulate n_f flips of a fair-coin by (a) generating realization values for n_f uniform pseudorandom numbers, and then (b) associating a head with each realization value that is less than 0.5. Microcomputer programs *FCOIN1* and *FCOIN2* employ this elementary algorithm to simulate flipping a fair coin n_f times. Microcomputer program *AVE1* extensively outputs the simulation-based proportion of heads for n_f up to 1000 flips, whereas microcomputer program *AVE2* extends n_f to 100,000, but only outputs the simulation-based proportion of heads for multiples of 10 flips. These two programs must employ the same three, three-digit odd seed numbers to be directly comparable.

Figure 3.1 is a plot of a typical outcome obtained by running program *FCOIN1*. The deviations of the simulation-based proportion of heads from its expected value (1/2) decrease statistically as n_f increases. Although plots such as Figure 3.1 may appear to differ markedly from simulation-based outcome to simulation-based outcome, especially for relatively small values of n_f, this trend is always evident when several replicate plots are compared.

Figure 3.2 is a plot of the maximum and minimum simulation-based proportions of heads obtained by a class of 20 students, each running microcomputer program *FCOIN1* (with a *different* set of seed numbers). The range between the maximum and minimum simulation-based proportions of heads plotted in Figure 3.2 is loosely akin to the width of a

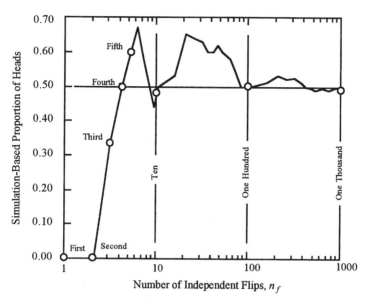

Figure 3.1 Simulation Example Two: a typical outcome obtained by running microcomputer program *FCOIN1*.

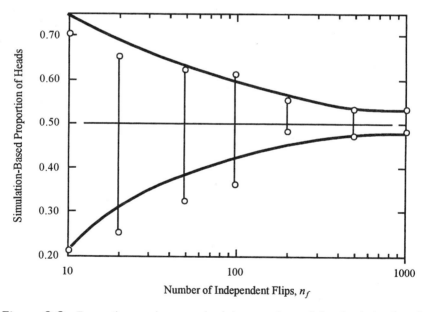

Figure 3.2 Respective maximum and minimum values of the simulation-based proportion of heads in 20 independent outcomes obtained by running microcomputer program *FCOIN1*. The respective faired maximum and minimum curves demonstrate the obvious decrease in statistical variability as n_f increases.

106

100(scp)% (two-sided) statistical confidence interval that allegedly includes the actual value of the probability that any future flip will produce a head. The width of this statistical confidence interval is inversely proportional to the square root of n_f. It decreases by a factor of 10 as n_f increases from 10 to 1000 and by another factor of 10 as n_f increases from 1000 to 100,000, etc. Note that this theoretical behavior is apparent in Figure 3.2.

```
C> FCOIN1

Input a new set of three, three-digit odd seed numbers

751 249 863

Simulation-Based Proportion of Heads Given nf Flips of a Fair Coin

                1 Flip  – Proportion of Heads = 1.000
                2 Flips – Proportion of Heads = 1.000
                3 Flips – Proportion of Heads = 1.000
                4 Flips – Proportion of Heads = 1.000
                5 Flips – Proportion of Heads = 0.800
                6 Flips – Proportion of Heads = 0.667
                7 Flips – Proportion of Heads = 0.714
                8 Flips – Proportion of Heads = 0.625
                9 Flips – Proportion of Heads = 0.556
               10 Flips – Proportion of Heads = 0.600
               11 Flips – Proportion of Heads = 0.636
               12 Flips – Proportion of Heads = 0.583
               13 Flips – Proportion of Heads = 0.615
               14 Flips – Proportion of Heads = 0.571
               15 Flips – Proportion of Heads = 0.533
               16 Flips – Proportion of Heads = 0.500
               17 Flips – Proportion of Heads = 0.471
               18 Flips – Proportion of Heads = 0.500
               19 Flips – Proportion of Heads = 0.526
               20 Flips – Proportion of Heads = 0.500
               21 Flips – Proportion of Heads = 0.476
               22 Flips – Proportion of Heads = 0.500
               23 Flips – Proportion of Heads = 0.478
               24 Flips – Proportion of Heads = 0.500
               25 Flips – Proportion of Heads = 0.480
```

```
 30 Flips – Proportion of Heads = 0.467
 35 Flips – Proportion of Heads = 0.457
 40 Flips – Proportion of Heads = 0.400
 45 Flips – Proportion of Heads = 0.400
 50 Flips – Proportion of Heads = 0.460
 55 Flips – Proportion of Heads = 0.455
 60 Flips – Proportion of Heads = 0.433
 65 Flips – Proportion of Heads = 0.415
 70 Flips – Proportion of Heads = 0.400
 75 Flips – Proportion of Heads = 0.400
 80 Flips – Proportion of Heads = 0.400
 85 Flips – Proportion of Heads = 0.424
 90 Flips – Proportion of Heads = 0.444
 95 Flips – Proportion of Heads = 0.463
100 Flips – Proportion of Heads = 0.450
200 Flips – Proportion of Heads = 0.480
300 Flips – Proportion of Heads = 0.510
400 Flips – Proportion of Heads = 0.498
500 Flips – Proportion of Heads = 0.496
600 Flips – Proportion of Heads = 0.505
700 Flips – Proportion of Heads = 0.504
800 Flips – Proportion of Heads = 0.499
900 Flips – Proportion of Heads = 0.497
1000 Flips – Proportion of Heads = 0.502
```

```
C> FCOIN2
```

Input the same set of three, three-digit odd seed numbers that was used as input to microcomputer program *FCOIN1*

```
751 249 863
```

Simulation-Based Proportion of Heads Given n_f Flips of a Fair Coin

```
    10 Flips – Proportion of Heads = 0.6
   100 Flips – Proportion of Heads = 0.45
  1000 Flips – Proportion of Heads = 0.502
 10000 Flips – Proportion of Heads = 0.4927
100000 Flips – Proportion of Heads = 0.49798
```

3.5.1.3. Discussion

It is demonstrated in Chapter 8 that the maximum likelihood (ML) statistical estimator for the actual probability that a head will occur as the outcome of any given flip is the observed proportion (relative frequency) of heads that occurred in n_f flips, viz., ML est$[p(\text{head})] = n_h/n_f$, where n_h is the number of heads that occurred in n_f independent flips. In turn, recall that every statistical estimate is equal to the actual value for the quantity being estimated plus an intrinsic statistical estimation error component. For example, the intrinsic statistical estimation error component for the maximum likelihood statistical estimate that is equal to 0.6 for 10 flips in our example output for microcomputer program *FCOIN2* is equal to $+0.1$, i.e., 0.500 plus the intrinsic statistical estimation error component is equal to 0.6. Figures 3.1 and 3.2 are intended to demonstrate that the magnitude of this intrinsic statistical estimation error component decreases (statistically) as n_f increases and thus asymptotically approaches zero as n_f increases without bound.

Exercise Set 6

(These exercises are intended to enhance your understanding of the fair-coin simulation example.)

1. (a) Run microcomputer programs *FCOIN1* and *FCOIN2* using your own set of three, three-digit odd seed numbers. (b) Plot the output of microcomputer program *FCOIN1*. (c) Plot the associated magnitude of the intrinsic statistical estimation error component. .

2. (Class exercise) Compile the maximum and minimum values for the simulation-based proportion of heads obtained by running microcomputer programs *FCOIN1* and *FCOIN2* in Example 1 *for the entire class* when $n_f = 10, 20, 50, 100, 200, 500, 1000,$ 10,000, and 100,000. Plot these maximum and minimum values to demonstrate the obvious decrease in the intrinsic statistical estimation error component as n_f increases.

3.5.1.4. Simulation Example Three: Arithmetic Average of Realization Values for n_a Uniform Pseudorandom Numbers

We now simulate the arithmetic average of the respective realization values for n_a uniform pseudorandom numbers. Microcomputer programs *AVE1*

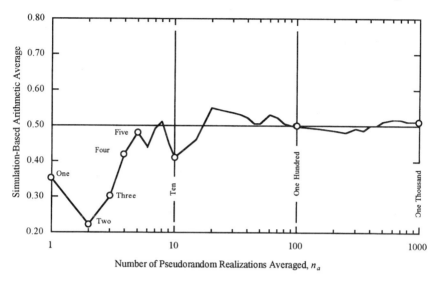

Figure 3.3 Simulation Example Three: a typical outcome obtained by running microcomputer program *AVE1*.

and *AVE2* are intended to demonstrate that, as n_a increases, the intrinsic statistical estimation error for the arithmetic average estimator asymptotically approaches zero as n_a increase without bound.

Figure 3.3 is a plot of a typical outcome obtained by running program *AVE1*. In turn, Figure 3.4 is a plot of the maximum and minimum simulation-based arithmetic averages in a class of size 20. The range between the maximum and minimum simulation-based simulation-based arithmetic averages plotted in Figure 3.4 is loosely akin to the width of a statistical confidence interval that allegedly includes the expected value for the arithmetic average estimator (0.5). The theoretical width of this statistical confidence interval is inversely proportional to the square root of n_a. Thus, this theoretical width decreases by a factor of 10 as n_a increases from 10 to 1000 and by another factor of 10 as n_a increases from 1000 to 100,000, etc. Note that this theoretical behavior is apparent in Figure 3.4.

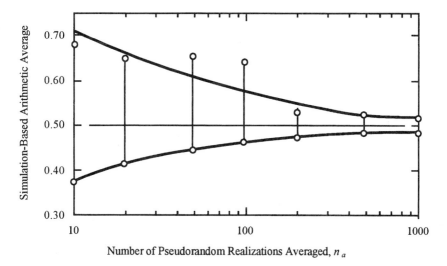

Figure 3.4 Respective maximum and minimum values of the simulation-based arithmetic averages in 20 independent outcomes obtained by running microcomputer program *AVE1*. The respective faired maximum and minimum curves demonstrate the obvious decrease in statistical variability as n_a increases.

```
C> AVE1

Input a new set of three, three-digit odd seed numbers

451 231 785
```
                    Arithmetic Average of Realization Values
                      for n_a Uniform Pseudorandom Numbers

                1 Realization Value – Arithmetic Average = 0.2598
                2 Realization Values – Arithmetic Average = 0.2968
                3 Realization Values – Arithmetic Average = 0.3354
                4 Realization Values – Arithmetic Average = 0.3216
                5 Realization Values – Arithmetic Average = 0.4398
                6 Realization Values – Arithmetic Average = 0.4813
                7 Realization Values – Arithmetic Average = 0.4297
                8 Realization Values – Arithmetic Average = 0.3913
                9 Realization Values – Arithmetic Average = 0.4032

10 Realization Values – Arithmetic Average = 0.4393
11 Realization Values – Arithmetic Average = 0.4029
12 Realization Values – Arithmetic Average = 0.4364
13 Realization Values – Arithmetic Average = 0.4797
14 Realization Values – Arithmetic Average = 0.4878
15 Realization Values – Arithmetic Average = 0.5213
16 Realization Values – Arithmetic Average = 0.5144
17 Realization Values – Arithmetic Average = 0.5380
18 Realization Values – Arithmetic Average = 0.5514
19 Realization Values – Arithmetic Average = 0.5478
20 Realization Values – Arithmetic Average = 0.5334
21 Realization Values – Arithmetic Average = 0.5093
22 Realization Values – Arithmetic Average = 0.5056
23 Realization Values – Arithmetic Average = 0.4951
24 Realization Values – Arithmetic Average = 0.5039
25 Realization Values – Arithmetic Average = 0.5019
30 Realization Values – Arithmetic Average = 0.4911
35 Realization Values – Arithmetic Average = 0.4737
40 Realization Values – Arithmetic Average = 0.4688
45 Realization Values – Arithmetic Average = 0.4774
50 Realization Values – Arithmetic Average = 0.4967
55 Realization Values – Arithmetic Average = 0.4986
60 Realization Values – Arithmetic Average = 0.4969
65 Realization Values – Arithmetic Average = 0.4886
70 Realization Values – Arithmetic Average = 0.4763
75 Realization Values – Arithmetic Average = 0.4718
80 Realization Values – Arithmetic Average = 0.4857
85 Realization Values – Arithmetic Average = 0.4849
90 Realization Values – Arithmetic Average = 0.4818
95 Realization Values – Arithmetic Average = 0.4920
100 Realization Values – Arithmetic Average = 0.4928
200 Realization Values – Arithmetic Average = 0.5060
300 Realization Values – Arithmetic Average = 0.4876
400 Realization Values – Arithmetic Average = 0.4909
500 Realization Values – Arithmetic Average = 0.5038
600 Realization Values – Arithmetic Average = 0.5065
700 Realization Values – Arithmetic Average = 0.5070
800 Realization Values – Arithmetic Average = 0.5063
900 Realization Values – Arithmetic Average = 0.5026
1000 Realization Values – Arithmetic Average = 0.4996

```
C> AVE2
```

Input the same set of three, three-digit odd seed numbers that was used as input to microcomputer program *AVE1*

```
451 231 785
```

Arithmetic Average of Realization Values
for n_a Uniform Pseudorandom Numbers

10 Realization Values – Arithmetic Average = 0.4393
100 Realization Values – Arithmetic Average = 0.4928
1000 Realization Values – Arithmetic Average = 0.4996
10000 Realization Values – Arithmetic Average = 0.5005
100000 Realization Values – Arithmetic Average = 0.5001

Exercise Set 7

(These exercises are intended to enhance your understanding of the probability behavior of averages.)

1. (a) Run microcomputer programs *AVE1* and *AVE2* using your own set of three, three-digit odd seed numbers. (b) Plot the output of microcomputer program *AVE1*. (c) Given that the arithmetic average is an unbiased estimator of the conceptual statistical (sampling) distribution mean, plot the intrinsic statistical estimation error for the arithmetic average estimator.

2. (Class exercise) Compile the maximum and minimum values generated by microcomputer programs *AVE1* and *AVE2* in Example 1 for your entire class for n_f = 10, 20, 50, 100, 200, 500, 1000, 10,000, and 100,000. Plot these maximum and minimum values to demonstrate the obvious decrease in the intrinsic statistical estimation error as n_f increases.

3.5.1.5. Simulation Example Four: Histogram for Proportions of Realization Values for Uniform Pseudorandom Numbers

We now extend Simulation Example One by constructing a histogram for the respective realization values for uniform pseudorandom numbers. Our

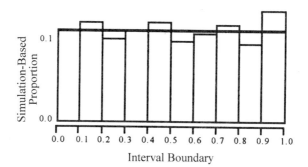

Figure 3.5 Typical histogram for proportions (relative frequencies) obtained by running microcomputer program *HISTPRO1.*

objective is to simulate the statistical variability of the proportion (relative frequency) of pseudorandom realization values that fall in each of the 10 equal-width intervals, 0.00000000 to 0.09999999, 0.10000000 to 0.20000000, . . . , and 0.90000000 to 0.99999999. Microcomputer program *HISTPRO1* generates 1000 realization values for uniform pseudorandom numbers and sorts these realization values into the 10 histogram intervals. See Figure 3.5. For comparison, program *HISTPRO2* generates and sorts 100,000 realization values for uniform pseudorandom numbers. Note that the magnitudes of the intrinsic statistical estimation errors generated by running program *HISTPRO2* are much smaller than the corresponding magnitudes of the intrinsic statistical estimation errors generated by running program *HISTPRO1*. In fact, based on the probability behavior demonstrated in Figures 3.2 and 3.4, it should be persuasive that the magnitudes of the respective intrinsic statistical estimation errors approach zero as the number of pseudorandom realization values used to construct this histogram increases without bound. Accordingly, in the limit as the number of pseudorandom realization values generated approaches infinity, this histogram intuitively becomes the conceptual histogram for proportions (relative frequencies) depicted in Figure 3.6. Moreover, as the number of pseudorandom realization values generated approaches infinity, the respective interval widths can be made infinitesimal in width (length), thereby permitting the conceptual histogram for proportions (relative frequencies) to be described by an appropriate continuous analytical expression, e.g., as in Figure 3.7. The analytical expression describing its geometric shape is called its *probability density function* (PDF).

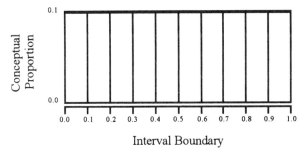

Figure 3.6 Conceptual (theory-based) histogram for proportions (relative frequencies) associated with Simulation Example Four. The conceptual proportion (relative frequency) in each interval is equal to its expected value (0.1). This conceptual behavior is theoretically attained as the respective intrinsic statistical estimation errors evident in Figure 3.5 asymptotically decrease to zero. (Recall that this asymptotic behavior was demonstrated in Figures 3.2 and 3.4 for unbiased estimators.)

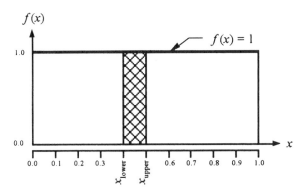

Figure 3.7 Probability density function (PDF) pertaining to the conceptual uniform distribution, zero to one.

```
C> HISTPRO1

Input a new set of three, three-digit odd seed numbers

359 917 431

     Interval   0.0 to 0.1   Simulation-Based Proportion   0.096
     Interval   0.1 to 0.2   Simulation-Based Proportion   0.098
```

Interval 0.2 to 0.3 Simulation-Based Proportion 0.112
Interval 0.3 to 0.4 Simulation-Based Proportion 0.095
Interval 0.4 to 0.5 Simulation-Based Proportion 0.100
Interval 0.5 to 0.6 Simulation-Based Proportion 0.098
Interval 0.6 to 0.7 Simulation-Based Proportion 0.100
Interval 0.7 to 0.8 Simulation-Based Proportion 0.098
Interval 0.8 to 0.9 Simulation-Based Proportion 0.106
Interval 0.9 to 1.0 Simulation-Based Proportion 0.097

```
C> HISTPRO2
```

Input the same set of three, three-digit odd seed numbers used as input to program *HISTPRO1*

```
359 917 431
```

Interval 0.0 to 0.1 Simulation-Based Proportion 0.09857
Interval 0.1 to 0.2 Simulation-Based Proportion 0.10088
Interval 0.2 to 0.3 Simulation-Based Proportion 0.10158
Interval 0.3 to 0.4 Simulation-Based Proportion 0.09966
Interval 0.4 to 0.5 Simulation-Based Proportion 0.09971
Interval 0.5 to 0.6 Simulation-Based Proportion 0.10114
Interval 0.6 to 0.7 Simulation-Based Proportion 0.09920
Interval 0.7 to 0.8 Simulation-Based Proportion 0.10025
Interval 0.8 to 0.9 Simulation-Based Proportion 0.10039
Interval 0.9 to 1.0 Simulation-Based Proportion 0.09862

Given a random variable X whose realization values are equally likely to lie in each infinitesimal interval along the continuous x metric from 0 to 1, under continual replication the respective realization values generate the *conceptual uniform distribution, 0 to 1*. The PDF that analytically defines this conceptual statistical distribution is expressed as

$$f(x) = 1 \quad 0 < x < 1 \quad \text{and} \begin{cases} f(x) = 0 & x \le 0 \\ f(x) = 0 & x \ge 1 \end{cases}$$

3.6. CONCEPTUAL STATISTICAL DISTRIBUTION THAT CONSISTS OF ALL POSSIBLE REPLICATE REALIZATION VALUES FOR A GENERIC RANDOM VARIABLE X WITH A CONTINUOUS METRIC

The conceptual statistical distribution that consists of all possible replicate realization values for a generic random variable X with a continuous metric can be analytically specified by stating its PDF. The enumeration-based probability that a randomly selected realization value for random variable X will fall in a given numerical interval, say from x_{lower} to x_{upper}, is established by the following ratio of areas

$$\text{probability}(x_{lower} < X < x_{upper}) = \frac{\int_{x_{lower}}^{x_{upper}} f(x)\,dx}{\int_{all\ x} f(x)\,dx}$$

in which the integral of $f(x)\,dx$ over all x is equal to one (by definition). Accordingly, this enumeration-based probability can be re-expressed simply as

$$\text{probability}(x_{lower} < X < x_{upper}) = \int_{x_{lower}}^{x_{upper}} f(x)\,dx$$

This generic expression is valid whenever the PDF associated with the random variable of specific interest can be expressed analytically in terms of a continuous metric.

There are several conceptual statistical distributions of specific interest in various mechanical reliability applications that employ a continuous metric. Each of these conceptual statistical distributions has a unique analytical PDF expression that establishes the value for $f(x)$ given a specific value of metric x. These conceptual statistical distributions, their associated PDF's, and their mechanical reliability and statistical applications are presented in subsequent chapters. However, to provide perspective regarding these presentations, it is important to understand that (a) all conceptual statistical distributions are mathematical abstractions, and (b) no conceptual statistical distribution ever exactly models the physical phenomena or the experiment test program $CRHDV_i$'s of specific interest.

The conceptual statistical distribution that consists of all possible replicate realization values for a generic random variable X with a continuous metric can also be specified analytically by stating its *cumulative distribution function* (CDF). The CDF function is defined as the integral of the area under the associated PDF from minus infinity to the metric value of specific interest, viz.,

$$\text{CDF} = F(x) = \int_{-\infty}^{x} f(u)du = \text{probability}(X < x)$$

in which u is the dummy variable of integration. The CDF is usually expressed explicitly by an analytical expression in which $F(x)$ is synonymous, for practical purposes, with the respective percentiles (p-tiles) of the associated conceptual statistical distribution.

Most CDF's of specific interest in mechanical reliability are sigmoidal when $F(x)$ is plotted along a linear ordinate and the associated metric value for x is plotted along a linear abscissa. For these CDF's, $F(x)$ asymptotically approaches zero at the smallest possible realization value for random variable X and asymptotically approaches unity at the largest possible realization value for random variable X.

> *Remark*: The *probability function* (PF) for a random variable with a discrete metric is analogous to the PDF for a random variable with a continuous metric. It is properly plotted as in Figure 1.5, but is more commonly plotted as in Figure 3.9. The *cumulative distribution* (CD) for a random variable with a discrete metric is analogous to the CDF for a random variable with a continuous metric. It plots as a step function, with its value equal to zero for all x less than the smallest possible realization value for random variable X and with its value equal to one for all x greater than the largest possible realization value for random variable X.

3.7. ACTUAL VALUES FOR THE MEAN AND VARIANCE OF THE CONCEPTUAL STATISTICAL DISTRIBUTION THAT CONSISTS OF ALL POSSIBLE REPLICATE REALIZATION VALUES FOR A GENERIC RANDOM VARIABLE X WITH A CONTINUOUS METRIC

The first two mechanics moments are intuitive descriptors of the two-parameter conceptual statistical distribution that consists of all possible replicate realization values for a generic random variable X with a continuous metric. The first moment of the area under the PDF for a conceptual statistical distribution establishes the centroid of this area. This centroid is technically termed the actual value for the mean of the conceptual statistical distribution. The second moment of the area under the PDF for this conceptual statistical distribution, evaluated about the area centroid (its mean), establishes the moment of inertia for this area. This area moment of inertia, divided by the (total) area under the PDF, is technically termed the actual value for the variance of the conceptual statistical distribution. In turn, the square root of a variance is termed the standard deviation. (Note that the units of the metric for the mean and the standard deviation are identical.)

Given a conceptual statistical distribution whose PDF $f(x)$ for the generic random variable X is stated using a continuous linear x metric, the actual values for its mean and variance are computed using the expressions (definitions):

$$\text{mean}(X) = \frac{\int_{\text{all } x} x f(x)\, dx}{\int_{\text{all } x} f(x)\, dx} = \int_{\text{all } x} x f(x)\, dx$$

$$\text{var}(X) = \frac{\int_{\text{all } x} [x - \text{mean}(X)]^2 f(x)\, dx}{\int_{\text{all } x} f(x)\, dx} = \int_{\text{all } x} [x - \text{mean}(X)]^2 f(x)\, dx$$

Mean(X) is technically verbalized as the actual value for the mean of the conceptual statistical distribution that consists of all possible replicate realization values for random variable X, but mean(X) is expeditiously verbalized as the mean of (random variable) X. Similarly, var(X) is technically verbalized as the actual value for the variance of the conceptual statistical distribution that consists of all possible replicate realization values for the random variable X, but var(X) is expeditiously verbalized as the variance of (random variable) X.

Mean(X) is also called the expected value of (random variable) X. Given some function of a generic random variable X, say $g(X)$, it is customary to refer to the expected value of $g(X)$ rather than to the actual value for the mean of $g(X)$. For example, when the actual value for the mean of the conceptual statistical distribution that consists of all possible replicate realization values for a generic random variable X is equal to zero, mean(X^2) is equal to var(X). Accordingly, we say that the expected value of X^2 is equal to var(X). Similarly, based on inspection of the analytical expressions (definitions) for mean(X) and var(X) given above, we say that the expected value of $[X - \text{mean}(X)]^2$ is equal to var(X).

When the metric for a random variable is linear, the actual value for the mean of its conceptual statistical distribution can be viewed as an intuitive location parameter for its PDF, and analogously, the actual value for the standard deviation of its conceptual statistical distribution can be viewed as an intuitive scale parameter for its PDF. However, the actual values for the mean and variance (or standard deviation) of a conceptual statistical distribution are not always the statistically recommended location and scale parameters for its PDF. In turn, when the metric for a random variable is logarithmic, the actual values for all PDF location and scale parameters have awkward geometric (and physical) interpretations.

Exercise Set 8

These exercises are intended to enhance your understanding of the mean and the variance of a conceptual distribution whose PDF is analytically expressed using a continuous metric x. Example 1 through 6 pertain specifically to the conceptual uniform distribution, but a generalization of the relationships developed in Exercises 5 and 6 will be used repeatedly in developing fundamental probability concepts in subsequent discussions.

1. Verify by integration that the actual value for the mean of the conceptual uniform distribution, zero to one, is $1/2$.
2. Verify by integration that the actual value for the variance of the conceptual uniform distribution, zero to one, is $1/12$.
3. Verify that the actual value for the mean of the conceptual statistical distribution that consists of all possible replicate realization values for the random variable X that is uniformly distributed over the interval from a to b is equal to $[(a + b)/2]$. (Remember that the total area under its PDF must, by definition, be equal to one.)
4. Verify that the actual value for the variance of the conceptual statistical distribution that consists of all possible replicate realization values for the random variable X that is uniformly distributed over the interval from a to b is equal to $\{[(b - a)^2]/12\}$.
5. (a) Verify that the actual value for the mean of the conceptual statistical distribution that consists of all possible replicate realization values for the random variable $(X + c)$ that is uniformly distributed over the *translated* interval from $(a + c)$ to $(b + c)$ is equal to $\{c + [(a + b)/2]\}$, where c is an arbitrary constant. This result can be generalized as follows:

   $$\text{mean}(X + c) = \text{mean}(X) + c$$

   In turn, (b) verify that the actual value for the variance of this translated conceptual uniform distribution is equal to $\{[(b - a)^2]/12\}$. This result can be generalized as follows:

   $$\text{var}(X + c) = \text{var}(X)$$

These two fundamental relationships are valid for all conceptual statistical distributions, whether the associated metric x is continuous or discrete.

Remark: An arbitrary constant c can be viewed as a random variable whose conceptual statistical distribution consists of the single numerical realization value c, viz., all possible outcomes of the associated experiment test program are identically equal to c. Then, clearly, mean(c) equals c and var(c) equals zero.

6. (a) Verify that the actual value for the mean of the conceptual statistical distribution that consists of all possible replicate realization values for a random variable $(c \cdot X)$ that is uniformly distributed over the *scaled* interval from $c \cdot a$ to $c \cdot b$ is equal to $[c \cdot (b + a)/2]$. This result can be generalized as follows:

$$\text{mean}(c \cdot X) = c \cdot \text{mean}(X)$$

Note that, when $c = -1$, then

$$\text{mean}(-X) = -\text{mean}(X)$$

In turn, (b) verify that the actual value for the variance of this scaled conceptual uniform distribution is equal to $\{c^2 \cdot [(b - a)^2]/12\}$. This result can be generalized as follows:

$$\text{var}(c \cdot X) = c^2 \cdot \text{var}(X)$$

Note that, when $c = -1$, then

$$\text{var}(-X) = \text{var}(X)$$

These fundamental relationships are valid for all conceptual statistical distributions, whether the associated metric x is continuous or discrete.

7. (a) Use the relationships (definitions) (i) that the actual value for the variance of the conceptual statistical distribution that consists of all possible replicate realization values for the generic random variable X is equal to the expected value (EV) of the random variable $[X - \text{mean}(X)]^2$, and (ii) that the expected value of a constant c is equal to c, to verify that

$$\text{var}(X) = \text{EV}(X^2) - [\text{mean}(X)]^2$$

(Hint: First expand the expression $[X - \text{mean}(X)]^2$ and then state the expected value of each of its terms.) Next, (b) rewrite this expression to obtain the fundamental relationship:

$$EV(X^2) = var(X) + [mean(x)]^2$$

Then, (c) verify that $EV(X^2)$ is equal to $var(X)$ when $EV(X)$ is equal to zero. In turn, (d) use the following expression for random variable X:

$$X = [X - mean(X)] + mean(X)$$

to verify that $EV(X^2)$ is given by the fundamental relationship developed in (b), viz.,

$$EV(X^2) = var(X) + [mean(X)]^2$$

In turn, (e) use the mechanics area moment of inertia concept and the associated parallel-axis theorem to explain this fundamental relationship in physical terms, viz., that the moment of inertia of the area under the conceptual statistical distribution PDF about the origin (zero) is equal to the moment of inertia of the area under the conceptual statistical distribution PDF about its centroid, [mean(X)], plus the area (equal to one) times the transfer distance [mean(X)] squared.

This fundamental relationship is also valid for conceptual statistical distributions whose metrics are discrete, but the mass centroid and the mass moment of inertia replace the area centroid and the area moment of inertia in the mechanics analogy.

Remark: The fact that the $EV(X^2)$ is equal to $var(X)$ when mean(X) is equal to zero, viz., when the $EV(X)$ is equal to zero, is fundamental to understanding why the *within*(MS) is used to estimate var(*APRCRHNDEE*'s) in classical ANOVA (Chapter 6).

3.8. SAMPLING DISTRIBUTIONS

3.8.1. Conceptual Sampling Distributions

Recall that the realization value for each statistic (test statistic) of specific interest is established by the outcome of the associated experiment test program. In turn, just as a conceptual statistical distribution consists of the collection of all possible realization values for its associated random variable, a conceptual sampling distribution consists of the collection of all possible realization values for its associated statistic (test statistic). In each case, the analytical expression for the PDF of the conceptual distribution is known. Thus, in each case, exact metric values pertaining to the respective percentiles of the conceptual distribution can be calculated (and tabulated if so desired). With few exceptions, conceptual sampling distributions are associated solely with classical statistical analyses (Chapters 5–7).

In a probability context there is no practical difference between a conceptual statistical distribution and a conceptual sampling distribution. However, in a mechanical reliability context, a specific conceptual statistical distribution is presumed to pertain to the physical phenomenon that generates the observed mode of failure. In turn, each experiment test program datum value is alleged to have been randomly selected from this presumed conceptual statistical distribution.

3.8.2. Empirical Sampling Distributions

There are numerous mechanical reliability applications where the analytical expression for the PDF of the conceptual sampling distribution of specific interest is not known and cannot be derived analytically. Accordingly, a simulation-based (randomization-based) methodology must be adopted in which pseudorandom numbers are used to construct n_{sbelo} (n_{rbelo}) equally likely outcomes for the experiment test program that was actually conducted. The associated empirical simulation-based (randomization-based) sampling distribution consists of the corresponding n_{sbelo} (n_{rbelo}) realization values for statistic (test statistic) of specific interest. In turn, as n_{sbelo} (n_{rbelo}) becomes larger and larger, the estimated metric values pertaining to the respective percentiles of simulation-based (randomization-based) empirical sampling distributions will display less and less variability (see Figures 3.2 and 3.4). Thus, the accuracy of the estimated metric values pertaining to the respective percentiles of the simulation-based (randomization-based) sampling distribution of specific interest is limited only by the run-time of the associated microcomputer program.

Table 3.3 is intended to illustrate the process of continually generating equally likely experiment test program outcomes using pseudorandom numbers and then computing the corresponding realization values for the statistic (test statistic) of specific interest. In turn, given $m = n_{sbelo}$ (n_{rbelo}), the number of simulation-based (randomization-based) experiment test program outcomes of specific interest, n_{oosi}, is established by the corresponding number of statistic (test statistic) realization values of specific interest. Table 3.3 is also intended to convey the notion that the limiting form of the empirical simulation-based (randomization-based) sampling distribution as $m = n_{sbelo}$ (n_{rbelo}) increases without bound is the conceptual sampling distribution, regardless of whether the analytical expression for its PDF is known or unknown. Thus, the same fundamental statistical abstractions pertain to empirical simulation-based (randomization-based) sampling distributions as pertain to conceptual sampling distributions. We statistically

Table 3.3   Table Illustrating Process of Continual Replication of Experiment Test Program, Either Conceptually or by Using a Simulation-Based (Randomization-Based) Methodology, to Generate *Equally–Likely* Outcomes and the Corresponding Realization Values for Statistic (Test Statistic) of Specific Interest

replicate experiment test program	associated *equally-likely* outcome	corresponding realization value for the statistic (test statistic) of specific interest[a]
1	$[x_1, x_2, x_3, \ldots, x_{n_{dv}}]_1$	$s_1(ts_1)$
2	$[x_1, x_2, x_3, \ldots, x_{n_{dv}}]_2$	$s_2(ts_2)$
3	$[x_1, x_2, x_3, \ldots, x_{n_{dv}}]_3$	$s_3(ts_3)$
$\vdots$	$\vdots$	$\vdots$
m	$[x_1, x_2, x_3, \ldots, x_{n_{dv}}]_m$	$s_m(ts_{n_{rep}})$

[a]A conceptual sampling distribution is the infinite collection of all possible statistic (test statistic) realization values, whereas an empirical sampling distribution is the finite collection of $m = n_{sbelo}\,(n_{rbelo})$ statistic (test statistic) realization values. However, probability values computed using empirical sampling distributions accurately approximate exact probability values computed using conceptual sampling distributions when $m = n_{sbelo}\,(n_{rbelo})$ is sufficiently large.

view the outcome for the experiment test program that is (was) actually conducted as being randomly selected from the collection of $m = n_{sbelo}$ (n_{rbelo}) equally likely simulation-based (randomization-based) experimental test program outcomes. Correspondingly, we statistically view its statistic (test statistic) realization value as being randomly selected from the empirical sampling distribution that consists of the collection of $m = n_{sbelo}\,(n_{rbelo})$ statistic (test statistic) realization values.

3.8.3. Asymptotic Sampling Distributions

In certain situations, e.g., maximum likelihood analysis (Chapter 8), an analytical expression for the PDF of the conceptual sampling distribution of specific interest can be derived, but it is only exact asymptotically, viz., as m increases without bound. Although asymptotic PDF's have been widely used in the statistical literature to calculate approximate probability values (or approximate values of the metric pertaining to given probability values), the accuracy of these calculations is always suspect for sample sizes that are practical in mechanical reliability applications. Thus, it is good statistical practice to avoid the use of asymptotic sampling distributions and to generate appropriate simulation-based (randomization-based) empirical sampling distributions instead.

3.9. TECHNICAL DEFINITION FOR THE NULL HYPOTHESIS THAT $B = A$ STATISTICALLY AND FOR THE SIMPLE (ONE-SIDED) ALTERNATIVE HYPOTHESIS THAT $B > A$ STATISTICALLY

We now define exactly what was intended to be connoted in our presentation of the sign test, the signed-rank test, and Fisher's enumeration-based and randomization-based tests (Sections 3.2–3.4) when the null hypothesis was stated as $B = A$ statistically and the associated simple (one-sided) alternative hypothesis was stated as $B > A$ statistically.

First, however, consider a CRD experiment test program with two treatments, A and B. Presume that the conceptual statistical distribution that consists of all possible treatment B datum values is identical to the conceptual statistical distribution that consists of all possible treatment A datum values, Figure 3.8(a). Then, *by definition*, $B = A$ statistically. In turn, presume that the conceptual statistical distribution that consists of all possible replicate treatment B datum values is identical to the conceptual statistical distribution that consists of all possible replicate treatment A datum values, except that it is translated by a positive increment *delta* along its (generic) x metric, Figure 3.8(b). Then, *by definition*, $B > A$ statistically. These definitions have two important consequences. First, the presumption of homoscedasticity for the respective treatment A and treatment B datum values is clearly valid. Second, given that the null hypothesis $B = A$ statistically, the conceptual sampling distribution for the statistic $(B - A)$ is symmetrical about its mean (median), and the x metric pertaining to its mean (median) is equal to zero. Accordingly, given the null hypothesis that $B = A$ statistically, the probability is exactly equal to $1/2$ that a randomly selected realization of the statistic $(B - A)$ will be positive. On the other hand, given the simple (one-sided) alternative hypothesis $B > A$ statistically, the conceptual sampling distribution for the statistic $(B - A)$ is symmetrical about its mean (median), but the x metric pertaining to its mean (median), is greater than zero. Accordingly, given the simple (one-sided) alternative hypothesis that $B > A$ statistically, the probability is greater than $1/2$ that a randomly selected realization of the statistic $(B - A)$ will be positive.

Remark: The symmetry of the conceptual sampling distribution for the statistic $(B - A)$ is intuitively obvious under the null hypothesis when realizations $(b - a)$, $(a - b)$, $(b_1 - b_2)$, and $(a_1 - a_2)$ are all regarded as interchangeable, where the subscripts for a and b refer to time order of selecting the respective realization values for A and B.

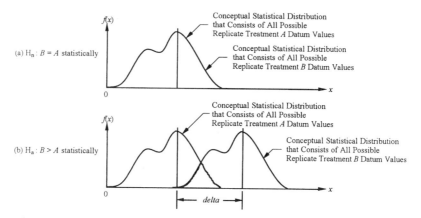

Figure 3.8 Technical definition of what is intended to be connoted when (a) the null hypothesis is stated as $B = A$ statistically, and (b) the simple one-sided) alternative hypothesis is stated as $B > A$ statistically. In (a), the null hypothesis, the conceptual statistical distribution that consists of all possible replicate treatment B datum values is identical to the conceptual statistical distribution that consists of all possible replicate treatment A datum values. In (b), the simple (one-sided) alternative hypothesis, the conceptual statistical distribution that consists of all possible replicate treatment B datum values is identical to the conceptual statistical distribution that consists of all possible replicate treatment A datum values, except that it is translated a positive distance *delta* along the x metric.

Now consider an unreplicated RCBD experiment test program with two treatments A and B. Presume that these two treatments do not interact with any of the infinite collection of available experiment test program blocks. If so, then the conceptual sampling distribution of specific interest consists of all possible realizations for the statistic $[(B - cbe_i) - (A - cbe_i)]$, but note that $[(B - cbe_i) - (A - cbe_i)] = (B - A)$. Thus, as above, given the null hypothesis that $B = A$ statistically, the conceptual sampling distribution that consists of all possible realizations for the statistic $[(B - cbe_i) - (A - cbe_i)]$ is symmetrical about its 50th percentile, and the x metric pertaining to its 50th percentile is equal to zero. Accordingly, given the null hypothesis that $B = A$ statistically, the probability is exactly equal to $1/2$ that a randomly selected realization for the statistic $[(B - cbe_i) - (A - cbe_i)]$ will be positive. On the other hand, given the simple (one-sided) alternative hypothesis that $B > A$ statistically, the conceptual sampling distribution that consists of all possible realizations for the statistic $[(B - cbe_i) - (A - cbe_i)]$ is also symmetrical about its 50th percentile, but the x metric pertaining to its 50th percentile is greater than zero. Accordingly, given the simple (one-sided) alternative hypothesis that $B = A$ statistically, the probability is greater than 1/2 that a randomly selected realization for the statistic $[(B - cbe_i) - (A - cbe_i)]$ will be positive.

3.9.1. Discussion

Note that if an explicit analytical expression for the PDF pertaining to the conceptual sampling distribution for the statistic $(B - A)$ were (alleged to be) known, then the exact probability that the ith realization value for the constructed random variable $(B - A)$ will be positive could be calculated. However, this PDF is never known in any mechanical metallurgy application. Thus, do not be deceived by simplistic reliability analyses published in engineering textbooks and statistical journals under the heading of "stress–strength interference theory." Such analyses only demonstrate an appalling naiveté regarding the enormous disparity between the actual service stress–time history and the allegedly corresponding material strength (resistance). The latter are always established by conducting a laboratory test with either an extremely elementary stress–time history or an amusingly fictitious stress–time history.

3.10. MECHANICAL RELIABILITY PERSPECTIVE

To provide perspective regarding the practical mechanical reliability application of a paired-comparison experiment test program, we now review the notions associated with the distribution-free (nonparametric) statistical test of the null hypothesis that $B = A$ statistically versus the simple (one-sided) alternative hypothesis that $B > A$ statistically. Perhaps surprisingly, these null and alternative hypotheses pertain to markedly different test objectives.

First, suppose that our test objective is to "assure" that proposed design B is at least as reliable as production design A statistically. Let the experiment test program consist of n_{pc} paired comparisons. (The issue of an adequate number of paired comparisons to limit the probability of committing a *Type II* error to an acceptable value will be considered later.) The resulting experiment test program datum values generate a numerical value for each of the respective n_{pc} $(b - a)$ differences. If $B > A$ statistically we expect a majority of these numerical values to be positive. However, if we do not obtain a sufficient majority of positive numerical values for the respective n_{pc} $(b - a)$ differences, microcomputer programs *EBST*, *EBSRT*, *FEBT*, and *FRBT* will compute enumeration-based probabilities that are not sufficiently small to warrant rejection of the null hypothesis that $A = B$ statistically. Then, we cannot rationally assert that proposed design B is at least as reliable as production design A statistically—even though we know that under the presumption that the null hypothesis is correct only 1/2 of all possible replicated experiment test program outcomes would actually exhibit a majority of positive $(b - a)$ differences. It is therefore evident that a paired-comparison experiment test program is not well suited to the test objective of "assuring" that B is at least as reliable as A.

Now suppose that our test objective is to "assure" that proposed design B is more reliable than production design A. Suppose also that the experiment test program outcome is such that so many $(b - a)$ differences are positive that we rationally opt to reject the null hypothesis that $B = A$ statistically in favor of the simple (one-sided) alternative hypothesis that $B > A$ statistically. (Recall that we rationally reject the null hypothesis *only* when the experiment test program outcome emphatically agrees with the alternative hypothesis.) Accordingly, we rationally assert that $B > A$ statistically (but by an unspecified and unknown amount akin to *delta* in Figure 3.8). Suppose, however, that we redefine the null and simple (one-sided) alternative hypotheses as follows:

$$H_n: B - mpd = A \text{ statistically}$$
$$H_a: B - mpd > A \text{ statistically} \qquad (mpd > 0)$$

or

$$H_n: \frac{B}{mpr} = A \text{ statistically}$$
$$H_a: \frac{B}{mpr} > A \text{ statistically} \qquad (mpr > 1)$$

in which *mpd* is the minimum practical difference of specific interest and *mpr* is the minimum practical ratio of specific interest. Correspondingly, suppose that we adjust the respective $(b - a)$ differences such that $(b' - a) = [(b - mpd)\text{-}a]$ or such that $(b' - a) = [(b/mpr) - a]$. If so, then we can statistically quantify the amount (ratio) that B excels A statistically (Supplemental Topic 3.C).

Hopefully, it is now clear that a paired-comparison experiment test program has practical application in improving reliability, but is not well suited to "assuring" reliability. The latter objective is usually associated with a so-called quality assurance experiment test program.

3.11. CLOSURE

The statistical abstraction of continually generating replicate experiment test programs is the key to understanding probability concepts of specific interest, whatever the application. This statistical abstraction establishes an explicit enumeration-based algorithm that, when coupled with a pseudorandom number generator for the relevant conceptual statistical distribution (Supplemental Topic 8.D), can be used to generate the empirical simulation-based sampling distribution for each mechanical reliability statistic of specific interest, however unusual or unique.

The primary criticism of the enumeration-based definition for probability is that it does not account for prior information. If this criticism is considered relevant, then we recommend the following procedure for combining prior information with actual experiment test program test datum values. First, using this prior information, state the PDF for the supposed conceptual statistical (or sampling) distribution of specific interest, along with the supposed conceptual parameter values. Next, estimate these conceptual parameter values using the actual experiment test program datum values. In turn, generate pseudorandom datum values from both the supposed and the estimated conceptual sampling distributions. Then, combine these two sets of pseudorandom datum values in that proportion which appears to be appropriate (based on the credibility of the prior information and the size of the experiment test program). Finally, establish the revised supposed conceptual statistical (or sampling) distribution empirically and estimate its conceptual sampling distribution parameters.

3.A. SUPPLEMENTAL TOPIC: STATISTICAL WEIGHTS OF DATUM VALUES AND ACTUAL VALUES FOR THE MEAN AND VARIANCE OF THE CONCEPTUAL STATISTICAL DISTRIBUTION THAT CONSISTS OF ALL POSSIBLE REPLICATE REALIZATION VALUES FOR A RANDOM VARIABLE WITH A DISCRETE METRIC

3.A.1. The Statistical Weights of Datum Values

Perhaps the most important statistical concept in data analysis is that each datum point (datum value) has a mass (a statistical weight) that is inversely proportional to the actual value for the variance of the conceptual statistical distribution from which that datum value was presumed to have been randomly selected. Accordingly, a datum value that exhibits a relatively large variability under continual replication of the experiment test program is assigned a relatively small statistical weight, whereas a datum value that exhibits a relatively small variability under continual replication of the experiment test program is assigned a relatively large statistical weight.

On the other hand, datum values that exhibit equal variabilities under continual replication of the experiment test program are assigned equal statistical weights (which are often conveniently presumed to be equal to one).

> *Remark*: The presumption that all experiment test program datum values have identical statistical weights (viz., are homoscedastic) is statistically equivalent to the presumption that the actual values for the variances (standard deviations) of their respective conceptual statistical distributions are all equal. This presumption so markedly simplifies data analysis that it is almost always tacitly accepted as being statistically credible without critical examination. (See Section 6.4.)

We now denote the ith weighted datum value as wdv_i and its associated statistical weight as sw_i. Then, for a specific collection of n_{wdv} weighted datum values, the mass centroid and mass moment of inertia about this mass centroid are respectively established by the expressions:

$$\text{mass centroid} = \frac{\displaystyle\sum_{i=1}^{n_{wdv}} sw_i \cdot wdv_i}{\displaystyle\sum_{i=1}^{n_{wdv}} sw_i}$$

and

$$\text{mass moment of inertia} = \sum_{i=1}^{n_{wdv}} sw_i \cdot (wdv_i - \text{mass centroid})^2$$

When the statistical weight is substituted for mass, the mechanics-based expression for the mass centroid is directly analogous to the statistics-based expression for the actual value for the mean of a conceptual statistical distribution whose metric is discrete. However, the mass moment of inertia about the mass centroid must be divided by the sum of the masses to have the mechanics-based expression be directly analogous to the statistics-based expression for the actual value for the variance of a conceptual statistical distribution whose metric is discrete. The respective statistics-based expressions are restated below.

> *Remark*: Just as conceptual statistical distributions and their associated random variables are commonly referred to as being continuous when the measurement metric is continuous, conceptual statistical distributions and their associated random variables are commonly referred to as being discrete when the measurement metric is discrete.

3.A.2. Actual Values for the Mean and Variance of the Conceptual Statistical Distribution that Consists of All Possible Replicate Realization Values for a Random Variable with a Discrete Metric

The analog to the continuous conceptual probability density function (PDF) is the discrete conceptual probability function (PF). The associated probability correspondence is

$$f(x_i)dx_i \Leftrightarrow p_i$$

in which p_i is directly analogous to the statistical weight sw_i for weighted (heteroscedastic) datum values, viz.,

$$p_i \Leftrightarrow sw_i$$

Accordingly, the actual value for the mean of the conceptual statistical distribution that consists of all possible realization values for a discrete random variable generically denoted Y is expressed as

$$\text{mean}(Y) = \frac{\sum_{i=1}^{n_{dyv}} p_i \cdot y_i}{\sum_{i=1}^{n_{dyv}} p_i} = \sum_{i=1}^{n_{dyv}} p_i \cdot y_i$$

in which (a) n_{dyv} is the number of the different discrete y_i metric values that the respective realization values for the random variable Y can take on, and (b) the sum of all p_i's is equal to one. Similarly, the actual value for the variance of the conceptual statistical distribution that consists of all possible realization values for a discrete random variable generically denoted Y is expressed as

$$\text{var}(Y) = \frac{\sum_{i=1}^{n_{dyv}} p_i \cdot [y_i - \text{mean}(Y)]^2}{\sum_{i=1}^{n_{dyv}} p_i} = \sum_{i=1}^{n_{dyv}} p_i \cdot [y_i - \text{mean}(Y)]^2$$

3.A.2.1. Perspective

We use these two fundamental expressions in Supplemental Topic 3.B to establish expressions for the actual values for the mean and variance of the conceptual statistical distribution that consists of the two mutually exclusive discrete outcomes of a binomial trial. These mean and variance expressions are then intuitively extended to pertain to a series of n_{bt} mutually independent binomial trials, viz., to the conceptual (one-parameter) binomial distribution. A series of n_{bt} independent individual binomial trials is statistically synonymous with (a) a series of n_{pc} mutually independent paired comparisons in a RCBD experiment test program, or (b) a series of n_{rt} mutually independent individual reliability tests in a CRD experiment test program. Thus, the conceptual (one-parameter) binomial distribution has direct application in either improving the reliability of a product or in estimating the reliability of a product. The statistical background for these applications is discussed next.

3.B. SUPPLEMENTAL TOPIC: CONCEPTUAL (ONE-PARAMETER) BINOMIAL DISTRIBUTION

The conceptual (one-parameter) binomial distribution describes the probability behavior of a random variable that takes on only discrete realization values, e.g., the number of favorable outcomes, n_{fo}, in n_{pc} mutually independent paired-comparisons or in n_{rt} mutually independent individual reliability tests, where the probability of a favorable outcome, p_{fo}, is invariant from paired comparison to paired comparison or from individual reliability test to individual reliability test. Each of the n_{pc} independent paired comparisons or n_{rt} individual reliability tests is generically termed a binomial trial.

Consider a hypothetical experiment test program with six (mutually independent) binomial trials, each with the same invariant probability pfo of a favorable outcome. The probability of a sequence of four favorable outcomes followed by two unfavorable outcomes occurring is $pfo \cdot pfo \cdot pfo \cdot pfo \cdot (1 - pfo) \cdot (1 - pfo)$. Similarly, a sequence with n_{fo} favorable outcomes followed by $(n_{bt} - n_{fo})$ unfavorable outcomes, where n_{bt} is the number of binomial trails that comprise the hypothetical experiment test program of specific interest, has a probability of occurring equal to $pfo^{n_{fo}} \cdot (1 - pfo)^{n_{bt} - n_{fo}}$. Note, however, that this probability expression is valid regardless of the specific order of the successes and failures in a sequence of n_{fo} favorable outcomes and $(n_{bt} - n_{fo})$ unfavorable outcomes. Moreover, there are exactly $(n_{pc})!/[(n_{fo})! \cdot (n_{bt} - n_{fo})!]$ experiment test program outcomes that have n_{fo} favorable outcomes and $(n_{bt} - n_{fo})$ unfavorable outcomes. Hence, given n_{bt} and pfo, the probability that a randomly selected equally likely reliability-based experiment test program will have exactly n_{fo} favorable outcomes is:

(enumeration-based) probability ($N_{FO} = n_{fo}$; given pfo and n_{bt})

$$= \frac{(n_{bt})!}{(n_{fo})! \cdot (n_{bt} - n_{fo})!} \cdot pfo^{n_{fo}} \cdot (1 - pfo)^{n_{bt} - n_{fo}}$$

This probability expression is called the binomial probability function. (Remember that, for this binomial probability function expression, binomial trials are synonymous with mutually independent paired comparisons and mutually independent individual reliability tests, viz., $n_{bt} = n_{pc} = n_{rt}$.)

We now present three microcomputer programs for the conceptual binomial distribution, each pertaining to a different way to tabulate cumulative binomial probability values. Microcomputer program *BINOM1* computes and tabulates the (exact) enumeration-based probability that $N_{FO} \geq n_{fo}$, given input values of *pfo* and n_{bt}. This tabulation is convenient when the classical sign test (without ties) is employed in a test of the null hypothesis that $B = A$ statistically versus the simple (one-sided) alternative hypothesis that $B > A$ statistically. In contrast, microcomputer program *BINOM2* computes and tabulates the (exact) enumeration-based probability that $N_{FO} < n_{fo}$, given input values of *pfo* and n_{bt}. This tabulation is convenient when the classical sign test (without ties) is employed in a test of the null hypothesis that $B = A$ statistically versus the simple (one-sided) alternative hypothesis that $B < A$ statistically. Finally, microcomputer program *BINOM3* computes the (exact) enumeration-based probability that $N_{FO} < n_{fo}$, given input values of *pfo* and n_{bt}. This tabulation is convenient

```
C> BINOM1

Input the invariant probability of a favorable outcome pfo and the
number of binomial trials nbt, e.g., 0.5 space 10

0.5 10
```

n_{fo}	probability that $N_{FO} \geq n_{fo}$
0	1.0000
1	0.9990
2	0.9893
3	0.9453
4	0.8281
5	0.6230
6	0.3770
7	0.1719
8	0.0547
9	0.0107
10	0.0010

```
C> BINOM2
```

Input the invariant probability of a favorable outcome *pfo* and the
number of binomial trials n_{bt}, e.g., 0.5 space 10

```
0.5 10
```

n_{fo}	probability that $N_{FO} \geq n_{fo}$
0	0.0010
1	0.0107
2	0.0547
3	0.1719
4	0.3770
5	0.6230
6	0.8281
7	0.9453
8	0.9893
9	0.9990
10	1.0000

when relating reliability to its associated statistical confidence probability
value (discussed later). Note that microcomputer programs *BINOM1* and
BINOM3 have their corresponding probabilities sum to 1.0000; the respective probabilities are complementary.

The maximum value for n_{bt} in programs *BINOM1*, *BINOM2*, and
BINOM3 is 32 (because it is very unlikely that n_{bt} will ever exceed 32 in a
mechanical reliability experiment test program.)

The respective derivations of the analytical expressions for the actual
values for the mean and variance of a conceptual binomial distribution are
tedious without the following change of variable to a single binomial trial.
Consider the random variable *Y* such that *y* takes on the discrete realization
value of 1 when a favorable outcome is observed and *y* takes on the discrete
realization value zero when an unfavorable outcome is observed. Then, the
actual value for the mean of the conceptual statistical distribution that
consists of both discrete realization values for random variable *Y*
(Supplemental Topic 3.A) is (summing over *y* = 1 and *y* = 0)

$$\text{mean}(Y) = pfo \cdot 1 + (1 - pfo) \cdot 0 = pfo$$

```
C> BINOM3

Input the invariant probability of a favorable outcome pfo and the
number of binomial trials n_bt, e.g., 0.5 space 10

0.5 10
```

n_{fo}	probability that $N_{FO} < n_{fo}$
0	0.0000
1	0.0010
2	0.0107
3	0.0547
4	0.1719
5	0.3770
6	0.6230
7	0.8281
8	0.9453
9	0.9893
10	0.9990

In turn, the actual value for the variance of the conceptual statistical distribution that consists of both discrete realization values for random variable Y is (again summing over $y = 1$ and $y = 0$)

$$\text{var}(Y) = pfo \cdot (1 - pfo)^2 + (1 - pfo) \cdot (0 - pfo)^2 = pfo \cdot (1 - pfo)$$

We assert in Chapter 4 (Exercise Set 1) that (a) the actual value for the mean
of the conceptual sampling distribution that consists of all possible replicate
realization values for the statistic (the sum of n_s realization values for the
random variable X) is equal to n_s times mean(X), and in turn that (b) the
actual value for the variance of the conceptual sampling distribution that
consists of all possible replicate realization values for the statistic (the sum
of n_s mutually independent realization values for the random variable X) is
equal to n_s times var(X). Accordingly, given a sequence of n_{bt} (mutually
independent) binomial trials, we assert that

$$\text{mean}(N_{FO}) = n_{bt} \cdot pfo$$

and

$$\text{var}(N_{FO}) = n_{bt} \cdot pfo \cdot (1 - pfo)$$

(Remember that binomial trials in a hypothetical experiment test program are synonymous with mutually independent paired comparisons in a paired-comparison experiment test program and with mutually independent individual reliability tests in a reliability-based experiment test program.)

The conceptual binomial distribution is often depicted in Figure 3.9(a) by plotting the numerical values of its probability function as ordinates, viz., the heights of vertical lines. This depiction is employed when the respective numbers of equally likely experiment test program outcomes are so large that it is inconvenient to construct a plot analogous to Figure 1.5(b). Note that the respective numerical binomial probability function values plotted as ordinates in Figure 3.9(a) can be redepicted in the histogram format as illustrated in Figure 3.9(b). Then, the resulting histogram also has a clear enumeration-based probability interpretation when it is restated in terms of proportions (relative frequencies). Advanced theory demonstrates that this histogram for proportions (relative frequencies) asymptotically takes on the exact bell-shaped form (Figure 3.10) of the conceptual (two-parameter) normal distribution (Chapter 5) as n_{bt} increases without bound.

Remark: We introduce the statistic (the sum of n_s uniform pseudor-andom numbers) in Chapter 4 and demonstrate that its conceptual sampling distribution also asymptotically (as n_s increases without bound) takes on the exact bell-shaped form of the conceptual (two-parameter) normal distribution. That demonstration is intended to support the presumption in classical statistical analysis that measurement-based random errors are normally distributed.

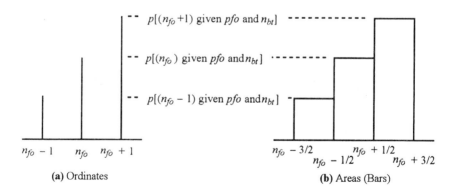

Figure 3.9 Depicting the conceptual binomial distribution by plotting the values of its probability function as ordinates, viz., the heights of the vertical lines in (a), or the heights of the vertical bars in (b).

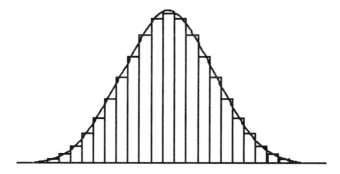

Figure 3.10 Conceptual (one-parameter) binomial distribution, when depicted in histogram format, becomes asymptotically identical to the bell-shaped conceptual (two-parameter) normal distribution, viz., as the number of binomial trials increases without bound.

Exercise Set 9

These exercises are intended (a) to acquaint you with the conceptual binomial distribution and its probability function, and with the respective outputs of microcomputer programs *BINOM1*, *BINOM2*, and *BINOM3*, and (b) to provide insight regarding microcomputer simulation of a sequence of binomial trials.

1. Given that $n_{bt} = 10$ and (a) $pfo = 0.1$, (b) $pfo = 0.5$, (c) $pfo = 0.9$, calculate the respective probabilities by hand that N_{fo} equals n_{fo} for n_{fo} from 0 to 10. Then, run microcomputer program *BINOM1* with $n_{bt} = 10$ and (a) $pfo = 0.1$, (b) $pfo = 0.5$, (c) $pfo = 0.9$ and *reinterpret* the respective outputs to verify your hand calculations. Following verification, plot the numerical values of the respective binomial probability functions in the formats of Figures 3.9(a) and (b).
2. Explain how you could use a uniform pseudorandom number generator to simulate the outcomes of n_{bt} mutually independent binomial trials.

Exercise Set 10

These exercises are intended to enhance your understanding of Figures 3.2 and 3.4.

1. Given that the actual value for the variance of the conceptual binomial distribution that consists of $(2)^{n_{bt}}$ discrete realization values for the random variable N_{FO} is equal to $n_{bt} \cdot pfo \cdot (1 - pfo)$, verify that the actual value for the variance of the conceptual binomial distribution that consists of $(2)^{n_{bt}}$ discrete realization values for the random variable $[(1/n_{bt}) \cdot N_{FO}]$ is equal to $\{[pfo \cdot (1 - pfo)]/n_{bt}\}$.

2. Consider the classical fair coin experiment. Verify that the actual value for the variance of the conceptual binomial sampling distribution that consists of $(2)^{n_f}$ discrete realization values for the statistic (the proportion of heads observed in n_f flips) is equal to $[1/(4 \cdot n_f)]$.

3. (a) Given that approximately 95% of all classical fair-coin experiments will generate a realization value for the statistic (the proportion of heads observed in n_f flips) that deviates less than 1.96 standard deviations from 0.5, extend Exercise 2 by computing *approximate* 95% (two-sided) statistical confidence intervals that allegedly include the actual value (0.5) for the mean of the conceptual binomial sampling distribution that consists of $(2)^{n_f}$ discrete equally likely realization values for the statistic (the proportion of heads observed in n_f flips)—when n_f is consecutively set equal to 10, 100, 1000, 10,000, and 100,000. Then, (b) compare the respective widths for these intervals to the simulation-based ranges for the curves faired in Figure 3.2.

4. (a) Given that the actual value for the variance of the conceptual sampling distribution that consists of all possible realization values for the statistic (the arithmetic average of n uniform pseudorandom numbers) is equal to $1/(12 \cdot n)$, what is the ratio of the actual value for this variance to the actual value for the variance of the conceptual binomial sampling distribution that consists of $(2)^{n_f}$ discrete realization values for the statistic (the proportion of heads observed in n_f flips) when $n_f = n$? Then, (b) is the corresponding standard deviation ratio consistent with the ratios of the widths of the simulation-based bands faired in Figures 3.2 and 3.4?

3.B.1. Reliability and Its Associated Statistical Confidence Probability Value

When the conceptual binomial distribution is used to summarize the equally likely outcomes of a hypothetical sequence of (mutually independent) binomial trials, there are two probabilities of specific interest: (a) the invariant probability *pfo* of a favorable outcome in each independent binomial trial, and (b) the computed value of the enumeration-based probability that $N_{FO} \geq n_{fo}$ given the number of binomial trials n_{bt}. Recall that the computed value of the enumeration-based probability that $N_{FO} \geq n_{fo}$ can be employed in the classical sign test of the null hypothesis that $B = A$ statistically versus the simple (one-sided) alternative hypothesis that $B > A$. Recall also that the probability of incorrectly rejecting the null hypothesis when it is presumed to be correct is the probability of committing a *Type I* error. We now assert that, in a reliability context, (a) *pfo* is synonymous with *reliability*, viz., the invariant probability of survival in each mutually independent reliability test, and (b) the complement of the probability of committing a *Type I* error is synonymous with the *statistical confidence probability*, subsequently denoted *scp*. Accordingly, if we had conducted 10 individual independent reliability tests and observed eight survivals, the example output of microcomputer program *BINOM3* indicates that the statistical confidence probability value is equal to 0.9453 when the reliability is equal to 0.5. However, this relationship is more properly interpreted as indicating that we can assert with $100 \cdot (0.9453)\%$ statistical confidence that the actual value for the reliability is at least equal to 0.50.

Statistical confidence probability values such as 0.9453 are not traditionally used in statistical analysis. Rather traditional values of the statistical confidence probability are 0.90, 0.95, and 0.99—because statistical analyses have traditionally been based on tabulated values for the 90th, 95th, and 99th percentiles of the conceptual (or empirical) sampling distribution for the test statistic of specific interest. (Recall that traditional values for the complementary probabilities of committing the associated *Type I* errors are 0.10, 0.05, and 0.01.)

Note that the input probability *pfo* in microcomputer program *BINOM3* must be iteratively adjusted to obtain the traditional *scp* value of specific interest. This iterative adjustment is accomplished in microcomputer program *MINREL* (minimum reliability), which computes the minimum actual value for the reliability given any *scp* value of specific interest, traditional or nontraditional. We limit $100(scp)\%$ to integer values in microcomputer program *MINREL* because values greater than 99% are not physically and statistically credible.

```
C> MINREL
```

Input the number of reliability tests n_{rt} that have been conducted

```
10
```

Input the number of items n_{is} that survived these n_{rt} reliability tests

```
10
```

Input the statistical confidence probability of specific interest, stated in percent

```
95
```

Given that 10 items survived in 10 reliability tests, we can assert that we have 95% statistical confidence that the actual value for the reliability is equal to at least 0.741.

Exercise Set 11

These exercises are intended to enhance your understanding of reliability and its associated statistical confidence probability value.

1. Let a paired-comparison experiment test program involve 10 blocks, each block with a single datum value each for components A and B. Then, (a) run microcomputer program *BINOM1* iteratively to establish the value of conceptual binomial probability *pfo* for which the probability of observing eight or more positive $(b - a)$ differences in 10 paired-comparisons is (almost exactly) equal to 0.05. Next, (b) run microcomputer program *MINREL* for 10 reliability tests with eight items surviving to the estimate of the minimum actual value for the reliability pertaining to 95% statistical confidence. Finally, (c) compare the results of (a) and (b).

2. Suppose we wish to select both a traditional *scp* value *and* the minimum value for reliability. How many reliability tests with no failures must be conducted to have 95% statistical confidence

that the reliability is at least equal to 0.90? Determine n_{rt} by iteratively running program *MINREL*.

3. (a) What is the fundamental difference (if any) between B is greater than A statistically and B is at least as good as A statistically? (b) Given 10 mutually independent reliability tests, how many items must survive these tests to have at least 95% statistical confidence that B is at least as good as A statistically? Is it practical to use this minimum number of survivals as the pass/fail criterion for a so-called quality assurance test? Discuss.

3.C. SUPPLEMENTAL TOPIC: LOWER 100(*SCP*)% (ONE-SIDED) STATISTICAL CONFIDENCE LIMITS

Recall that microcomputer program *MINREL* (Supplemental Topic 3.B) can be used to compute a lower $100(scp)\%$ (one-sided) statistical confidence limit that allegedly bounds the minimum actual value for the reliability, given (a) any (traditional or nontraditional) value of the *scp* value of specific interest, and (b) the number of items n_{is} that survived in n_{rt} independent reliability tests. We now present an analogous lower $100(scp)\%$ (one-sided) statistical confidence limit that allegedly bounds the actual value of the minimum practical difference (*mpd*) between the respective means (or medians, or any other corresponding percentiles) of B and A in Figure 3.8(b).

Consider Fisher's enumeration-based test and the paired-comparison data used to compute our example output for microcomputer program *FEBT*. Given the null hypothesis that $B = A$ statistically, the probability that a randomly selected value of Fisher's enumeration-based conceptual sampling distribution will be equal to or greater than 138 is equal to 0.0039. Thus, we could rationally assert that we have $100(1 - 0.0039)\% = 99.61\%$ statistical confidence that $B > A$ statistically. However, there is a practical difficulty associated with this reinterpretation of the null hypothesis rejection probability value in terms of its nontraditional complementary value for the *scp*. When we conceptually replicate our paired-comparison experiment test program, it is clear that the resulting null hypothesis rejection probability and its nontraditional *scp* complement will change from replicated experiment test program to replicated experiment test program. Thus, our reinterpretation of the null hypothesis rejection probability in terms of its nontraditional *scp* complement is generally considered awkward. The statistical confidence value is traditionally preselected before the experiment test program is conducted to be either 90, 95, or 99%, just as the acceptable probability of committing a *Type I* error is traditionally preselected before the experiment test program is conducted to be either 0.10, 0.05, or 0.01.

Now suppose that (a) the null hypothesis is that $(B - mpd) = B' = A$ statistically, and (b) the simple (one-sided) alternative hypothesis is that $(B - mpd) = B' > A$ statistically, where mpd connotes minimum practical difference. Then, as illustrated below, microcomputer program *FEBMPDT* (Fisher's enumeration-based minimum practical difference test) can be run iteratively to establish the mpd value that is subsequently reinterpreted as the lower $100(scp)\%$ (one-sided) statistical confidence limit for the actual value for the difference between the respective means (medians) of the respective conceptual statistical distributions for B and A.

mpd	enumeration-based probability that a randomly selected experiment test program will have its sum of $(b' - a)$ differences equal to or greater than the data-based sum of $(b' - a)$ differences for the experiment test program that was actually conducted, given the null hypothesis that $B' = A$ statistically	corresponding nontraditional *scp*, stated in %
0	0.0039	99.61
⋮	⋮	⋮
9	0.0039	99.61
10	0.0078	99.22
11	0.0156	98.44
12	0.0273	97.27
13	0.0547	94.53
14	0.1016	89.84

Given this iteratively established mpd value we can subsequently assert with $100(scp)\%$ statistical confidence that the actual value for the difference between the respective means (medians) of the conceptual statistical distributions for B and A is at least equal to mpd (units). For example, we can assert with more than 95% statistical confidence that the actual value of the mean (median) of the conceptual statistical distribution for B excels the actual value of the median of the conceptual statistical distribution for A by at least 12 units. Clearly, this assertion is either correct, or it is not. Accordingly, the arbitrarily selected scp value is the probability that the associated lower $100(scp)\%$ (one-sided) statistical confidence limit assertion is actually correct.

Given any scp value of specific interest, say 0.95, we can conduct a simulation study of the probability behavior of the proportion (percentage) of lower $100(scp)\%$ (one-sided) statistical confidence limit assertions that are actually correct. Microcomputer program *SSLOSSCL* (simulation study of a lower one-sided statistical confidence limit, pages 144 and 145) has the scp of interest, stated in percentage, as input data. It clearly demonstrates that,

```
C> COPY FEMPDDTA DATA


   1 file(s) copied


C> FEBMPDT
```

Given a minimum practical difference *mpd* equal to 14, the data-based sum of the actual values for the $[(b - mpd) - a]$ differences that constitute the outcome of the paired-comparison experiment test program that was actually conducted is equal to 26.

Given the null hypothesis that $(B - mpd) = A$ statistically, this microcomputer program constructed exactly 256 equally-likely outcomes for this paired-comparison experiment test program by using Yate's enumeration algorithm to reassign positive and negative signs to its $[(b - mpd) - a]$ differences. The number of these outcomes that had its sum of the actual values for its $[(b - mpd) - a]$ differences equal to or greater than 26 is equal to 26. Thus, given the null hypothesis that $(B - mpd) = A$ statistically, the enumeration-based probability that a randomly selected outcome of this paired-comparison experiment test program when continually replicated will have its sum of the actual values for its $[(b - mpd) - a]$ differences equal to or greater than 26 is equal to 0.1016. When this probability is sufficiently small, reject the null hypothesis in favor of the simple (one-sided) alternative hypothesis that $(B - mpd) > A$ statistically.

as illustrated in Figures 3.2 and 3.4, the variability in the actual proportion (percentage) of lower $100(scp)\%$ (one-sided) statistical confidence limit assertions that are actually correct decreases as the number of simulated experiment test program outcomes increases. Accordingly, it is intuitively clear that, in the limit as the number of simulated experiment test program outcomes increases without bound, the proportion (percentage) of statistical confidence assertions that are actually correct asymptotically approaches the preselected (input) *scp* value. Remember, however, that when we preselect the *scp* value of specific interest we simultaneously select the acceptable probability $(1 - scp)$ that the lower $100(scp)\%$ (one-sided) statistical confidence limit assertion will actually be incorrect.

Microcomputer program *SSLOSSCL* is also intended to demonstrate that it is impossible to state *a priori* whether any specific lower $100(scp)\%$ (one-sided) statistical confidence limit assertion is or is not actually correct. In particular, given any future simulation study pertaining to a new set of three, three-digit seed numbers, it is impossible to state *a priori* whether the assertion associated with any randomly selected experiment test program outcome will be correct or not.

Finally, a traditional (exact) value for the *scp* can be obtained, if required, by using the realization of a uniform pseudorandom number to select between the two nontraditional lower $100(scp)\%$ (one-sided) statistical confidence limits whose *scp* values immediately straddle the traditional (exact) *scp* value of specific interest. This random selection process requires that the respective nontraditional values have probabilities of being selected as follows:

probability of selecting nontraditional scp_{low}

$$= \frac{\text{nontraditional } scp_{\text{high}} - \text{traditional } scp}{\text{nontraditional } scp_{\text{high}} - \text{nontraditional } scp_{\text{low}}}$$

and

probability of selecting nontraditional scp_{high}

$$= \frac{\text{traditional } scp - \text{nontraditional } scp_{\text{low}}}{\text{nontraditional } scp_{\text{high}} - \text{nontraditional } scp_{\text{low}}}$$

3.D. SUPPLEMENTAL TOPIC: RANDOMIZATION-BASED TEST OF THE NULL HYPOTHESIS THAT $B = A$ STATISTICALLY VERSUS THE SIMPLE (ONE-SIDED) ALTERNATIVE HYPOTHESIS THAT $B > A$ STATISTICALLY PERTAINING TO A CRD EXPERIMENT TEST PROGRAM

It may seem ironic that we present and discuss alternative computer-intensive statistical tests of hypotheses before we have developed the probability background theory (Chapters 4 and 5) underlying the classical statistical tests of hypotheses (Chapter 6). However, the reason for this syllabus reversal is simply that these alternative computer-intensive analyses involve minimal statistical presumptions and therefore require minimal probability

(continued on page 145)

```
C> SSLOSSCL
```

Input a new set of three, three-digit odd seed numbers

```
227 191 359
```

Input the statistical confidence probability desired, stated in percent (integer values only)

```
95
```

Percentage of Simulation-Based Lower 95% (One-Sided) Statistical
Confidence Limit Assertions that are Actually Correct

```
 1 Assertion  – 100.000 Percent Correct
 2 Assertions – 100.000 Percent Correct
 3 Assertions –  66.667 Percent Correct
 4 Assertions –  75.000 Percent Correct
 5 Assertions –  80.000 Percent Correct
 6 Assertions –  83.333 Percent Correct
 7 Assertions –  71.429 Percent Correct
 8 Assertions –  75.000 Percent Correct
 9 Assertions –  77.778 Percent Correct
10 Assertions –  70.000 Percent Correct
11 Assertions –  72.727 Percent Correct
12 Assertions –  75.000 Percent Correct
13 Assertions –  76.923 Percent Correct
14 Assertions –  71.429 Percent Correct
15 Assertions –  73.333 Percent Correct
16 Assertions –  75.000 Percent Correct
17 Assertions –  76.471 Percent Correct
18 Assertions –  77.778 Percent Correct
19 Assertions –  78.947 Percent Correct
20 Assertions –  80.000 Percent Correct
21 Assertions –  80.952 Percent Correct
22 Assertions –  81.818 Percent Correct
23 Assertions –  82.609 Percent Correct
24 Assertions –  83.333 Percent Correct
25 Assertions –  84.000 Percent Correct
```

```
  30 Assertions -   83.333 Percent Correct
  35 Assertions -   82.857 Percent Correct
  40 Assertions -   85.000 Percent Correct
  45 Assertions -   84.444 Percent Correct
  50 Assertions -   86.000 Percent Correct
  55 Assertions -   87.273 Percent Correct
  60 Assertions -   88.333 Percent Correct
  65 Assertions -   89.231 Percent Correct
  70 Assertions -   90.000 Percent Correct
  75 Assertions -   90.667 Percent Correct
  80 Assertions -   91.250 Percent Correct
  85 Assertions -   91.765 Percent Correct
  90 Assertions -   92.222 Percent Correct
  95 Assertions -   92.632 Percent Correct
 100 Assertions -   93.000 Percent Correct
 200 Assertions -   95.000 Percent Correct
 300 Assertions -   94.000 Percent Correct
 400 Assertions -   94.250 Percent Correct
 500 Assertions -   94.200 Percent Correct
 600 Assertions -   94.500 Percent Correct
 700 Assertions -   94.429 Percent Correct
 800 Assertions -   94.500 Percent Correct
 900 Assertions -   94.667 Percent Correct
1000 Assertions -   94.600 Percent Correct
```

background theory to support their presentation and discussion. Clearly, the fewer and the simpler the presumptions underlying a statistical test of hypothesis, the more credible the statistical conclusions drawn (statistical inferences made).

Consider the (comparative) CRD experiment test program in Figure 3.11 (a repeat of Figure 2.2) with two treatments, viz., treatments A and B, where treatment A is final mechanical polishing in the circumferential direction and treatment B is final mechanical polishing in the longitudinal direction. Given that the null hypothesis is that no direction-of-final-mechanical-polishing effect exists, each possible reassignment of the respective 16 datum values to treatments A and B constitutes an equally likely outcome for this comparative CRD experiment test program. Accordingly, given this null hypothesis and an appropriate test statistic, we can employ a randomization-based test of hypotheses methodology to calculate the associated null

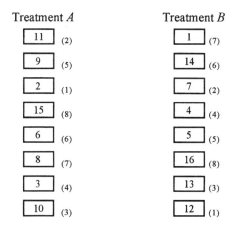

Figure 3.11 Schematic of the organizational structure of a comparative CRD experiment test program involving two treatments, viz., treatments A and B, each with eight replicates.

hypothesis rejection probability. (Recall that the empirical randomization-based null hypothesis rejection probability value asymptotically approaches the exact enumeration-based null hypothesis rejection probability value as n_{rbelo} becomes larger and larger.)

Randomization-based tests of hypotheses employ uniform pseudorandom numbers to construct equally likely outcomes for the experiment test program that was actually conducted. Then, the test statistic value for each of these constructed equally likely outcomes is calculated and compared to the data-based test statistic value for the outcome of the experiment test program that was actually conducted. In turn, the number n_{oosi} of constructed equally likely outcomes that exceed the data-based test statistic value is used to calculate the randomization-based null hypothesis rejection probability.

The rearrangements of the datum values for the CRD experiment test program that was actually conducted is conveniently accomplished in a two-step procedure using uniform pseudorandom numbers (Section 3.5). First, the integers 1 through n_{dv} are generated in random order (as in microcomputer program *RANDOM1*). These integers are then reordered from smallest to largest while maintaining their association with the datum values for the CRD experiment test program that was actually conducted. This two-step procedure, when repeated, generates equally likely outcomes for a CRD experiment test program (regardless of the number of its treatments or treatment levels).

Remark: The appropriate number of constructed equally likely experiment test program outcomes, n_{rbelo}, is subjective. It is best established by demonstrating that the randomization-based null hypothesis rejection probability value does not change for practical purposes as n_{rbelo} is increased beyond the value for n_{rbelo} employed in your randomization-based analysis. Typically n_{rbelo} equal 9999 suffices.

Microcomputer program *RBBVACRD* (randomization-based *B* versus *A* in a CRD experiment test program) pertains to a CRD experiment test program with only two treatments, denoted *A* and *B*. The null hypothesis is that $B = A$ statistically, whereas the simple (one-sided) alternative hypothesis is that $B > A$ statistically. The test statistic employed in this microcomputer program to test the null hypothesis is the arithmetic

```
COPY RBBVADTA DATA

   1 file(s) copied

C> RBBVACRD
```

The data-based value of the (arithmetic average of the *b*'s minus the arithmetic average of the *a*'s) test statistic for the CRD experiment test program that was actually conducted is equal to 15.60.

Given the null hypothesis that $B = A$ statistically, this microcomputer program constructed 9999 equally-likely outcomes for this experiment test program by using uniform pseudorandom numbers to reorder its datum values. The number of these outcomes that had its (arithmetic average of the *b*'s minus the arithmetic average of the *a*'s) test statistic value equal to or greater than 15.60 is equal to 48. Thus, given the null hypothesis that $B = A$ statistically, the randomization-based probability that a randomly selected outcome of this experiment test program when continually replicated will have its (arithmetic average of the *b*'s minus the arithmetic average of the *a*'s) test statistic value equal to or greater than 15.60 is equal to 0.0049. When this probability is sufficiently small, reject the null hypothesis in favor of the simple (one-sided) alternative hypothesis that $B > A$ statistically.

average of the b experiment test program datum values minus the arithmetic average of the a experiment test program datum values. The constructed experiment test program outcomes of specific interest are those outcomes that have their test statistic value equal to or greater than the data-based test statistic value for the experiment test program that was actually conducted.

> *Remark*: The test statistic employed in microcomputer program *RBBVACRD* is used in the classical *independent t test* that is also based on the presumption that the $CRHDV_i$'s are normally distributed. If this additional presumption is valid, then the classical alternative lower $100(scp)\%$ (one-sided) statistical confidence limit that allegedly bounds the actual value for the difference [mean(B) − mean(A)] is also valid.

```
COPY RBMPDDTA DATA


  1 file(s) copied


C> RBBVAMPD
```

The data-based value of the [(arithmetic average of the b's minus the mpd) − the arithmetic average of the a's] test statistic for the CRD experiment test program that was actually conducted is equal to 5.60.

Given the null hypothesis that $(B - mpd) = A$ statistically, this microcomputer program constructed 9999 equally-likely outcomes for this experiment test program by using uniform pseudorandom numbers to reorder its datum values. The number of these outcomes that had its [(arithmetic average of the b's minus the mpd)—the arithmetic average of the a's] test statistic value equal to or greater than 5.60 is equal to 336. Thus, given the null hypothesis that $(B - mpd) = A$ statistically, the randomization-based probability that a randomly selected outcome of this experiment test program when continually replicated will have its [(arithmetic average of the b's minus the mpd)—the arithmetic average of the a's] test statistic value equal to or greater than 15.60 is equal to 0.0337. When this probability is sufficiently small, reject the null hypothesis in favor of the simple (one-sided) alternative hypothesis that $(B - mpd) > A$ statistically.

3.D.1. Discussion

Microcomputer *RBBVACRD* is much more practical when it is revised to include a minimum practical difference (*mpd*) between the mean (median) of the conceptual statistical distribution that consists of *APRCRHDV*'s for treatment *B* and the mean (median) of the conceptual statistical distribution that consists of *APRCRHDV*'s for treatment *A*. Then, a lower 100(*scp*)% (one-sided) statistical confidence limit that allegedly bounds the actual value for the difference [mean(*B*) − mean(*A*)] can be established as discussed in Supplemental Topic 3.C. This lower 100(*scp*)% (one-sided) statistical confidence limit is established by iteratively running microcomputer program *RBBVAMPD* with different input values for the *mpd* to establish a nontraditional *scp* value that is approximately equal to the *scp* value of specific interest. Recall, however, that a traditional (exact) value for the *scp* of specific interest can be obtained, if required, by using the realization of a uniform pseudorandom number to select between the two nontraditional lower 100(*scp*)% (one-sided) statistical confidence limits whose *scp* values immediately straddle the traditional *scp* value of specific interest. (The associated equations are also given in Supplemental Topic 3.C.)

4

The Classical Statistical Presumption of Normally Distributed Experiment Test Program Datum Values

4.1. INTRODUCTION

The fundamental presumption underlying classical statistical analyses is that each experiment test program datum value is randomly selected from an appropriately scaled conceptual (two-parameter) normal distribution. This fundamental presumption is supported in this chapter by intuitively examining the statistical behavior of the numerous minor sources of unavoidable variability that are included in each experiment test program datum value. We use a combination of elementary analysis and simulation in this intuitive examination.

4.2. CONCEPTUAL SAMPLING DISTRIBUTION THAT CONSISTS OF ALL POSSIBLE REPLICATE REALIZATION VALUES FOR THE STATISTIC (THE SUM OF n_s UNIFORM PSEUDORANDOM NUMBERS)

Consider the conceptual sampling distribution that consists of all possible replicate realization values for the statistic (the sum of n_s uniform pseudorandom numbers). We now demonstrate that this conceptual sampling distribution can be accurately approximated by a conceptual (two-parameter) normal distribution when n_s is sufficiently large. However, to keep this demonstration tractable, we initially limit n_s to be equal to only 1, 2, and 3.

4.2.1. Conceptual Sampling Distribution that Consists of All Possible Replicate Realization Values for the Statistic (the Sum of One Uniform Pseudorandom Number)

Recall that uniform pseudorandom numbers are statistically synonymous with (mutually independent) realization values randomly selected from the conceptual uniform distribution, zero to one (Figure 3.7). Thus the conceptual uniform distribution, zero to one, is identical to the conceptual sampling distribution that consists of all possible replicate realization values for the statistic (the sum of one uniform pseudorandom number).

4.2.2. Conceptual Sampling Distribution that Consists of All Possible Replicate Realization Values for the Statistic (the Sum of Two Uniform Pseudorandom Numbers)

To develop the PDF expression for the conceptual sampling distribution that consists of all possible replicate realization values for the statistic (the sum of two pseudorandom numbers), we plot x_1 along the abscissa and x_2 along the ordinate in Figure 4.1.

The unit square in Figure 4.1(a) establishes all possible locations of the respective realization values x_1 and x_2 for the two uniform pseudorandom numbers, X_1 and X_2, when plotted as the point (x_1, x_2). This unit square is generically termed a *probability space*. Note that (a) the probability that X_1 will lie inside an interval of infinitesimal width dx_1 is equal to dx_1 regardless of the value of x_1, and (b) the probability that X_2 will lie inside an interval of infinitesimal width dx_2 is equal to dx_2 regardless of the value of x_2. Thus, because the respective realization values for x_1 and x_2 are independent, the probability that the plotted realization value point (x_1, x_2) will lie inside of an infinitesimal rectangular area with dimensions dx_1 by dx_2 is equal to dx_1 times dx_2, regardless of the actual location of this infinitesimal area in the unit square probability space. Accordingly, we can use the enumeration-based definition for probability to establish the exact value for the probability that the plotted realization value point (x_1, x_2) will lie inside any area of specific interest. Note that when $(x_1 + x_2)$ takes on a specific value, say x^*, the equation $x_1 + x_2 = x^*$ defines a line as

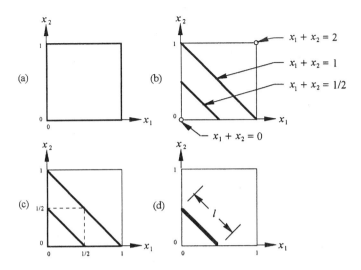

Figure 4.1 Diagrams useful in developing the PDF defining the conceptual sampling distribution that consists of all possible replicate realization values for the statistic (the sum of two uniform pseudorandom numbers).

illustrated in Figure 4.1(b). This line intersects the unit square probability space and establishes an area that is defined by the inequality $x_1 + x_2 < x^*$. This area is numerically equal to the enumeration-based probability that $0 < X_1 + X_2 < x^*$ (because the area of the unit square probability space is equal to one). For example, as illustrated in Figure 4.1(c), the enumeration-based probability that $0 < X_1 + X_2 < 1/2$ is equal to 1/8, and the enumeration-based probability that $1/2 < X_1 + X_2 < 1$ is equal to 3/8. Next, consider the event $[0 < X_1 + X_2 < x^*]$. Note that, for x^* between 0 and 1 and for x^* between 1 and 2, the area that is defined by the inequality $x_1 + x_2 < x^*$ changes linearly with a change in x^*. These changes can be visualized by successively adding an elemental area with length l, Figure 4.1(d), to the area previously included in the inequality. Note that length l starts at 0, increases linearly with x^* up to $x^* = 1$, and then decreases linearly with x^* up to $x^* = 2$, where $l = 0$ again. Thus, the PDF defining the conceptual sampling distribution that consists of all possible replicate realization values for the statistic (the sum of two uniform pseudorandom numbers) must increase linearly for $(x_1 + x_2)$ between 0 and 1, and then decrease linearly for $(x_1 + x_2)$ between 1 and 2. Accordingly $f(x)$ has two linear (first-order) segments, viz.,

$$f(x) = x \qquad \text{for } 0 < x < 1$$
$$f(x) = 2 - x \qquad \text{for } 1 < x < 2$$
$$f(x) = 0 \qquad \text{for all other values of } x$$

This PDF is plotted in Figure 4.2 (along with the PDF's that define the analogous conceptual sampling distributions pertaining to n_s equal to 1, 2, 3, and 4).

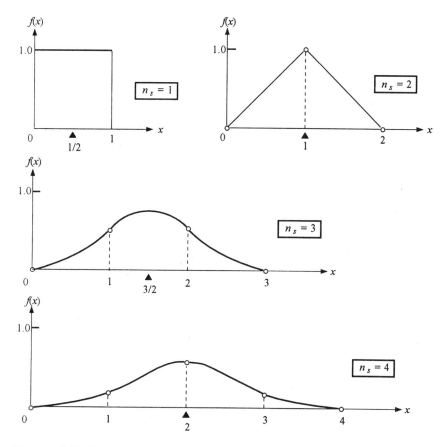

Figure 4.2 Comparison of the respective conceptual sampling distributions for the statistics (the sum of 1, 2, 3, and 4 uniform pseudorandom numbers). Note that the "bell-shaped contour" of the PDF's for the conceptual sampling distributions pertaining to n_s is equal to 3 and 4.

4.2.3. Conceptual Sampling Distribution that Consists of All Possible Replicate Realization Values for the Statistic (the Sum of Three Uniform Pseudorandom Numbers)

The conceptual sampling distribution that consists of all possible replicate realization values for the statistic (the sum of three uniform pseudorandom numbers) must be deduced because its probability space is three-dimensional, viz., a unit cube with its inboard edges along mutually orthogonal axes x_1, x_2, and x_3. Nevertheless, it is clear that, because the respective realization values x_1, x_2, and x_3 for the associated uniform pseudorandom numbers are mutually independent, all infinitesimal parallelepipeds with dimensions dx_1 by dx_2 by dx_3 are equally likely to contain the realization value point (x_1,x_2,x_3) regardless of their actual locations in the unit cube probability space. Accordingly, the enumeration-based probability that $X_1 + X_2 + X_3 < x^*$ is exactly computed as the ratio of the volume of the probability space defined by the inequality $x_1 + x_2 + x_3 < x^*$ to the total volume of the probability space (one). The intersection of the octahedral plane $x_1 + x_2 + x_3 = x^*$ with the three inboard edges of the unit cube is an equilateral triangle for x^* between 0 and 1. The associated inequality defines a tetrahedral volume with this equilateral triangle as its base. For x^* between 1 and 2, the basal area of the sample space volume defined by the inequality $x_1 + x_2 + x_3 < x^*$ is more complex geometrically because the octahedral plane associated with the equality $x_1 + x_2 + x_3 = x^*$ intersects the unit cube probability space along its three inboard edges and its three intermediate edges. If the probability space cube were larger, the intersection of this octahedral plane and the probability space would still be an equilateral triangle. However, restricting the size of the probability space to a unit cube has the geometric consequence of cutting off each apex of the equilateral triangle defining the basal area for a larger probability space cube. Accordingly, for the unit cube probability space, the basal area of the probability space volume defined by the inequality $x_1 + x_2 + x_3 < x^*$ continues to increase as x^* exceeds 1, but at a slower and slower rate until x^* reaches 3/2. This basal area then starts to decrease, slowly at first, and then at a faster and faster rate until x^* reaches 2. This change in the basal area of the probability space volume is symmetrical for x^* between 1 and 3/2 and x^* between 3/2 and 1. For x^* greater than 2, the octahedral plane intersects the probability space unit cube only along its outboard edges. Accordingly, the change in basal area for x^* between 2 and 3 is symmetrical to the change in basal area for x^* between 0 and 1. The PDF defining the conceptual sampling distribution that consists of all possible replicate realization values for the statistic (the sum of three uniform pseudorandom numbers) can be deduced by examining the quadratic nature of the basal areas of the elemental volumes that are successively added to generate the resulting probability space volumes as x^* successively increases from 0 to 1, from 1 to 2, and from 2 to 3. This PDF has three quadratic (second-order) segments, viz.,

$$
\begin{aligned}
f(x) &= \tfrac{1}{2}x^2 & 0 < x < 1 \\
f(x) &= -x^2 + 3x - \tfrac{3}{2} & 1 < x < 2 \\
f(x) &= \tfrac{1}{2}(3 - x)^2 & 2 < x < 3 \\
f(x) &= 0 & \text{for all other } x
\end{aligned}
$$

and recall that it is depicted in Figure 4.2 (along with the PDF's that define the respective conceptual sampling distributions pertaining to n_s equal to 1, 2, 3, and 4).

4.2.4 Conceptual Sampling Distribution that Consists of All Possible Replicate Realization Values for the Statistic (the Sum of n_s Pseudorandom Numbers)

When the statistic of specific interest is the sum of more than three uniform pseudorandom numbers, the relevant probability space is a unit hypercube and mathematical induction must be used to deduce the PDF's that define the associated conceptual sampling distributions (Parzen, 1960). However, as evident in Table 4.1 for n_s from 1 to 4, each successive PDF

Table 4.1 Expressions for the PDFs that Define the Respective Conceptual Sampling Distributions Depicted in Figure 4.2

n_s	$f(x)$	for x between
1	1	0 and 1
2	x	0 and 1
	$2 - x$	1 and 2
3	$\tfrac{1}{2}x^2$	0 and 1
	$-x^2 + 3x - \tfrac{3}{2}$	1 and 2
	$\tfrac{1}{2}(3 - x)^2$	2 and 3
4	$\tfrac{1}{6}x^3$	0 and 1
	$-\tfrac{1}{2}x^3 + 2x^2 - 2x + \tfrac{2}{3}$	1 and 2
	$\tfrac{1}{2}x^3 - 4x^2 + 10x - \tfrac{44}{6}$	2 and 3
	$\tfrac{1}{6}(4 - x)^3$	3 and 4

consists of n_s segments of $(n_s - 1)^{th}$ order polynomials and is symmetrical relative to its midpoint. Thus, as n_s increases, the n_s segments of $(n_s - 1)^{th}$ order polynomials provide an increasingly more accurate approximation to the PDF for a conceptual (two-parameter) normal distribution (Chapter 5). In fact, this normal approximation becomes so accurate that it has been widely used with $n_s = 12$ to generate normally distributed pseudorandom numbers.

We now need to demonstrate that the statistical behavior displayed in Figure 4.2 is also valid for pseudorandom numbers generated from diverse conceptual statistical distributions. However, we temporarily postpone this demonstration to develop analytical expressions in Exercise Set 1 for establishing the actual values for the mean and variance of the conceptual sampling distributions in Figure 4.3. (These expressions have important statistical application when generalized appropriately.)

Exercise Set 1

These exercises are intended to use the PDF expressions given in Table 4.1 to provide insight regarding the fundamental statistical notion that (a) the actual value for the mean of the conceptual sampling distribution that consists of all possible replicate realization values for the statistic (the sum of n_s pseudorandom numbers) is equal to n_s times the actual value for the mean of the conceptual statistical distribution that consists of all possible replicate realization values for these pseudorandom numbers, and (b) the actual value of the variance of the conceptual sampling distribution that consists of all possible replicate realization values for the statistic (the sum of n_s pseudorandom numbers) is equal to n_s times the actual value for the variance of the conceptual statistical distribution that consists of all possible replicate realization values for these pseudorandom numbers. However, (b) is valid only when the respective pseudorandom numbers are mutually independent.

1. For $n_s = 1$ to 4, verify by integration that the respective areas under the PDF's in Table 4.1 are equal to 1.

2. For $n_s = 1$ to 4, compute the actual values for the means of the respective conceptual sampling distributions defined by the PDF's in Table 4.1. (a) Then, use these results to demonstrate that

$$\text{mean}(X_1 + X_2 + X_3 + \cdots + X_{n_s}) = \text{mean}\left(\sum_{i=1}^{n_s} X_i\right)$$

$$= n_s \cdot \text{mean}(X)$$

In turn, (b) use the result (Chapter 3, Exercise Set 4):

$$\text{mean}(c \cdot X) = c \cdot \text{mean}(X)$$

to demonstrate that

$$\text{mean[arithmetic average } (X_1 + X_2 + X_3 + \cdots + X_{n_s})]$$

$$= \text{mean}\left[\left(\frac{1}{n_s}\right) \cdot \left(\sum_{i=1}^{n_s} X_i\right) \right] = \text{mean}(X)$$

These relationships are valid regardless of the actual form for the conceptual statistical distribution that consists of all possible replicate experiment test program realization values for the *generic* random variable X, whether its metric is discrete or continuous.

3. For $n_s = 1$ to 4, compute the actual values for the variances of the respective conceptual sampling distributions defined by PDFs given in Table 4.1. (a) Then, use these results to demonstrate that

$$\text{var}(X_1 + X_2 + X_3 + \cdots + X_{n_s}) = \text{var}\left(\sum_{i=1}^{n_s} X_i\right) = n_s \cdot \text{var}(X)$$

In turn, (b) use the result (Chapter 3, Exercise Set 4):

$$\text{var}(c \cdot X) = c^2 \cdot \text{var}(X)$$

to demonstrate that

$$\text{var[arithmetic average}(X_1 + X_2 + X_3 + \cdots + X_{n_s})]$$

$$= \text{var}\left[\left(\frac{1}{n_s}\right) \cdot \left(\sum_{i=1}^{n_s} X_i\right) \right] = \frac{\text{var}(X)}{n_s}$$

These relationships are valid regardless of the form for the conceptual statistical distribution that consists of all possible replicate experiment test program realization values for the *generic* random variable X, whether its metric is discrete or continuous—*provided that these replicate experiment test program realizations are mutually independent.* (See Supplemental Topic 7.A for more information in this regard.)

4. The variance relationship developed in Exercise 3 has direct application in the classical statistical analysis of variance (Chapter 6). Recall that, for the orthogonal conceptual statistical models presented in Chapter 2, the column-vector-based least-squares est(sc_j) is

$$\text{est}(sc_j) = \frac{\sum_{i=1}^{n_{dv}} c_{j,i} \cdot (\text{experiment test program datum value})_i}{\sum_{i=1}^{n_{dv}} c_{j,i}^2}$$

$$= \sum_{i=1}^{n_{dv}} \left(\frac{c_{j,i}}{\sum_{i=1}^{n_{dv}} c_{j,i}^2} \right) \cdot (\text{experiment test program datum value})_i$$

in which $c_{j,i}$ is the integer value for the ith element in the jth column vector in the n_{dv} by n_{dv} orthogonal augmented contrast array associated with the experiment test program. Presume that each experiment test program datum value is randomly selected from its associated conceptual statistical distribution with (homoscedastic) variance var($APRCRHEE's$), then verify that

$$\text{var}[\text{est}(sc_j)] = \frac{\text{var}(APRCRHEE's)}{\sum_{i=1}^{n_{dv}} c_{j,i}^2}$$

Remark: Since the least-squares est(sc_j) is a linear function of the respective experiment test program datum values, the associated least-squares estimator is unbiased when the presumed conceptual statistical model is correct.

5. Given that the EV[est(sc_j)] is equal to zero, demonstrate that the expected value for the sum of squares of the respective elements that comprise the column vector in the estimated complete analytical model is equal to var($APRCRHEE's$). Since each column vector in the estimated complete analytical model has one statistical degree of freedom, this result can be used to establish a general rule for computing an unbiased estimate of var($APRCRHEE's$). See Section 5.2.

6. (a) Use the analytical expression for each PDF in Table 4.1 to verify that the probability that the sum of n_s uniform pseudorandom numbers will lie in a central interval whose width is one-half of the PDF range is equal to 1/2 (0.500) for $n_s = 1$, 3/4 (0.750) for $n_s = 2$, 55/64 (0.859) for $n_s = 3$, and 11/12 for (0.917) for $n_s = 4$. Is it intuitively obvious that this probability value asymptotically approaches one as n_s increases without bound? (b) Explain why exactly the same probability values and asymptotic probability behavior as in (a) pertain to the arithmetic average of $n_a = n_s$ uniform pseudorandom numbers.

(This asymptotic probability behavior pertains to every central interval regardless of how small its width. It is the direct consequence of the outcome developed in Exercise 3 that the variance of an arithmetic average is inversely proportional to n_a. Thus the width of the associated statistical confidence interval is proportional to the square root of n_a. Accordingly, its width decreases by a factor of ten as n_a increases by a factor of one hundred, which confirms the assertions made in discussing the simulation outcome summarized in Figures 3.2 and 3.4.)

7. (a) Use the probability space in Figure 4.1(a) to develop the analytical expression for the PDF that defines the conceptual sampling distribution that consists of all possible replicate realization values for the statistic $(X_1 - X_2)$. Then, (b) use this analytical expression verify that

$$\text{mean}(X_1 - X_2) = 0 = \text{mean}(X - X) = \text{mean}(X) - \text{mean}(X)$$

and

$$\text{var}(X_1 - X_2) = 2 \cdot \text{var}(X) = \text{var}(X - X) = \text{var}(X) + \text{var}(X)$$

In turn, (c) let X_1 be denoted X and X_2 be denoted Y, then verify that

$$\text{mean}(X \pm Y) = \text{mean}(X) \pm \text{mean}(Y)$$

and

$$\text{var}(X \pm Y) = \text{var}(X) + \text{var}(Y)$$

4.3. ACCURACY OF THE CONCEPTUAL (TWO-PARAMETER) NORMAL DISTRIBUTION APPROXIMATION TO THE CONCEPTUAL SAMPLING DISTRIBUTION THAT CONSISTS OF ALL POSSIBLE REPLICATE REALIZATION VALUES FOR THE STATISTIC (THE SUM OF n_s PSEUDORANDOM NUMBERS)

Advanced statistical theory indicates that (a) the conceptual sampling distribution that consists of all possible replicate realization values for the statistic (the sum of n_s normal pseudorandom numbers) is an appropriately scaled conceptual (two-parameter) normal distribution regardless of the value for n_s, and (b) the conceptual sampling distribution that consists of all possible replicate realization values for the statistic (the sum of n_s pseudorandom numbers) can be accurately approximated by the conceptual (two-parameter) normal distribution regardless of the conceptual statistical distribution used to generate these pseudorandom numbers, provided that n_s is sufficiently large. We now provide insight regarding the accuracy of this normal approximation by conducting three simulation studies pertaining to the statistic (the sum of n_s pseudorandom numbers). The first simulation study pertains to normal (normally distributed) pseudorandom numbers. It provides perspective by establishing the magnitudes of inherent (unavoidable) simulation errors. The second simulation study pertains to uniform pseudorandom numbers. It is intended to confirm and to extend the analytical results developed (and deduced) in Section 4.2. The third simulation study pertains to exponential pseudorandom numbers. It is intended to establish the magnitudes of the "worst-case" normal approximation errors. These three simulations studies are then augmented by a discussion of cognate simulation studies.

4.3.1. Simulation Study One—Magnitudes of Typical Simulation Errors

In this simulation study, we translate and scale the conceptual (two-parameter) normal distribution such that the resulting actual values for the means and variances of the conceptual sampling distributions that consist of all possible realization values for the statistic (the sum of n_s normal pseudorandom numbers) are zero and one, respectively. Thus, 90% of our simulation-based sums will theoretically lie in the numerical interval from -1.645 to $+1.645$, 95% will theoretically lie in the numerical interval from -1.960 to $+1.960$, and 99% will theoretically lie in the numerical interval from -2.576 to $+2.576$. In turn, we (a) generate 10,000 replicate experiment test programs with n_s normal pseudorandom numbers, for n_s from 3 to 15, and (b) compute the proportions of the respective statistics that actually lie in the three probability intervals of specific interest. Then, by appropriately subtracting $p = 0.90$, 0.95, or 0.99 from these proportions, we generate a typical set of simulation errors. As evidenced by the example output from microcomputer program $AVE3A$, typical simulation errors are less than 1% and do not appear to decrease noticeably as n_s increases.

```
C> AVE3A

Input a new set of three, three-digit odd seed numbers

371 583 915
```

Sum of n_s Normal Pseudorandom Numbers	Simulation Error for the 0.90 Interval	Simulation Error for the 0.95 Interval	Simulation Error for the 0.99 Interval
3	−0.0019	0.0007	0.0000
4	0.0000	0.0000	0.0006
5	−0.0019	−0.0028	−0.0006
6	−0.0020	0.0003	0.0018
7	−0.0071	−0.0036	−0.0012
8	0.0049	0.0017	0.0020
9	0.0018	0.0012	0.0003
10	−0.0026	−0.0007	0.0003
11	−0.0003	−0.0002	−0.0018
12	0.0005	0.0003	0.0004
13	−0.0034	−0.0035	0.0002
14	0.0058	0.0003	0.0007
15	−0.0005	−0.0008	0.0005

Remark One: Microcomputer programs $AVE3A$ and $AVE3A2$ employ the theoretically exact polar method (Knuth, 1969) to generate normal pseudorandom numbers (whose actual behavior depends on the behavior of the underlying uniform pseudorandom number generator). Comparative microcomputer programs $AVE3D$

and $AVE3D2$ employ a normal pseudorandom number generator that was used in the original IBM scientific subroutine package. It is based on the approximate normality of the sum of twelve uniform pseudorandom numbers.

Remark Two: Given 10,000 replicate experiment test programs, approximately 95% of the microcomputer program $AVE3A$ simulation errors will lie in the statistical confidence interval ± 0.006 for $p = 0.90$, ± 0.004 for $p = 0.95$, and ± 0.002 for $p = 0.99$ (Exercise Set 10). However, if one million replicate experiment test programs were used in this simulation study, the width of these statistical confidence intervals would be reduced by a factor of 10 (and the microcomputer execution time would increase by a factor of 100). Thus, we establish the size of the simulation errors (and the associated microcomputer execution time) by the selection of the number of replicate experiment test programs employed in our simulation study.

Microcomputer program $AVE3A2$ pertains to one million replicate experiment test programs for each value of n_s. Its example output appears in microcomputer file $AVE3A2$. It is prudent to run program $AVE3A$ before attempting to run program $AVE3A2$.

4.3.2. Simulation Study Two—Accuracy of the Normal Approximation to the Conceptual Sampling Distribution that Consists of All Possible Replicate Realization Values for the Statistic (the Sum of n_s Uniform Pseudorandom Numbers)

In this simulation study, we translate and scale the conceptual uniform distribution, 0 to 1, such that the for the actual values for the means and variances of the conceptual sampling distributions that consist of all possible realization values for the statistics (the sum of n_s uniform pseudorandom numbers) are zero and one, respectively. Thus, if the normal approximations to these conceptual sampling distributions are exact, 90% of our simulation-based sums will theoretically lie in the numerical interval from -1.645 to $+1.645$, 95% will theoretically lie in the numerical interval from -1.960 to $+1.960$, and 99% will theoretically lie in the numerical interval from 2.576 to $+2.576$. In turn, we (a) generate 10,000 replicate experiment test programs with n_s uniform pseudorandom numbers, for n_s from 3 to 15, and (b) compute the proportions of the respective statistics that actually lie in the three probability intervals of specific interest. Then, by appropriately subtracting $p = 0.90$, 0.95, or 0.99 from these proportions, we obtain a typical set of simulation-based estimates for the normal approximation errors. As evidenced by the example output from microcomputer program *AVE3B*, the magnitudes of the normal approximation errors are remarkably similar to the magnitudes of the simulation errors in the example output from micro-computer program *AVE3A*. Thus, for all three probability intervals of spe-cific interest, the magnitudes of the systematic (bias) components of these approximation errors are obscured by the magnitudes of their simulation components—even for n_s as small as 3. In fact, the number of replicate experiment test programs must be at least one million to obtain reasonably accurate estimates of the actual values for the systematic (bias) components of the respective approximation errors. (Microcomputer program *AVE3B2* pertains to one million replicate sets of n_s uniform pseudorandom numbers. Its example output appears in microcomputer file *AVE3B2*.)

```
C> AVE3B
```

Input a new set of three, three-digit odd seed numbers

```
371 583 915
```

Sum of n_s Uniform Pseudorandom Numbers	Approximation Error for the 0.90 Interval	Approximation Error for the 0.95 Interval	Approximation Error for the 0.99 Interval
3	−0.0019	0.0044	0.0069
4	0.0027	0.0040	0.0038
5	0.0001	0.0021	0.0028
6	0.0016	0.0037	0.0020
7	0.0022	0.0032	0.0016
8	−0.0028	−0.0020	−0.0008
9	−0.0043	0.0008	0.0026
10	−0.0017	−0.0016	−0.0004
11	0.0006	0.0001	0.0009
12	0.0046	0.0058	0.0025
13	0.0039	0.0001	−0.0002
14	−0.0041	0.0001	0.0019
15	−0.0030	−0.0029	0.0014

4.3.3. Simulation Study Three—Accuracy of the Normal Approximation for the Conceptual Sampling Distribution that Consists of All Possible Replicate Realization Values for the Statistic (the Sum of n_s Exponential Pseudorandom Numbers)

The PDF for the conceptual exponential distribution is expressed as $f(x) = \exp[(x + 1)]$ for $x > -1$ and $f(x) = 0$ elsewhere. As in Simulation Study Two, we translate and scale the conceptual exponential distribution such that the resulting actual values for the means and variances of the conceptual sampling distributions for the statistic (the sum of n_s exponential pseudorandom numbers) are zero and one, respectively. Thus, if the normal approximations to these conceptual statistical distributions are exact, 90% of our simulation-based sums will theoretically lie in the numerical interval from -1.645 to $+1.645$, 95% will theoretically lie in the numerical interval from -1.960 to $+1.960$, and 99% will theoretically lie in the numerical interval from -2.576 to $+2.576$. In turn, we (a) generate 10,000 replicate experiment test programs with n_s uniform pseudorandom numbers, for n_s from 3 to 15, and (b) compute the proportions of the respective statistics that actually lie in the three probability intervals of specific interest. Then, by appropriately subtracting $p = 0.90$, 0.95, or 0.99 from these proportions, we obtain a typical set of simulation-based estimates for the normal approximation errors. As evidenced by the example output from microcomputer program *AVE3C*, the magnitudes of the normal approximation errors are too large to be obscured by the magnitudes of their corresponding simulation components. Nevertheless it is clear that, for $p = 0.95$ and 0.99, the magnitudes of the normal approximation errors are less than 1%, even for n_s as small as 5. The actual values for the systematic (bias) components of these approximation errors can be estimated by examining several independent outputs from microcomputer program *AVE3C2* which pertains to one million replicate experiment test programs. An example output from microcomputer program *AVE3C2* appears in microcomputer file *AVE3C2*.

```
C> AVE3C
```

Input a new set of three, three-digit odd seed numbers

```
371 583 915
```

Sum of n_s Exponential Pseudorandom Numbers	Approximation Error for the 0.90 Interval	Approximation Error for the 0.95 Interval	Approximation Error for the 0.99 Interval
3	0.0315	0.0058	−0.0094
4	0.0315	0.0118	−0.0062
5	0.0224	0.0075	−0.0079
6	0.0187	0.0054	−0.0056
7	0.0172	0.0058	−0.0074
8	0.0133	0.0042	−0.0066
9	0.0118	0.0070	−0.0012
10	0.0158	0.0067	−0.0046
11	0.0129	0.0075	−0.0039
12	0.0139	0.0056	−0.0028
13	0.0091	0.0036	−0.0029
14	0.0074	0.0011	−0.0048
15	0.0022	0.0002	−0.0027

4.3.4. Cognate Simulation Studies

Given any of the alternative conceptual statistical distributions commonly employed in mechanical reliability, the normal approximation errors pertaining to the conceptual sampling distributions for the statistic (the sum of n_s replicate pseudorandom numbers) are intermediate in magnitude to those generated by running microcomputer programs *AVE3B* and *AVE3C*. Similarly, given the collection of alternative conceptual statistical distributions commonly employed in mechanical reliability, the normal approximation errors pertaining to the conceptual sampling distributions that consist of all possible replicate realization values for the statistic (the sum of n_s *assorted* pseudorandom numbers) are also intermediate in magnitude to the simulation errors generated by running microcomputer programs *AVE3B* and *AVE3C*. Moreover, this collection of alternative conceptual statistical distributions can even have markedly different actual values for their respective means and variances without noticeably affecting the magnitudes of the normal approximation errors.

4.3.5. Perspective

Whatever the underlying conceptual statistical distribution or distributions used to generate the pseudorandom numbers, our simulation studies demonstrate that the conceptual sampling distribution that consists of all possible replicate realization values for the statistic (the sum of n_s pseudorandom numbers) is accurately approximated by an appropriately scaled conceptual (two-parameter) normal distribution—even when n_s is as small as 6 to 8. This statistical behavior supports the fundamental presumption of normally distributed experiment test program datum values (experimental errors) that underlies classical statistical analyses.

This normality presumption has an important consequence in the classical statistical analysis of variance (Chapter 6). Recall that the respective [est(sc_j)]'s in the complete analytical model are computed using the (linear) expression:

$$est(sc_j) = \frac{\sum_{i=1}^{n_{dv}} c_{j,i} \cdot (\text{experiment test program datum value})_i}{\sum_{i=1}^{n_{dv}} c_{j,i}^2}$$

in which $c_{j,i}$ is the integer value for the ith element of the jth column vector in the n_s by n_s orthogonal augmented contrast array associated with the experiment test program. Thus, based on the presumption that each of the respective experiment test program datum values is normally distributed, under continual replication of the experiment test program the conceptual sampling distribution for the statistic [est(sc_j)] is an appropriately scaled conceptual (two-parameter) normal distribution. This assertion (presumption) provides a rational basis for calculating the classical (shortest) $100(scp)\%$ (two-sided) statistical confidence interval that allegedly includes the actual value for each sc_j of specific interest. The associated calculation is developed in Chapter 5 and is illustrated in Chapter 6. It is based on the fundamental variance expression:

$$var[est(sc_j)] = \frac{var(APRCRHNEE\text{'s})}{\sum_{i=1}^{n_{dv}} c_{j,i}^2}$$

in which *APRCRHNDEE*'s connotes all possible replicate conceptual random homoscedastic *normally distributed* (experiment test program) experimental errors.

4.4. CLOSURE

The experiment test program $CRHDV_i$'s are presumed to be normally distributed in the classical statistical analysis, termed analysis of variance, in Chapter 6. Thus, we first consider the conceptual (two-parameter) normal distribution and its associated conceptual sampling distributions in Chapter 5.

4.A. SUPPLEMENTAL TOPIC: RANDOMIZATION-BASED TEST OF THE NULL HYPOTHESIS OF INDEPENDENCE FOR PAIRED DATUM VALUES VERSUS THE ALTERNATIVE HYPOTHESIS OF A MONOTONIC ASSOCIATION

We now present a test of the null hypothesis that the paired datum values of specific interest are independent versus the composite alternative hypothesis that these paired datum values actually exhibit either a concordant or a discordant monotonic association. This statistical test is particularly relevant for a proper perspective regarding the respective simulation and approximation errors in our Section 4.3 simulation studies. Since, for each value of n_s, the respective simulation errors associated with the $p = 0.90, 0.95,$ and 0.99 intervals are based on pseudorandom numbers that were generated using the *same* set of three, three-digit odd seed numbers, it is reasonable to suspect that these errors are not independent, but rather are concordant pair-wise (the simple alternative hypothesis). Note that a concordant association is strongly suggested by a plot of the paired example microcomputer program *AVE3D* simulation errors pertaining to the $p = 0.95$ and 0.99 intervals, Figure 4.3.

This statistical test can also be used to examine the statistical credibility of the presumption that the respective est($CRHEE_i$'s) of specific interest are random, as opposed to being monotonically associated with (or systematically influenced by) some experiment test program variable or condition.

Suppose that the two sets of n_{pdv} paired datum values are indeed independent. Then, when the n_{pdv} datum values of data set one are reordered from smallest to largest, all sequences of the correspondingly reordered n_{pdv} datum values for data set two are equally likely to occur. Next, suppose that data set one is reordered from smallest to largest and that data set two is correspondingly reordered by maintaining the respective associations

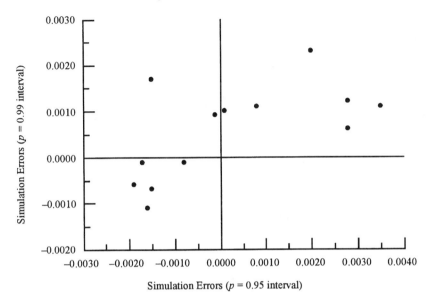

Figure 4.3 Plot of the paired example microcomputer program *AVE3D* simulation errors pertaining to the $p = 0.95$ and 0.99 intervals. Visual examination suggests a strong concordant association between the respective paired simulation errors. Given the null hypothesis that the signs and magnitudes of the respective paired simulation errors are independent versus the simple alternative hypothesis that their signs and magnitudes are concordant, randomization-based microcomputer program *RBKTAU* (Section 4.A.1) calculates the null hypothesis rejection probability as being equal to 0.0076 (0.0152/2). Thus, we statistically conclude that the respective simulation errors are not independent, but rather are indeed concordant.

between the paired datum values. If so, then Kendall's positive score test statistic, *kps,* is defined as the number of reordered datum values in data set two that exceed each respective reordered datum value. For example, consider the following hypothetical paired datum values:

data set one	data set two
1	2
4	3
3	4
2	1
5	5

As illustrated below, when the datum values of data set one are reordered from smallest to largest and the respective associations between the paired datum values in data sets one and two are maintained, the reordered datum values in data set two form a sequence that is convenient for calculating the data-based value of *kps*:

data set one	data set two
1	2
2	1
3	4
4	3
5	5

Now consider only reordered data set two. Datum value 2 is exceeded by subsequent ordered datum values 4, 3, and 5 (incremental $kps = 3$); datum value 1 is exceeded by subsequent ordered datum values 4, 3, and 5 (incremental $kps = 3$); datum value 4 is exceeded by subsequent ordered datum value 5 (incremental $kps = 1$); and datum value 3 is exceeded by subsequent ordered datum value 5 (incremental $kps = 1$). Accordingly, for this illustrative example, the data-based value of *kps* is equal to $3 + 3 + 1 + 1 = 8$. In turn, we calculate the corresponding value of the more intuitive Kendall's *tau* test statistic using the expression:

$$ktau = \frac{4 \cdot kps}{npdv \cdot (npdv - 1)} - 1$$

where *ktau*, by definition, lies in the interval from -1 to $+1$. When the datum values (or ranks) of data set two ascend in strict concordance with data set one, then $ktau = +1$. When the datum values (or ranks) of data set two descend in strict discordance with data set one, then $ktau = -1$.

> *Perspective*: Kendall's *tau* test statistic is defined such that its range, -1 to $+1$, is identical to the range of the well-known conceptual correlation coefficient (*ccc*) test statistic.

The absolute value of *ktau*, denoted abs(*ktau*), is equal to 0.6000 for our illustrative example. The exact enumeration-based probability that a randomly selected value of abs(*ktau*) for five paired datum values under continual replication will be greater than or equal to 0.6000 is equal to 0.2333 (28/120), whereas microcomputer program *RBKTAU* calculates the corresponding randomization-based probability as being equal to 0.2420 (see microcomputer computer file *RBKTDTA1*). Accordingly, we cannot rationally reject the null hypothesis of independence for our five example paired datum values.

> *Caveat*: It was presumed for convenience in writing microcomputer program *RBKTAU* that neither data set contains ties.

4.A.1. Numerical Example

Suppose that the paired datum values of specific interest are the respective example microcomputer program *AVE3D* simulation errors pertaining to the $p = 0.95$ and 0.99 intervals. These paired simulations errors (conveniently rescaled) appear in microcomputer file *RBKTDATA*. Microcomputer program *RBKTAU* calculates the randomization-based probability (0.0152) that a randomly selected set of 13 paired datum values will have its value of the abs(*ktau*) test statistic equal to or greater than the data-based value of the abs(*ktau*) test statistic for these paired datum values (0.5128). However, its null hypothesis rejection probability pertains to the composite alternative hypothesis that includes both concordant and discordant monotonic associations. In contrast, the simple alternative hypothesis relevant to these simulation errors pertains only to a concordant association. Thus, the appropriate randomization-based null hypothesis rejection probability is equal to 0.0076 (0.0152/2). Accordingly, we rationally opt to reject the null hypothesis that the respective simulation errors are independent and assert instead that these errors exhibit a concordant association.

```
C> TYPE RBKTDATA

13                Number of Paired Datum Values in Each Data Set

+28  +12          Corresponding Paired Datum Values
−08  −01
−15  +17
−01  +09
−16  −11
+28  +06
−19  −06
+20  +23
−17  −01
+35  +11
+01  +10
−15  −07
+08  +11
```

9999   Number of Randomly Reordered Sequences of the
                  Data Set Two Datum Values

237 755 913A   A New Set of Three, Three-Digit Odd Seed Numbers

(These paired datum values pertain to microcomputer program
AVE3D simulation errors for *scp* equal to 0.95 and 0.99.)

```
C> COPY RBKTDATA DATA

  1 file(s) copied

C> RBKTAU
```

This microcomputer program reordered the datum values in data set
two by reordering the datum values in data set one from smallest to
largest and maintaining the pairings of the respective datum values in
data sets one and two. The data-based value of Kendall's abs(*tau*) test
statistic for the 13 reordered datum values in data set two is equal to
0.5128.

Given the null hypothesis of independence for the paired datum values
of specific interest, this microcomputer program constructed 9999
equally-likely sets of 13 paired datum values by using uniform pseudorandom numbers to re-order only the datum values in data set two.
The number of these sets that had its Kendall's abs(*tau*) test statistic
value equal to or greater than 0.5128 is equal to 151. Thus, given the
null hypothesis of independence for the paired datum values of specific
interest, the randomization-based probability that a randomly selected
set of 13 paired datum values will have its Kendall's abs(*tau*) test
statistic value equal to or greater than 0.5128 is equal to 0.0152.
When this probability is sufficiently small, reject the null hypothesis
in favor of the composite alternative hypothesis that the paired datum
values of specific interest exhibit either a monotonic concordant or
discordant association.

Exercise Set 2

These exercises are intended to familiarize you with running program microcomputer *RBKTAU*.

    1. Run microcomputer program *RBKTAU* to test the independence of the example simulation-based normal approximation errors that were generated by running microcomputer program *AVE3A*. Discuss your results.

    2. Run microcomputer program *RANDOM1* with $n_{digit} = 8$ and $n_{elpri} = 8$ with two different sets of three, three-digit odd seed numbers to generate two *independent* data sets, each with eight pseudorandom two-digit integer numbers. Then, (a) examine the respective paired datum values relative to a monotonic association, and (b) each pseudorandom data set relative to a time-order-of-generation trend.

4.A.2. Additional Examples

The est($CRSIEE_i$'s) for the three respective numerical examples pertaining to the quantitative CRD experiment test program depicted in Figure 2.1 are plotted in Figure 4.4. Given the sequence of est($CRSIEE_i$'s) in Figure 4.4(a), microcomputer program *RBKTAU* calculates the randomization-based probability that a randomly selected equally-likely outcome of the example experiment test program when continually replicated will have its abs(*ktau*) equal to or greater than 1.0000 as being equal to 0.0001 (see microcomputer file *RBKTDTA2*). The corresponding enumeration-based probability is exactly equal to 0.00005 (2/40320). These probability values are so small that we must reject the null hypothesis that there is no time-order-of-testing effect for the est($CRSIEE_i$'s) in Figure 4.4(a). Rather, we assert that the respective magnitudes of these est($CRSIEE_i$'s) depend on the time order of testing. (This conclusion would be the subject of considerable test conduct concern if the example experiment test program datum values were not hypothetical.) Next, given the sequence of est($CRSIEE_i$'s) in Figure 4.4(b), microcomputer program *RBKTAU* calculates the randomization-based probability that a randomly selected equally-likely outcome of the example experiment test program when continually replicated will have its abs(*ktau*) equal to or greater than 0.1429 as being equal to 0.7128 (see microcomputer file *RBKTDTA3*). This probability value is so large that we have no objective reason to conclude that a time-order-of-testing association exists. Recall that pseudorandom numbers were used to generate the datum values whose est($CRSIEE_i$'s) are shown in Figure 4.4(b). Finally, given the sequence of est($CRSIEE_i$'s) in Figure 4.4(c), microcomputer program *RBKTAU* calcu-

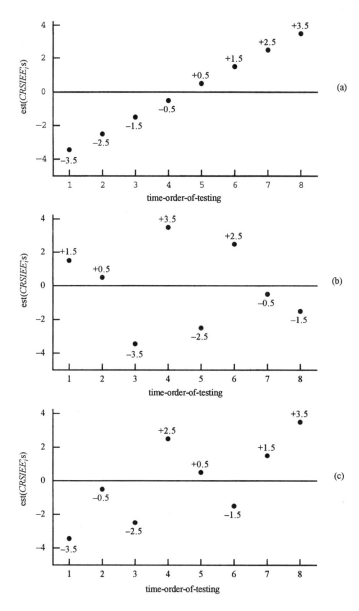

Figure 4.4 Plots of the est($CRSIEE_i$'s) versus time-order-of-testing for the three example (hypothetical) sets of datum values pertaining to the quantitative CRD experiment test program depicted schematically in Figure 2.1.

lates the randomization-based probability that a randomly selected equally-likely outcome of this example experiment test program when continually replicated will have its abs($ktau$) equal to or greater than 0.5714 as being equal to 0.0624 (see microcomputer computer file *RBKTDTA4*). This probability is small enough to warrant some concern that the test-specimen-blank location in the rod actually affects the test outcome. If so, there must also be concern regarding the physical interpretation of the data-based values for est($csdm$) and est[stddev($APRCRSIEE$'s)]. However, for this example experiment test program, recall that the test specimen blanks were randomly selected from the 16 test specimen blank locations in the rod. Accordingly, the experiment test program data-based value of Kendall's abs($ktau$) test statistic actually corresponds to a situation where, if we were to reject the null hypothesis of no test-specimen-blank-location effect on the est($CRSIEE_i$'s) in Figure 4.4(c), we would do so incorrectly, viz., we would commit a *Type I* error.

5

The Conceptual (Two-Parameter) Normal Distribution and the Associated Conceptual Sampling Distributions for Pearson's Central χ^2 (Chi Square), Snedecor's Central F, and Student's Central t Test Statistics

5.1. CONCEPTUAL (TWO-PARAMETER) NORMAL DISTRIBUTION

The PDF for the conceptual (two-parameter) normal distribution that consists of all possible replicate realization values for the random variable X is generically expressed as

$$f(x) = \frac{1}{\sqrt{2\pi} \cdot csp} \cdot \exp\left[-\frac{1}{2} \cdot \left(\frac{x - clp}{csp}\right)^2\right]$$

in which clp and csp denote conceptual location and scale parameters, and x is the continuous linear measurement metric. However, for a conceptual (two-parameter) normal distribution, the clp and the csp are, respectively, its mean and its standard deviation. Accordingly, the PDF for the conceptual (two-parameter) normal distribution that consists of all possible replicate realization values for the random variable X is almost always expressed as

$$f(x) = \frac{1}{\sqrt{2\pi} \cdot [\text{stddev}(X)]} \cdot \exp\left[-\frac{1}{2} \cdot \left(\frac{x - \text{mean}(X)}{\text{stddev}(X)}\right)^2\right]$$

This conceptual (two-parameter) normal distribution is depicted in Figure 5.1, where the magnitude stddev(X) establishes uniformly spaced tick-marks along the linear x abscissa.

Tedious numerical computations would be required to calculate each probability of specific interest for the conceptual (two-parameter) normal distribution (Figure 5.1). Rather, for computational purposes, this distribution is shifted and scaled to obtain the standardized conceptual normal distribution that consists of all possible replicate realization values for random variable Y. This conceptual statistical distribution is displayed in Figure 5.2. Then, when the linear X, Y relationship is used to establish equivalent events, the exact probability computations that pertain to the standardized conceptual normal distribution that consists of all possible replicate realization values for random variable Y also pertain to the conceptual (two-parameter) normal distribution that consists of all possible replicate realization values for random variable X.

The derivation of the standardized conceptual normal distribution that consists of all possible replicate realization values for random variable Y is straightforward. First, we subtract mean(X) from random variable X to generate a conceptual normal distribution that consists of all possible replicate realization values for the random variable $[X - \text{mean}(X)]$, whose mean is (now) equal to zero and whose standard deviation is (still) equal to the stddev(X). Then, this normally distributed random variable $[X - \text{mean}(X)]$ is divided by stddev(X) to generate the standardized conceptual normal distribution that consists of all possible replicate realization values for

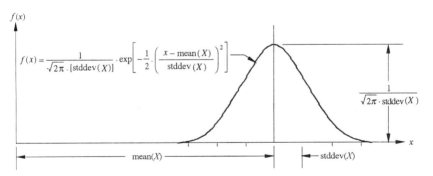

Figure 5.1 PDf for the conceptual (two-parameter) normal distribution that consists of all possible replicate realization values for random variable X.

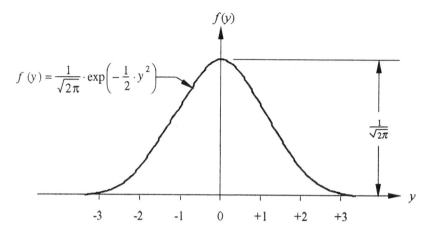

Figure 5.2   PDF for the standardized conceptual normal distribution that consists of all possible replicate realization values for random variable Y. The actual values for the mean and the variance of the standardized conceptual normal distribution are, respectively, zero and one.

random variable $Y = [X - \text{mean}(X)]/\text{stddev}(X)\}$, where mean($Y$) is equal to zero and stddev(Y) is equal to one. The associated X, Y relationship is clearly linear, viz.,

$$\frac{X - \text{mean}(X)}{\text{stddev}(X)} = Y = a + bX$$

in which $b = \{(1)/[\text{stddev}(X)]\}$ and $a = \{[\text{mean}(X)]/[\text{stddev}(X)]\}$. Micro-computer program PY, given any numerical input value for standardized conceptual normal distribution variate y_p of specific interest, computes the numerical value for the probability p that a randomly selected realization value for the standardized conceptual normal distribution random variable Y will be less than y_p. Analogously, microcomputer program YP, given any numerical input value for the probability p of specific interest, computes the numerical value of the standardized conceptual normal distribution variate y_p such that the probability is exactly equal to p that a randomly selected realization value for the standardized normal distribution random variable Y will be less than y_p.

```
C> PY
```

Input the value of specific interest for the variate $y(p)$ of the standardized conceptual normal distribution that consists of all possible replicate realization values for the random variable Y

```
0.0
```

The probability that a realization value randomly selected from the standardized conceptual normal distribution that consists of all possible replicate realization values for the random variable Y will be less than 0.0000 equals 0.5000

```
C> PY
```

Input the value of specific interest for the variate $y(p)$ of the standardized conceptual normal distribution that consists of all possible replicate realization values for the random variable Y

```
1.0
```

The probability that a realization value randomly selected from the standardized conceptual normal distribution that consists of all possible replicate realization values for the random variable Y will be less than 1.0000 equals 0.8413

```
C> PY
```

Input the value of specific interest for the variate $y(p)$ of the standardized conceptual normal distribution that consists of all possible replicate realization values for the random variable Y

```
2.0
```

The probability that a realization value randomly selected from the standardized conceptual normal distribution that consists of all possible replicate realization values for the random variable Y will be less than 2.0000 equals 0.9772

```
C> PY
```

Input the value of specific interest for the variate $y(p)$ of the standardized conceptual normal distribution that consists of all possible replicate realization values for the random variable Y

```
3.0
```

The probability that a realization value randomly selected from the standardized conceptual normal distribution that consists of all possible replicate realization values for the random variable Y will be less than 3.0000 equals 0.9987

```
C> YP
```

Input the probability p of specific interest that a realization value randomly selected from the standardized conceptual normal distribution that consists of all possible replicate realization values for the random variable Y will be less than $y(p)$

```
0.5
```

The value for $y(0.5000)$ equals 0.0000

```
C> YP
```

Input the probability p of specific interest that a realization value randomly selected from the standardized conceptual normal distribution that consists of all possible replicate realization values for the random variable Y will be less than $y(p)$

```
0.9
```

The value for $y(0.9000)$ equals 1.2816

```
C> YP
```

Input the probability p of specific interest that a realization value randomly selected from the standardized conceptual normal distribution that consists of all possible replicate realization values for the random variable Y will be less than $y(p)$

```
0.95
```

The value for $y(0.9500)$ equals 1.6449

```
C> YP
```

Input the probability p of specific interest that a realization value randomly selected from the standardized conceptual normal distribution that consists of all possible replicate realization values for the random variable Y will be less than $y(p)$

```
0.99
```

The value for $y(0.9900)$ equals 2.3263

5.1.1. Example Probability Calculation Pertaining to the Conceptual Normal Distribution that Consists of All Possible Replicate Realization Values for Random Variable X

First, we assert that the conceptual normal distribution that consists of all possible replicate realization values for random variable X and the associated standardized conceptual normal distribution that consists of all possible replicate realization values for random variable Y pertain to equivalent x,y events when

$$y = \frac{x - \text{mean}(X)}{\text{stddev}(X)} = a + bx$$

where

$$a = -\frac{\text{mean}(X)}{\text{stddev}(X)} \quad \text{and} \quad b = \frac{1}{\text{stddev}(X)}$$

Accordingly, when

$$y_{lower} < Y < y_{upper}$$

then

$$x_{lower} < X < x_{upper}$$

The equivalence of these events is easily demonstrated geometrically by plotting the straight line $y = a + bx$ relationship and noting that x_{lower} determines y_{lower}, and x_{upper} determines y_{upper}.

Now suppose that (a) mean(X) is equal to 11.15 and that stddev(X) is equal to 6.51, and (b) the probability of specific interest is the probability that a randomly selected realization value for random variable X will lie in numerical interval [10.07,13.59]. Then, $x_{lower} = 10.07$ and $x_{upper} = 13.59$. Accordingly,

$$y_{lower} = \frac{10.07 - 11.15}{6.51} = -0.1659$$

and

$$y_{upper} = \frac{13.59 - 11.15}{6.51} = +0.3748$$

In turn, successively running microcomputer program PY generates the results:

$$\text{probability}[Y < -0.1659] = 0.4341$$

and

$$\text{probability}[Y < +0.3748] = 0.6461$$

Hence,

$$\text{probability}[-0.1659 < Y < +0.3748] = 0.2120$$

Finally, because the respective probabilities pertaining to equivalent events are identical, the probability that a randomly selected realization value for the random variable Y will lie in the numerical interval [$-0.1659, +0.3748$] exactly equals the probability that a randomly selected realization value for random variable X will lie in the numerical interval [10.07,13.59]. Accordingly,

$$\text{probability}[10.07 < X < 13.59] = 0.2120$$

5.2. PLOTTING REPLICATE DATUM VALUES ON PROBABILITY PAPER

Probability papers are widely used to plot replicate datum values that are presumed (alleged) to have been randomly selected from a known conceptual statistical distribution. However, there is no actual need for commercial probability paper. Any plot of the straight-line relationship $y = a + bx$ suffices. Accordingly, it is not the construction of probability paper that is an issue, rather it is the method of plotting replicate datum values on probability paper that requires explanation.

Both the respective abscissa (x) and ordinate (y) metric values must be known to plot replicate datum values on probability paper. The respective abscissa metric values are the replicate datum values. The corresponding *ordered* ordinate metric values are called plotting positions. Approximate $p(pp)_i$ plotting positions were originally used because exact $y(pp)_i$ plotting positions could not be computed (except in special situations). However, experience now indicates that, for practical values of n_{dv}, no plotting position, exact or approximate, is adequate to assert on the basis of visual inspection that the given replicate datum values can rationally be viewed as having been randomly selected from some alleged conceptual statistical distribution. Accordingly, the choice among alternative plotting positions is not a major issue.

Because the $p(pp)_i$ plotting position metric is nonlinear for all probability papers other than for the uniform distribution, the direct use of $p(pp)_i$ plotting positions introduces two problems. First and foremost, the quantitative assessment of the estimated slope of the CDF is clouded. (This assessment requires understanding how the given probability paper is actually constructed.) Second, the use of $p(pp)_i$ plotting positions requires interpolation along a nonlinear metric. Thus, we strongly recommend against the use of commercial probability paper whatever the presumed conceptual statistical distribution.

5.2.1. Normal Probability Paper

The construction (and proper documentation) of normal probability paper is illustrated in Figure 5.4. (The construction of probability papers for other distributions is discussed later.) For replicate datum values that are presumed to be normally distributed, we recommend using Blom's $p(pp)_i$ plotting position expression:

$$p(pp)_i = \frac{i - 3/8}{n_{dv} + 1/4}$$

and successively converting the respective $p(pp)_i$ plotting positions to their corresponding $y(pp)_i$ plotting positions by running microcomputer program *YP*.

5.3. SIMULATION STUDY OF NORMALLY DISTRIBUTED PSEUDORANDOM DATUM VALUES PLOTTED ON NORMAL PROBABILITY PAPER

Microcomputer *SIMNOR* (presented later) was successively run with n_{dv} equal to 9, 19, 49, and 99 to generate the pseudorandom normally distributed datum values plotted on commercial normal distribution probability paper in Figures 5.3(a)–(d), respectively. Clearly, these normally distributed pseudorandom datum values do not plot exactly on the conceptual CDF. However, in the limit as n_{dv} increases without bound, the deviations of the respective plotted datum values from the conceptual CDF will theoretically diminish to zero. Note, however, that for n_{dv} as large as 99, the plotted pseudorandom datum values at the respective tails (extreme percentiles) still deviate markedly from the conceptual CDF. This statistical behavior prevails even for much larger values for n_{dv}.

Next, consider the 12 sets of $n_{dv} = 10$ normally distributed pseudorandom datum values plotted on properly documented normal probability paper in Figure 5.4. Note that, given the respective approximate $y(pp)_i$ plotting positions computed by microcomputer program *SIMNOR*, neither the nonlinear p scale nor its directly associated y scale have intrinsic value and thus need not be included when constructing normal probability paper.

The pseudorandom datum values plotted in Figure 5.4 illustrate typical random variability that occurs when $n_{dv} = 10$ replicate datum values are randomly selected from a conceptual (two-parameter) normal distribution. Accordingly, given (a) the conceptual distribution, (b) its parameter values, and (c) the sample sizes of specific interest, simulation-based plots of this type provide valuable perspective regarding how much the estimated CDF can differ from the conceptual CDF and how much plotted datum values can differ from either the conceptual or the estimated CDF. No other statistical methodology generates as much information with so little effort.

The estimated CDF's plotted in Figure 5.4 are computed using data-based estimates of the actual values for the mean and the standard deviation of the conceptual (two-parameter) normal distribution that consists of all possible realization values for the random variable X. These data-based estimates were computed using the following expressions (page 189):

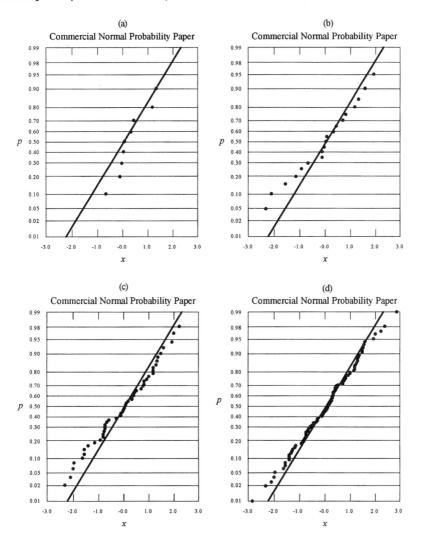

Figure 5.3 Plots of 9, 19, 49, and 99 pseudorandom normally distributed datum values generated by running microcomputer program *SIMNOR* with mean(X) equal to zero and stddev(X) equal to one (using the *same* set of three, three-digit odd seed numbers). The so-called mean plotting position $p(pp)_i = 1/(n_{dv} + 1)$ was used in plotting these datum values on commercial probability paper because of its convenience. The *sigmoidal* CDF for the conceptual (two-parameter) normal distribution plots as a *straight line* on normal probability paper, when abscissa x is the *linear* metric for random variable X and p is the *nonlinear* ordinate, where p is the probability that a randomly selected realization value for the normally distributed random variable X will be less than x.

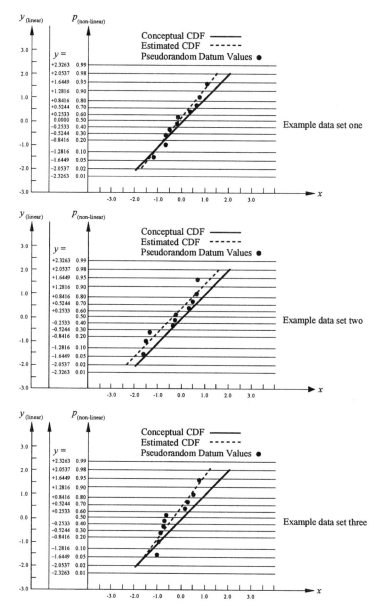

Figure 5.4 Twelve sets of 10 normally distributed pseudorandom datum values plotted on properly documented normal probability paper. Each pseudorandom data set was generated by running microcomputer program *SIMNOR* with mean(X) equal to zero and stddev(X) equal to one, but using a *different* set of three, three-digit odd seed numbers.

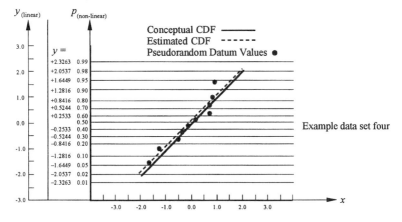

Example data set four

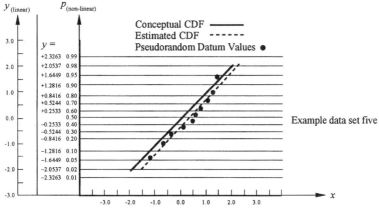

Example data set five

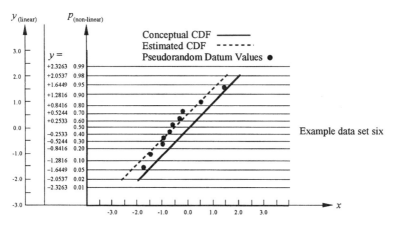

Example data set six

Figure 5.4 (*continued*)

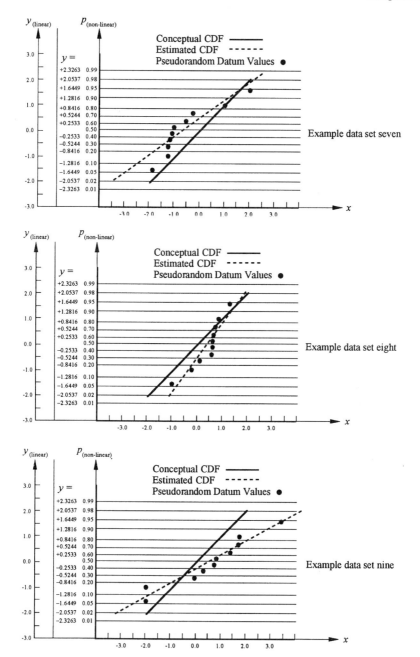

Figure 5.4 (*continued*)

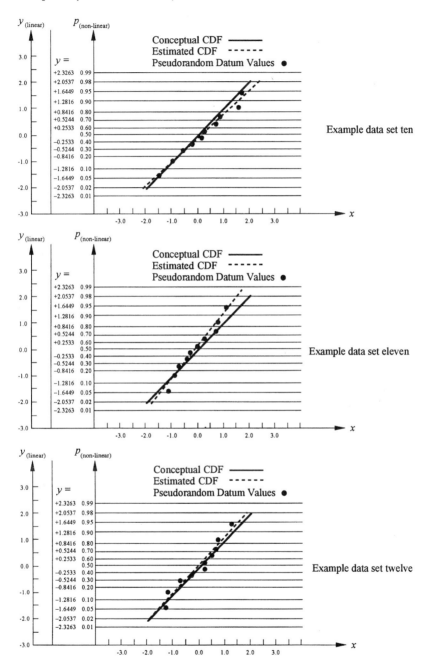

Figure 5.4 (*continued*)

$$\text{est[mean}(X)] = \text{arithmetic average}\,(x\text{'s}) = \text{ave}(x\text{'s}) = \frac{\displaystyle\sum_{i=1}^{n_{dv}} x_i}{n_{dv}}$$

and

$$\text{est[stddev}\,(X)] = \left(\sum_{i=1}^{n_{dv}} \frac{[x_i - \text{ave}(x\text{'s})]^2}{n_{dv} - 1}\right)^{1/2}$$

Section 5.4 presents a discussion of the probability behavior of the associated statistical estimators for the parameters of a conceptual (two-parameter) normal distribution.

Figure 5.5 summarizes the 12 estimated CDF's plotted in Figure 5.4. When numerous replicate estimated CDF's are summarized in a plot similar to that of Figure 5.5, their envelope resembles a hyperbolic-shaped region. Clearly, the variability of these replicate estimated CDF's is smallest at mean(X) and is largest at their extreme percentiles. This variability characterization is valid for all estimated CDF's of specific interest in mechanical reliability. It is the fundamental reason why we compare the respective actual values for the means of conceptual statistical and sampling distributions rather than comparing their respective extreme percentiles, viz., the variability of their respective estimated CDF's at extreme percen-

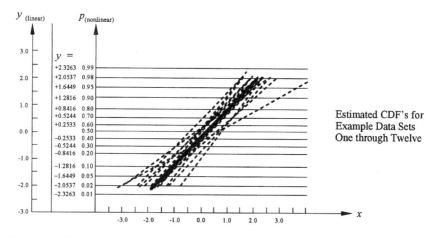

Figure 5.5 Summary of the respective estimated CDF's pertaining to 12 sets of pseudorandom normally distributed datum values plotted in Figure 5.4. Observe that the variability of the percentiles of the respective estimated CDF's is smallest near $p = 0.5$ and is largest at very high and low values of $p(y)$.

tile is too large to permit precise statistical comparisons for practical experiment test program sizes.

5.3.1. Discussion

The typical variabilities displayed in Figures 5.3(a) through 5.5 for normally distributed datum values are intended to convey another fundamental notion. The variability among numerous small sets of replicate datum values is much too large to allow us to use a plot of replicate datum values on normal probability paper to discern visually whether these datum values can rationally be presumed to be normally distributed. In fact, the relevant issue is not whether experiment test program datum values are normally distributed. (Recall that conceptual statistical distributions are merely mathematical abstractions.) Rather, the relevant issue is whether the apparent departures from normality are sufficiently pronounced to affect the statistical credibility of a probability calculation that is based on the presumption of normality. Accordingly, we should always test the null hypothesis of normality (Section 5.5). Then, when the null hypothesis of normality is rationally rejected, we have a statistical basis for concern regarding the credibility of a probability calculation that is based on the presumption of normality.

```
C> SIMNOR

Input a new set of three, three-digit odd seed numbers

273 697 353

Input the sample size of specific interest (maximum size = 1000)

10

Input the conceptual normal distribution mean of specific interest

0.0

Input the conceptual normal distribution standard deviation of speci-
fic interest
```

1.0

Ordered normal datum values	Blom's $p(pp)_i$ plotting position	Corresponding $y(pp)_i$ plotting position
−0.8748	0.0610	−1.5466
−0.6126	0.1585	−1.0005
−0.1048	0.2561	−0.6554
−0.0220	0.3537	−0.3755
0.0654	0.4512	−0.1226
0.0839	0.5488	0.1226
0.3268	0.6463	0.3755
0.4248	0.7439	0.6554
1.1445	0.8415	1.0005
1.2803	0.9390	1.5466

The estimated conceptual normal distribution mean is equal to 0.1712

The estimated conceptual normal distribution standard deviation is equal to 0.6755

Exercise Set 1

These exercises are intended to enhance your intuition regarding the obvious discrepancies between plotted normally distributed pseudorandom datum values and the associated conceptual (two-parameter) normal distribution.

In the exercises that follow, run microcomputer program *SIMNOR* to generate normally distributed pseudorandom datum values. Then, plot these datum values on ordinary graph paper using $y(pp)_i$ plotting positions corresponding to Blom's $p(pp)_i$ approximate plotting positions.

1. (a) Construct 12 plots similar to the 12 plots in Figure 5.4, by successively running program *SIMNOR* 12 times to generate 12 sets of 10 pseudorandom normal datum values. For each set of these pseudorandom normal datum values, let mean(X) be equal to zero and stddev(X) be equal to one. Then, (b) construct a summary plot for the respective estimated CDF's as illustrated in Figure 5.5.
2. (a) Randomly select one of the plots in Exercise 1. Then, (b) using the same set of three, three-digit odd seed numbers that was used

to generate the datum values for this plot, select markedly different values for mean(X) and stddev(X) and generate 10 pseudo-random normal datum values. In turn, (c) plot these datum values as in Exercise 1. Finally, (d) discuss the effect of the values selected for mean(X) and stddev(X) on the deviations of these datum values from their respective estimated CDFs.

5.4. ESTIMATING MEAN(X) AND STDDEV(X) GIVEN REPLICATE DATUM VALUES RANDOMLY SELECTED FROM A CONCEPTUAL (TWO-PARAMETER) NORMAL DISTRIBUTION

Statistical theory indicates that given n_{dv} datum values randomly selected from a conceptual (two-parameter) normal distribution that consists of all possible replicate realizations of random variable X, the best statistical estimator for mean(X) is the arithmetic average of the respective experiment test program realization values, viz.,

$$\text{best statistical estimator for mean}(X) = \text{est}[\text{mean}(X)]$$

$$= \text{arithmetic average}(X\text{'s})$$

$$= \text{ave}(X\text{'s}) = \frac{\displaystyle\sum_{i=1}^{n_{dv}} X_i}{n_{dv}}$$

in which the term best explicitly connotes that est[mean(X)] is unbiased and that the actual value for the variance of its conceptual sampling distribution is less than the actual value for variance of the conceptual sampling distribution for any alternative statistical estimator.

The selection of the *appropriate* statistical estimator for stddev(X) depends on the application of specific interest. In mechanical reliability applications, stddev(X) is estimated as the square root of the unbiased generic statistical estimator for var(X), where

$$\text{unbiased generic statistical estimator for var}(X) = \text{est}[\text{var}(X)]$$

$$= \frac{\displaystyle\sum_{i=1}^{n_{dv}} [X_i - \text{ave}(X\text{'s})]^2}{n_{dv} - 1}$$

Unfortunately, because a square root is a nonlinear transformation, the corresponding generic est[stddev(X)] is biased.

We now try to provide intuition regarding the reason that its $(n_{dv} - 1)$ divisor makes the generic statistical estimator for var(X) unbiased (regardless of the actual conceptual statistical distribution for X). Accordingly, we first establish the actual value for the variance of the conceptual sampling distribution that consists of all possible replicate realization values for the statistic $[X_i - \text{ave}(X\text{'s})]$. Consider the ith row in the following estimated conceptual statistical model stated in hybrid column vector notation:

$$
\begin{vmatrix} X_1 \\ X_2 \\ \vdots \\ X_i \\ \vdots \\ X_{n_{dv}} \end{vmatrix} = \begin{vmatrix} \text{ave}(X\text{'s}) \\ \text{ave}(X\text{'s}) \\ \vdots \\ \text{ave}(X\text{'s}) \\ \vdots \\ \text{ave}(X\text{'s}) \end{vmatrix} + \begin{vmatrix} [X_1 - \text{ave}(X\text{'s})] \\ [X_2 - \text{ave}(X\text{'s})] \\ \vdots \\ [X_i - \text{ave}(X\text{'s})] \\ \vdots \\ [X_{n_{dv}} - \text{ave}(X\text{'s})] \end{vmatrix}
$$

Clearly, the actual value for the variance of the conceptual statistical distribution that consists of all possible replicate realization values for the random variable X is equal to var(X). In turn, the actual value for the variance of the conceptual sampling distribution that consists of all possible replicate realization values for the statistic $[\text{ave}(X\text{'s})]$ is equal to $[\text{var}(X)]/n_{dv}$. However, since the ave(X's) column vector is orthogonal to the $[X_i - \text{ave}(X\text{'s})]$ column vector, the associated statistics are independent and the variances of their respective conceptual sampling distributions sum algebraically. Thus, the actual value for the variance of the conceptual sampling distribution that consists of all possible replicate realization values for the statistic $[X_i - \text{ave}(X\text{'s})]$ must be equal to var(X) times the (statistical bias) factor $[(n_{dv} - 1)/n_{dv}]$.

Next, if mean(X) were known, we would estimate the actual value for the variance of the conceptual statistical distribution that consists of all possible replicate realization values for the random variable X as the sum of squares of all $[X_i - \text{mean}(X)]$'s divided by n_{dv}, where n_{dv} is the number of statistical degrees of freedom for the X_i column vector. However, because mean(X) is unknown we have no rational alternative other than to estimate the actual value for the variance of the conceptual statistical distribution that consists of all possible replicate realization values for the random variable X as the sum of squares of the realization values for the respective $\{X_i - \text{est}[\text{mean}(X)]\}$'s divided by n_{dv}, in which ave(X's) is substituted for est[mean(X)]. However, by definition of ave(X's), the sum of squares of the realization values for the respective $[X_i - \text{ave}(X\text{'s})]$'s is always less than the sum of squares of the realization values for the respective $[X_i - \text{mean}(X)]$'s. Accordingly, var(X) is always statistically *underestimated* (statistically biased) unless we multiply it by the inverse of the (statistical bias) factor $[(n_{dv}-1)/n_{dv}]$. The algebraic consequence of employing this multiplicative statistical-bias-correction factor is that an unbiased statistical estimator for var(X) is obtained by dividing the sums of squares of the realization values for the respective $[X_i - \text{ave}(X\text{'s})]$'s by $(n_{dv} - 1)$, the number of statistical degrees of freedom for the $[X_i - \text{ave}(X\text{'s})]$'s column vector. This result is a special case of the following general rule that pertains to each of the orthogonal estimated statistical models considered herein:

> *General Rule*: When the expected value for each element of a column vector is equal to zero, the sum of the squares of the respective elements in this column vector, divided by its number of statistical degrees of freedom, is a statistically unbiased estimator of var($APRCRHNDEE's$).

5.5. TESTING NORMALITY FOR REPLICATE DATUM VALUES

There are several statistical tests for which the null hypothesis is that the replicate (presumed replicate) datum values of specific interest were randomly drawn from a conceptual (two-parameter) normal distribution and the omnibus alternative hypothesis is that these replicate (presumed replicate) datum values pertain to some other (unspecified) conceptual statistical distribution. Unfortunately, even the best of these tests (with greatest statistical power) cannot reliably detect non-normality for the (small) sample sizes typically used in mechanical reliability tests. Nevertheless, whenever normality is presumed in statistical analysis, we are obliged to examine the credibility of this presumption. Accordingly, we now present a microcomputer program employing a test statistic that can be used to test the null hypothesis that the given replicate (presumed replicate) datum values were randomly selected from a conceptual (two-parameter) normal distribution. This test statistic was chosen because it can also be employed to test the null hypothesis of normality for the respective $CRHNDEE_i's$ associated with statistical models pertaining to equally replicated CRD and unreplicated RCBD and SPD experiment test programs (Chapter 6).

Perhaps the most intuitive way to test normality is to plot the presumed normally distributed datum values on normal probability paper as in Figure 5.4 and then use the magnitude of the maximum deviation (md) of a plotted datum value from the estimated conceptual normal CDF as the test statistic. The problem with this intuitive test statistic is that, as evident in Figure 5.4, the largest deviation seldom occurs near the middle of the estimated distribution, but rather almost always occurs at or near the extremes of the estimated distribution. Michael (1983) used an empirical arc-sine transformation to generate a stabilized probability plot (*spp*) that mitigates this problem.

Microcomputer program *NORTEST* pertains to a modified version of Michael's D_{spp} test statistic that employs Blom's approximate plotting position $p(pp)_i = (i - 3/8)/(n_{dv} + 1/4)$ and the classical estimators for the mean and variance of a conceptual (two-parameter) normal distribution (Section 5.4). This simulation-based microcomputer program first computes the value for the modified *MDSPP* test statistic given the n_{rep} replicate datum values of specific interest. It then generates n_{sim} sets of n_{rep} replicate datum values from a conceptual normal distribution and establishes the proportion of these n_{sim} data sets whose modified *MDSPP* test statistic realization value exceeds the modified *MDSPP* test statistic value pertaining to the n_{rep} replicate datum values of specific interest. Presuming that this proportion is reduced when the n_{rep} replicate datum values of specific interest are not normally distributed, we rationally opt to reject the presumption of normality when the simulation-based probability computed by running microcomputer program *NORTEST* is sufficiently small.

5.5.1. Normality Test Example

First, run microcomputer program *NOR* (page 196) to generate 10 pseudorandom normally distributed replicate datum values. Then, augment these data as indicated in microcomputer file *ANORDATA*. In turn, run microcomputer program *NORTEST* to compute the data-based value for the modified *MDSPP* test statistic (0.0541) and the simulation-based probability that this test statistic value will be equal to or larger than 0.0541 (0.9488) given the null hypothesis of normality. Since this probability is not small, viz., less than either 0.10, 0.05, or 0.01, we must rationally opt not to reject the null hypothesis of normality. Clearly, our decision is correct—because all of the pseudorandom datum values generated by program *NOR* are normally distributed. However, we cannot know for certain whether our test of hypothesis decision is correct when the replicate (presumed replicate) datum values of specific interest are not known to be normally distributed. Rather, we must view the correctness of our test of hypothesis decision in the context of a continually replicated experiment test program, viz., in terms of the probability that our decision is actually correct.

Recall that a *Type I* error is committed when we incorrectly opt to reject a null hypothesis that is correct. Recall also that it is not practical to select the probability of committing a *Type I* error to be extremely small because this selection reduces the probability of correctly rejecting the null hypothesis when the alternative hypothesis is correct. The latter probability is called statistical power. It is important to estimate (and compare) the statistical power for each test statistic of potential interest. Fortunately, the simulation-based empirical sampling distribution that consists of all possible replicate realization values for the test statistic of specific interest can also be generated for the alternative hypothesis of specific interest. Thus, the simulation-based statistical power of the test of the null hypothesis can be established relative to this alternative hypothesis. Typically, the statistical power of a test of the null hypothesis is low for small values of n_{dv}, sometimes barely larger than the acceptable probability of committing a *Type I* error, but it increases markedly as n_{dv} increases.

```
C> NOR

Input a new set of three, three-digit odd seed numbers

345 761 799

Input the sample size of specific interest (maximum size = 1000)

10

Input the conceptual normal distribution mean of specific interest

10.0

Input the conceptual normal distribution standard deviation of speci-
fic interest

2.0

       6.86091450498240
       7.89512237885230
       8.82680050201190
       9.31976099687290
       9.52103697451820
      10.26841061562700
      10.41102005946200
      11.34970238520000
      11.75034014152200
      13.91231724763700
```

```
C>COPY ANORDATA DATA

  1 file(s) copied

C>NORTEST
```

The data-based value of the modified *MDSPP* test statistic for the 10 replicate datum values of specific interest is equal to 0.0541.

This microcomputer program generated 10,000 sets of 10 normally distributed replicate datum values. The number of these sets that had its modified *MDSPP* test statistic value equal to or greater than 0.0541 is equal to 9488. Thus, given the null hypothesis of normality, the simulation-based probability that a randomly selected set of 10 replicate datum values will have its modified *MDSPP* test statistic value equal to or greater than 0.0541 is equal to 0.9488. When this probability is sufficiently small, reject the null hypothesis in favor of the alternative hypothesis of non-normality.

```
C> TYPE ANORDATA
```

10	n_{rep}, the Number of Replicate Datum Values of Specific Interest

6.86091450498240
7.89512237885230
8.82680050201190
9.31976099687290
9.52103697451820
10.26841061562700
10.41102005946200
11.34970238520000
11.75034014152200
13.91231724763700

10000	n_{sim}, the Number of Pseudorandom Replicate Normally Distributed Data Sets, Each of Size n_{rep}

879	247	751	A New Set of Three, Three-Digit Odd Seed Numbers

These datum values are the datum values generated by microcomputer program *NOR*, augmented as indicated

Exercise Set 2

This exercise is intended to familiarize you with testing the null hypothesis of normality for a set of n_{rep} replicate (presumed replicate) normally distributed datum values using the modified *MDSPP* test statistic. When the normality of normally distributed datum values is tested, we commit a *Type I* error each time that we (incorrectly) opt to reject normality.

> Run microcomputer program *NOR* to generate 5, 10, 20, 50, and 100 pseudorandom replicate datum values from a conceptual normal distribution, using the same set of three, three-digit odd seed numbers. Then, augment each respective set of normally distributed pseudorandom replicate datum values as illustrated in microcomputer file *ANORDATA*. In turn, run microcomputer program *NORTEST* to test the normality of each respective set of normally distributed pseudorandom replicate datum values using an acceptable probability of committing a *Type I* error equal to 0.05. (a) Let mean(X) be equal to 10 and stddev(X) equal to 2. (b) Let mean(X) be equal to 0 and stddev(X) equal to 1. Compare the respective sets of results. What do you conclude regarding the effect of different values for mean(X) and var(X) for this statistical test of normality?

Exercise Set 3

These exercises are intended to provide perspective regarding the statistical power of the modified *MDSPP* test statistic. Statistical power is the complement of the probability of committing a *Type II* error, where a *Type II* error is defined as (incorrectly) failing to reject the null hypothesis when the alternative hypothesis is correct. Accordingly, when the normality of replicate non-normally distributed datum values is tested, we commit a *Type II* error each time that we incorrectly fail to reject normality.

Microcomputer programs *UNI*, *LOG*, *SEV*, and *LEV* respectively generate n_{rep} replicate pseudorandom datum values from the conceptual (two-parameter) uniform distribution (symmetrical PDF), the conceptual (two-parameter) logistic distribution (symmetrical PDF), the conceptual (two-parameter) smallest-extreme-value distribution (asymmetrical PDF, skewed to the left), and the conceptual (two-parameter) largest-extreme-value distribution (asymmetrical PDF, skewed to the right). In addition, microcomputer programs *LNOR* and *WBL* respectively generate n_{rep} replicate pseudorandom datum values from the conceptual two-parameter $\log_e$-normal distribution and the conceptual two-parameter Weibull distribution. (We explain how to generate pseudorandom datum values from these conceptual two-parameter distributions in Supplemental Topic 8.D.)

1. Run either (a) microcomputer program *UNI* or (b) microcomputer program *LOG* with the same set of three, three-digit odd seed numbers to generate 5, 10, 20, 50, and 100 replicate pseudorandom datum values from (a) a conceptual (two-parameter) uniform distribution or (b) a conceptual (two-parameter) logistic distribtion with its mean equal to 10 and its standard deviation equal to 2. Run microcomputer program *NORTEST* to test the null hypothesis of normality for each respective set of pseudorandom datum values using an acceptable probability of committing a *Type I* error equal to 0.05. Is normality properly rejected in each case? Discuss your results.

2. Run microcomputer program *SEV* with the same set of three, three-digit odd seed numbers to generate 5, 10, 20, 50, and 100 replicate pseudorandom datum values from a conceptual (two-parameter) smallest-extreme-value distribution with its mean equal to 10 and its standard deviation equal to 2. Run microcomputer program *NORTEST* to test the null hypothesis of normality for each respective set of pseudorandom datum values using an acceptable probability of committing a *Type I* error equal to 0.05. Is normality properly rejected in each case? Discuss your results.

3. Run microcomputer program *LEV* with the same set of three, three-digit odd seed numbers to generate 5, 10, 20, 50, and 100 replicate pseudorandom datum values from a conceptual (two-parameter) largest-extreme-value distribution with its mean equal to ten and its standard deviation equal to 2. Run microcomputer program *NORTEST* to test the null hypothesis of normality for each respective set of pseudorandom datum values using an

acceptable probability of committing a *Type I* error equal to 0.05. Is normality properly rejected in each case? Discuss your results.

4. Run microcomputer program *LNOR* with the same set of three, three-digit odd seed numbers to generate the 5, 10, 20, 50, and 100 replicate pseudorandom datum values from a conceptual (two-parameter) $\log_e$-normal distribution whose mean is equal to 10 and whose standard deviation is equal to 2. Run microcomputer program *NORTEST* to test the null hypothesis of normality for each respective set of pseudorandom datum values using an acceptable probability of committing a *Type I* error equal to 0.05. Is normality properly rejected in each case? Discuss your results.

5. Run microcomputer program *WBL* with the same set of three, three-digit odd seed numbers to generate 5, 10, 20, 50, and 100 replicate pseudorandom datum values from a conceptual (two-parameter) Weibull distribution whose mean is equal to 10 and whose standard deviation is equal to 2. Run microcomputer program *NORTEST* to test the null hypothesis of normality for each respective set of pseudorandom datum values using an acceptable probability of committing a *Type I* error equal to 0.05. Is normality properly rejected in each case? Discuss your results.

5.5.2. Discussion

Hopefully, it is now clear that, given a test statistic whose simulation-based empirical sampling distribution does not depend of the actual values for the parameters of the conceptual distribution used to generate the n_{sim} replicate sets of n_{rep} replicate pseudorandom datum values, we can test the null hypothesis that any given (uncensored) set of replicate (presumed replicate) datum values was randomly selected from some known conceptual statistical distribution. Moreover, we can establish the statistical power of this test of the null hypothesis relative to the specific alternative hypothesis that the given (uncensored) set of n_{rep} replicate (presumed replicate) datum values was actually randomly selected from some other known conceptual statistical distribution.

5.6. CONCEPTUAL SAMPLING DISTRIBUTIONS FOR STATISTICS BASED ON DATUM VALUES RANDOMLY SELECTED FROM A CONCEPTUAL (TWO-PARAMETER) NORMAL DISTRIBUTION

Exact analytical expressions are known for the conceptual sampling distribution PDF's presented in this section. Nevertheless, it is informative to adopt a simulation-based perspective regarding these sampling distributions—because this simulation process intuitively underlies the proper statistical interpretation of data-based realization values that are presumed to have been randomly selected from these conceptual sampling distributions.

5.6.1. Pearson's Central χ^2 (Chi Square) Conceptual Sampling Distribution

Consider a hypothetical quantitative CRD experiment test program that consists of a single realization value randomly selected from a conceptual (two-parameter) normal distribution whose mean and standard deviation are known. Let the statistic of specific interest be Pearson's central χ^2 statistic with one statistical degree of freedom, viz.,

$$\text{Pearson's central } \chi^2_{n_{sdf}=1}\text{ statistic} = \left[\frac{X - \text{mean}(X)}{\text{stddev}(X)}\right]^2 = Y^2$$

Under continual replication of this hypothetical experiment test program, the respective realization values of Pearson's central χ^2 statistic with one statistical degree of freedom generate a conceptual sampling distribution that is referred to as Pearson's central χ^2 conceptual sampling distribution with one statistical degree of freedom.

Analogously, consider a hypothetical quantitative CRD experiment test program that consists of n_s independent replicate realization values randomly selected from a conceptual (two-parameter) normal distribution whose mean and standard deviation are known. Let the statistic of specific interest be Pearson's central χ^2 statistic with $n_{sdf} = n_s$ statistical degree of freedom, viz.,

$$\text{Pearson's central } \chi^2_{n_{sdf}=n_s}\text{ statistic} = \sum_{i=1}^{n_s}\left[\frac{X_i - \text{mean}(X)}{\text{stddev}(X)}\right]^2 = \sum_{i=1}^{n_s} Y_i^2$$

Under continual replication of this hypothetical experiment test program, the respective realization values of Pearson's central χ^2 statistic with $n_{sdf} = n_s$ statistical degrees of freedom generate a conceptual sampling distribution that is referred to as Pearson's central χ^2 conceptual sampling distribution with n_{sdf} statistical degrees of freedom. Figure 5.6 depicts Pearson's central χ^2 conceptual sampling distribution given three illustrative values for its number of statistical degrees of freedom n_{sdf}.

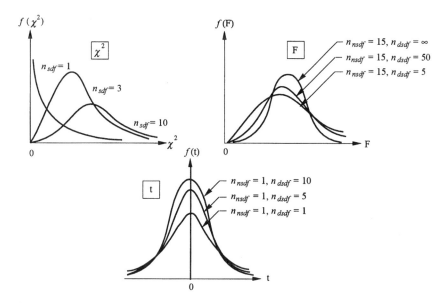

Figure 5.6 Plot of Pearson's central χ^2, Snedecor's central F, and Student's central t conceptual sampling distributions, each with three illustrative values for its statistical degrees of freedom.

There is also an associated Pearson's noncentral χ^2 statistic. It can be succinctly expressed as

$$\text{Pearson's noncentral } \chi^2_{n_{sdf}=n_s} \text{ statistic} = \sum_{i=1}^{n_s} [Y_i + \text{mean}(X)]^2$$

Pearson's noncentral χ^2 conceptual sampling distribution is analytically much more complex than Pearson's central χ^2 conceptual sampling distribution because its PDF includes a noncentrality parameter whose value depends on mean(X).

5.6.2. Snedecor's Central F Conceptual Sampling Distribution

Consider a hypothetical quantitative CRD experiment test program that consists of $(n_{nsdf} + n_{dsdf})$ independent replicate realization values randomly selected from a conceptual (two-parameter) normal distribution whose mean and standard deviation are known. Let the statistic of specific interest be Snedecor's central F statistic with n_{nsdf} numerator statistical degrees of freedom and n_{dsdf} denominator statistical degrees of freedom, viz.,

Snedecor's central $F_{n_{nssf}, n_{dsdf}}$ statistic

$$= \frac{\sum_{i=1}^{n_{nsdf}} [X_i - \text{mean}(X)/\text{stddev}(X)]^2 / n_{nsdf}}{\sum_{i=1 n_{nsdf}+1}^{n_{nsdf}+n_{dsdf}} [X_i - \text{mean}(X)/\text{stddev}(X)]^2 / n_{dsdf}}$$

$$= \frac{\chi^2_{n_{nsdf}} / n_{nsdf}}{\chi^2_{n_{dsdf}} / n_{dsdf}}$$

Under continual replication of this hypothetical experiment test program, the respective realization values of Snedecor's central F statistic with n_{nsdf} numerator statistical degrees of freedom and n_{dsdf} denominator statistical degrees of freedom generate a conceptual sampling distribution that is referred to as Snedecor's central F conceptual sampling distribution with n_{nsdf} numerator statistical degrees of freedom and n_{dsdf} denominator statistical degrees of freedom. Figure 5.6 depicts Snedecor's central F conceptual sampling distribution given three sets of illustrative values for its numbers of numerator and denominator statistical degrees of freedom.

There is also an associated Snedecor's noncentral F statistic. It can be succinctly expressed as

Snedecor's noncentral $F_{n_{nsdf}, n_{dsdf}}$ statistic $= \dfrac{\sum_{i=1}^{n_{nsdf}} [Y_i + \text{mean}(X)]^2 / n_{nsdf}}{\chi^2_{n_{dsdf}} / n_{dsdf}}$

Snedecor's noncentral F conceptual sampling distribution is analytically much more complex than Snedecor's central F conceptual sampling distribution because its PDF includes a noncentrality parameter whose value depends on mean(X). Probability values based on noncentral F conceptual sampling distributions are used in microcomputer programs PCRD and PRCBD to compute statistical power (Section 6.5).

5.6.2.1. Application

Data-based realization values of Snedecor's central F statistic are traditionally used to test the null hypotheses of specific interest in analysis of variance (Chapter 6) and to test the statistical adequacy of the presumed conceptual model in linear regression analysis (Chapter 7). It is convenient in these applications to state the data-based value of Snedecor's central F statistic as

data-based value of Snedecor's central $F_{between_{n_{sdf}}, within_{n_{sdf}}}$ statistic

$$= \frac{[between(\text{SS})]/between_{n_{sdf}}}{[within(\text{SS})/within_{n_{sdf}}} = \frac{between(\text{MS})}{within(\text{MS})}$$

in which the *within* and *between* sum of squares, respectively denoted *within*(SS) and *between*(SS), are computed as the sum of squares of the elements of appropriate orthogonal column vectors in the associated estimated conceptual statistical model. Then, under continual replication of the given experiment test program and subject to certain additional presumptions described later, the respective realization values for the statistic [*between*(MS)/*within*(MS)] generate Snedecor's central F conceptual sampling distribution.

> *Remark*: G. W. Snedecor named the central F conceptual sampling distribution to honor R. A. Fisher. Thus, it is always written using a capital F, regardless of whether F is viewed as being a random variable or as being the realization value for this random variable.

5.6.3. Student's Central t Conceptual Sampling Distribution

Consider a hypothetical quantitative CRD experiment test program that consists of $(1 + n_{dsdf})$ independent replicate realization values randomly selected from a conceptual (two-parameter) normal distribution whose mean and standard deviation are known. Let the statistic of specific interest be Student's central t statistic with one numerator statistical degree of freedom and n_{dsdf} denominator statistical degrees of freedom, viz.,

Student's central $T_{1, n_{dsdf}}$ statistic

$$= \left\{ \frac{[X_1 - \text{mean}(X)/\text{stddev}(X)]^2/1}{\sum_{i=2}^{1+n_{dsdf}} [X_i - \text{mean}(X)/\text{stddev}(X)]^2/n_{dsdf}} \right\}^{1/2}$$

$$= \frac{Y}{(\chi^2_{n_{dsdf}}/n_{dsdf})^{1/2}}$$

Under continual replication of this hypothetical experiment test program, the respective realization values for Student's central T statistic with one numerator statistical degrees of freedom and n_{dsdf} denominator statistical degrees of freedom generate a conceptual sampling distribution that is referred to herein as Student's central t conceptual sampling distribution with one numerator statistical degree of freedom and n_{dsdf} denominator statistical degrees of freedom. Figure 5.6 depicts Student's central t conceptual sampling distribution given three sets of illustrative values for its number of denominator statistical degrees of freedom. Note that Student's central t conceptual sampling distribution is merely a special case of Snedecor's central F conceptual sampling distribution with one numerator statistical degree of freedom. Nevertheless it is traditionally employed in classical statistical analyses, e.g., the calculation of classical statistical confidence intervals (Applications One and Two below).

There is also an associated Student's noncentral T statistic. It can be succinctly expressed as

$$\text{Student's noncentral } T_{1, n_{dsdf}} \text{ statistic} = \frac{Y + \text{mean}(X)}{(\chi^2_{n_{dsdf}}/n_{dsdf})^{1/2}}$$

Student's noncentral t conceptual sampling distribution is analytically more complex than Student's central t conceptual sampling distribution because its PDF includes a noncentrality parameter whose value depends on mean(X). (It is a special case of Snedecor's noncentral F conceptual sampling distribution with one numerator statistical degree of freedom, just as Student's central t conceptual sampling distribution is a special case of Snedecor's central F conceptual sampling distribution with one numerator statistical degree of freedom.) Probability values based on Student's noncentral t conceptual sampling distributions are used in microcomputer programs *ABNSTL*, *BBNSTL*, *ABLNSTL*, and *BBLBSTL* to compute *A*-basis and *B*-basis statistical tolerance limits (Supplemental Topic 8.A).

5.6.3.1. Application One

Given a quantitative CRD experiment test program that consists of n_{dv} independent replicate realization values randomly selected from a conceptual (two-parameter) normal distribution, consider the following traditional expression for Student's central t statistic with one numerator statistical degree of freedom and $n_{dv} - 1$ denominator statistical degrees of freedom:

$$\text{traditional Student's central } T_{1, n_{dv}-1} \text{ statistic} = \frac{\text{ave}(X\text{'s}) - \text{mean}(X)}{\text{est[stddev}(X)]/\sqrt{n_{dv}}}$$

in which 1 is the number of (numerator) statistical degrees of freedom pertaining to ave(X's) and ($n_{dv}-1$) is the number of (denominator) statistical degrees of freedom pertaining to est[stddev(X)]. This traditional

Student's central $T_{1,n_{dv-1}}$ statistical expression and a *scp*-based value established by its associated Student's central t_1, n_{dv-1} conceptual sampling distribution is employed in Section 5.7 to establish the classical (shortest) $100(scp)\%$ (two-sided) statistical confidence interval that allegedly includes mean(X).

5.6.3.2. Application Two

First recall (a) that the linear least-squares estimation expression for the actual value of each scalar coefficient in the complete analytical models presented in Chapter 2 is statistically unbiased, and that (b) each resulting scalar coefficient estimate is normally distributed (under continual replication of the experiment test program) when the respective experiment test program datum values are normally distributed. Then, consider the following generic expression for Student's central t statistic with one numerator statistical degree of freedom and $(n_{dv}-1)$ denominator statistical degrees of freedom:

generic Student's central $T_{1,n_{dsdf}}$ statistic

$$= \frac{\text{statistically unbiased normally distributed estimator} - \text{its expected value}}{\text{est[stddev (statistically unbiased normally distributed estimator)]}}$$

in which 1 is the number of (numerator) statistical degrees of freedom pertaining to the statistically unbiased normally distributed estimator and n_{dsdf} is the number of (denominator) statistical degrees of freedom pertaining to the estimated standard deviation of the statistically unbiased normally distributed estimator. This generic Student's central $T_{1,n_{dsdf}}$ statistic expression and a *scp*-based value established by its associated Student's central $t_{1,n_{dsdf}}$ conceptual sampling distribution is used in Chapter 6 to compute classical (shortest) $100(scp)\%$ (two-sided) statistical confidence intervals that allegedly include the actual values for the scalar coefficients of specific interest. (See also Chapter 7, Exercise Set 2, Exercise 7.)

5.6.4. Probability Relationships Among Standardized Conceptual Normal Distribution and Associated Pearson's Central χ^2, Snedecor's Central F, and Student's Central t Conceptual Sampling Distributions

Figure 5.7 attempts to illustrate the fundamental relationships among the standardized conceptual normal distribution and its three associated central conceptual sampling distributions. Note that Snedecor's central F conceptual sampling distribution contains all of the probability informa-

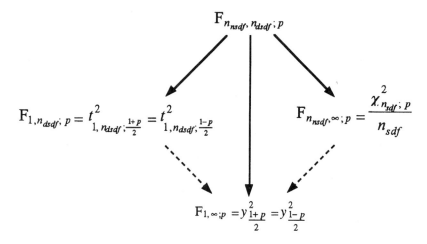

Figure 5.7 Relationships among the standardized conceptual normal distribution and Pearson's central χ^2, Snedecor's central F, and Student's central t conceptual sampling distributions.

tion found in the other three conceptual distributions. These relationships can easily be verified numerically using the microcomputer programs enumerated below.

It is important to recognize that both the standardized conceptual normal distribution variate y and Student's central t test statistic can take on either positive or negative values, whereas Pearson's central χ^2 test statistic and Snedecor's central F test statistic can only take on positive values. Accordingly, the probabilities associated with both the upper (positive) and lower (negative) tails of the standardized conceptual normal distribution and Student's central t conceptual sampling distribution are aggregated to establish the probabilities pertaining to the upper tail of Pearson's central χ^2 conceptual sampling distribution and the upper tail of Snedecor's central F conceptual sampling distribution. These relationships are evident when the associated PDF's in Figure 5.7 are examined.

5.6.5 Computer Programs for Pearson's Central χ^2, Snedecor's Central F, and Student's Central t Conceptual Sampling Distributions

We now present three sets of microcomputer programs that are analogous to microcomputer programs PY and YP pertaining to the standardized conceptual normal distribution:

1. Microcomputer program *PCS* is used to compute probability values pertaining to Pearson's central χ^2 conceptual sampling distribution, whereas microcomputer program *CSP* is used to compute values of Pearson's central χ^2 test statistic that correspond to specific percentiles of Pearson's central χ^2 conceptual sampling distribution.

2. Microcomputer program *PF* is used to compute probability values pertaining to Snedecor's central F conceptual sampling distribution, whereas microcomputer program *FP* is used to compute values of Snedecor's central F test statistic that correspond to specific percentiles of Snedecor's central F conceptual sampling distribution.

3. Microcomputer program *PT* is used to compute probability values for Student's central t conceptual sampling distribution, whereas microcomputer program *TP* is used to compute values of Student's central t test statistic that correspond to specific percentiles of Student's central t test statistic conceptual sampling distribution.

5.7. CLASSICAL (SHORTEST) 100(*SCP*)% (TWO-SIDED) STATISTICAL CONFIDENCE INTERVAL THAT ALLEGEDLY INCLUDES MEAN(*X*)— COMPUTED USING STUDENT'S CENTRAL t CONCEPTUAL SAMPLING DISTRIBUTION.

We now develop the classical (shortest) 100(*scp*)% (two-sided) statistical interval that allegedly includes the actual value for the mean of the conceptual (two-parameter) normal distribution from which it is presumed that the quantitative CRD experiment test program replicate datum values were randomly selected.

Consider the outcome of a quantitative CRD experiment test program that consists of n_{dv} measurement values presumed to be replicate realizations of a normally distributed random variable X. Suppose that we compute the arithmetic average of these replicate realizations and then we compute est[stddev(X)] using the square root of the unbiased generic variance estimator given in Section 5.4. In turn, suppose that we state Student's central $T_{1,n_{dv}-1}$ statistic using its traditional expression, viz.,

$$\text{traditional Student's central } T_{1,n_{dv}-1} \text{ statistic} = \frac{\text{ave}(X\text{'s}) - \text{mean}(X)}{\text{est[stddev}(X)]/\sqrt{n_{dv}}}$$

Recall that the experiment test program that is (was) actually conducted is statistically viewed as being randomly selected from the collection of all possible replicate experiment test programs. Accordingly, the probability that a randomly selected experiment test program will have its data-based realization value for $T_{1,n_{dv}-1}$ lie in the numerical interval from $t_{1,n_{dv}-1;p_{lower}}$ to $t_{1,n_{dv}-1;p_{upper}}$ is

$$\text{probability}\left\{ t_{1,n_{dv}-1;p_{lower}} < T_{1,n_{dv}-1} = \frac{\text{ave}(X\text{'s}) - \text{mean}(X)}{\text{est[stddev}(X)]/\sqrt{n_{dv}}} < t_{1,n_{dv}-1;p_{upper}} \right\}$$

$$= p_{upper} - p_{lower}$$

in which p_{upper} and p_{lower} are established by the selected values for $t_{1,n_{dv}-1;p_{lower}}$ and $t_{1,n_{dv}-1;p_{upper}}$. This probability expression can be *reinterpreted* as

$$\text{probability}\left[\text{ave}(X\text{'s}) - t_{1,n_{dv}-1;p_{upper}} \cdot \left\{ \frac{\text{est[stddev}(X)]}{\sqrt{n_{dv}}} \right\} < \text{mean }(X) \right.$$

$$\left. < \text{ave}(X\text{'s}) - t_{1,n_{dv}-1;p_{lower}} \cdot \left\{ \frac{\text{est[stddev}(X)]}{\sqrt{n_{dv}}} \right\} \right] = scp$$

in which $p_{upper} - p_{lower} = scp$. When $t_{1,n_{dv}-1;p_{lower}}$ is deliberately selected such that the probability that $(T_{1,n_{dv}-1} < t_{1,n_{dv}-1;p_{lower}})$ equals $(1 - scp)/2$, and $t_{1,n_{dv}-1;p_{upper}}$ is deliberately selected such that the probability that $(T_{1,n_{dv-1}} > t_{1,n_{dv}-1;p_{upper}})$ equals $(1 - scp)/2$, the corresponding classical $100(scp)\%$ (two-sided) statistical confidence interval takes on its minimum width (because Student's central t conceptual sampling distribution is symmetrical about zero, its mean).

5.7.1. Numerical Example

Suppose the following five datum values are presumed to have been randomly selected from a conceptual (two-parameter) normal distribution with an unknown mean and variance: 10.1, 9.3, 9.9, 10.2, and 9.7.

Then the data-based realization value for the statistical estimator $\text{est[mean}(X)] = \text{ave}(X\text{'s})$ is

$$\text{est[mean}(X)] = \text{ave}(x\text{'s}) = (10.1 + 9.3 + 9.9 + 10.2 + 9.7)/5 = 9.84$$

Similarly, the data-based realization value for statistical estimator $\text{est[var}(X)]$ is

$$\text{est[var}(X)] = \sum_{i=1}^{n_{dv}} \frac{[x_i - \text{ave}(x\text{'s})]^2}{n_{dv} - 1}$$

$$= \frac{\begin{array}{c}(10.1 - 9.84)^2 + (9.3 - 9.84)^2 + (9.9 - 9.84)^2 + \\ (10.2 - 9.84)^2 + (9.7 - 9.84)^2\end{array}}{5 - 1} = 0.128$$

In turn,

$$\frac{\text{est[stddev}(X)]}{\sqrt{n_{dv}}} = \frac{\sqrt{0.128}}{\sqrt{5}} = 0.16$$

To compute the associated classical (shortest) 95% (two-sided) statistical confidence interval ($scp = 0.95$) that allegedly includes mean(X), we next run microcomputer program *TP* to obtain values for $-t_{1,4;0.025}$ and for $t_{1,4;0.975}$, viz., $+2.7764$. We then substitute into the *reinterpreted* probability expression:

$$\text{probability}\left[\text{ave}(X\text{'s}) - t_{1,n_{dv}-1;p_{\text{upper}}} \cdot \left\{ \frac{\text{est[stddev}(X)]}{\sqrt{n_{dv}}} \right\} < \text{mean } (X) \right.$$

$$\left. < \text{ave } (X\text{'s}) - t_{1,n_{dv}-1;p_{\text{upper}}} \cdot \left\{ \frac{\text{est[stddev}(X)]}{\sqrt{n_{dv}}} \right\} \right] = scp$$

as follows to obtain the result:

$$\text{probability}[9.84 - (2.7764) \cdot (0.16) < \text{mean}(X)$$

$$< 9.84 + (2.7764) \cdot (0.16)] = 0.95$$

Thus,

$$\text{probability}[9.40 < \text{mean}(X) < 10.28] = 0.95$$

Accordingly, the corresponding classical (shortest) 95% (two-sided) statistical confidence interval that allegedly includes mean(X) is [9.40, 10.28].

5.7.2. Proper Statistical Interpretation of the Classical (Shortest) 100(*scp*)% (Two-Sided) Statistical Confidence Interval that Allegedly Includes Mean(X)

Recall that the first fundamental concept underlying statistical analysis is that the outcome for an experiment test program that is (was) actually conducted can be viewed as having been randomly selected from the infinite collection of all possible replicate experiment test program outcomes. Analogously, the corresponding classical (shortest) 100(*scp*)% (two-sided) statistical confidence interval that allegedly includes mean(X) can be viewed as having been randomly selected from the infinite collection of corresponding classical (shortest) 100(*scp*)% (two-sided) statistical confidence intervals that allegedly include mean(X). Clearly, the data-based numerical limits of the associated classical (shortest) 100(*scp*)% (two-sided) statistical confidence interval depends on which experiment test program outcome was randomly selected from this infinite collection of all possible replicate experiment test program outcomes. Now suppose that mean(X) is known (as in a simulation-based study). Then, we can state positively whether or not the data-based numerical limits of any given classical (shortest) $100(scp)\%$ (two-sided) statistical confidence interval actually include mean(X). However, if mean(X) is unknown, we can never state positively whether or not these numerical limits actually include mean(X). Rather, we can only say that, under the conceptual process of continually replicating the experiment test program, each successive test experiment program outcome will generate a (different) data-based associated classical (shortest) $100(scp)\%$ (two-sided) statistical confidence interval which, *a priori*, has the same probability, viz., the specified *scp*, that it actually includes mean(X). This statistical behavior can be simulated by repeatedly running microcomputer program *SIMNOR* to generate n_{sim} sets of n_{dv} normally distributed pseudorandom datum values with any mean(X) and stddev(X) of specific interest. Then, when these associated classical (shortest) $100(scp)\%$ (two-sided) statistical confidence intervals are successively computed and subsequently summarized relative to the proportion that actually include mean(X), the outcome will resemble the simulation-based result obtained by running microcomputer program *SSTSSCI1* (simulation study for a two-sided statistical confidence interval—version 1). See page 212.

5.7.3. Proper Quantitative Interpretation (Precision) of the Classical (Shortest) 100(scp)% (Two-Sided) Statistical Confidence Interval that Allegedly Includes Mean(X)

The proper quantitative interpretation (precision) of a classical (shortest) $100(scp)\%$ (two-sided) statistical confidence interval can be deduced by appropriate simulation. If mean(X) and stddev(X) are presumed known, then n_{rep} replicate normally distributed pseudorandom data sets of size n_{dv} can be generated, e.g., by running microcomputer program *SIMNOR*. In turn, classical (shortest) $100(scp)\%$ (two-sided) statistical confidence intervals can be computed and compared for each of these replicate data sets. This comparison is always remarkably informative, even for n_{rep} as small as 12. However, because mean(X) and stddev(X) are unknown in mechanical test and mechanical reliability applications, est[mean(X)] must be substituted for mean(X) and est[stddev(X)] must be substituted for stddev(X) in the generation of the desired n_{rep} "replicate" normally distributed pseudorandom data sets. Then, the sampling distributions that are comprised of the n_{rep} "replicate" realizations of the statistics of specific interest are termed pragmatic rather than empirical.

Microcomputer program *SSTSSCI2* (page 214) computes 12 replicate classical (shortest) $100(scp)\%$ (two-sided) statistical confidence intervals and displays their variability. Microcomputer program *SSTSSCI3* (page 216) generates the empirical sampling distribution for the ratio of the half-width of the classical (shortest) $100(scp)\%$ (two-sided) statistical confidence interval to its associated midpoint based on 30,000 replicate data sets, each with n_{dv} normally distributed pseudorandom datum values, and then computes its mean, median, and selected percentiles. Analogously, micro-computer program *SSTSSCI4* computes 12 "replicate" classical (shortest) $100(scp)\%$ (two-sided) statistical confidence intervals and displays their variability. In turn, microcomputer program *SSTSSCI5* generates the prag-matic sampling distribution for the ratio of the half-width of the classical (shortest) $100(scp)\%$ (two-sided) statistical confidence interval to its asso-ciated midpoint based on 30,000 "replicate" data sets, each with n_{dv} nor-mally distributed pseudorandom datum values, and then computes its mean, median, and selected percentiles. Microcomputer programs *SSTSSCI4* and *SSTSSCI5* employ an exact statistical basis correction for est[stddev(X)].

```
C> SSTSSCI1
```

Input a new set of three, three-digit odd seed numbers

```
933 449 175
```

Input the statistical confidence probability of specific interest, stated in per cent (integer value)

```
95
```

Percentage of Simulation-Based 95% (Two-Sided) Statistical
Confidence Intervals that Correctly Include Mean(X)

Interval   1 – Percent Correct 100.000
Interval   2 – Percent Correct 100.000
Interval   3 – Percent Correct 100.000
Interval   4 – Percent Correct 100.000
Interval   5 – Percent Correct 100.000
Interval   6 – Percent Correct 100.000
Interval   7 – Percent Correct 100.000
Interval   8 – Percent Correct 100.000
Interval   9 – Percent Correct 100.000

```
Interval   10 – Percent Correct 100.000
Interval   11 – Percent Correct 100.000
Interval   12 – Percent Correct 100.000
Interval   13 – Percent Correct 100.000
Interval   14 – Percent Correct 100.000
Interval   15 – Percent Correct 100.000
Interval   16 – Percent Correct 100.000
Interval   17 – Percent Correct 100.000
Interval   18 – Percent Correct 100.000
Interval   19 – Percent Correct 100.000
Interval   20 – Percent Correct 100.000
Interval   21 – Percent Correct 100.000
Interval   22 – Percent Correct 100.000
Interval   23 – Percent Correct 100.000
Interval   24 – Percent Correct 100.000
Interval   25 – Percent Correct 100.000
Interval   30 – Percent Correct 100.000
Interval   35 – Percent Correct 100.000
Interval   40 – Percent Correct 100.000
Interval   45 – Percent Correct 100.000
Interval   50 – Percent Correct 100.000
Interval   55 – Percent Correct 100.000
Interval   60 – Percent Correct 100.000
Interval   65 – Percent Correct  98.462
Interval   70 – Percent Correct  97.143
Interval   75 – Percent Correct  96.000
Interval   80 – Percent Correct  96.250
Interval   85 – Percent Correct  95.294
Interval   90 – Percent Correct  95.556
Interval   95 – Percent Correct  95.789
Interval  100 – Percent Correct  96.000
Interval  200 – Percent Correct  94.000
Interval  300 – Percent Correct  94.000
Interval  400 – Percent Correct  94.000
Interval  500 – Percent Correct  94.000
Interval  600 – Percent Correct  94.333
Interval  700 – Percent Correct  94.429
Interval  800 – Percent Correct  94.875
Interval  900 – Percent Correct  95.222
Interval 1000 – Percent Correct  95.400
```

```
C> SSTSSCI2
```

Input the number of replicate (presumed) normally distributed datum values of specific interest, where $3 < n_{dv} < 33$

```
5
```

Input the conceptual (two-parameter) normal distribution mean of specific interest

```
10.0
```

Input the conceptual (two-parameter) normal distribution standard deviation mean of specific interest

```
1.0
```

Input the statistical confidence probability of specific interest (integer value)

```
95
```

Input a new set of three, three-digit odd seed numbers

```
489 739 913
```

Replicate Classical (Shortest) 95%
(Two-Sided) Statistical Confidence Intervals

Interval 1: from 8.55 to 11.03
Interval 2: from 8.02 to 10.62
Interval 3: from 9.21 to 10.75
Interval 4: from 8.73 to 10.55
Interval 5: from 10.03 to 10.47
Interval 6: from 9.19 to 11.15
Interval 7: from 9.61 to 12.01
Interval 8: from 9.12 to 11.44
Interval 9: from 8.58 to 11.78
Interval 10: from 9.23 to 11.22
Interval 11: from 9.60 to 10.76
Interval 12: from 9.65 to 12.14

5.7.4. Relationship Between Classical Shortest (Two-Sided) Statistical Confidence Interval that Allegedly Includes Mean(X) and Corresponding Classical Statistical Test of Hypothesis

The experiment test program datum values in our numerical example can also be analyzed using the classical test of hypothesis methodology. Suppose that the null hypothesis of specific interest is that mean(X) = 10.0 and that the associated composite (two-sided) alternative hypothesis is that mean(X) ≠ 10.0. Then, given an acceptable probability of committing a *Type I* error equal to 0.05, viz., equal to $(1 - scp) = (1 - 0.95)$, the classical test of hypothesis methodology will indicate that the null hypothesis cannot rationally be rejected. (The null hypothesis cannot rationally be rejected in a classical statistical test of hypothesis unless the experiment test program data-based value for Snedecor's central F test statistic exceeds the value of Snedecor's central F conceptual sampling distribution pertaining to the acceptable probability of committing a *Type I* error.) Recall that the classical (shortest) 95% (two-sided) statistical confidence interval for mean(X) is [9.40, 10.28]. Since mean(X) = 10.0 lies in this interval, the outcome of the classical test of hypothesis methodology is clearly consistent with the corresponding classical (shortest) 95% (two-sided) statistical confidence interval for mean(X). Next, suppose that the null hypothesis of specific interest is that mean(X) = 9.0 and that the associated composite (two-sided) alternative hypothesis is that mean(X) ≠ 9.0. Then, given the same acceptable probability of committing a *Type I* error equal to 0.05, the classical test of hypothesis methodology will indicate that the null hypothesis can rationally be rejected. This outcome is also clearly consistent with the corresponding classical shortest 95% (two-sided) statistical confidence interval that allegedly includes mean(X). Note that the interval [9.40, 10.28] does not include mean(X) = 9.0.

These two examples can be summarized as follows: the classical (shortest) $100(scp)$% (two-sided) statistical confidence interval that allegedly includes mean(X) is the collection of all possible candidate values for mean(X) such that the null hypothesis that mean(X) is equal to this candidate value cannot rationally be rejected in a classical statistical test of hypothesis, given that the associated composite (two-sided) alternative hypothesis that mean(X) is not equal to this candidate value. However, remember that for this relationship to pertain, the acceptable probability of committing a *Type I* error must be set equal to the complement of the corresponding value for the *scp*.

```
C> SSTSSCI3
```

Input the number of replicate (presumed) normally distributed datum values of specific interest, where $3 < n_{dv} < 33$

```
5
```

Input the conceptual (two-parameter) normal distribution mean of specific interest

```
10.0
```

Input the conceptual (two-parameter) normal distribution standard deviation mean of specific interest

```
1.0
```

Input the statistical confidence probability of specific interest (integer value)

```
95
```

Input a new set of three, three-digit odd seed numbers

```
843 511 745
```

This microcomputer program generated 30,000 replicate normally distributed data sets, each of size 5, and computed the associated classical (shortest) 95% (two-sided) statistical confidence intervals. The mean of the empirical sampling distribution that consists of 30,000 replicate realizations for the statistic [the ratio of the half-width of the classical (shortest) 95% (two-sided) statistical confidence interval to its midpoint] is equal to 0.1172. The median of this empirical sampling distribution is equal to 0.1137. Thus, approximately 50% of all possible replicate classical (shortest) 95% (two-sided) statistical confidence intervals will bound its midpoint within 11.37%. The corresponding median empirical classical (shortest) 95% (two-sided) statistical confidence interval is [8.86, 11.14].

Approximately p% of all possible replicate classical (shortest) 95% (two-sided) statistical confidence intervals will bound its mid-point within $p*$ %:

p%	$p*$%
75	14.45
90	17.50
95	19.38
99	23.17

The relationship between the classical (shortest) $100(scp)$% (two-sided) statistical confidence interval that allegedly includes mean(X) and the corresponding classical statistical test of hypothesis, given a composite (two-sided) alternative hypothesis, is based on the relationship between Snedecor's central F test statistic and Student's central T test statistic. To emphasize this relationship we use Snedecor's central F test statistic to compute the classical (shortest) $100(scp)$% (two-sided) statistical confidence interval that allegedly includes mean(X) in Section 5.8.

5.7.4.1. Discussion

Statistical confidence intervals are typically more informative than classical statistical tests of hypotheses. The latter usually serve primarily to ascertain which experiment test program variables (treatments) are actually important and which are apparently benign. The former displays intuitive information about both the magnitude and precision of the estimate of specific interest. A classical statistical test of hypothesis for which the null hypothesis is rationally rejected in favor of the composite (two-sided) alternative hypothesis should always be followed by a computation of the relevant classical (shortest) $100(scp)$% (two-sided) statistical confidence interval. In turn, the variability (precision) of "replicate" classical (shortest) $100(scp)$% (two-sided) statistical confidence intervals should be assessed by running either microcomputer program *SSTSSCI4* or *SSTSSCI5* (or both).

When the proportion of statistical assertions that are actually correct asymptotically approaches the *scp* value of specific interest under continual replication of the experiment test program the associated statistical confidence interval or limit is said to be exact (unbiased). Exact (unbiased) statistical confidence intervals and limits are based either on

```
C> SSTSSCI4
```

For the given set of n_{dv} (presumed) normally distributed datum values of specific interest:

Input n_{dv}, where $3 < n_{dv} < 33$

```
5
```

Input est[mean(X)]

```
9.84
```

Input est[stddev(X)]

```
0.35771
```

Input the statistical confidence probability of specific interest (integer value)

```
95
```

Input a new set of three, three-digit odd seed numbers

```
727 349 173
```

"Replicate" Classical (Shortest) 95%
(Two-Sided) Statistical Confidence Intervals

Interval   1: from 9.40 to 10.20
Interval   2: from 9.55 to 10.23
Interval   3: from 9.61 to 9.92
Interval   4: from 9.06 to 9.90
Interval   5: from 9.80 to 10.17
Interval   6: from 9.12 to 10.19
Interval   7: from 9.43 to 10.27
Interval   8: from 9.61 to 10.30
Interval   9: from 9.15 to 9.96
Interval 10: from 8.68 to 10.53
Interval 11: from 9.35 to 10.25
Interval 12: from 9.72 to 10.17

known conceptual sampling distributions or on accurate empirical sampling distributions. In contrast, approximate (biased) statistical confidence intervals and limits are typically based on asymptotic sampling distributions. This distinction is important in maximum likelihood analysis (Chapter 8) where empirical (pragmatic) statistical bias corrections must be established and appropriately employed to reduce the difference between the actual value for the *scp* and its desired (specification) value. If the empirical (pragmatic) bias correction make this difference negligible for practical purposes, then the resulting 100(*scp*)% (two-sided) statistical confidence interval can be viewed as being exact (unbiased) for practical purposes.

Exercise Set 4

These exercises are intended to provide experience in computing classical (shortest) 100(*scp*)% (two-sided) confidence intervals for mean(*X*), given datum values that are (or are alleged to be) normally distributed. Run microcomputer program *SIMNOR* to generate your datum values.

1. Generate 10 pseudorandom datum values from a conceptual normal distribution whose mean(*X*) = 10.0 and whose stddev(*X*) = 1.0 and compute the associated classical (shortest) 95% (two-sided) statistical confidence interval that allegedly includes mean(*X*). Then, run microcomputer program *SSTSSCI2* and comment appropriately.

2. Using the same set of three, three-digit odd seed numbers as in Exercise 1, generate 10 pseudorandom datum values from a conceptual normal distribution whose mean(*X*) = 10.0 and whose stddev(*X*) = 2.0. Then, compute the classical (shortest) 95% (two-sided) statistical confidence interval that allegedly includes mean(*X*) and comment appropriately.

3. Using the same set of three, three-digit odd seed numbers as in Exercise 1, generate 10 pseudorandom datum values from a conceptual normal distribution whose mean(*X*) = 10.0 and whose stddev(*X*) = 0.5. Then, compute the classical (shortest) 95% (two-sided) statistical confidence interval that allegedly includes mean(*X*) and comment appropriately.

4. Repeat Exercise 1 for classical (shortest) 50, 75, 90, 99, and 99.9% (two-sided) statistical confidence intervals that allegedly include mean(*X*). Plot the respective half-widths versus the corresponding *scp* value and comment appropriately.

C> SSTSSCI5

For the given set of n_{dv} (presumed) normally distributed datum values of specific interest:

Input n_{dv}, where $3 < n_{dv} < 33$

5

Input est[mean(X)]

9.84

Input est[stddev(X)]

0.35771

Input the statistical confidence probability of specific interest (integer value)

95

Input a new set of three, three-digit odd seed numbers

913 283 779

This microcomputer program generated 30,000 "replicate" normally distributed data sets, each of size 5, and computed the associated classical (shortest) 95% (two-sided) statistical confidence intervals. The mean of the pragmatic sampling distribution that consists of 30,000 "replicate" realizations for the statistic (the ratio of the half-width of the classical (shortest) 95% (two-sided) statistical confidence interval to its midpoint is equal to 0.0453. The median of this pragmatic sampling distribution is equal to 0.0443. Thus, approximately 50% of all possible "replicate" classical (shortest) 95% (two-sided) statistical confidence intervals will bound its midpoint within 4.13%. The corresponding median pragmatic classical (shortest) 95% (two-sided) statistical confidence interval is [9.40, 10.28].

Approximately $p\%$ of all possible "replicate" classical (shortest) 95% (two-sided) statistical confidence intervals will bound its midpoint within $p^*\%$:

$p\%$	$p^*\%$
75	5.60
90	6.69
95	7.40
99	8.78

5.8. SAME CLASSICAL (SHORTEST) 100(*SCP*)% (TWO-SIDED) STATISTICAL CONFIDENCE INTERVAL THAT ALLEGEDLY INCLUDES MEAN(*X*)— COMPUTED USING SNEDECOR'S CENTRAL F CONCEPTUAL SAMPLING DISTRIBUTION

We now use Snedecor's central F conceptual sampling distribution to compute the same classical (shortest) 100(scp)% (two-sided) statistical confidence interval that allegedly includes mean(X) that was computed for our numerical example in Section 5.7.1. First, we express our estimated statistical model in hybrid column vector format as

$$|X_i| = |\text{mean}(X)| + |\text{ave}(X\text{'s}) - \text{mean}(X)| + |X_i - \text{ave}(X\text{'s})|$$

and, in turn, more explicitly as

$$\begin{vmatrix} X_1 \\ X_2 \\ X_3 \\ \vdots \\ X_{n_{dv}} \end{vmatrix} = \begin{vmatrix} \text{mean}(X) \\ \text{mean}(X) \\ \text{mean}(X) \\ \vdots \\ \text{mean}(X) \end{vmatrix} + \begin{vmatrix} \text{ave}(X\text{'s}) - \text{mean}(X) \\ \text{ave}(X\text{'s}) - \text{mean}(X) \\ \text{ave}(X\text{'s}) - \text{mean}(X) \\ \\ \text{ave}(X\text{'s}) - \text{mean}(X) \end{vmatrix} + \begin{vmatrix} X_1 - \text{ave}(X\text{'s}) \\ X_2 - \text{ave}(X\text{'s}) \\ X_3 - \text{ave}(X\text{'s}) \\ \\ X_{n_{dv}} - \text{ave}(X\text{'s}) \end{vmatrix}$$

It is then clear that the sum of squares for the data-based realizations of the elements in the $|X_i - \text{ave}(X\text{'s})|$ column vector is the *within*(SS). Recall that the *within*(MS) is equal to the *within*(SS) divided by its number of statistical degrees of freedom and that the *within*(MS) is an unbiased statistical estimator for var(X) under continual replication of the experiment test program.

Recall also that est[var(X)] for our numerical example is equal to 0.128 with $n_{dv} - 1$ ($5 - 1 = 4$) statistical degrees of freedom. Hence, the *within*(MS) is equal to 0.128 with four statistical degrees of freedom.

Next, consider the sum of squares for the elements of the |ave(X's) $-$ mean(X)| column vector. The sum of squares of these elements (differences) is clearly the *between*(SS), viz.,

$$between(\text{SS}) = N_{dv} \cdot [\text{ave}(X\text{'s}) - \text{mean}(X)]^2$$

Substituting 5 for n_{dv} and ave(x's) $= 9.84$ for ave(Xs) gives

$$between(\text{SS}) = 5 \cdot [9.84 - \text{mean}(X)]^2$$

In turn, the *between*(MS) is equal to the *between*(SS) divided by its number of statistical degrees of freedom, viz., 1. Thus, the *between*(MS) is equal to the *between*(SS) for our numerical example.

Finally, under continual replication of the quantitative CRD experiment test program underlying our numerical example, we assert that the respective realization values for the test statistic [*between*(MS)/*within*(MS)] generate Snedecor's central F conceptual sampling distribution with $n_{nsdf} = 1$ numerator statistical degrees of freedom and $n_{dsdf} = 4$ denominator statistical degrees of freedom. In turn, running microcomputer program *FP* indicates that $F_{1,4;0.95} = 7.7086$. Accordingly, we assert that

$$\text{probability}(F_{1,4} \leq 7.7086) = \text{probability}\left\{ \frac{(5) \cdot [9.84 - \text{mean}(X)]^2}{0.128} \leq 7.7086 \right\} = 0.95$$

Thus,

$$\text{probability}\left\{[9.84 - \text{mean}(X)]^2 \leq 0.1973\right\} = 0.95$$

Taking the square root of both sides of the inequality gives

$$\text{probability}[9.84 - 0.4442 \leq \text{mean}(X) \leq 9.84 + 0.4442] = 0.95$$

The corresponding classical (shortest) 95% (two-sided) statistical confidence interval that allegedly includes mean(X) is [9.40,10.28]—which is exactly the same as the classical (shortest) 95% (two-sided) statistical confidence interval that allegedly includes mean(X) that was computed using Student's central t conceptual sampling distribution.

5.9. CLASSICAL 100(*SCP*)% (TWO-SIDED) STATISTICAL CONFIDENCE INTERVAL THAT ALLEGEDLY INCLUDES VAR(X)—COMPUTED USING PEARSON'S CENTRAL χ^2 CONCEPTUAL SAMPLING DISTRIBUTION

The development of the classical 100(*scp*)% (two-sided) statistical confidence interval that allegedly includes var(X) is straightforward. It is based on the notion that est[var(X)] would be exactly equal to var(X) if its number of statistical degrees of freedom were infinite. Accordingly, under continual replication of a quantitative CRD experiment test program, the respective realization values for the statistic {est[var(X)]}/[var(X)] generate Snedecor's central F conceptual sampling distribution with $n_{nsdf} - n_{dv} - 1$ numerator statistical degrees of freedom and an infinite number of denominator statistical degrees of freedom, viz.,

$$F_{n_{dv}-1,\infty} = \frac{\text{est}[\text{var}(X)]}{\text{var}(X)} = \frac{\chi^2_{n_{dv}-1}}{n_{dv}-1}$$

In turn, after solving algebraically for var(X), we can write the following the probability interval expression:

$$\text{Probability}\left\{\frac{(n_{dv}-1)\cdot\text{est}[\text{var}(X)]}{\chi^2_{n_{dv}-1;p_{\text{upper}}}}\right\} < \text{var}(X)$$

$$< \left\{\frac{(n_{dv}-1)\cdot\text{est}[\text{var}(X)]}{\chi^2_{n_{dv}-1;p_{\text{lower}}}}\right\} = p_{\text{upper}} - p_{\text{lower}}$$

This probability interval expression is subsequently reinterpreted as the classical 100(*scp*)% (two-sided) statistical confidence interval that allegedly includes var(X) when $(p_{\text{upper}} - p_{\text{lower}}) = scp$. The values for p_{upper} and p_{lower} are almost always selected as $(1+scp)/2$ and $(1-scp)/2$ because the associated $\chi^2_{n_{dv}-1;p_{\text{upper}}}$ and $\chi^2_{n_{dv}-1;p_{\text{lower}}}$ values are almost always obtained from standard statistical tables for Pearson's central χ^2 conceptual sampling distribution.

> *Remark One*: The classical 100(*scp*)% (two-sided) statistical confidence interval that allegedly includes var(X) pertains only to normally distributed replicate datum values generated in a quantitative CRD experiment test program. Unfortunately, the actual probability content of this statistical confidence interval is relatively sensitive to non-normality.

Remark Two: The number of replicate datum values required to estimate var(X) precisely is typically too large to be practical (Tukey, 1986). Moreover, since test specimens (test items, experimental units) are processed in batches, the presumption that the appropriate conceptual statistical model pertains to a quantitative CRD experiment test program is dubious. In addition, when the batch-to-batch effects are improperly ignored in the presumed conceptual statistical model, these batch-to-batch effects are statistically confounded with the actual values for the respective *CRHNDEE$_i$*'s. If so, then the resulting data-based value for est[var(*APRCRHNDEE*'s)] is inflated.

Remark Three: The traditional selections of $p_{upper} = (1 + scp)/2$ and $p_{lower} = (1 - scp)/2$ do not generate the shortest classical statistical confidence interval that allegedly includes var(X). (Recall that Pearson's central χ^2 conceptual sampling distribution is not symmetrical about its mean.) However, it is seldom deemed worth the additional effort to compute the values for p_{upper} and p_{lower} that establish the shortest classical $100(scp)\%$ (two-sided) statistical confidence interval that allegedly includes var(X).

5.9.1. Numerical Example

Consider the same hypothetical datum values (allegedly generated in a quantitative CRD experiment test program) that were used in the two previous statistical confidence interval examples. Recall that est[var(X)] = 0.128 and that $n_{dv} = 5$. Hence, after running microcomputer program *CSP* with inputs $p_{lower} = 0.025$ and $p_{upper} = 0.975$ to obtain the respective values for $\chi^2_{n_{dv}-1;p_{lower}}$ and $\chi^2_{n_{dv}-1;p_{upper}}$ pertaining to $n_{dv} = 5$, the classical probability interval expression for var(X) can be stated as

$$\text{probability}\left[\frac{(4)(0.128)}{(11.1433)} < \text{var}(X) < \frac{(4)(0.128)}{(0.4844)}\right] = 0.975 - 0.025 = 0.95$$

Thus

$$\text{probability}[0.046 < \text{var}(X) < 1.057] = 0.95$$

Accordingly, the corresponding classical 95% (two-sided) statistical confidence interval that allegedly includes var(X) is [0.046, 1.057].

If the corresponding classical 95% (two-sided) statistical confidence interval that allegedly includes stddev(X) is desired, it is computed by respectively taking the square root of the upper and lower limits of the classical 95% (two-sided) statistical confidence interval that allegedly includes var(X). Accordingly, the classical 95% (two-sided) statistical confidence interval that allegedly includes stddev(X) is [0.214, 1.028].

5.10. CLOSURE

The conceptual (two-parameter) normal distribution is so widely known and used in classical statistical analyses that it has attained a mystical status. However, it is important to understand that all conceptual statistical distributions are merely mathematical abstractions that are used to model (approximate) actual physical behaviors. Accordingly, given the conceptual statistical model of specific interest (Chapter 2) and the associated experiment test program datum values, our analytical concern in classical statistical analyses is whether the est($CRHNDEE_i$'s) undermine the fundamental presumptions of randomness, homoscedasticity, and normality.

5.A. SUPPLEMENTAL TOPIC: STATISTICAL ESTIMATORS

The purpose of this section is to demonstrate that the arithmetic average is a poor statistical estimator of the actual value for the mean of a conceptual uniform distribution. Thus, it follows that the appropriate statistical estimator of the actual value for the mean of each different conceptual statistical distribution (or for any other conceptual parameter of a conceptual statistical distribution) must be established by an appropriate comparison of alternative statistical estimators.

5.A.1. Estimating Actual Value for the Mean of the Conceptual Uniform Distribution, Zero to One

Suppose we consider three alternative statistical estimators for the actual value of the mean of a conceptual uniform distribution, zero to one: (a) the arithmetic average, denoted ave(X's), (b) the median, denoted X_{median}, and (c) the midpoint of the maximum range, denoted $(X_{smallest} + X_{largest})/2$. What criteria should be used to compare these three alternative statistical estimators? The results of the following simulation study is intended to provide both insight regarding appropriate criteria and perspective regarding the choice among alternatives of statistical estimators.

5.A.1.1.   Simulation Study

Figure 5.8 summarizes a simulation study of the three proposed alternative statistical estimators, given 999 replicate sets of $n_{dv} = 5$ uniform pseudo-random numbers, zero to one.

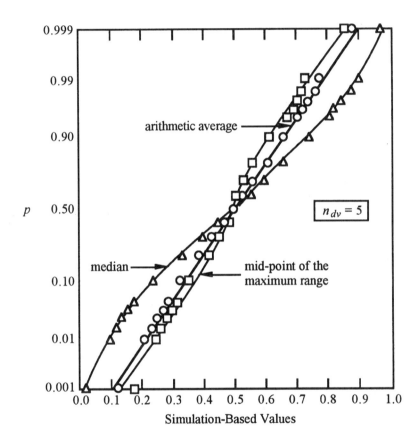

Figure 5.8   Results of a simulation study of the arithmetic average, median, and midpoints of the maximum range as statistical estimators of the actual value for the mean of the conceptual uniform distribution, zero to one. The respective empirical sampling distributions for these three proposed alternative statistical estimators are based on 999 simulation-based realization values for each statistical estimator. These realization values were then plotted on commercial normal probability paper using the so-called mean plotting position $p(pp)_i = i/(n_{dv} + 1)$ and the associated empirical CDF's were faired.

Visual examination of the respective simulation-based empirical sampling distributions plotted on normal probability paper suffices for comparison purposes. Each of the faired CDFs passes through the point, $p = 0.50$ and $x = 0.50$. Thus, it appears (correctly) that each of these three alternative statistical estimators is unbiased. However, among unbiased alternative statistical estimators, we opt for the statistical estimator whose conceptual sampling distribution has the smallest variance. Accordingly, reconsider the three respective faired CDFs in Figure 5.8. The steeper the CDF of a conceptual sampling distribution when plotted on probability paper, the smaller the actual value for the standard deviation (and variance) of its conceptual sampling distribution. Thus, the midpoint of the maximum range statistical estimator is preferred over the alternative statistical estimators because the actual value for the variance of its conceptual sampling distribution is smaller than the actual values for variances of the conceptual sampling distributions pertaining to the two alternative statistical estimators.

One way of rationalizing our preference for an unbiased minimum variance statistical estimator over alternative unbiased statistical estimators is to compare the respective probabilities that a realization value randomly selected from the conceptual sampling distribution that consists of all possible replicate realization values for each alternative statistical estimator will lie in a narrow central interval around the actual value for the mean of the underlying conceptual statistical distribution. For example, draw vertical lines through $x = 0.45$ and 0.55 in Figure 5.8 and observe that the empirical simulation-based probability that a randomly selected realization value for the midpoint of the maximum range statistical estimator will lie in this interval is larger than the corresponding probability for either alternative statistical estimator. Accordingly, given alternative unbiased statistical estimators whose conceptual sampling distribution CDF's have different standard deviations (slopes), the minimum variance statistical estimator generates the largest probability that a randomly selected realization value will lie in a narrow central interval around the actual value for the mean of the underlying conceptual statistical distribution.

Remark: The statistical objective is (almost) always to have the estimated value lie as close, in a statistical sense, as possible to the actual value for the conceptual parameter. It is possible that a biased statistical estimator excels an unbiased statistical estimator in this regard. If so, then the biased statistical estimator is generally preferred (especially when a statistical bias correction is available).

5.A.2. Relative Efficiency of Statistical Estimators

The relative efficiency of statistical estimators is defined in terms of the ratio of the actual values of the variances of the respective conceptual sampling distributions for the statistical estimators of specific interest. If we take the actual value for the variance of the conceptual sampling distribution pertaining to the midpoint of the maximum range as our reference (numerator) value, the respective theoretical variance ratios (derivations omitted) generate the relative efficiency curves shown in Figure 5.9. Clearly, neither the arithmetic average nor the median is a statistically efficient estimator of the actual value for the mean of the conceptual uniform distribution, zero to one, especially when n_{dv} is fairly large.

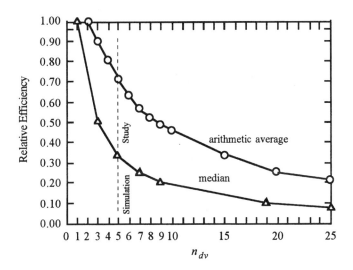

Figure 5.9 Relative efficiency of the arithmetic average and the median as statistical estimators of the actual value for the mean of the conceptual uniform distribution, zero to one—using the actual value for the variance of the conceptual sampling distribution that consists of all possible replicate realization values for the midpoint of the maximum range estimator as the reference (numerator) value. (A similar plot can be developed either analytically or empirically for alternative estimators of the actual value for each parameter of the conceptual statistical distribution of specific interest.

Conclusion: The arithmetic average (or the median) is not necessarily a statistically efficient (effective) estimator of the actual value for the mean of any conceptual statistical distribution of specific interest in mechanical reliability.

Corollary: The statistical literature should be consulted regarding the appropriate statistical estimator (and its conceptual or simulation-based empirical sampling distribution) for each parameter of the conceptual statistical distribution of specific interest.

6

Statistical Analysis of Variance (ANOVA)

6.1 CLASSICAL STATISTICAL ANALYSIS OF VARIANCE (FIXED EFFECTS MODEL)

We now present the statistical test of hypothesis called classical (fixed effects) analysis of variance (ANOVA) for the comparative experiment test programs presented in Chapter 2. The omnibus null hypothesis that the actual values for the respective $ctesc_j$'s are all equal to zero is traditionally tested versus its associated omnibus composite alternative hypothesis that the actual values for the respective $ctesc_j$'s are not all equal to zero. The test statistic of specific interest is Snedecor's central F test statistic, viz., the ratio of the *between*(MS) to the *within*(MS). The *within*(MS) is equal to the sum of squares of the elements of the $|est(CRHNDEE_i\text{'s})|$ column vector, divided by its number of statistical degrees of freedom; whereas the *between*(MS) is equal to the sum of squares of the elements of the aggregated $|est(cte_i\text{'s})|$ column vector, divided by its number of statistical degrees of freedom. Under continual replication of the experiment test program, the respective realizations values for Snedecor's central F test statistic generate Snedecor's central F conceptual sampling distribution *provided that the omnibus null hypothesis is correct*. However, if the actual values for the respective $ctesc_j$'s are not all equal to zero as asserted by the omnibus alternative hypothesis, then the expected value of the *between*(MS) is inflated to

231

account for the magnitudes of the actual values for the $ctesc_j$'s that are not equal to zero. Small actual values for these $ctesc_j$'s will inflate the expected value of the *between*(MS) only a small amount, but large actual values for these $ctesc_j$'s will markedly inflate the expected value of the *between*(MS). Thus, if the experiment test program data-based value for Snedecor's central F test statistic is sufficiently large, we can (must) rationally reject the omnibus null hypothesis and assert instead that the actual values for the $ctesc_j$'s are not all equal to zero.

This classical ANOVA methodology can also be used to test a specific null hypothesis, e.g., the specific null hypothesis that the actual value for a given $ctesc_j$ is equal to zero versus the specific composite (two-sided) alternative hypothesis that the actual value for this $ctesc_j$ is not equal to zero. The associated $ctec_j$ can pertain either to a simple comparison of $ctKm$'s or a compound comparison of $ctKm$'s, e.g., an interaction effect.

> *Remark*: The terminology *fixed effects* model in ANOVA connotes that treatments (treatment levels) are specifically selected (as opposed to being randomly selected to be statistically representative of a population of treatments of specific interest).

6.2 A DEMONSTRATION OF VALIDITY OF SNEDECOR'S CENTRAL F TEST STATISTIC IN ANOVA USING FOUR CONSTRUCTED DATA SETS

We demonstrate in this section that the data-based value for Snedecor's central F test statistic is independent of (a) the actual values for the scalar coefficients in the conceptual statistical model, and (b) the actual value for the variance of the conceptual statistical distribution that is comprised of $APRCRHNDEE$'s. These demonstrations are intended to validate the use of Snedecor's central F test statistic in ANOVA. We start, however, by providing background information regarding the computation of the classical (shortest) $100(scp)\%$ (two-sided) statistical confidence intervals that are of specific interest when the null hypothesis employed in ANOVA is rejected.

Consider the least-square estimator of the actual value for the $ctesc_j$ of specific interest. Presume that the conceptual statistical model is correct. When the null hypothesis that the actual value for this $ctesc_j$ is equal to zero is also presumed to be correct, then the est($ctesc_j$) is equal to its intrinsic statistical estimation error component. Accordingly, we assert that the conceptual sampling distribution that consists of all possible replicate realization values for est($ctesc_j$) under continual replication of the experiment test program is normally distributed, has its mean equal to zero, and has an estimated standard deviation given by the expression:

$$\text{est}\{\text{stddev}[\text{est}(ctesc_j)]\} = \left\{ \frac{\text{est}[\text{var}(APRCRHNDEE\text{'s})]}{\sum_{i=1}^{n_{dv}} c_{j,i}^2} \right\}^{1/2}$$

On the other hand, given the composite (two-sided) alternative hypothesis that the actual value for this $ctesc_j$ is not equal to zero, the mean of the conceptual sampling distribution that consists of all possible replicate realization values for est($ctesc_j$) is equal to the actual value for the $ctesc_j$. (Recall that the least-squares estimator for the actual value of each $ctesc_j$ is unbiased.)

If we reject the null hypothesis that the actual value for the $ctesc_j$ is equal to zero in ANOVA, then it is good statistical practice to compute the classical (shortest) $100(scp)\%$ (two-sided) statistical confidence interval that allegedly includes the actual value for the $ctesc_j$. The relevant generic expression for Student's central $T_{1,n_{dsdf}}$ test statistic is $\{[\text{est}(ctesc_j) - ctesc_j]/ (\text{est}\{\text{stddev}[\text{est}(ctesc_j)]\})\}$, where n_{dsdf} is the number of statistical degrees of freedom pertaining to est$\{$stddev[est($ctesc_j$)]$\}$, viz., pertaining to est[var($APRCRHNDEE$'s). This test statistic can be used in conjunction with Student's $t_{1,n_{dsdf}}$ conceptual sampling distribution to establish the following classical (shortest) $100(scp)\%$ (two-sided) statistical confidence interval:

$$\Big[\text{est}(ctecs_j) - t_{1,n_{dsdf};(1+scp)/2} \cdot \text{est}\{\text{stddev}[\text{est}(ctecs_j)]\},$$

$$\text{est}(ctesc_j) - t_{1,n_{dsdf};(1-scp)/2} \cdot \text{est}\{\text{stddev}[\text{est}(ctecs_j)]\}\Big]$$

that allegedly includes the actual value for the $ctesc_j$, where n_{dsdf} is the number of denominator statistical degrees of freedom.

Next, suppose that the omnibus null hypothesis that all of the $ctesc_j$'s are equal to zero is of specific interest. If we reject this omnibus null hypothesis in ANOVA, then we may be tempted to compute the analogous classical (shortest) $100(scp)\%$ (two-sided) statistical confidence intervals that allegedly include the actual values for the respective $ctKm$'s, viz.,

$$\Big[\text{est}(ctKm) - t_{1,n_{dsdf};(1+scp)/2} \cdot \text{est}\{\text{stddev}[\text{est}(ctKm)]\},$$

$$\text{est}(ctKm) - t_{1,n_{dsdf};(1-scp)/2} \cdot \text{est}\{\text{stddev}[\text{est}(ctKm)]\}\Big]$$

in which the arithmetic average of the n_a treatment K datum values is substituted for est($ctKm$) and

$$\text{est}\{\text{stddev}[\text{est}(ctKm)]\} = \left\{ \frac{\text{est}[\text{var}(APRCRHNDEE\text{'s})]}{n_a} \right\}^{1/2}$$

However, as discussed later, the associated classical (shortest) $100(scp)\%$ (two-sided) statistical confidence intervals are valid individually (separately), but not collectively (simultaneously).

6.2.1. Constructed Data Sets

We now construct four experiment test program data sets to demonstrate that the experiment test program value of Snedecor's central F test statistic does not depend on the actual values of the parameters of the conceptual statistical model. We arbitrarily employ an unreplicated RCBD experiment test program with $n_t = 4$ treatments in each of $n_b = 5$ blocks in these four demonstration examples. However, an equally replicated CRD or an unreplicated SPD experiment test program could just as well have been used in this demonstration.

Table 6.1 presents the 20 by 20 orthogonal array for the unreplicated RCBD experiment test program employed in each of our four demonstration examples. The actual values of the conceptual parameters, generically denoted cp_i's, for the first two demonstration examples were arbitrarily selected as follows: $csmm = 20$, $cb1e = 1$, $cb2e = 2$, $cb3e = 3$, $cb4e = 4$, $cb5e = -10$, $ctAe = -4$, $ctBe = 3$, $ctCe = 2$, and $ctDe = -1$, where the actual values of the cbe_i's and the cte's have been deliberately constrained to sum to zero. These constraints are selected so that estimates generated by microcomputer program $AGESTCV$ are exact when var($APRCRHNDEE$'s) is deliberately set equal to zero in our first demonstration example. Our first demonstration example can then be interpreted as verifying that the intrinsic statistical estimation error component of each est(sc_j) is equal to zero when var($APRCRHNDEE$'s) is equal to zero.

6.2.1.1. Constructed Data Set One

All $CRHNDEE_i$'s were deliberately set equal to zero for Constructed Data Set One. The resulting constructed experiment test program datum values are thus computed simply as the sums of the respective selected values for the $csmm$, the cbe_i's, and the cte_i's. In turn, given these constructed datum values and the array in Table 6.1, appropriately running microcomputer program $AGESTCV$ demonstrates that the least-squares algorithm acting on these datum values exactly estimates the actual values for respective conceptual parameters. Thus, Constructed Data Set One serves to establish reference experiment test program datum values that are subsequently

Table 6.1 Transpose of a Nonunique Orthogonal Augmented Contrast Array for an Unreplicated RCBD Experiment Test Program with Four Treatments in Each of Five Blocks[a]

#	Label	Contrast values
1	|+1's|	+1 +1 +1 +1 +1 +1 +1 +1 +1 +1 +1 +1 +1 +1 +1 +1 +1 +1 +1 +1 +1 +1 +1 +1 +1 +1 +1 +1 +1 +1
2	|cbec(1)'s|	+1 +1 +1 +1 +1 +1 +1 +1 −2 −2 −2 −2 0 0 0 0 0 0 0 0 0
3	|cbec(2)'s|	+1 +1 +1 +1 +1 +1 +1 +1 −2 −2 −2 −2 −3 −3 −3 −3 −3 −3 −3 −3 −3
4	|cbec(3)'s|	+1 +1 +1 +1 +1 +1 +1 +1 −2 −2 −2 −2 −3 −3 −3 −3 −3 −3 −3 −3 −4
5	|cbec(4)'s|	+1 +1 +1 +1 +1 +1 +1 +1 +1 +1 +1 +1 +1 +1 +1 +1 +1 +1 +1 +1 −4 −4 −4 −4
6	|ctec(1)'s|	−1 −1 −1 −1 +1 +1 +1 +1 +1 +1 +1 +1 +1 +1 +1 +1 −1 −1 −1 −1 −1 −1 −1 −1
7	|ctec(2)'s|	+1 +1 +1 +1 −1 −1 −1 −1 +1 +1 +1 +1 +1 +1 +1 +1 +1 +1 +1 +1 +1 +1 +1 +1
8	|ctec(3)'s|	−1 −1 −1 −1 +1 +1 +1 +1 −1 −1 −1 −1 +1 +1 +1 +1 +1 +1 +1 +1 +1 +1 +1 +1
9	|cbtiec(1,1)'s|	+1 +1 +1 +1 0 0 0 0 0 0 0 0 0 0 0 0 0 0 0 0 0
10	|cbtiec(2,1)'s|	−1 −1 −1 −1 +2 +2 +2 +2 −2 −2 −2 −2 0 0 0 0 0 0 0 0 0
11	|cbtiec(3,1)'s|	+1 +1 +1 +1 −1 −1 −1 −1 +1 +1 +1 +1 −3 −3 −3 −3 0 0 0 0 0
12	|cbtiec(4,1)'s|	+1 +1 +1 +1 −1 −1 −1 −1 +1 +1 +1 +1 +1 +1 +1 +1 −4 −4 −4 −4 0
13	|cbtiec(1,2)'s|	+1 +1 +1 +1 0 0 0 0 0 0 0 0 0 0 0 0 0 0 0 0 0
14	|cbtiec(2,2)'s|	−1 −1 −1 −1 +2 +2 +2 +2 −2 −2 −2 −2 0 0 0 0 0 0 0 0 0
15	|cbtiec(3,2)'s|	+1 +1 +1 +1 −1 −1 −1 −1 +1 +1 +1 +1 −3 −3 −3 −3 0 0 0 0 0
16	|cbtiec(4,2)'s|	+1 +1 +1 +1 −1 −1 −1 −1 +1 +1 +1 +1 +1 +1 +1 +1 −4 −4 −4 −4 0
17	|cbtiec(1,3)'s|	+1 +1 +1 +1 0 0 0 0 0 0 0 0 0 0 0 0 0 0 0 0 0
18	|cbtiec(2,3)'s|	−1 −1 −1 −1 +2 +2 +2 +2 −2 −2 −2 −2 0 0 0 0 0 0 0 0 0
19	|cbtiec(3,3)'s|	+1 +1 +1 +1 −1 −1 −1 −1 +1 +1 +1 +1 −3 −3 −3 −3 +1 +1 +1 +1 0
20	|cbtiec(4,3)'s|	+1 +1 +1 +1 −1 −1 −1 −1 +1 +1 +1 +1 +1 +1 +1 +1 −4 −4 −4 −4 −4

[a]This array pertains to all four constructed data sets in this section and appears in microcomputer file *C6ARRAY1*.

aggregated with pseudorandom $CRHNDEE_i$'s to establish the experiment test program datum values for Constructed Data Sets Two, Three, and Four. The respective pseudorandom $CRHNDEE_i$'s for these constructed data sets are generated by running microcomputer program $ANOVADTA$ and thus can be viewed as having been randomly selected from a conceptual (two-parameter) normal distribution whose mean is zero and whose standard deviation is known (is input information to microcomputer program $ANOVADTA$). Thus, the realization of the intrinsic statistical estimation error component of each least-squares estimate can be computed because the actual values for the conceptual parameters are known.

*****Constructed Data Set One*****

$\|$datum value$_i$'s$\|$	=	$\|csmm\|$	+	$\|cbe_i$'s$\|$	+	$\|cte_i$'s$\|$	+	$\|CRHNDEE_i$'s$\|$
17	=	20	+	(+1)	+	(−4)	+	0
24	=	20	+	(+1)	+	(+3)	+	0
23	=	20	+	(+1)	+	(+2)	+	0
20	=	20	+	(+1)	+	(−1)	+	0
18	=	20	+	(+2)	+	(−4)	+	0
25	=	20	+	(+2)	+	(+3)	+	0
24	=	20	+	(+2)	+	(+2)	+	0
21	=	20	+	(+2)	+	(−1)	+	0
19	=	20	+	(+3)	+	(−4)	+	0
26	=	20	+	(+3)	+	(+3)	+	0
25	=	20	+	(+3)	+	(+2)	+	0
22	=	20	+	(+3)	+	(−1)	+	0
20	=	20	+	(+4)	+	(−4)	+	0
27	=	20	+	(+4)	+	(+3)	+	0
26	=	20	+	(+4)	+	(+2)	+	0
23	=	20	+	(+4)	+	(−1)	+	0
6	=	20	+	(−10)	+	(−4)	+	0
13	=	20	+	(−10)	+	(+3)	+	0
12	=	20	+	(−10)	+	(+2)	+	0
9	=	20	+	(−10)	+	(−1)	+	0

Constructed Data Set One is stored in microcomputer file $C6DATA1$. Microcomputer program $AGESTCV$ (with microcomputer file $C6DATA1$ copied into microcomputer file $DATA$ and with microcomputer file $C6ARRAY1$ copied into microcomputer file $ARRAY$) computes the following column vectors in the estimated statistical model:

$\lvert$datum value$_i$'s$\rvert$	$=$	$\lvert$est($csmm$)$\rvert$	$+$	$\lvert$est(cbe_i's)$\rvert$	$+$	$\lvert$est(cte_i's)$\rvert$	$+$	$\lvert$est($CRHNDEE_i$'s$\rvert$
17	$=$	20	$+$	$(+1)$	$+$	(-4)	$+$	0
24	$=$	20	$+$	$(+1)$	$+$	$(+3)$	$+$	0
23	$=$	20	$+$	$(+1)$	$+$	$(+2)$	$+$	0
20	$=$	20	$+$	$(+1)$	$+$	(-1)	$+$	0
18	$=$	20	$+$	$(+2)$	$+$	(-4)	$+$	0
25	$=$	20	$+$	$(+2)$	$+$	$(+3)$	$+$	0
24	$=$	20	$+$	$(+2)$	$+$	$(+2)$	$+$	0
21	$=$	20	$+$	$(+2)$	$+$	(-1)	$+$	0
19	$=$	20	$+$	$(+3)$	$+$	(-4)	$+$	0
26	$=$	20	$+$	$(+3)$	$+$	$(+3)$	$+$	0
25	$=$	20	$+$	$(+3)$	$+$	$(+2)$	$+$	0
22	$=$	20	$+$	$(+3)$	$+$	(-1)	$+$	0
20	$=$	20	$+$	$(+4)$	$+$	(-4)	$+$	0
27	$=$	20	$+$	$(+4)$	$+$	$(+3)$	$+$	0
26	$=$	20	$+$	$(+4)$	$+$	$(+2)$	$+$	0
23	$=$	20	$+$	$(+4)$	$+$	(-1)	$+$	0
6	$=$	20	$+$	(-10)	$+$	(-4)	$+$	0
13	$=$	20	$+$	(-10)	$+$	$(+3)$	$+$	0
12	$=$	20	$+$	(-10)	$+$	$(+2)$	$+$	0
9	$=$	20	$+$	(-10)	$+$	(-1)	$+$	0

Clearly, microcomputer program *AGESTCV* exactly computes the elements in each column vector in the conceptual statistical model when all $CRHNDEE_i$'s are arbitrarily set equal to zero.

6.2.1.2. Constructed Data Set Two

Microcomputer program *ANOVADTA* was run with an input standard deviation equal to 2 to generate the pseudorandom $CRHNDEE_i$'s for Constructed Data Set Two. (The reason for selecting this value for the standard deviation is discussed later.)

*****Constructed Data Set Two*****

$\lvert$datum value$_i$'s$\rvert$	$=$	$\lvert csmm \rvert$	$+$	$\lvert cbe_i$'s$\rvert$	$+$	$\lvert cte_i$'s$\rvert$	$+$	$\lvert CRHNDEE_i$'s$\rvert$
20.912	$=$	20	$+$	$(+1)$	$+$	(-4)	$+$	$(+3.912)$
24.411	$=$	20	$+$	$(+1)$	$+$	$(+3)$	$+$	$(+0.411)$
19.861	$=$	20	$+$	$(+1)$	$+$	$(+2)$	$+$	(-3.139)
21.350	$=$	20	$+$	$(+1)$	$+$	(-1)	$+$	$(+1.350)$
18.268	$=$	20	$+$	$(+2)$	$+$	(-4)	$+$	$(+0.268)$
23.827	$=$	20	$+$	$(+2)$	$+$	$(+3)$	$+$	(-1.173)
23.521	$=$	20	$+$	$(+2)$	$+$	$(+2)$	$+$	(-0.479)

22.750	=	20	+	(+2)	+	(−1)	+	(+1.750)
16.895	=	20	+	(+3)	+	(−4)	+	(−2.105)
25.320	=	20	+	(+3)	+	(+3)	+	(−0.680)
24.283	=	20	+	(+3)	+	(+2)	+	(−0.717)
24.828	=	20	+	(+3)	+	(−1)	+	(+2.828)
16.730	=	20	+	(+4)	+	(−4)	+	(−3.270)
26.648	=	20	+	(+4)	+	(+3)	+	(−0.352)
27.013	=	20	+	(+4)	+	(+2)	+	(+1.013)
24.135	=	20	+	(+4)	+	(−1)	+	(+1.135)
5.997	=	20	+	(−10)	+	(−4)	+	(−0.003)
12.229	=	20	+	(−10)	+	(+3)	+	(−0.771)
9.263	=	20	+	(−10)	+	(+2)	+	(−2.737)
8.312	=	20	+	(−10)	+	(−1)	+	(−0.688)

Constructed Data Set Two is stored in microcomputer file *C6DATA2*. Microcomputer program *AGESTCV* (with microcomputer *C6DATA2* copied into microcomputer file *DATA* and with microcomputer file *C6ARRAY1* copied into microcomputer file *ARRAY*) computes the following column vectors for the estimated statistical model:

|datum value$_i$'s|	=	|est($csmm$)|	+	|est(cbe_i's)|	+	|est(cte_i's)|	+	|est($CRHNDEE_i$'s)|
20.912	=	19.82765	+	(+1.80585)	+	(−4.06725)	+	(+3.34575)
24.411	=	19.82765	+	(+1.80585)	+	(+2.65935)	+	(+0.11815)
19.861	=	19.82765	+	(+1.80585)	+	(+0.96055)	+	(−2.73305)
21.350	=	19.82765	+	(+1.80585)	+	(+0.44735)	+	(−0.73085)
18.268	=	19.82765	+	(+2.26385)	+	(−4.06725)	+	(+0.24375)
23.827	=	19.82765	+	(+2.26385)	+	(+2.65935)	+	(−0.92385)
23.521	=	19.82765	+	(+2.26385)	+	(+0.96055)	+	(+0.46895)
22.750	=	19.82765	+	(+2.26385)	+	(+0.44735)	+	(+0.21115)
16.895	=	19.82765	+	(+3.00385)	+	(−4.06725)	+	(−1.86925)
25.320	=	19.82765	+	(+3.00385)	+	(+2.65935)	+	(−0.17085)
24.283	=	19.82765	+	(+3.00385)	+	(+0.96055)	+	(+0.49095)
24.828	=	19.82765	+	(+3.00385)	+	(+0.44735)	+	(+1.54915)
16.730	=	19.82765	+	(+3.80385)	+	(−4.06725)	+	(−2.83425)
26.648	=	19.82765	+	(+3.80385)	+	(+2.65935)	+	(+0.35715)
27.013	=	19.82765	+	(+3.80385)	+	(+0.96055)	+	(+2.42095)
24.135	=	19.82765	+	(+3.80385)	+	(+0.44735)	+	(+0.05615)
5.997	=	19.82765	+	(−10.87740)	+	(−4.06725)	+	(+1.11400)
12.229	=	19.82765	+	(−10.87740)	+	(+2.65935)	+	(+0.61940)
9.263	=	19.82765	+	(−10.87740)	+	(+0.96055)	+	(−0.64780)
8.312	=	19.82765	+	(−10.87740)	+	(+0.44735)	+	(−1.08560)

Clearly, each conceptual parameter estimate has its realization value for the intrinsic statistical estimation error confounded with its actual value. We next change the values of these conceptual parameters to examine how (if) the respective realization values for this intrinsic statistical estimation error change.

6.2.1.3. Constructed Data Set Three

We now arbitrarily change all parameters in the conceptual statistical model to be equal to zero, but we do not change the $CRHNDEE_i$'s. Our objective is to compare the realizations of the intrinsic statistical estimation error components for Constructed Data Set Three to the corresponding realization components for Constructed Data Set Two to determine how (if) these intrinsic statistical estimation error components change with changes in the parameters of the conceptual statistical model.

*****Constructed Data Set Three*****

$|$datum value$_i$'s$|$	=	$|csmm|$	+	$|cbe_i$'s$|$	+	$|cte_i$'s$|$	+	$|CRHNDEE_i$'s$|$
+3.912	=	0	+	0	+	0	+	(+3.912)
+0.411	=	0	+	0	+	0	+	(+0.411)
−3.139	=	0	+	0	+	0	+	(−3.139)
+1.350	=	0	+	0	+	0	+	(+1.350)
+0.268	=	0	+	0	+	0	+	(+0.268)
−1.173	=	0	+	0	+	0	+	(−1.173)
−0.479	=	0	+	0	+	0	+	(−0.479)
+1.750	=	0	+	0	+	0	+	(+1.750)
−2.105	=	0	+	0	+	0	+	(−2.105)
−0.680	=	0	+	0	+	0	+	(−0.680)
−0.717	=	0	+	0	+	0	+	(−0.717)
+2.828	=	0	+	0	+	0	+	(+2.828)
−3.270	=	0	+	0	+	0	+	(−3.270)
−0.352	=	0	+	0	+	0	+	(−0.352)
+1.013	=	0	+	0	+	0	+	(+1.013)
+1.135	=	0	+	0	+	0	+	(+1.135)
−0.003	=	0	+	0	+	0	+	(−0.003)
−0.771	=	0	+	0	+	0	+	(−0.771)
−2.737	=	0	+	0	+	0	+	(−2.737)
−0.688	=	0	+	0	+	0	+	(−0.688)

Constructed Data Set Three is stored in microcomputer file *C6DATA3*. Microcomputer program *AGESTCV* (with microcomputer file *C6DATA3* copied into microcomputer file *DATA* and with microcomputer file *C6ARRAY1* copied into microcomputer file *ARRAY*) computes the following column vectors for the estimated statistical model:

$|\text{datum value}_i\text{'s}|$	$=$	$|\text{est}(csmm)|$	$+$	$|\text{est}(cbe_i\text{'s})|$	$+$	$|\text{est}(cte_i\text{'s})|$	$+$	$|\text{est}(CRHNDEE_i\text{'s})|$
$+3.912$	$=$	-0.17235	$+$	$(+0.80585)$	$+$	(-0.06725)	$+$	$(+3.34575)$
$+0.411$	$=$	-0.17235	$+$	$(+0.80585)$	$+$	(-0.34065)	$+$	$(+0.11815)$
-3.139	$=$	-0.17235	$+$	$(+0.80585)$	$+$	(-1.03945)	$+$	(-2.73305)
$+1.350$	$=$	-0.17235	$+$	$(+0.80585)$	$+$	$(+1.44735)$	$+$	(-0.73085)
$+0.268$	$=$	-0.17235	$+$	$(+0.26385)$	$+$	(-0.06725)	$+$	$(+0.24375)$
-1.173	$=$	-0.17235	$+$	$(+0.26385)$	$+$	(-0.34065)	$+$	(-0.92385)
-0.479	$=$	-0.17235	$+$	$(+0.26385)$	$+$	(-1.03945)	$+$	$(+0.46895)$
$+1.750$	$=$	-0.17235	$+$	$(+0.26385)$	$+$	$(+1.44735)$	$+$	$(+0.21115)$
-2.105	$=$	-0.17235	$+$	$(+0.00385)$	$+$	(-0.06725)	$+$	(-1.86925)
-0.680	$=$	-0.17235	$+$	$(+0.00385)$	$+$	(-0.34065)	$+$	(-0.17085)
-0.717	$=$	-0.17235	$+$	$(+0.00385)$	$+$	(-1.03945)	$+$	$(+0.49095)$
$+2.828$	$=$	-0.17235	$+$	$(+0.00385)$	$+$	$(+1.44735)$	$+$	$(+1.54915)$
-3.270	$=$	-0.17235	$+$	(-0.19615)	$+$	(-0.06725)	$+$	(-2.83425)
-0.352	$=$	-0.17235	$+$	(-0.19615)	$+$	(-0.34065)	$+$	$(+0.35715)$
$+1.013$	$=$	-0.17235	$+$	(-0.19615)	$+$	(-1.03945)	$+$	$(+2.42095)$
$+1.135$	$=$	-0.17235	$+$	(-0.19615)	$+$	$(+1.44735)$	$+$	$(+0.05615)$
-0.003	$=$	-0.17235	$+$	(-0.87740)	$+$	(-0.06725)	$+$	$(+1.11400)$
-0.771	$=$	-0.17235	$+$	(-0.87740)	$+$	(-0.34065)	$+$	$(+0.61940)$
-2.737	$=$	-0.17235	$+$	(-0.87740)	$+$	(-1.03935)	$+$	(-0.64780)
-0.688	$=$	-0.17235	$+$	(-0.87740)	$+$	$(+1.44735)$	$+$	(-1.08560)

Given the least-squares estimates for Constructed Data Sets Two and Three, we now compare the corresponding values for the respective $[\text{est}(cp_i)]$'s and est($CRHNDEE_i$'s) in these data sets to deduce the effect (if any) of changes in the selected values for the conceptual parameters on the associated intrinsic statistical estimation error components. These comparisons establish and highlight the fundamental statistical behavior that underlies classical ANOVA. First, observe that both sets of constructed datum values have exactly the same numerical values for the est($CRHNDEE_i$'s). Thus, the est($CRHNDEE_i$'s) do not depend on the actual values for the cp_i's in the conceptual statistical model. Second, observe that the respective values for corresponding $[\text{est}(cp_i)]$'s are directly related, viz., each pair of corresponding $[\text{est}(cp_i)]$'s have values that differ by exactly the differences between their actual (selected) values. For example,

$$(19.82765) - (-0.17235) = (20) - (0),$$

$$(1.80585) - (0.80585) = (1) - (0),$$

$$(-4.06725) - (-0.06725) = (4) - (0),$$

$$\text{etc.}$$

These relationships can also be stated as

$$(19.82765) - (20) = (-0.17235) - (0),$$
$$(1.80585) - (1.0) = (+0.80585) - (0),$$
$$\text{etc.}$$

Thus, the intrinsic statistical estimation error component of an est(cp_i) does not depend on the actual value for that cp_i, or on the actual values for any of the other cp_i's in the conceptual statistical model. The results of these two comparisons lead to the following conclusions and perspective.

Conclusion One: Since the least-squares est($CRHNDEE_i$'s) do not depend on the actual values for any of the cp_i's in the conceptual statistical model, the experiment test program data-based value of the *within*(MS) does not depend on the actual values for these cp_i's.

Conclusion Two: When the actual value for a $ctesc_j$ in the conceptual statistical model is equal to zero, the intrinsic statistical estimation error component of this $ctesc_j$ is exactly equal to est($ctesc_j$). Thus, given the null hypothesis that the actual value for this $ctesc_j$ is equal to zero, the experiment test program data-based value for its associated *between*(MS) does not depend on the actual value for either this $ctesc_j$ or for any other $cpsc_j$ in the conceptual statistical model.

Perspective: The fundamental problem in ANOVA is that, even if the actual value for any given $ctesc_j$ in the conceptual statistical model is alleged to be known, it is still statistically confounded with its intrinsic statistical estimation error component. However, the null hypothesis assertion that the actual value for a $ctesc_j$ is equal to zero circumvents this fundamental problem—thereby providing a rational basis for computing the associated null hypothesis rejection probability.

6.2.1.4. Constructed Data Set Four

We now demonstrate that the intrinsic statistical estimation error components of the respective [est(cp_i)]'s and the respective est($CRHNDEE_i$'s) are proportional to the value of the standard deviation used in running microcomputer program *ANOVADTA* to generate the pseudorandom $CRHNDEE_i$'s. Accordingly, we next run program *ANOVADTA* with its input standard deviation equal to 1 (instead of 2) to generate $CRHNDEE_i$'s in Constructed Data Set Four that are only one-half as large as the $CRHNDEE_i$'s in Constructed Data Set Three. (To avoid differences arising from round-off errors, we state the Constructed Data Set Four pseudorandom $CRHNDEE_i$'s to four decimal places.)

*****Constructed Data Set Four*****

$|\text{datum value}_i\text{'s}|$	$=$	$|csmm|$	$+$	$|cbe_i\text{'s}|$	$+$	$|cte_i\text{'s}|$	$+$	$|CRHNDEE_i\text{'s}|$
$+1.9560$	$=$	0	$+$	0	$+$	0	$+$	$(+1.9560)$
$+0.2055$	$=$	0	$+$	0	$+$	0	$+$	$(+0.2055)$
-1.5695	$=$	0	$+$	0	$+$	0	$+$	(-1.5695)
$+0.6750$	$=$	0	$+$	0	$+$	0	$+$	$(+0.6750)$
$+0.1340$	$=$	0	$+$	0	$+$	0	$+$	$(+0.1340)$
-0.5865	$=$	0	$+$	0	$+$	0	$+$	(-0.5865)
-0.2395	$=$	0	$+$	0	$+$	0	$+$	(-0.2395)
$+0.8750$	$=$	0	$+$	0	$+$	0	$+$	$(+0.8750)$
-1.0525	$=$	0	$+$	0	$+$	0	$+$	(-1.0525)
-0.3400	$=$	0	$+$	0	$+$	0	$+$	(-0.3400)
-0.3585	$=$	0	$+$	0	$+$	0	$+$	(-0.3585)
$+1.4140$	$=$	0	$+$	0	$+$	0	$+$	$(+1.4140)$
-1.6350	$=$	0	$+$	0	$+$	0	$+$	(-1.6350)
-0.1760	$=$	0	$+$	0	$+$	0	$+$	(-0.1760)
$+0.5065$	$=$	0	$+$	0	$+$	0	$+$	$(+0.5065)$
$+0.5675$	$=$	0	$+$	0	$+$	0	$+$	$(+0.5675)$
-0.0015	$=$	0	$+$	0	$+$	0	$+$	(-0.0015)
-0.3855	$=$	0	$+$	0	$+$	0	$+$	(-0.3855)
-1.3685	$=$	0	$+$	0	$+$	0	$+$	(-1.3685)
-0.3440	$=$	0	$+$	0	$+$	0	$+$	(-0.3440)

Constructed Data Set Four is stored in microcomputer file *C6DATA4*. Microcomputer program *AGESTCV* (with microcomputer *C6DATA4* copied into microcomputer file *DATA* and with microcomputer file *C6ARRAY1* copied into microcomputer file *ARRAY*) computes the following column vectors for the estimated statistical model:

$|\text{datum value}_i\text{'s}|$	$=$	$|\text{est}(csmm)|$	$+$	$|\text{est}(cbe_i\text{'s})|$	$+$	$|\text{est}(cte_i\text{'s})|$	$+$	$|\text{est}(CRHNDEE_i\text{'s})|$
$+1.9560$	$=$	-0.086175	$+$	$(+0.402925)$	$+$	(-0.033625)	$+$	$(+1.672875)$
$+0.2055$	$=$	-0.086175	$+$	$(+0.402925)$	$+$	(-0.170325)	$+$	$(+0.059075)$
-1.5695	$=$	-0.086175	$+$	$(+0.402925)$	$+$	(-0.519725)	$+$	(-1.366525)
$+0.6750$	$=$	-0.086175	$+$	$(+0.402925)$	$+$	$(+0.723675)$	$+$	(-0.365425)
$+0.1340$	$=$	-0.086175	$+$	$(+0.131925)$	$+$	(-0.033625)	$+$	$(+0.121875)$
-0.5865	$=$	-0.086175	$+$	$(+0.131925)$	$+$	(-0.170325)	$+$	(-0.461925)
-0.2395	$=$	-0.086175	$+$	$(+0.131925)$	$+$	(-0.519725)	$+$	$(+0.234475)$
$+0.8750$	$=$	-0.086175	$+$	$(+0.131925)$	$+$	$(+0.723675)$	$+$	$(+0.105575)$
-1.0525	$=$	-0.086175	$+$	$(+0.001950)$	$+$	(-0.033625)	$+$	(-0.934625)
-0.3400	$=$	-0.086175	$+$	$(+0.001950)$	$+$	(-0.170325)	$+$	(-0.085425)
-0.3585	$=$	-0.086175	$+$	$(+0.001950)$	$+$	(-0.519725)	$+$	$(+0.245475)$
$+1.4140$	$=$	-0.086175	$+$	$(+0.001950)$	$+$	$(+0.723675)$	$+$	$(+0.774575)$

-1.6350	$=$	-0.086175	$+$	(-0.098075)	$+$	(-0.033625)	$+$	(-1.417125)
-0.1760	$=$	-0.086175	$+$	(-0.098075)	$+$	(-0.170325)	$+$	$(+0.178575)$
$+0.5065$	$=$	-0.086175	$+$	(-0.098075)	$+$	(-0.519725)	$+$	$(+1.210475)$
$+0.5675$	$=$	-0.086175	$+$	(-0.098075)	$+$	$(+0.723675)$	$+$	$(+0.028075)$
-0.0015	$=$	-0.086175	$+$	(-0.438700)	$+$	(-0.033625)	$+$	$(+0.557000)$
-0.3855	$=$	-0.086175	$+$	(-0.438700)	$+$	(-0.170325)	$+$	$(+0.309700)$
-1.3685	$=$	-0.086175	$+$	(-0.438700)	$+$	(-0.519725)	$+$	(-0.323900)
-0.3440	$=$	-0.086175	$+$	(-0.438700)	$+$	$(+0.723675)$	$+$	(-0.542800)

First note that the est($CRHNDEE_i$'s) pertaining to Constructed Data Set Four are exactly equal to one-half of the corresponding est($CRHNDEE_i$'s) pertaining to Constructed Data Set Three. Then note that the intrinsic statistical estimation error component of each est(cp_i) in Constructed Data Set Four is also exactly equal to one-half of the intrinsic statistical estimation error component of its corresponding est(cp_i) in Constructed Data Set Three. On the other hand, if the $CRHNDEE_i$'s in Constructed Data Set Four had been generated by running microcomputer program *ANOVADTA* with an input standard deviation equal to 4, then its est($CRHNDEE_i$'s) would have been exactly twice as large as the corresponding est($CRHNDEE_i$'s) in Constructed Data Set Three. Moreover, the intrinsic statistical estimation error component of each est(cp_i) in Constructed Data Set Four would have been twice as large as the intrinsic statistical estimation error component of its corresponding est(cp_i) in Constructed Data Set Three. (The latter two assertions can be verified by modifying Constructed Data Set Four to pertain to the pseudorandom $CRHNDEE_i$'s that appear in microcomputer computer file *C6DATA5*.)

> *Conclusion Three*: Given either (a) the omnibus null hypothesis that the actual values for all of the respective $ctesc_j$'s are equal to zero or (b) the specific null hypothesis that the actual value for a given $ctesc_j$ is equal to zero, the *ratio* of the experiment test program data-based value for the *between*(MS) to the experiment test program data-based value for the *within*(MS) does not depend on the magnitude of var($APRCRHNDEE$'s). Accordingly, the experiment test program data-based value for Snedecor's central F test statistic does not depend on the magnitude of var($APRCRHNDEE$'s).

6.2.1.5. Summary and Perspective

The respective comparisons of the [est(cp_i)]'s in Constructed Data Sets One through Four demonstrate that, given the relevant null hypothesis, the experiment test program data-based value for Snedecor's central F test statistic does not depend either on the actual values for the cp_i's or on the magnitude of var($APRCRHNDEE$'s). Thus, subject to its underlying presumptions being valid (statistically credible), these comparisons demonstrate the validity of using Snedecor's central F test statistic in classical ANOVA for the experiment test programs presented in Chapter 2.

Exercise Set 1

These exercises are intended to confirm the text Conclusions One, Two, and Three and to examine the necessity of the fundamental presumption of normality in classical ANOVA.

1. Run microcomputer program *ANOVADTA* with the same set of three, three-digit odd seed numbers to construct data sets similar to Constructed Data Sets Two and Three. Use the same arbitrarily selected magnitude for var(*APRCRHNDEE*'s), but use different arbitrarily selected values for the *csmm*, the cbe_i's, and the cte_i's. Then, given these two constructed data sets, demonstrate that the respective intrinsic statistical estimation error components do not depend on your arbitrarily selected values. In turn, construct a data set similar to Constructed Data Set Four using the same arbitrarily selected values for the *csmm*, the *cbe*'s, the *cte*'s, and the same set of three, three-digit odd seed numbers as used for Constructed Data Set Three, but use a different arbitrarily selected magnitude for var(*APRCRHNDEE*'s). Then, given these two constructed data sets, demonstrate that the respective experiment test program data-based values for Snedecor's central F test statistic are identical.

2. Rework Exercise 1 using microcomputer program (a) *UNI*, (b) *SEV*, or (c) *LEV* to generate pseudorandom $CRHEE_i$'s for three similar sets of constructed datum values. Do the conclusions confirmed in Exercise 1 depend on the presumption that the $CRHDV_i$'s are normally distributed?

3. If the experiment test program data-based value for Snedecor's central F test statistic does not depend on the presumption of normality, explain (a) why normality must be presumed in classical ANOVA, then (b) why a randomization-based ANOVA is feasible. What presumptions must be made to validate a randomization-based ANOVA?

6.3. CLASSICAL ANOVA USING SNEDECOR'S CENTRAL F TEST STATISTIC

The classical analysis of variance methodology is typically summarized in a tabular format, e.g., Table 6.2 pertains to an equally replicated CRD experiment test program, and Table 6.3 pertains to an unreplicated RCBD experiment test program.

The expected value of the *within*[est($CRHNDEE_i$'s)]MS in Table 6.2 is equal to var($APRCRHNDEE$'s). In turn, given the omnibus null hypothesis that the actual values for the respective $ctesc_j$'s are all equal to zero, the expected value of the *between*[est(cte_i's)]MS is also equal to var($APRCRHNDEE$'s). However, given the omnibus composite alternative hypothesis that the actual values for the respective $ctesc_j$'s are not all equal to zero, the expected value of the *between*[est(cte_i's)]MS is inflated to account for the differences among the actual values for the respective $ctesc_j$'s. Thus, when a sufficiently large data-based value of Snedecor's central F test statistic is computed for the ANOVA experiment test program that is actually conducted, viz., such that the corresponding probability value that pertains to the upper tail of Snedecor's central F conceptual sampling distribution is sufficiently small, then we can (must) rationally reject the omnibus null hypothesis and assert instead that the actual values for the respective $ctesc_j$'s are not all equal to zero.

The omnibus null hypothesis and its associated omnibus composite alternative hypothesis can also be stated in terms of the respective $ctKm$'s. However, when the ANOVA experiment test program involves several treatments, the omnibus composite alternative hypothesis that the actual values of the respective $ctKm$'s are all equal to one another is often too inclusive to be practical. If so, then it is generally preferable to avoid this omnibus null hypothesis and its associated composite alternative hypothesis and work instead with set of $ctec_i$'s that correspond directly to the objectives of the ANOVA experiment test program (Supplemental Topic 2.A).

Given the specific null hypothesis that the actual value for a particular $ctesc_j$ is equal to zero, the associated alternative hypothesis can be: (a) the specific composite (two-sided) alternative hypothesis that the actual value for this particular $ctesc_j$ is not equal to zero, (b) the specific simple (one-sided) alternative hypothesis that the actual value for this particular $ctesc_j$ is greater than zero, or (c) the specific simple (one-sided) alternative hypothesis that the actual value for this particular $ctesc_j$ is less than zero. For (b) and (c) use only one-half of the upper tail probability of Snedecor's central F conceptual sampling distribution as the null hypothesis rejection probability.

The classical (shortest) 100(scp)% (two-sided) confidence interval that allegedly includes the actual value for the given $ctesc_j$ is computed when the

Table 6.2 ANOVA for an Equally Replicated CRD Experiment Test Program[a]

Type of Variability	Sums of Squares (SS)	Number of Statistical Degrees of Freedom (n_{sdf})	Mean Square (MS)
between (estimated conceptual treatment effect$_i$'s) = *between*[est(cte_i's)]	*between*[est(cte_i's)]SS = Σ[est(cte_i's)]^2	$(n_t - 1)$	*between*[est(cte_i's)]MS = *between*[est(cte_i's)]SS/$(n_t - 1)$
within (estimated conceptual random homoscedastic normally distributed experiment error$_i$'s) = *within*[est($CRHNDEE_i$'s)]	*within*[est($CRHNDEE_i$'s)]SS = Σ[est($CRHNDEE_i$'s)]^2	$n_t \cdot (n_r - 1)$	*within*[est($CRHNDEE_i$'s)]MS = *within*[est($CRHNDEE_i$'s)]SS/ $n_t \cdot (n_r - 1)$

Snedecor's central $F_{n_t - 1, n_r \cdot (n_r - 1)}$ test statistic = *between*[est(cte_i's)]MS/*within*[est($CRHNDEE_i$'s)]MS

[a](with n_r replicate datum values for each of the n_t treatments), given the omnibus null hypothesis that the actual values for the respective $ctesc_j$'s are all equal to zero.

Table 6.3 ANOVA for an Unreplicated RCBD Experiment Test Program[a]

Type of Variability	Sums of Squares (SS)	Number of Statistical Degrees of Freedom (n_{sdf})	Mean Square (MS)
between (estimated conceptual block effect$_i$'s) = *between*[est(cbe$_i$'s)]	*between*[est(cbe$_i$'s)]SS = Σ[est(cbe_i's)]²	$(n_b - 1)$	—
between (estimated conceptual treatment effect$_i$'s) = *between*[est(cte$_i$'s)]	*between*[est(cte$_i$'s)]SS = Σ[est(cte$_i$'s)]²	$(n_t - 1)$	*between*[est(cte$_i$'s)]MS = *between*[est(cte$_i$'s)]SS/$(n_t - 1)$
within (estimated presumed random homoscedastic normally distributed experimental error$_i$'s) = *within*[est(*CRHNDEE$_i$*'s)]	*within*[est(*CRHNDEE$_i$*'s)]SS = Σ[est(cbtie$_i$'s)]²	$(n_b - 1)\cdot(n_t - 1)$	*within*[est(*CRHNDEE$_i$*'s)]MS = *within*[est(*CRHNDEE$_i$*'s)]SS/[$(n_b - 1)\cdot(n_t - 1)$]

Snedecor's central $F_{n_t-1,\,(n_b-1)\cdot(n_t-1)}$ test statistic = *between*[est(cte$_i$'s)]MS/*within*[est(*CRHNDEE$_i$*'s)]MS

[a]With n_b blocks, each n_t treatments, given the *omnibus* null hypothesis that the actual values for all of the respective *ctsec*'s are equal to zero.

specific null hypothesis that the actual value for the given $ctesc_j$ is equal to zero is rejected in favor of the specific composite (two-sided) alternative hypothesis that the actual value for the given $ctesc_j$ is not equal to zero. Similarly, a lower (upper) $100(scp)\%$ (one-sided) statistical confidence limit that allegedly bounds the actual value of the given $ctesc_j$ is computed when the specific null hypothesis that the actual value for the given $ctesc_j$ is equal to zero is rejected in favor of the specific simple (one-sided) alternative hypothesis that the actual value for the given $ctesc_j$ is greater (less) than zero.

Now consider Table 6.3. The discussion of Table 6.2 is also relevant here. In addition, we note that the *between*[est(cbe_i's)]MS is not explicitly stated herein because (a) blocks must be specifically selected so that no block,treatment interaction effect is physically credible in an unreplicated RCBD experiment test program, and (b) blocks pertain to nuisance variables and, therefore, the respective est($cbesc_j$'s) are not of specific interest. (Recall that batch-to-batch effects are viewed as treatment effects.) On the other hand, suppose that the blocks in an unreplicated RCBD experiment test program are merely time blocks used as a precautionary measure against possible equipment breakdown. Then, if no equipment breakdown occurs, the *between*[est(cbe_i's)]SS is aggregated with the *within*[est($CRHNDEE_i$'s)]SS and the number of est(cbe_i's) statistical degrees of freedom is aggregated with the number of est($CRHNDEE_i$'s) statistical degrees of freedom, viz., the un-replicated RCBD experiment test program statistically reduces to a CRD experiment test program with $n_b = n_r$ replicates of each of its n_t treatments.

6.3.1. Classical ANOVA for an Unreplicated RCBD Experiment Test Program

We now run microcomputer program *ANOVA* with the orthogonal aug-mented contrast array transpose in Table 6.1 copied in microcomputer file *ARRAY* and Constructed Data Set Two copied in microcomputer file *DATA*. This program first computes the experiment test program data-based value for Snedecor's central F test statistic and then it computes the probability that a value of Snedecor's central F test statistic that is randomly selected from Snedecor's central F conceptual sampling distribution will be greater than or equal to the experiment test program data-based value of Snedecor's central F test statistic.

6.3.1.1. Example One

Consider the omnibus null hypothesis that the actual values for all of the respective $ctesc_j$'s are equal to zero (or that the actual values for all of the respective $ctKm$'s are identical) versus the omnibus composite alternative hypothesis that the actual values for the respective $ctesc_j$'s are not all equal to zero (or that the actual values for the respective $ctKm$'s are not all identical). The *between* [est(cte_i's)]SS thus pertains to columns (rows) 6–8 in Table 6.1, whereas the corresponding *within*[est($CRHNDEE_i$'s)]SS pertains to columns (rows) 9–20.

```
C> COPY C6ARRAY1 ARRAY

     1 file(s) copied

C> COPY C6DATA2 DATA

     1 file(s) copied

C> ANOVA
```

This program presumes that the transpose of the relevant orthogonal augmented contrast array appears in microcomputer file *ARRAY* and the corresponding appropriately ordered experiment test program datum values appear in microcomputer file *DATA*.

Input the adjacent column vectors in the estimated complete analytical model that are aggregated to compute the desired *between*[est(cte_i's)]SS (e.g., for columns 6 through 8, type 6 space 8)

```
6 8
```

Input the adjacent column vectors in the estimated complete analytical model that are aggregated to compute the desired *within*[est($CRHNDEE_i$'s)]SS (e.g., for columns 9 through 20, type 9 space 20)

```
9 20
```

between[est(cte_i's)]SS	=	0.1236872166D + 03
[est(cte_i's)]n_{sdf}	=	3
between[est(cte_i's)]MS	=	0.4122907218D + 02
within[est($CRHNDEE_i$'s)]SS	=	0.4380080620D + 02
[est($CRHNDEE_i$'s)]n_{sdf}	=	12
within[est($CRHNDEE_i$'s)]MS	=	0.3650067183D + 01

The data-based value of Snedecor's central $F_{3,12}$ test statistic for the outcome of the experiment test program that was actually conducted is equal to $0.112954D + 02$.

Given the null hypothesis that the actual value(s) for the $ctesc_j(s)$ of specific interest is (are) equal to zero and given that the experiment test program datum values are random, homoscedastic, and normally distributed, this data-based value of Snedecor's central $F_{3,12}$ test statistic can statistically be viewed as having been randomly selected from Snedecor's central $F_{3,12}$ conceptual sampling distribution. Thus, the probability that a randomly selected outcome of this experiment test program when continually replicated will have its data-based value of Snedecor's central $F_{3,12}$ test statistic equal to or greater than $0.112954D + 02$ is equal to 0.0008. When this probability is sufficiently small, reject the null hypothesis in favor of the alternative hypothesis that the actual value(s) for the $ctesc_j(s)$ of specific interest is (are) not equal to zero.

Given the omnibus null hypothesis that the actual values for the respective $ctesc_j$'s are all equal to zero, the probability that a randomly selected replicate ANOVA experiment test program will have its data-based value for Snedecor's central $F_{3,12}$ test statistic equal to or greater than 11.2954 equals 0.0008. This probability is sufficiently small to warrant rejection of the null hypothesis. Thus, we (correctly) reject the omnibus null hypothesis that the actual values for the respective $ctesc$'s are all equal to zero in favor of the omnibus alternative hypothesis that the actual values for the respective $ctesc_j$'s are not all equal to zero. (The actual values for the respective $ctesc_j$'s are 1, 0.5, and -2.5.)

When the omnibus null hypothesis is rejected, the classical (shortest) $100(scp)\%$ (two-sided) statistical confidence intervals that allegedly include the actual values for the respective $ctKm$'s should be computed. First, we assert that $est[var(APRCRHNDEE\text{'s})] = within[est(CRHNDEE_i\text{'s})]$, where the value for $within[est(CRHNDEE_i\text{'s})]$ is obtained from the output of microcomputer program ANOVA. Thus, $est[var(APRCRHNDEE\text{'s})] = 3.6501$. We then estimate the actual values for the variances of the conceptual sampling distributions for the respective $[est(ctKm)]$'s using the expression $est\{var[est(ctKm)]\} = \{est[var(APRCRHNDEE\text{'s})]\}/n_a$, in which $n_a = n_b$, where n_b ($= 5$) is the number of datum values whose arithmetic average was used to estimate the actual value for each respective $ctKm$. Accordingly, est{var[est($ctKm$)]} = (3.6501/5) = 0.7300 and est{stddev[est($ctkm$)]} = (3.6501/5)^{1/2} = 0.8544. Next, we select the desired value for the scp, say 0.95; then, $(1+0.95)/2 = 0.975$ and $(1-0.95)/2 = 0.025$. In turn, we run microcomputer program TP to establish that $-t_{1,12;0.025} = t_{1,12;0.975} = 2.1788$. Note that the denominator number of statistical degrees of freedom for Student's central t conceptual sampling distribution is the number of statistical degrees of freedom pertaining to the est[var($APRCRHNDEE$'s)]. In turn, we compute the half-widths for the respective classical (shortest) $100(scp)\%$ (two-sided) statistical confidence intervals that allegedly include the actual values for the respective $ctKm$'s as $(2.1788) \cdot (0.8544) = 1.8616$. Accordingly, the classical (shortest) 95% (two-sided) statistical confidence intervals that allegedly include the actual values for the respective $ctKm$'s are equal to their associated estimated values, 15.7604 (A), 22.4870 (B), 20.7882 (C), and 20.2750 (D), ∓ 1.8616. These four intervals, [13.8988, 17.6220], [20.6254, 24.3486], [18.9262, 22.6504], and [18.4134, 22.1366], are valid individually (separately), but not collectively (simultaneously). See Section 6.4.

6.3.1.2. Example Two

Suppose that the four treatments in the unreplicated RCBD experiment test program actually pertain to the four treatment combinations associated with two treatments, each at two levels, in a $(2)^2$ factorial arrangement. Then, the three respective $ctec_i$'s in Table 6.1 must be reinterpreted and reidentified as $ctlec_i$'s, $ct2ec_i$'s, and $ct1t2iec_i$'s, as in Table 6.4. Accordingly, there are now three specific null hypotheses of interest in ANOVA: $H_n(1)$: $ct1esc = 0$, $H_n(2)$: $ct2esc = 0$, and $H_n(3)$: $ct1t2iesc = 0$.

1. Consider the specific null hypothesis that the actual value for the $ct1esc$ is equal to zero versus the specific composite (two-sided) alternative hypothesis that the actual value for the $ct1esc$ is not equal to zero. Running microcomputer program $ANOVA$ in which the respective appropriate adjacent estimated column vectors are aggregated, generates the following results:

```
C> COPY C6ARRAY2 ARRAY


      1 file(s) copied


C> ANOVA
```

This program presumes that the transpose of the relevant orthogonal augmented contrast array appears in microcomputer file *ARRAY* and the corresponding appropriately ordered experiment test program datum values appear in microcomputer file *DATA*.

Input the adjacent column vectors in the estimated complete analytical model that are aggregated to compute the desired *between-*[est(cte_i's)]SS (e.g., for columns 6 through 8, type 6 space 8)

6 6

Input the adjacent column vectors in the estimated complete analytical model that are aggregated to compute the desired *within-*[est($CRHNDEE_i$'s)]SS (e.g., for columns 9 through 20, type 9 space 20)

9 20

between[est(cte_i's)]SS	=	0.4825792445D + 02
[est(cte_i's)]n_{sdf}	=	1
between[est(cte_i's)]MS	=	0.4825792445D + 02
within[est($CRHNDEE_i$'s)]SS	=	0.4380080620D + 02
[est($CRHNDEE_i$'s)]n_{sdf}	=	12
within[est($CRHNDEE_i$'s)]MS	=	0.3650067183D + 01

The data-based value of Snedecor's central $F_{1,12}$ test statistic for the outcome of the experiment test program that was actually conducted is equal to 0.132211D + 02.

Given the null hypothesis that the actual value(s) for the $ctesc_j$(s) of specific interest is (are) equal to zero and given that the experiment test program datum values are random, homoscedastic, and normally distributed, this data-based value of Snedecor's central $F_{1,12}$ test statistic can statistically be viewed as having been randomly selected from Snedecor's central $F_{1,12}$ conceptual sampling distribution. Thus, the probability that a randomly selected outcome of this experiment test program when continually replicated will have its data-based value of Snedecor's central $F_{1,12}$ test statistic equal to or greater than 0.132211D + 02 is equal to 0.0034. When this probability is sufficiently small, reject the null hypothesis in favor of the alternative hypothesis that the actual value(s) for the $ctesc_j$(s) of specific interest is (are) not equal to zero.

Table 6.4 Transpose of a Nonunique Orthogonal Augmented Contrast Array for an Example Unreplicated RCBD Experiment Test Program with Four Treatment Combinations in a $(2)^2$ Factorial Arrangement Within Each of Five Blocks[a]

1	2	3	4	5	6	7	8	9	10	11	12	13	14	15	16	17	18	19	20	No.	Label
+1	+1	+1	+1	+1	+1	+1	+1	+1	+1	+1	+1	+1	+1	+1	+1	+1	+1	+1	+1	1	$\lvert +1$'s$\rvert$
+1	+1	+1	+1	−1	−1	−1	−1	0	0	0	0	0	0	0	0	0	0	0	0	2	$\lvert cbec(1)$'s$\rvert$
+1	+1	+1	+1	+1	+1	+1	+1	−2	−2	−2	−2	0	0	0	0	0	0	0	0	3	$\lvert cbec(2)$'s$\rvert$
+1	+1	+1	+1	+1	+1	+1	+1	+1	+1	+1	+1	−3	−3	−3	−3	0	0	0	0	4	$\lvert cbec(3)$'s$\rvert$
+1	+1	+1	+1	+1	+1	+1	+1	+1	+1	+1	+1	+1	+1	+1	+1	−4	−4	−4	−4	5	$\lvert cbec(4)$'s$\rvert$
−1	+1	−1	+1	−1	+1	−1	+1	−1	+1	−1	+1	−1	+1	−1	+1	−1	+1	−1	+1	6	$\lvert ct1ec$'s$\rvert$
−1	−1	+1	+1	−1	−1	+1	+1	−1	−1	+1	+1	−1	−1	+1	+1	−1	−1	+1	+1	7	$\lvert ct2ec$'s$\rvert$
+1	−1	−1	+1	+1	−1	−1	+1	+1	−1	−1	+1	+1	−1	−1	+1	+1	−1	−1	+1	8	$\lvert ct12iec$'s$\rvert$
−1	+1	−1	+1	+1	−1	+1	−1	0	0	0	0	0	0	0	0	0	0	0	0	9	$\lvert cbt1iec(1)$'s$\rvert$
−1	+1	−1	+1	−1	+1	−1	+1	+2	−2	+2	−2	0	0	0	0	0	0	0	0	10	$\lvert cbt1iec(2)$'s$\rvert$
−1	+1	−1	+1	−1	+1	−1	+1	−1	+1	−1	+1	+3	−3	+3	−3	0	0	0	0	11	$\lvert cbt1iec(3)$'s$\rvert$
−1	+1	−1	+1	−1	+1	−1	+1	−1	+1	−1	+1	−1	+1	−1	+1	+4	−4	+4	−4	12	$\lvert cbt1iec(4)$'s$\rvert$
−1	−1	+1	+1	+1	+1	−1	−1	0	0	0	0	0	0	0	0	0	0	0	0	13	$\lvert cbt2iec(1)$'s$\rvert$
−1	−1	+1	+1	−1	−1	+1	+1	+2	+2	−2	−2	0	0	0	0	0	0	0	0	14	$\lvert cbt2iec(2)$'s$\rvert$
−1	−1	+1	+1	−1	−1	+1	+1	−1	−1	+1	+1	+3	+3	−3	−3	0	0	0	0	15	$\lvert cbt2iec(3)$'s$\rvert$
−1	−1	+1	+1	−1	−1	+1	+1	−1	−1	+1	+1	−1	−1	+1	+1	+4	+4	−4	−4	16	$\lvert cbt2iec(4)$'s$\rvert$
+1	−1	−1	+1	−1	+1	+1	−1	0	0	0	0	0	0	0	0	0	0	0	0	17	$\lvert cbt1t2iec(1)$'s$\rvert$
+1	−1	−1	+1	+1	−1	−1	+1	−2	+2	+2	−2	0	0	0	0	0	0	0	0	18	$\lvert cbt1t2iec(2)$'s$\rvert$
+1	−1	−1	+1	+1	−1	−1	+1	+1	−1	−1	+1	−3	+3	+3	−3	0	0	0	0	19	$\lvert cbt1t2iec(3)$'s$\rvert$
+1	−1	−1	+1	+1	−1	−1	+1	+1	−1	−1	+1	+1	−1	−1	+1	−4	+4	+4	−4	20	$\lvert cbt1t2iec(4)$'s$\rvert$

[a]This array appears in microcomputer file *C6ARRAY2*. Note that the three respective $ctec_i$'s in Table 6.1 are reinterpreted and reidentified. Each of the three reinterpreted and reidentified $ctec_i$'s now has a distinct physical interpretation that is consistent with the organizational structure of the experiment test program, its objective, and its conceptual statistical model (Chapter 2). Note also that the remaining elements in this orthogonal augmented contrast array do not change.

Given the specific null hypothesis that the actual value for the *ctlesc* is equal to zero, the probability that a randomly selected replicate ANOVA experiment test program will have its data-based value for Snedecor's central $F_{1,12}$ test statistic equal to or greater than 13.2211 equals 0.0034. This probability is sufficiently small to warrant rejection of the null hypothesis. Thus, we (correctly) reject the specific null hypothesis that the actual value for the *ctlesc* is equal to zero in favor of the specific alternative (two-sided) hypothesis that the actual value for the *ctlesc* is not equal to zero. (The actual value for the *ctlesc* is equal to 1 and its least-squares estimate is equal to 1.55335.) The issue of concern now is whether to compute 100(*scp*)% (two-sided) statistical confidence intervals that allegedly include the actual values for $(ctlm)_{\text{low}}$ and $(ctlm)_{\text{high}}$, or to compute a 100(*scp*)% (two-sided) statistical confidence interval that allegedly includes the actual value for the associated *ctlesc*.

First, consider the classical (shortest) 95% (two-sided) statistical confidence interval that allegedly includes the actual value for the *ctlesc*. The half-width of this interval is equal to $(2.1788) \cdot (3.6501/20)^{1/2} = 0.9308$ in which $-t_{1,12;0.025} = t_{1,12;0.975} = 2.1788$ and the divisor $(= 20)$ of est[var(*APRCRHNDEE*'s)] is the sum of the squares of the 20 elements of the $|ctlec_i$'s$|$ column vector (Table 6.4). Thus, the classical (shortest) 95% (two-sided) statistical confidence interval that allegedly includes the actual value for the *ctlesc* is equal to 1.55335 ∓ 0.9308, viz., [0.6226, 2.4841]. Note that this interval does not include the value zero and is therefore consistent with our (correct) decision to reject the associated specific null hypothesis. The classical (shortest) 100(*scp*)% (two-sided) statistical confidence interval that just barely includes the value zero has its required half-width when $(1.55335)/[(3.6501/20)^{1/2}] = t_{1,12;p} = 3.6361$, viz., when $p = 0.9983$ and $scp = 1 - 2 \cdot (1 - p) = 0.9966$. Accordingly, *scp* can be increased to 0.9966 (99.66%) before this classical (shortest) 100(*scp*)% (two-sided) statistical confidence interval just barely includes the value zero at its lower limit. Note that the value $2 \cdot (1 - p) = 0.0034$ exactly agrees with the corresponding ANOVA null hypothesis rejection probability. Obviously, this ANOVA outcome could also have been stated in terms of a 100(*scp*)% (two-sided) statistical confidence interval from zero to twice est(*ctlesc*), where this *scp* value is equal to 0.9966.

Now consider the classical (shortest) 95% (two-sided) statistical confidence intervals that allegedly include the actual values for $ctlm_{\text{low}}$ and $ctlm_{\text{high}}$, where est($ctlm_{\text{low}}$) = 18.2743 and est($ctlm_{\text{high}}$) = 21.3810. The half-widths of each of these intervals is equal to $2.1788 \cdot [(3.6501/10)^{1/2}] = 1.3163$, in which the divisor $(= 10)$ of est[var(*APRCRHNDEE*'s)] is the number of datum values whose arithmetic average was used to compute est($ctlm_{\text{low}}$) and est($ctlm_{\text{high}}$). Thus, the classical (shortest) 95% (two-sided) statistical confidence interval that allegedly includes the actual value for $ctlm_{\text{low}}$ is [16.9580, 19.5906], whereas the classical (shortest) 95% (two-sided) statistical confidence interval that allegedly includes the actual value for $ctlm_{\text{high}}$ is [20.0647, 22.6973]. The respective upper and lower limits of these two statistical confidence intervals just barely meet when their half-widths are equal to 1.55335, viz., when $t_{1,12;p} = (1.55335)/[(3.6501/10)^{1/2}] = 2.5711$ and $p = 0.9877$. Then, $2 \cdot (1 - p) = 0.0246$ instead of the correct value, viz., 0.0034 as established by ANOVA and the associated classical (shortest) 99.66% (two-sided) statistical confidence interval that allegedly includes the actual value for the $ctlesc$. This discrepancy highlights the problem of computing separate $100(scp)\%$ (two-sided) statistical confidence intervals that allegedly include the actual values for $ctlm_{\text{low}}$ and $ctlm_{\text{high}}$. These two intervals are valid individually (separately), but not collectively (simultaneously).

Caveat: Be careful when comparing two or more $ctKm$'s because these $ctKm$'s can differ statistically even when their respective classical (shortest) $100(scp)\%$ (two-sided) statistical confidence intervals slightly overlap. The appropriate expression to use for Student's central $T_{1,n_{dsdf}}$ test statistic in developing the classical (shortest) $100(scp)\%$ (two-sided) statistical confidence interval that allegedly includes the actual value for the difference $(ctlm_{\text{high}} - ctlm_{\text{low}})$ is

Student's central $T_{1,n_{dsdf}}$ test statistic

$$= \frac{[\text{est}(ctlm_{\text{high}}) - \text{est}(ctlm_{\text{low}})] - [ctlm_{\text{high}} - ctlm_{\text{low}}]}{\text{est}\{stddev[\text{est}(ctlm_{\text{high}}) - \text{est}(ctlm_{\text{low}})]\}}$$

in which the number of denominator statistical degrees of freedom is equal to the number of statistical degrees of freedom pertaining to est[var($APRCRHNDEE$'s). Accordingly, the half-width of the classical (shortest) $100(scp)\%$ (two-sided) statistical confidence interval that allegedly includes the actual value for the difference $(ctlm_{\text{high}} - ctlm_{\text{low}})$ is numerically equal to $[(1.55335) - (-1.55335)]/\{(3.6501) \cdot [(1/10) + (1/10)]\}^{1/2} = 3.6361$ when this interval just barely includes the value zero. Then, $p = 0.9983$ and $2 \cdot (1 - p) = 0.0034$, which exactly checks with our example ANOVA pertaining to the $ctlesc$ and the associated classical (shortest) $100(scp)\%$ (two-sided) statistical confidence interval that just barely includes the value zero. Note that $3.6361 = (2.5711) \cdot (2)^{1/2}$. See Section 6.4 when statistically comparing the actual values for more than two $ctKm$'s.

2. Consider the specific null hypothesis that the actual value for the *ct2esc* is equal to zero versus the specific composite (two-sided) alternative hypothesis that the actual value for the *ct2esc* is not equal to zero. Running microcomputer program *ANOVA* in which the respective appropriate adjacent estimated column vectors are aggregated, generates the following results:

C> ANOVA

This program presumes that the transpose of the relevant orthogonal augmented contrast array appears in microcomputer file *ARRAY* and the corresponding appropriately ordered experiment test program datum values appear in microcomputer file *DATA*.

Input the adjacent column vectors in the estimated complete analytical model that are aggregated to compute the desired *between-*[est(cte_i's)]SS (e.g., for columns 6 through 8, type 6 space 8)

7 7

Input the adjacent column vectors in the estimated complete analytical model that are aggregated to compute the desired *within-*[est($CRHNDEE_i$'s)]SS (e.g., for columns 9 through 20, type 9 space 20)

9 20

between[est(cte_i's)]SS	=	0.9910912050D + 01
[est(cte_i's)]n_{sdf}	=	1
between[est(cte_i's)]MS	=	0.9910912050D + 01
within[est($CRHNDEE_i$'s)]SS	=	0.4380080620D + 02
[est($CRHNDEE_i$'s)]n_{sdf}	=	12
within[est($CRHNDEE_i$'s)]MS	=	0.3650067183D + 01

The data-based value of Snedecor's central $F_{1,12}$ test statistic for the outcome of the experiment test program that was actually conducted is equal to 0.271527D + 01.

Given the null hypothesis that the actual value(s) for the *ctesc$_j$*(s) of specific interest is (are) equal to zero and given that the experiment test program datum values are random, homoscedastic, and normally dis-

---

tributed, this data-based value of Snedecor's central $F_{1,12}$ test statistic can statistically be viewed as having been randomly selected from Snedecor's central $F_{1,12}$ conceptual sampling distribution. Thus, the probability that a randomly selected outcome of this experiment test program when continually replicated will have its data-based value of Snedecor's central $F_{1,12}$ test statistic equal to or greater than $0.271527D + 01$ is equal to 0.1253. When this probability is sufficiently small, reject the null hypothesis in favor of the alternative hypothesis that the actual value(s) for the $ctesc_j$(s) of specific interest is (are) not equal to zero.

Given the specific null hypothesis that the actual value for the $ct2esc$ is equal to zero, the probability that a randomly selected ANOVA experiment test program will have its data-based value for Snedecor's central $F_{1,12}$ test statistic equal to or greater than 2.71527 equals 0.1253. This probability is not sufficiently small to warrant rejection of the null hypothesis. Accordingly, we (incorrectly) do not reject the specific null hypothesis that the actual value for the $ct2esc$ is equal to zero in favor of the specific composite (two-sided) alternative hypothesis that the actual value for the $ct2esc$ is not equal to zero. (The actual value for the $ct2esc$ is equal to 0.5 and its estimated value is equal to 0.70395.) The classical (shortest) 95% (two-sided) statistical confidence interval that allegedly includes the actual value for the $ct2esc$ is $[-0.2268, +1.6347]$. Note that this interval does include the value zero and is therefore consistent with our (incorrect) decision not to reject the specific null hypothesis that the actual value for the $ct2esc$ is equal to zero in favor of the specific composite (two-sided) alternative that the actual value for the $ct2esc$ is not equal to zero.

3. Consider the specific null hypothesis that the actual value for the $ct1t2iesc$ is equal to zero versus the specific composite (two-sided) alternative hypothesis that the actual value for the $ct1t2iesc$ is not equal to zero. Running microcomputer program *ANOVA* in which the respective appropriate adjacent estimated column vectors are aggregated, generates the following results:

C> ANOVA

This program presumes that the transpose of the relevant orthogonal augmented contrast array appears in microcomputer file *ARRAY* and the corresponding appropriately ordered experiment test program datum values appear in microcomputer file *DATA*.

Input the adjacent column vectors in the estimated complete analytical model that are aggregated to compute the desired *between-*[est(cte_i's)]SS (e.g., for columns 6 through 8, type 6 space 8)

8 8

Input the adjacent column vectors in the estimated complete analytical model that are aggregated to compute the desired *within-*[est($CRHNDEE_i$'s)]SS (e.g., for columns 9 through 20, type 9 space 20)

9 20

$$
\begin{aligned}
between[\text{est}(cte_i\text{'s})]\text{SS} &= 0.6551838005\text{D}+02 \\
[\text{est}(cte_i\text{'s})]n_{sdf} &= 1 \\
between[\text{est}(cte_i\text{'s})]\text{MS} &= 0.6551838005\text{D}+02 \\
within[\text{est}(CRHNDEE_i\text{'s})]\text{SS} &= 0.4380080620\text{D}+02 \\
[\text{est}(CRHNDEE_i\text{'s})]n_{sdf} &= 12 \\
within[\text{est}(CRHNDEE_i\text{'s})]\text{MS} &= 0.3650067183\text{D}+01
\end{aligned}
$$

The data-based value of Snedecor's central $F_{1,12}$ test statistic for the outcome of the experiment test program that was actually conducted is equal to 0.179499D + 02.

Given the null hypothesis that the actual value(s) for the $ctesc_j$(s) of specific interest is (are) equal to zero and given that the experiment test program datum values are random, homoscedastic, and normally distributed, this data-based value of Snedecor's central $F_{1,12}$ test statistic can statistically be viewed as having been randomly selected from

Snedecor's central $F_{1,12}$ conceptual sampling distribution. Thus, the probability that a randomly selected outcome of this experiment test program when continually replicated will have its data-based value of Snedecor's central $F_{1,12}$ test statistic equal to or greater than $0.179499D + 02$ is equal to 0.0012. When this probability is sufficiently small, reject the null hypothesis in favor of the alternative hypothesis that the actual value(s) for the $ctesc_j(s)$ of specific interest is (are) not equal to zero.

Given the specific null hypothesis that the actual value for the $ct1t2iesc$ is equal to zero, the probability that a randomly selected ANOVA experiment test program will have its data-based value for Snedecor's central $F_{1,12}$ test statistic equal to or greater than 17.9499 equals 0.0012. This probability is sufficiently small to warrant rejection of the null hypothesis. Thus, we (correctly) reject the specific null hypothesis that the actual value for the $ct1t2esc$ is equal to zero in favor of the specific composite (two-sided) alternative hypothesis that the actual value for the $ct2esc$ is not equal to zero. (The actual value for the $ct1t2iesc$ is equal to -2.5 and its estimated value is equal to -1.80995). The classical (shortest) 95% (two-sided) statistical confidence interval that allegedly includes the actual value for the $ct1t2iesc$ is $[-2.7407, -0.8792]$. This interval does not include the value zero and is therefore consistent with our (correct) decision to reject the specific null hypothesis that the actual value for the $ct1t2iesc$ is equal to zero in favor of the specific composite (two-sided) alternative that the actual value for the $ct1t2iesc$ is not equal to zero.

Remark: Recall that Constructed Data Sets One and Two were based on the arbitrary cp_i selections: $ct1e = -4$, $ct2e = 3$, $ct3e = 2$, and $ct4e = -1$. These selections, when restated to pertain to a $(2)^2$ factorial arrangement of four treatment combinations, are such that $ct1esc = 1$, $ct2esc = 0.5$, and $ct1t2iesc = -2.5$. Accordingly, for our hypothetical example, the two-factor interaction term dominates its associated main effects. Although such a dominance is rare, it occurs occasionally (e.g., in high strain rate, high-temperature tension testing of median-carbon steels). Sometimes a strictly monotonic transformation of the experiment test program datum values (e.g., a logarithmic transformation) will cause the estimated main effects to dominate the estimated interaction term.

6.3.1.3.   Example Three

Suppose that our unreplicated RCBD experiment test program had included time blocks that were used only to avoid analytical problems if the test equipment failed during the experiment test program. Suppose further that it was subsequently decided to ignore time blocks and to reinterpret this experiment test program as a CRD so that the resulting $|\text{est}(CRHNDEE_i\text{'s})|$ column vector thus had 16 statistical degrees of freedom. The resulting orthogonal augmented contrast array appears in Table 6.5.

Exercise Set 2

These exercises are intended to acquaint you with the use of microcomputer program $ANOVA$.

1. Verify Example Two by running microcomputer program $ANOVA$ appropriately to generate the same data-based values for Snedecor's central F test statistic with the experiment test program datum values found in Constructed Data Sets Three and Four, respectively, copied into microcomputer file $DATA$.

2. Given the CRD experiment test program datum values found in Constructed Data Set Two and the transpose of the orthogonal augmented contrast array found in Table 6.5, run program ANOVA appropriately to test (a) the specific null hypotheses that the actual value for the $ct1esc$ is equal to zero versus the specific composite (two-sided) alternative hypothesis that the actual value for the $ct1esc$ is not equal to zero, (b) the specific null hypotheses that the actual value for the $ct1esc$ is equal to zero versus the specific composite (two-sided) alternative hypothesis that the actual value for the $ct2esc$ is not equal to zero, and (c) the specific null hypotheses that the actual value for the $ct1t2iesc$ is equal to zero versus the specific composite (two-sided) alternative hypothesis that the actual value for the $ct1t2iesc$ is not equal to zero.

3. Compare the respective ANOVA's in Exercise 1 and 2 and discuss what happens when the actual values for the cbe_i's are not negligible, but nevertheless are ignored in ANOVA.

Table 6.5 Transpose of a Nonunique Orthogonal Augmented Contrast Array for an Example CRD Experiment Test Program with Four Treatment Combinations in a $(2)^2$ Factorial Arrangement, Each Treatment Combination with Five Replicates[a]

#	Vector	C1	C2	C3	C4	C5	C6	C7	C8	C9	C10	C11	C12	C13	C14	C15	C16	C17	C18	C19	C20
1	|+1's	+1	+1	+1	+1	+1	+1	+1	+1	+1	+1	+1	+1	+1	+1	+1	+1	+1	+1	+1	+1
2	|ct1ec's	−1	+1	−1	+1	−1	+1	−1	+1	−1	+1	−1	+1	−1	+1	−1	+1	−1	+1	−1	+1
3	|ct2ec's	−1	−1	+1	+1	−1	−1	+1	+1	−1	−1	+1	+1	−1	−1	+1	+1	−1	−1	+1	+1
4	|ct1t2iec's	+1	−1	−1	+1	+1	−1	−1	+1	+1	−1	−1	+1	+1	−1	−1	+1	+1	−1	−1	+1
5	|crhndeec(1)'s	+1	+1	+1	+1	−1	−1	−1	−1	0	0	0	0	0	0	0	0	0	0	0	0
6	|crhndeec(2)'s	+1	+1	+1	+1	+1	+1	+1	+1	−2	−2	−2	−2	0	0	0	0	0	0	0	0
7	|crhndeec(3)'s	+1	+1	+1	+1	+1	+1	+1	+1	+1	+1	+1	+1	−3	−3	−3	−3	0	0	0	0
8	|crhndeec(4)'s	+1	+1	+1	+1	+1	+1	+1	+1	+1	+1	+1	+1	+1	+1	+1	+1	−4	−4	−4	−4
9	|crhndeec(5)'s	−1	+1	−1	+1	+1	−1	+1	−1	0	0	0	0	0	0	0	0	0	0	0	0
10	|crhndeec(6)'s	−1	+1	−1	+1	−1	+1	−1	+1	+2	−2	+2	−2	0	0	0	0	0	0	0	0
11	|crhndeec(7)'s	−1	+1	−1	+1	−1	+1	−1	+1	−1	+1	−1	+1	+3	−3	+3	−3	0	0	0	0
12	|crhndeec(8)'s	−1	+1	−1	+1	−1	+1	−1	+1	−1	+1	−1	+1	−1	+1	−1	+1	+4	−4	+4	−4
13	|crhndeec(9)'s	−1	−1	+1	+1	+1	+1	−1	−1	0	0	0	0	0	0	0	0	0	0	0	0
14	|crhndeec(10)'s	−1	−1	+1	+1	−1	−1	+1	+1	+2	+2	−2	−2	0	0	0	0	0	0	0	0
15	|crhndeec(11)'s	−1	−1	+1	+1	−1	−1	+1	+1	−1	−1	+1	+1	+3	+3	−3	−3	0	0	0	0
16	|crhndeec(12)'s	−1	−1	+1	+1	−1	−1	+1	+1	−1	−1	+1	+1	−1	−1	+1	+1	+4	+4	−4	−4
17	|crhndeec(13)'s	+1	−1	−1	+1	−1	+1	+1	−1	0	0	0	0	0	0	0	0	0	0	0	0
18	|crhndeec(14)'s	+1	−1	−1	+1	+1	−1	−1	+1	−2	+2	+2	−2	0	0	0	0	0	0	0	0
19	|crhndeec(15)'s	+1	−1	−1	+1	+1	−1	−1	+1	+1	−1	−1	+1	−3	+3	+3	−3	0	0	0	0
20	|crhndeec(16)'s	+1	−1	−1	+1	+1	−1	−1	+1	+1	−1	−1	+1	+1	−1	−1	+1	−4	+4	+4	−4

[a]This array appears in microcomputer file C6ARRAY3. It is the same array as presented in Table 6.4, except that the column vectors that pertained to the blocks and to the block,treatment interaction effects in an unreplicated RCBD test program array pertain to $CRHNDEE_i$'s in an equally replicated CRD experiment test program.

6.3.2. Classical ANOVA for an Unreplicated SPD Experiment Test Program with a $(2)^2$ Factorial Arrangement

Analysis of variance for an unreplicated SPD experiment test program differs from that for CRD and RCBD experiment test programs in that no omnibus null hypothesis that pertains to all treatment combinations is relevant. This distinction is due to the hierarchical nature of SPD experiment test programs, viz., split-plots always pertain to nominally identical experimental units within (relative to) each of the respective main-plot treatment experimental units.

As for unreplicated RCBD experiment test programs, unreplicated SPD experiment test programs require the presumption that no block,treatment interaction has a physical basis. The respective $|cbtie_i\text{'s}|$ column vectors in the complete analytical model can then be selectively aggregated and reinterpreted as relevant $|CRHNDEE_i\text{'s}|$ column vectors in the conceptual statistical model: (a) the aggregated $|\text{est}(cbmptie_i\text{'s})|$ column vector is reinterpreted as the $|\text{est}(CRHNDMPTEEE_i\text{'s})|$ column vector in Table 6.6; (b) the aggregated $|\text{est}(cbsptie_i\text{'s})|$ column vector is reinterpreted as the $|\text{est}(CRHNDSPTEEE_i\text{'s})|$ column vector in Table 6.7; and (c) the aggregated $|\text{est}(cbmptsptie_i\text{'s})|$ column vector is reinterpreted as the $|\text{est}(CRHNDMPTSPTIEEE_i\text{'s})|$ column vector in Table 6.8. However, the traditional conceptual statistical model for an unreplicated SPD experiment test program (Snedecor and Cochran, 1967) does not include a $|CRHNDMPTSPTIEEE_i\text{'s}|$ column vector. Rather, the est($CRHNDSPTEEE_i\text{'s}$) in Table 6.7 are aggregated with the est($CRHNDMPTSPTIEEE_i\text{'s}$) in Table 6.8 and interpreted as est($CRHNDSubPlotEE\text{'s}$) in Table 6.9. The $within[\text{est}(CRHNDSubPlotEE_i\text{'s})]$MS is then used in classical ANOVA for this traditional conceptual statistical model (as numerically illustrated below).

We do not recommend using Snedecor and Cochran's traditional model *unless* the null hypothesis that var($APRCRHNDSPTEEE\text{'s}$) = var($APRCRHNDMPTSPTIEEE\text{'s}$) is not rejected in favor of the (one-sided) alternative hypothesis that var($APRCRHNDSPTEEE\text{'s}$) > var($APRCRHNDMPTSPTIEEE\text{'s}$). Accordingly, we recommend performing the ANOVA's in Tables 6.7 and 6.8 and then testing the null hypothesis that var($APRCRHNDSPTEEE_i\text{'s}$) = var($APRCRHNDMPTSPTIEEE_i\text{'s}$) using

Snedecor's central $F_{(n_b-1)\cdot(n_{spt}-1),(n_b-1)\cdot(n_{mpt}-1)\cdot(n_{spt}-1)}$ test statistic

$$\frac{within[\text{est}(CRH\bar{N}DSPTEEE_i\text{'s})]\text{MS}}{within[\text{est}(CRHNDMPTSPTIEEE_i\text{'s})]\text{MS}}$$

continued on page 267

Table 6.6 ANOVA for an Unreplicated SPD Experiment Test Program[a]

Type of Variability	Sums of Squares (SS)	Number of Statistical Degrees of Freedom (n_{sdf})	Mean Square (MS)
between (estimated conceptual block effects) = $between[est(cbe_i\text{'s})]$	$between[est(cbe_i\text{'s})]SS = \Sigma[est(cbe_i\text{'s})]^2$	$(n_b - 1)$	—
between (estimated conceptual main-plot treatment effect$_i$'s) = $between[est(cmpte_i\text{'s})]$	$between[est(cmpte_i\text{'s})]SS = \Sigma[est(cmpte_i\text{'s})]^2$	$(n_{mpt} - 1)$	$between[est(cmpte_i\text{'s})]MS = between[est(cmpte_i\text{'s})]SS/(n_{mpt} - 1)$
within (estimated presumed random homoscedastic normally distributed main-plot treatment effect experimental error$_i$'s) = $within[CRHNDMPTEEE_i\text{'s}]$	$within[est(CRHNDMPTEEE_i\text{'s})]SS = \Sigma[est(cbmptie_i\text{'s})]^2$	$(n_b - 1) \cdot (n_{mpt} - 1)$	$within[est(CRHNDMPTEEE_i\text{'s})]MS = within[est(CRHNDMPTEEE_i\text{'s})]SS/(n_b - 1) \cdot (n_{mpt} - 1)$

Snedecor's central $F_{n_{mpt}-1,(n_b-1)\cdot(n_{mpt}-1)}$ test statistic $= between[est(cmpte_i\text{'s})]MS/within[est(CRHNDMPTEEE_i\text{'s})]MS$

[a]With n_b blocks, given the omnibus null hypothesis that the actual values for the respective $cmptesc_j$'s are all equal to zero.

Table 6.7 Technically Proper ANOVA for an Unreplicated SPD Experiment Test Program[a]

Type of Variability	Sums of Squares (SS)	Number of Statistical Degrees of Freedom (n_{sdf})	Mean Square (MS)
between (estimated conceptual split-plot treatment effect$_i$'s) = *between*[est($cspte_i$'s)]	*between*[est($cspte_i$'s)]SS = Σ[est($cspte_i$'s)]^2	$(n_{spt} - 1)$	*between*[est($cspte_i$'s)]MS = *between*[est($cspte_i$'s)]SS/ $(n_{spt} - 1)$
within (estimated presumed random homoscedastic normally distributed split-plot treatment effect experimental error$_i$'s) = *within*[est($CRHNDSPTEEE_i$'s)]	*within*[est ($CRHNDSPTEEE_i$'s)]SS = Σ[est($cbsptie_i$'s)]^2	$(n_b - 1) \cdot (n_{spt} - 1)$	*within*[est ($CRHNDSPTEEE_i$'s)]MS = *within*[est ($CRHNDSPTEEE_i$'s)]SS/ $(n_b - 1) \cdot (n_{spt} - 1)$

Snedecor's central $F_{n_{spt}-1, (n_b-1) \cdot (n_{spt}-1)}$ test statistic = *between*[est($cspte_i$'s)]MS/*within*[est($CRHNDSPTEEE_i$'s)]MS

[a]With n_b blocks, given the omnibus null hypothesis that the actual values for the respective $cspte_i$'s are all equal to zero.

Table 6.8 Technically Proper ANOVA for an Unreplicated SPD Experiment Test Program[a]

Type of variability	Sums of Squares (SS)	Number of Statistical Degrees of Freedom (n_{sdf})	Mean Square (MS)
between (estimated conceptual main-plot treatment, split-plot treatment interaction effect$_i$'s) = *between*[est($csptie_i$'s)]	*between*[est($cmptsptie_i$'s)]SS = Σ[est($cmptsptie_i$'s)]^2	$(n_{mpt} - 1) \cdot (n_{spt} - 1)$	*between*[est($cmptsptie_i$'s)]MS = *between*[est($cmptsptie_i$'s)]SS/ $(n_{mpt} - 1) \cdot (n_{spt} - 1)$
within (estimated presumed random homoscedastic normally distributed main-plot treatment, split-plot treatment interaction effect experimental error$_i$'s) = *within*[est($CRHNDMPTSPTIEEE_i$'s)]	*within*[est($CRHNDMPTSPTIEEE_i$'s)] SS = Σ[est($cbmptsptie_i$'s)]^2	$(n_b - 1) \cdot (n_{mpt} - 1) \cdot (n_{spt} - 1)$	*within*[est($CRHNDMPTSPTIEEE_i$'s)]MS = *within*[est($CRHNDMPTSPTIEEE_i$'s)]SS/ $(n_b - 1) \cdot (n_{mpt} - 1) \cdot (n_{spt} - 1)$

Snedecor's central $F_{(n_{mpt}-1) \cdot (n_{spt}-1),\, (n_b-1) \cdot (n_{mpt}-1) \cdot (n_{spt}-1)}$ test statistic = *between*[est($cmptsptie_i$'s)]MS/*within*[est($CRHNDMPTSPTIEEE_i$'s)]MS

[a]With n_b blocks, given the omnibus null hypothesis that the actual values for the respective $cmptsptiesc_i$'s are all equal to zero.

Table 6.9 The *within*(MS) Expression Pertaining to the Traditional Conceptual Statistical Model for an Unreplicated SPD Experiment Test Program[a]

Type of Variability	Sums of Squares (SS)	Number of Statistical Degrees of Freedom (n_{sdf})	Mean Square (MS)
within [presumed random homoscedastic normally distributed *SubPlot* experimental error$_i$'s] = *within*[est ($CRHNDSubPlotEE_i$'s)]	*within*[est ($CRHNDSubPlotEE_i$'s)]SS = Σ[est($cbsptie_i$'s)]^2 + Σ[est($cbmptsptie_i$'s)]^2	$(n_b - 1) \cdot (n_{spi} - 1)$ $[1 + (n_{mpt} - 1)]$	*within*[est ($CRHNDSubPlotEE_i$'s)]MS = *within*[est ($CRHNDSubPlotEE_i$'s)]SS/ $(n_b - 1) \cdot (n_{spt} - 1)$ $[1 + (n_{mpt} - 1)]$

[a]This *within*(MS) expression replaces the respective [*within*(MS)] expressions in the technically proper ANOVAs presented in Tables 6.7 and 6.8.

This test can be performed by running microcomputer program $ANOVA$ with the $within$[est($CRHNDSPTEEE_i$'s)]MS serving as the surrogate $between$[est(cte_i's)]MS and the $within$[est ($CRHNDMPTSPTIEEE_i$'s)]MS serving as the surrogate $within$[est($CRHNDEE_i$'s)]MS.

Note that the null hypothesis that var($APRCRHNDSPTEEE$'s) = var($APRCRHNDMPTSPTIEEE$'s) is essentially a null hypothesis of homoscedasticity, except that only the (one-sided) alternative hypothesis that var($APRCRHNDSPTEEE$'s) > var($APRCRHNDMPTSPTIEEE$'s) is relevant in this application. Recall that the est($CRHNDSPTEEE_i$'s) are based on a two-factor interaction, whereas the est($CRHNDMPTSPTIEEE_i$'s) are based on a three-factor interaction. Thus, it is rational to expect var($APRCRHNDSPTEEE$'s) to exceed var($APRCRHNDMPTSPTIEE$'s). Accordingly, if the associated null hypothesis rejection probability is sufficiently small, we rationally reject this null hypothesis (and decline to adopt Snedecor and Cochran's traditional statistical model for an unreplicated SPD experiment test program).

The $within$[est($CRHNDSubPlotEE_i$'s)]MS is used in the following unreplicated SPD experiment test program example. Note that microcomputer program $ANOVA$ requires that the orthogonal augmented contrast array be deliberately rearranged as illustrated in Table 6.9 so that the estimated column vectors that must be aggregated to obtain the |est($CRHNDSubPlotEE_i$'s)| column vector are adjacent to one another.

6.3.2.1. Example Four

Consider an unreplicated split-plot experiment test program with five blocks, two main plots (within each block), and two split-plots (within each main-plot). The main-plot treatments (treatment levels) could be the longitudinal and transverse directions in a sheet or plate. If so, the paired longitudinally oriented and transversely oriented split-plot test specimens must be machined from blanks cut from adjacent locations in the sheet or plate. Then, whatever the split-plot treatments (treatment levels), the respective main-plot treatments (treatment levels) will be as uniform as possible relative to the split-plot treatments (treatment levels). Alternatively, the main-plot treatment (treatment levels) could be two grades of machine screws, where all machine screws of each grade were selected from the same box and thus presumably are as nominally identical as possible (practical) relative to the two split-plot treatments (treatment levels), whatever their nature. The transpose of a nonunique orthogonal augmented contrast array for this unreplicated split-plot experiment test program appears in Table 6.10. This array is subsequently used in three numerical example ANOVA's that are analogous the three numerical example ANOVA's in Example Two.

Table 6.10 Transpose of a Nonunique Orthogonal Augmented Contrast Array for an Unreplicated SPD Experiment Test Program with Five Blocks, Two Main-Plots (Within Each Block), and Two Split-Plots (Within Each Main-Plot)[a]

1	2	3	4	5	6	7	8	9	10	11	12	13	14	15	16	17	18	19	20		
+1	+1	+1	+1	+1	+1	+1	+1	+1	+1	+1	+1	+1	+1	+1	+1	+1	+1	+1	+1	1	|+1's|
+1	+1	+1	+1	−1	−1	−1	−1	0	0	0	0	0	0	0	0	0	0	0	0	2	|cbec(1)'s|
+1	+1	+1	+1	+1	+1	+1	+1	−2	−2	−2	−2	0	0	0	0	0	0	0	0	3	|cbec(2)'s|
+1	+1	+1	+1	+1	+1	+1	+1	+1	+1	+1	+1	−3	−3	−3	−3	0	0	0	0	4	|cbec(3)'s|
+1	+1	+1	+1	+1	+1	+1	+1	+1	+1	+1	+1	+1	+1	+1	+1	−4	−4	−4	−4	5	|cbec(4)'s|
+1	+1	−1	−1	+1	+1	−1	−1	+1	+1	−1	−1	+1	+1	−1	−1	+1	+1	−1	−1	6	|cmptec's|
+1	+1	−1	−1	−1	−1	+1	+1	0	0	0	0	0	0	0	0	0	0	0	0	7	|cmpteeec(1)'s|
+1	+1	−1	−1	+1	+1	−1	−1	−2	−2	+2	+2	0	0	0	0	0	0	0	0	8	|cmpteeec(2)'s|
+1	+1	−1	−1	+1	+1	−1	−1	+1	+1	−1	−1	−3	−3	+3	+3	0	0	0	0	9	|cmpteeec(3)'s|
+1	+1	−1	−1	+1	+1	−1	−1	+1	+1	−1	−1	+1	+1	−1	−1	−4	−4	+4	+4	10	|cmpteeec(4)'s|
+1	−1	+1	−1	+1	−1	+1	−1	+1	−1	+1	−1	+1	−1	+1	−1	+1	−1	+1	−1	11	|csptec's|
+1	−1	−1	+1	+1	−1	−1	+1	+1	−1	−1	+1	+1	−1	−1	+1	+1	−1	−1	+1	12	|cmptsptiec's|
+1	−1	+1	−1	−1	+1	−1	+1	0	0	0	0	0	0	0	0	0	0	0	0	13	|cspteeec(1)'s|
+1	−1	+1	−1	+1	−1	+1	−1	−2	+2	−2	+2	0	0	0	0	0	0	0	0	14	|cspteeec(2)'s|
+1	−1	+1	−1	+1	−1	+1	−1	+1	−1	+1	−1	−3	+3	−3	+3	0	0	0	0	15	|cspteeec(3)'s|
+1	−1	+1	−1	+1	−1	+1	−1	+1	−1	+1	−1	+1	−1	+1	−1	−4	+4	−4	+4	16	|cspteeec(4)'s|
+1	−1	−1	+1	−1	+1	+1	−1	0	0	0	0	0	0	0	0	0	0	0	0	17	|cmptsptieeec(1)'s|
+1	−1	−1	+1	+1	−1	−1	+1	−2	+2	+2	−2	0	0	0	0	0	0	0	0	18	|cmptsptieeec(2)'s|
+1	−1	−1	+1	+1	−1	−1	+1	+1	−1	−1	+1	−3	+3	+3	−3	0	0	0	0	19	|cmptsptieeec(3)'s|
+1	−1	−1	+1	+1	−1	−1	+1	+1	−1	−1	+1	+1	−1	−1	+1	−4	+4	+4	−4	20	|cmptsptieeec(4)'s|

[a]This array appears in microcomputer file C6ARRAY4. It is the same as the array in Table 6.4, except that the $|ct2ec_i's|$ column vector is now reinterpreted as the $|cmptec_i's|$ column vector, the $|ct1ec_i's|$ column vector is now reinterpreted as the $|csptec's|$ column vector, and the $|ct12ec_i's|$ column vector is now reinterpreted as the $|cmptsptiec_i's|$ column vector—and the associated $|cmpteeec(k)'s|$, $|cspteeec(k)'s|$ and $|cmptsptieeec(k)'s|$ column vectors are appropriately reinterpreted and rearranged (relocated).

1. Consider the specific null hypothesis that the actual value for the *cmptesc* is equal to zero versus the specific composite (two-sided) alternative hypothesis that the actual value for the *cmptesc* is not equal to zero. Running microcomputer program *ANOVA* in which the respective appropriate adjacent estimated column vectors are aggregated, gives the following results:

```
C> COPY C6ARRAY4 ARRAY

    1 file(s) copied


C> ANOVA
```

This program presumes that the transpose of the relevant orthogonal augmented contrast array appears in microcomputer file *ARRAY* and the corresponding appropriately ordered experiment test program datum values appear in microcomputer file *DATA*.

Input the adjacent column vectors in the estimated complete analytical model that are aggregated to compute the desired *between-*[est(cte_i's)]SS (e.g., for columns 6 through 8, type 6 space 8)

```
6 6
```

Input the adjacent column vectors in the estimated complete analytical model that are aggregated to compute the desired *within-*[est($CRHNDEE_i$'s)]SS (e.g., for columns 9 through 20, type 9 space 20)

```
7 10
```

$$\begin{array}{lll}
between[\text{est}(cte_i\text{'s})]SS & = & 0.9910912050D+01 \\
[\text{est}(cte_i\text{'s})]n_{sdf} & = & 1 \\
between[\text{est}(cte_i\text{'s})]MS & = & 0.9910912050D+01 \\
within[\text{est}(CRHNDEE_i\text{'s})]SS & = & 0.2576384720D+02 \\
[\text{est}(CRHNDEE_i\text{'s})]n_{sdf} & = & 4 \\
within[\text{est}(CRHNDEE_i\text{'s})]MS & = & 0.6440961800D+01
\end{array}$$

The data-based value of Snedecor's central $F_{1,4}$ test statistic for the outcome of the experiment test program that was actually conducted is equal to $0.153873D + 01$.

Given the null hypothesis that the actual value(s) for the $ctesc_j$(s) of specific interest is (are) equal to zero and given that the experiment test program datum values are random, homoscedastic, and normally distributed, this data-based value of Snedecor's central $F_{1,4}$ test statistic can statistically be viewed as having been randomly selected from Snedecor's central $F_{1,4}$ conceptual sampling distribution. Thus, the probability that a randomly selected outcome of this experiment test program when continually replicated will have its data-based value of Snedecor's central $F_{1,4}$ test statistic equal to or greater than $0.153873 + 01$ is equal to 0.2826. When this probability is sufficiently small, reject the null hypothesis in favor of the alternative hypothesis that the actual value(s) for the $ctesc_j$(s) of specific interest is (are) not equal to zero.

Given the specific null hypothesis that the actual value for the *cmptesc* is equal to zero, the probability that a randomly selected replicate ANOVA experiment test program will have its data-based value for Snedecor's central $F_{1,4}$ test statistic equal to or greater than 1.53873 equals 0.2826. This probability is not sufficiently small to warrant rejection of the null hypothesis. Accordingly, we (incorrectly) do not reject the specific null hypothesis that the actual value for the *cmptesc* is equal to zero in favor of the specific composite (two-sided) alternative hypothesis that the actual value for the *cmptesc* is not equal to zero. (The actual value for the *cmptesc* is equal to 0.5 and its least-squares estimate is equal to 0.70395.) Nevertheless, we now compute the classical (shortest) 95% (two-sided) statistical confidence interval that allegedly includes the actual value for the *cmptesc*. Its half-width is equal to $[6.4410/20)^{1/2}] \cdot (-t_{1,4;0.025} = t_{1,4;0.975} = 2.7764) = 1.5756$. Accordingly, the classical (shortest) 95% (two-sided) statistical confidence interval that allegedly includes the actual value for the *cmptesc* is $[-0.8716, 2.2795]$. Note that this interval includes the value zero and is thus consistent with our (incorrect) decision not to reject the specific null hypothesis in favor of the composite (two-sided) alternative hypothesis that the actual value for the *cmptesc* is not equal to zero.

2. Consider the specific null hypothesis that the actual value for the *csptesc* is equal to zero versus the specific composite (two-sided) alternative

hypothesis that the actual value for the *csptesc* is not equal to zero. We "validate" the use of the *within*[est($CRHNDSubPlotEE_i$'s)]MS in this numerical example in Section 6.5 by running microcomputer program *RBBHT* to demonstrate that the null hypothesis that var($APRCRHNDSPTEEE$'s) = var($APRCRHNDMPTSPTIEEE$'s) cannot rationally be rejected. Accordingly, the following numerical results are obtained by running microcomputer program *ANOVA*:

C> ANOVA

This program presumes that the transpose of the relevant orthogonal augmented contrast array appears in microcomputer file *ARRAY* and the corresponding appropriately ordered experiment test program datum values appear in microcomputer file *DATA*.

Input the adjacent column vectors in the estimated complete analytical model that are aggregated to compute the desired *between*-[est(cte_i's)]SS (e.g., for columns 6 through 8, type 6 space 8)

11 11

Input the adjacent column vectors in the estimated complete analytical model that are aggregated to compute the desired *within*-[est($CRHNDEE_i$'s)]SS (e.g., for columns 9 through 20, type 9 space 20)

13 20

$$
\begin{array}{lcl}
between[\text{est}(cte_i\text{'s})]\text{SS} & = & 0.4825792445\text{D} + 02 \\
[\text{est}(cte_i\text{'s})]n_{sdf} & = & 1 \\
between[\text{est}(cte_i\text{'s})]\text{MS} & = & 0.4825792445\text{D} + 02 \\
within[\text{est}(CRHNDEE_i\text{'s})]\text{SS} & = & 0.1803695900\text{D} + 02 \\
[\text{est}(CRHNDEE_i\text{'s})]n_{sdf} & = & 8 \\
within[\text{est}(CRHNDEE_i\text{'s})]\text{MS} & = & 0.2254619875\text{D} + 01
\end{array}
$$

The data-based value of Snedecor's central $F_{1,8}$ test statistic for the outcome of the experiment test program that was actually conducted is equal to 0.214040D + 02.

Given the null hypothesis that the actual value(s) for the $ctesc_j$(s) of specific interest is (are) equal to zero and given that the experiment test program datum values are random, homoscedastic, and normally distributed, this data-based value of Snedecor's central $F_{1,8}$ test statistic can statistically be viewed as having been randomly selected from Snedecor's central $F_{1,8}$ conceptual sampling distribution. Thus, the probability that a randomly selected outcome of this experiment test program when continually replicated will have its data-based value of Snedecor's central $F_{1,8}$ test statistic equal to or greater than $0.214040D + 02$ is equal to 0.0017. When this probability is sufficiently small, reject the null hypothesis in favor of the alternative hypothesis that the actual value(s) for the $ctesc_j$(s) of specific interest is (are) not equal to zero.

Given the specific null hypothesis that the actual value for the $csptesc$ is equal to zero is correct, the probability that a randomly selected replicate ANOVA experiment test program will have its data-based value for Snedecor's central $F_{1,8}$ test statistic equal to or greater than 21.4040 equals 0.0017. This probability is sufficiently small to warrant rejection of the null hypothesis. Thus, we (correctly) reject the specific null hypothesis that the actual value for the $csptesc$ is equal to zero in favor of the specific composite (two-sided) alternative hypothesis that the actual value for the $csptesc$ is not equal to zero. (The actual value for the $csptesc$ is equal to 1 and its least-squares estimate is equal to 1.55335.) The classical (shortest) 95% (two-sided) statistical confidence interval that allegedly includes the actual value for the $csptesc$ is $[0.7791, 2.3276]$. The half-width of the interval is equal to $[(2.2546/20)^{1/2}] \cdot (> -t_{1,8;0.025} = t_{1,8;0.975} = 2.3060) = 0.7742$.

Suppose, in contrast, that the specific simple (one-sided) alternative hypothesis that the actual value for $csptesc$ is greater than zero is physically relevant. If so, the probability of randomly selecting a replicate ANOVA experiment test program that has it data-based value for Snedecor's central $F_{1,8}$ test statistic equal to 21.4040 or larger equals 0.00085 (viz., $0.0017/2$). Thus, we (correctly) reject the specific null hypothesis that the actual value for the $csptesc$ is equal to zero in favor of the specific simple (one-sided) alternative hypothesis that the actual value for the $csptesc$ is greater than zero. Then, since the null hypothesis is rejected, we compute the associated lower $100(scp)$% (one-sided) statistical confidence limit that allegedly bounds the actual value for the $csptesc$. For $scp = 0.95$, this lower limit is equal to 0.9290, viz., $1.55335 - t_{1,8;0.95} \cdot (2.2546/20)^{1/2}$, where $t_{1,8;0.95} = 1.8595$. Note that this lower limit exceeds zero and is therefore consistent with our (correct) decision to reject the specific null hypothesis that the actual value for the *csptesc* is equal to zero in favor of the specific simple (one-sided) alternative hypothesis that the actual value for the *csptesc* is greater than zero.

3. Consider the specific null hypothesis that the actual value for the *cmptsptiesc* is equal to zero versus the specific composite (two-sided) alternative hypothesis that the actual value for the *cmptsptiesc* is not equal to zero. The following results are obtained by running microcomputer program *ANOVA*:

C> ANOVA

This program presumes that the transpose of the relevant orthogonal augmented contrast array appears in microcomputer file *ARRAY* and the corresponding appropriately ordered experiment test program datum values appear in microcomputer file *DATA*.

Input the adjacent column vectors in the estimated complete analytical model that are aggregated to compute the desired *between-*[est(cte_i's)]SS (e.g., for columns 6 through 8, type 6 space 8)

12 12

Input the adjacent column vectors in the estimated complete analytical model that are aggregated to compute the desired *within-*[est($CRHNDEE_i$'s)]SS (e.g., for columns 9 through 20, type 9 space 20)

13 20

between[est(cte_i's)]SS	=	0.6551838005D + 02
[est(cte_i's)]n_{sdf}	=	1
between[est(cte_i's)]MS	=	0.6551838005D + 02
within[est($CRHNDEE_i$'s)]SS	=	0.1803695900D + 02
[est($CRHNDEE_i$'s)]n_{sdf}	=	8
within[est($CRHNDEE_i$'s)]MS	=	0.2254619875D + 01

The data-based value of Snedecor's central $F_{1,8}$ test statistic for the outcome of the experiment test program that was actually conducted is equal to 0.290596D + 02.

Given the null hypothesis that the actual value(s) for the $ctesc_j$(s) of specific interest is (are) equal to zero and given that the experiment test program datum values are random, homoscedastic, and normally distributed, this data-based value of Snedecor's central $F_{1,8}$ test statistic can statistically be viewed as having been randomly selected from Snedecor's central $F_{1,8}$ conceptual sampling distribution. Thus, the probability that a randomly selected outcome of this experiment test program when continually replicated will have its data-based value of Snedecor's central $F_{1,8}$ test statistic equal to or greater than $0.290596D + 02$ is equal to 0.0007. When this probability is sufficiently small, reject the null hypothesis in favor of the alternative hypothesis that the actual value(s) for the $ctesc_j$(s) of specific interest is (are) not equal to zero.

Given the specific null hypothesis that the actual value for the *cmptsptiesc* is equal to zero, the probability that a randomly selected replicate ANOVA experiment test program will have its data-based value for Snedecor's central $F_{1,8}$ test statistic equal to or greater than 29.0596 equals 0.0007. This probability is sufficiently small to warrant rejection of the null hypothesis. Thus, we (correctly) reject the specific null hypothesis the actual value for the *cmptsptiesc* is equal to zero in favor of the specific composite (two-sided) alternative hypothesis that the actual value for the *cmptsptiesc* is not equal to zero. (The actual value for the *cmptsptiesc* is equal to -2.5 and its least-squares estimate is equal to -1.80995.)

Exercise Set 3

These exercises are intended to provide perspective regarding classical (shortest) $100(scp)\%$ (two-sided) statistical confidence intervals pertaining to the respective treatment and interaction effect scalar coefficients in a $(2)^2$ factorial arrangement.

1. Rationalize the magnitude of the divisor ($= 20$) for est[var($APRCRHNDEE$'s)] in computing est{var[est($ct1esc$)]} in Example Two by recalling that est($ctle$'s) can be algebraically computed as one-half the difference between the corresponding respective treatment arithmetic averages. (a) How many datum values are used to compute each of the two respective treatment arithmetic averages? (b) What is the actual value for the variance of the conceptual sampling distributions for each of these arith-

metic averages? (c) What is the actual value for the variance of the conceptual sampling distribution that consists of all possible replicate realizations values for the algebraic difference of two of these arithmetic averages? (d) What is the actual value for the variance of the conceptual sampling distribution that consists of all possible replicate realization values for one-half of the algebraic difference for two of these arithmetic averages? (e) What divisor is relevant for the actual value for the variance of the conceptual sampling distribution that consists of all possible replicate realizations values for the statistic est($ct1esc$)?

2. (a) Reconsider the three classical (shortest) 100(scp)% (two sided) statistical confidence intervals presented in Example Four that allegedly include the actual value for the *cmptesc*, the *csptesc*, and the *cmptsptiesc*. Compute the respective scp's for those statistical confidence intervals whose half-widths are just large enough to include the value zero. Then, (b) compare these scp values to the null hypothesis rejection probability values pertaining to the corresponding ANOVA's.

6.3.3. Classical ANOVA Summary and Perspective

It is easy to construct experiment test program datum values that will warrant rejection of the null hypothesis of specific interest. When the magnitudes of typical elements of the |est($CRHNDEE_i$'s)| column vector are much smaller than the elements of the |est(cte_i's)| column vector of specific interest, the experiment test program data-based value for the associated Snedecor's central F test statistic will likely be sufficiently large to warrant rejection of the null hypothesis. However, even if the actual values for the $ctesc_j$'s are all exactly equal to zero, as in Constructed Data Sets Three and Four, the respective est($ctesc_j$'s) will not be equal to zero and thus there is a possibility (however small) that this null hypothesis will be incorrectly rejected. Thus, the probability of correctly or incorrectly rejecting a null hypothesis of specific interest is akin to the probability that a classical (shortest) 100(scp)% (two-sided) statistical confidence interval does or does not include the actual value for the quantity of specific interest. It is never known in any practical situation whether we have correctly rejected or failed to reject the null hypothesis. We can only assert that, under continual replication of the given experiment test program, we will correctly reject the null hypothesis in $100 \cdot (1-p)$% of the respective ANOVA's, where p is the null hypothesis rejection probability. Accordingly, when we reject a null hypothesis, we can assert with $100 \cdot (1-p)$% statistical confidence that we have correctly rejected this null hypothesis.

Remark: A simulation study of a continually replicated statistical test of hypothesis will generate an outcome that is akin to that generated by running microcomputer program *SSTSSCI1*.

When the objective of the experiment test program is to detect cte_i's whose actual values are at least equal to some minimum practical value it is imperative to have a high probability of rejecting the null hypothesis when the alternative hypothesis is correct. This probability is called statistical power (Section 6.6).

6.4. COMPARING ALL PAIRS OF CONCEPTUAL TREATMENT MEANS

We do not recommend the traditional statistical practice of comparing all pairs of conceptual treatment means. Rather, we recommend selecting sets of $cetc_i$'s that are consistent with the experiment test program objectives. Nevertheless, if it is desired to make all $(n_t) \cdot [(n_t - 1)/2]$ comparisons of paired $ctKm_i$'s, then a plot of *corrected* experiment test program datum values similar to Figure 6.1 should be prepared for all experiment test program with blocks. Visual examination of this plot should support subsequent statistical conclusions.

Several statistical procedures are available for simultaneously comparing the actual values for all $(n_t) \cdot [(n_t - 1)/2]$ combinations of paired $ctKm$'s. However, Fisher's protected t test, which is conducted only after the omnibus null hypothesis has been rejected in classical ANOVA, is adequate for most reliability applications. Consider, for example, the ANOVA for the unreplicated RCBD experiment test program in Example One where the omnibus null hypothesis that the actual values for all four $ctKm$'s are equal was rejected in favor of the omnibus composite alternative hypothesis that not all of the actual values for these four $ctKm$'s are equal. Recall that est[var($APRCRHNDEE$'s)] had 12 statistical degrees of freedom and was equal to 3.6501. Suppose that our null hypothesis rejection decision was based on an acceptable probability of committing a *Type I* error equal to 0.05. Then, the appropriate value of Student's central t test statistic for Fisher's protected t test is $t_{1,12;(1+scp)/2} = t_{1,12;0.975} = 2.1788$. The resulting value for the least significant difference, *lsd*, is $(2.1788) \cdot \{[(3.6501/5) + (3.6501/5)]^{1/2}\} = 2.6327$ (units). Accordingly, we assert that we have approximately 95% statistical confidence that the actual values for any given pair of $ctKm$'s differ if their associated est($ctKm$'s) differ by more than 2.6327 (units). Since the respective est($ctKm$'s) in Example One are 15.7604 (A), 22.4870 (B), 20.7882 (C), and 20.2750 (D), we (correctly) assert with approximately 95% statistical

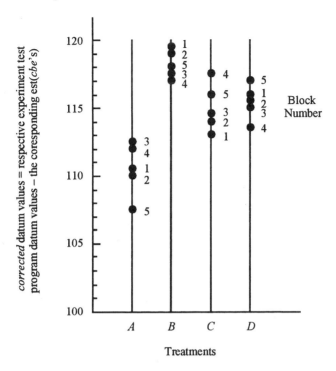

Figure 6.1 Plot of hypothetical unreplicated RCBD experiment test program datum values, *corrected* by subtracting the corresponding est(*cbe*'s). This plot often suffices to draw obvious conclusions regarding which *ctKm*'s differ.

confidence that *ctAm* differs from *ctBm*, that *ctAm* differs from *ctCm*, and that *ctAm* differs from *ctDm*. However, we (incorrectly) fail to detect three additional differences that actually occur, viz., that *ctBm* differs from *ctCm*, that *ctBm* differs from *ctDm*, and that *ctCm* differs from *ctDm*.

> *Remark*: Recall that good statistical practice requires keeping the number of treatments of specific interest to as few as practical while increasing the amount of (equal) replication for these treatments as much as practical.

6.5. CHECKING THE PRESUMPTIONS UNDERLYING CLASSICAL ANOVA

The respective ANOVA-based est($CRHNDEE_i$'s) should always be examined relative to normality, homoscedasticity, and randomness (independence, viz., lack of associations).

6.5.1. Normality

The null hypothesis of normality for the $CRHNDEE_i$'s of specific interest in ANOVA can be tested by running simulation-based microcomputer program *ANOVANT*. This program first computes the ANOVA-based value of the generalized modified Michael's *MDSPP* test statistic for the experiment test program that was actually conducted. It then (a) generates n_{sim} additional ANOVA's for this experiment test program, each pertaining to datum values constructed using pseudorandom normally distributed $CRHNDEE_i$'s, and (b) computes the associated n_{sim} ANOVA-based generalized modified Michael's *MDSPP* test statistic values. In turn, it counts the number of these n_{sim} values that are equal to or greater than the ANOVA-based value for the experiment that was actually conducted and computes the corresponding simulation-based null hypothesis rejection probability.

Note that, to run microcomputer program *ANOVANT*, the experiment test program datum values must be augmented to have the format illustrated in microcomputer file *AANOVDTA*.

```
C>COPY C6ARRAY1 ARRAY

    1 file(s) copied

C>COPY AANOVDTA DATA

    1 file(s) copied

C>ANOVANT
```

The ANOVA-based value of the generalized modified Michael's *MDSPP* test statistic for the experiment test program that was actually conducted is equal to 0.1010.

This microcomputer program constructed 10,000 sets of normally distributed datum values for this experiment test program and then calculated the associated ANOVA-based values of the generalized modified Michael's $MDSPP$ test statistic. The number of these values that were equal to or greater than 0.1010 was equal to 5287. Thus, given the null hypothesis of normality, the simulation-based probability that a randomly selected outcome of this experiment test program when continually replicated will have its generalized modified Michael's $MDSPP$ test statistic value equal to or greater than 0.1010 is equal to 0.5287. When this probability is sufficiently small, reject the null hypothesis in favor of the alternative hypothesis of nonnormality.

```
C>TYPE AANOVDTA
```

20	Number of Experiment Test Program Datum Values
20.912	
24.411	
19.861	
21.350	

⋮

9	Adjacent Estimated Column Vectors that Are Aggregated to Form the $|est(CRHNDEE_i\text{'s})|$ Column
20	Vector of Specific Interest, First to Last (Inclusive)
10000	n_{sim}, the Number of Simulation-Based Replicate Experiment Test Programs
277 911 819	A New Set of Three, Three-Digit Odd Seed Numbers

These datum values are the datum values that appear in microcomputer file $C6DATA2$, augmented as indicated.

6.5.2. Homoscedasticity (and Testing the Null Hypothesis of the Equality of Variances When Appropriate)

The presumption of homoscedasticity for the respective $CRHNDEE_i$'s cannot be examined statistically for a paired-comparison experiment test program. However, with this exception, the presumption of homoscedasticity should always be statistically examined in classical ANOVA. Microcomputer program *BARTLETT* employs Bartlett's likelihood-ratio-based test statistic to test the null hypothesis of homoscedasticity. This test statistic has two shortcomings: (a) it pertains only to CRD experiment test programs, and (b) its conceptual sampling distribution is asymptotic. However, shortcoming (a) can be circumvented by examining the respective ANOVA-based est($CRHNDEE_i$'s) for the experiment test program design of specific interest. In turn, shortcoming (b) can be circumvented by running randomization-based microcomputer program *RBBHT* to calculate the associated null hypothesis rejection probability. However, when the ANOVA-based est($CRHNDEE_i$'s) contain groups of repeated values (ignoring signs), the respective ANOVA-based est($CRHNDEE_i$'s) must be appropriately edited to be statistically equivalent to the (nonrepeated) [est($CRHNDEE_i$'s)] that pertain to a CRD experiment test program with n_b replicates for each of its n_t treatments (treatment levels, treatment combinations). For example, microcomputer file *RBBHTDTA* contains the appropriately edited Example Four unreplicated SPD experiment test program ANOVA-based est($CRHNDSPTEEE_i$'s) and est($CRHNDMPTSPTIEEE_i$'s) that are subsequently employed in microcomputer programs *BARTLETT* and *RBBHT* to test the null hypothesis that var($APRCRHNDSPTEEE$'s) = var($APRCRHNDMPTSPTIEEE$'s). Recall, however, that the appropriate alternative hypothesis for combining the est($CRHNDSPTEEE_i$'s) with the est($CRHNDMPTSPTIEEE_i$'s) to form the est($CRHNDSub$-$PlotEE_i$'s) is simple (one-sided), viz., is var($APRCRHNDSPTEEE$'s) > var(APR $CRHNDMPTSPTIEEE$'s), whereas programs *BARTLETT* and *RBBHT* pertain to the composite (two-sided) alternative hypothesis that var($APRCRHNDSPTEEE$'s) $\neq$ var($APRCRHND$-$MPTSPTIEEE$'s).

C> RBBHTDTA

| 10 | Total Number of Nonrepeated est($CRHNDEE_i$'s) |
| 2 | Number of Treatments, n_t, Each With an Equal Number of Nonrepeated est($CRHNDEE_i$'s) |

```
+0.30635
+0.35635
−0.68915
−0.20665
+0.23310      End of Treatment One Nonrepeated est(CRHNDEEᵢ's)
+1.30745
+0.22745
−0.16005
−1.38905
+0.01420      End of Treatment Two Nonrepeated est(CRHNDEEᵢ's),
              etc.
9999          Number of Randomly Reordered Sequences of Non-
              repeated Experiment Test Program est(CRHNDEEᵢ's)
615 993 179   A New Set of Three, Three-Digit Odd Seed Numbers
```

```
C> COPY RBBHTDTA DATA

    1 file(s) copied

C> RBBHT
```

The data-based value of Bartlett's homoscedasticity test statistic for the 2 sets of 5 nonrepeated ANOVA-based est($CRHNDEE_i$'s) of specific interest is equal to 1.9391.

Given the null hypothesis of homoscedasticity, this microcomputer program constructed 9999 equally-likely sequences for these 10 est($CRHNDEE_i$'s) by using uniform pseudorandom numbers to rearrange their order. The number of these sequences that, when repartitioned into 2 sets of 5 est($CRHNDEE_i$'s), had its Bartlett's homoscedasticity test statistic value equal to or greater than 1.9391 is equal to 4328. Thus, given the null hypothesis of homoscedasticity, the randomization-based probability that a randomly selected sequence, when repartitioned into 2 sets of 5 est($CRHNDEE_i$'s), will have its Bartlett's homoscedasticity test statistic value equal to or greater than 1.9391 is equal to 0.4329. When this probability is sufficiently small, reject the null hypothesis in favor of the alternative hypothesis of heteroscedasticity.

6.5.2.1. Discussion

Microcomputer program *ANOVA* calculates the null hypothesis rejection probability value for homoscedasticity as being equal to 0.9181. This value can be confirmed by running microcomputer program *BARTLETT* and properly reinterpreting its null hypothesis rejection probability value. When the data-based value for the test statistic of specific interest is not consistent with its associated simple (one-sided) alternative hypothesis, the correct null hypothesis rejection probability value is equal to $[1 - (p/2)]$, which in this example is $[1 - (0.1638/2)]$ or 0.9181. On the other hand, when the data-based value for the test statistic of specific interest is consistent with its associated simple (one-sided) alternative hypothesis, then the correct null hypothesis rejection probability value is equal to $(p/2)$.

> *Remark*: When this procedure is used to reinterpret the null hypothesis rejection probability value calculated by running microcomputer program *ANOVA* for the example paired-comparison experiment test program datum values in Chapter 3, it will be evident that the correct null hypothesis rejection probability value is consistent with the respective null hypothesis rejection probability values pertaining to the sign test, the signed-rank test, and Fisher's enumeration test.

Microcomputer program *RBBHT* can also be run to calculate this null hypothesis rejection probability. When properly reinterpreted, the correct null hypothesis rejection probability value is equal to 0.7835, viz., $[1 - (0.4329/2)]$. A discrepancy of this magnitude is unusual for datum values that are indeed normally distributed. It occurs for our unreplicated SPD numerical example because two of the aggregated nonrepeated est($CRHNDSPTEEE_i$'s) and est($CRHNDMPTSPTIEEE_i$'s) are much larger than the other eight. Thus, their locations in each randomly rearranged sequence generated in microcomputer program *RBBHT* effectively dictates the magnitude of the associated data-based value for Bartlett's homoscedasticity test statistic. Accordingly, approximately one-half of these randomly rearranged sequences generates a large calculated value for this test statistic.

> *Remark*: Figure 6.1 should also be examined relative to time-order-of-testing effects and the variabilities of datum values pertaining to the respective treatments (treatment levels, treatment combinations). Hopefully, your intuition indicates that these hypothetical datum values exhibit unusually uniform variabilities.

6.5.3. Randomness (Independence, viz., Lack of Associations)

The respective ANOVA-based est($CRHNDEE_i$'s) are presumed to be independent of the time-order-of-testing and all experiment test program conditions. Accordingly, run microcomputer program *RBKTAU* to examine the respective ANOVA-based est($CRHNDEE_i$'s) relative to all possible monotonic associations. Consider, for example, our Example Four unreplicated SPD experiment test program. The nonrepeated est($CRHNDMPTEEE_i$'s) are plotted in Figure 6.2 versus the corresponding nonrepeated est($CRHNDSPTEEE_i$'s). If RBKTAO microcomputer program indicates that the null hypothesis that the $CRHNDMPTEEE_i$'s and the $CRHNDSPTEEE_i$'s are independent must rationally be rejected, then the credibility of the conceptual statistical model underlying ANOVA is undermined.

It is particularly important to test the null hypothesis that the respective $CRHNDSPTEEE_i$'s and $CRHNDMPTSPTIEEE_i$'s do not exhibit a

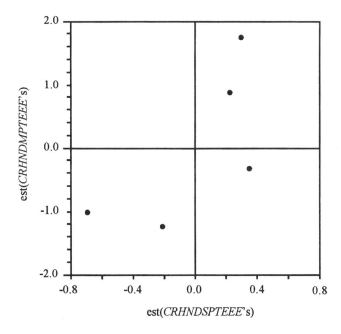

Figure 6.2 Plot of the non-repeated est($CRHNDMPTEEE_i$'s) versus the non-repeated est($CRHNDSPTEEE_i$'s) for the split-plot experiment test program in ANOVA Example Four. (These datum values appear in microcomputer file *RBKTDTA6*.)

monotonic association before the est($CRHNDSPTEEE_i$'s) and the est($CRHNDMPTSPTIEEE_i$'s) are aggregated to form the est($CRHND$-$SubPlotEE_i$'s). If this null hypothesis is rejected in favor of the alternative hypothesis that either a concordant or discordant association exists between these two conceptual experimental errors, then the respective estimated errors should not be combined regardless of the outcome of the test for homoscedasticity.

Exercise Set 4

These exercises are intended to enhance your understanding of the tests relevant to checking the presumptions underlying classical ANOVA.

1. (a) Test the null hypothesis of normality for the $CRHNDEE_i$'s in ANOVA Example One. Then, (b) test the null hypothesis of homoscedasticity. In turn, (c) construct a plot analogous to Figure 6.1 for the respective est($CRHNDEE_i$'s). Is this plot intuitively consistent with the outcome of your homoscedasticity test?

2. (a) Test the null hypotheses of normality for the $CRHNDSPTEEE_i$'s and the $CRHNDMPTSPTIEEE_i$'s in ANOVA Example Four. Then, (b) construct plots of the *corrected* datum values (analogous to Figure 6.1) for the high and low levels of the *mpt*'s and the *spt*'s. Are these plots informative relative to the respective presumptions of homoscedasticity? (If not, why not?)

3. Examine the possibility of a monotonic association among the respective est($CRHNDMPTEEE_i$'s), est($CRHNDSPTEEE_i$'s), and est($CRHNDMPTSPTIEEE_i$'s) in ANOVA Example Four. (a) Do the respective est($CRHNDMPTEEE_i$'s) appear to be statistically independent of their corresponding est($CRHNDSP$-$TEEE_i$'s)? (b) Do the respective est($CRHNDMPTEEE_i$'s) appear to be statistically independent of their corresponding est(CRH-$NDMPTSPTIEEE_i$'s)? (c) Do the respective est($CRHNDS$-$PTEEE_i$'s) appear to be statistically independent of their corresponding est($CRHNDMPTSPTIEEE_i$'s)?

6.6. STATISTICAL POWER

Recall that statistical power is the probability of (correctly) rejecting the null hypothesis when the alternative hypothesis is correct. For a comparative experiment test program this probability is directly related to its organizational structure and size, and to the magnitude of the minimum practical value of specific interest for the maximum difference among the respective *ctKm*'s. Recall also that our mechanical reliability objective in conducting a comparative experiment test program is to detect, with a high probability, differences among the respective *ctKm*'s that are sufficiently large to be of practical interest. Therefore, a vital consideration in planning a comparative experiment test program is to assure (as well as possible) that it will have satisfactory statistical power. Accordingly, before the experiment test program is formalized, a decision must be made regarding the minimum practical value of the maximum difference (*mpvmd*) among the *ctKm*'s pertaining to the respective n_t treatments, n_{tc} treatment combinations, or n_{tl} treatment levels. In addition, we must guestimate stddev(*APRCRHNDEE*'s) with reasonable accuracy. Accordingly, experiment test program planning involves engineering judgment and prior information (that hopefully is based on either extensive experience or preliminary testing).

Microcomputer programs *PCRD* and *PRCBD* should be used in experiment test program planning to guestimate the statistical power of a proposed experiment test program and thereby decide whether it appears to be practical. These programs have the guestimated value of the *standardized* minimum practical value of the maximum difference (*smpvmd*) among the respective *ctKm*'s as input information, where

guestimated(*smpvmd*)

$$= \frac{\text{minimum practical value of } [\max(ctKm) - \min(ctKm)]}{\text{guestimated[stddev } (APRCRHNDEE\text{'s})]}$$

Suppose an unreplicated RCBD experiment test program is proposed with four treatments and five blocks. Suppose further that (a) the *mpvmd* of engineering interest is equal to 20 (units) and (b) the guestimated[stddev(*APRCRHNDDV*'s)] is equal to 10 (units); then, guestimated *smpvmd* is equal to 2.0 (unitless). In turn, given an acceptable probability of committing a *Type I* error equal to 0.05, microcomputer program *PRCBD* computes the statistical power as being equal to at least 0.606. This statistical power is definitely not adequate. However, given an an unreplicated RCBD experiment test program with four treatments and 10 blocks, microcomputer program *PRCBD* computes the statistical power to be at least equal to 0.951. Thus, it is not advisable to undertake this unreplicated RCBD experiment test program with less than approximately 10 blocks. On the other hand, even more than 10 blocks should be employed in this unreplicated RCBD experiment test program when it is convenient to do so.

Suppose that the experiment test program of specific interest has already been conducted and the null hypothesis of specific interest has been rejected. Then, the *post hoc* statistical power for this experiment test program should be computed using the ANOVA-generated values for the maximum est($ctKm$'s) and minimum est($ctKm$'s) values and the associated est[stddev($APRCRHNDEE$'s)] value. In turn, if the acceptable guestimated *smpvmd* is smaller than the *post hoc smpvmd*, we can be reasonably confident that the statistical power for this experiment test program is quite adequate.

$$post\ hoc\ smpvmd = \frac{\max[est(ctKm)] - \min[est(ctKm)]}{est[stddev\ (APRCRHNDEE's)]}$$

Remark: Microcomputer programs *PCRD* and *PRCBD* respectively compute the exact statistical power for equally replicated CRD and unreplicated RCBD experiment test programs with only two treatments (Kastenbaum, et al., 1970). However, for experiment test programs with more than two treatments, all *ctKm*'s except the maximum *ctKm* and the minimum *ctKm* must be equal to the arithmetic average of the maximum *ctKm* and the minimum *ctKm*. Otherwise, the actual statistical power exceeds its computed value by an unknown amount.

The statistical power for an unreplicated SPD experiment test program is much more complex because of its organizational structure. Recall that split-plots are always nested within main-plots and main-plots are always nested within blocks. Thus, SPD experiment test programs always involve treatment combinations applied in a specific order. Nevertheless, microcomputer programs *PCRD* and *PRCBD* can be employed in an *ad hoc* manner to guestimate the statistical power for a test of the omnibus null hypothesis

```
C> PCRD
```

Input the acceptable probability of committing a *Type I* error in per cent (integer value)

```
5
```

Input the number of equally replicated treatments for the CRD experiment test program of specific interest

```
4
```

Input the number of replicate treatment datum values for the CRD experiment test program of specific interest

5

Input the *smpvmd* of specific interest

2.0

The statistical power for this CRD experiment test program is equal to at least 0.644.

C> PRCBD

Input the acceptable probability of committing *Type I* error in per cent (integer value)

5

Input the number of treatments for the RCBD experiment test program of specific interest

4

Input the number of blocks for the RCBD experiment test program of specific interest

5

Input the *smpvmd* of specific interest

2.0

The statistical power for this RCBD experiment test program is equal to at least 0.606.

that the actual values for the *cmptesc$_j$*'s or for the *csptesc$_j$*'s are all equal to zero. Accordingly, stddev(*APRCRHNDMPTEEE*'s) is guestimated and divided into the *mpvmd* of specific interest for the respective *cmptKm*'s to obtain the guestimated value for the *smpvmd* that pertains only to the main-plot treatment, whereas stddev(*APRCRHNDSPTEEE*'s) or stddev(*APR-CRHNDSubPlotEE*'s) is guestimated and divided into the *mpvmd* of specific interest for the respective *csptKm*'s to obtain a guestimated value for the *smpvmd* that pertains only to the split-plot treatment.

> *Remark*: Typically the power of the test of the omnibus null hypothesis that the actual value for the *csptesc$_i$*'s are all equal to zero exceeds the power of the test of the omnibus null hypothesis that the actual value for the *cmptesc$_i$*'s are all equal to zero because the split-plot treatment has greater replication than the main-plot treatment.

6.7. ENUMERATION-BASED AND RANDOMIZATION-BASED ANOVA's

Given the speed of the present generation of personal microcomputers, enumeration-based and randomization-based ANOVA's are clearly competitive with classical ANOVA's, and computer-intensive ANOVA's do not require the presumption of normality. Exact enumeration-based (empirical randomization-based) ANOVA null hypothesis rejection probability values can be calculated by constructing all (n_{rbelo}) equally-likely outcomes for the experiment test program of specific interest. We now illustrate enumeration-based and randomization-based ANOVA's based on the distribution-free (non-parametric) F test statistic.

$$\text{distribution-free } F_{between_{n_{sdf}}, within_{n_{sdf}}} \text{ test statistic} = \frac{between(\text{MS})}{within(\text{MS})}$$

Given a randomly rearranged sequence of experiment test program datum values, the data-based value of the distribution-free F test statistic is computed using the same algorithm as is used to compute the data-based value of Snedecor's central F test statistic in classical ANOVA. Recall that this algorithm does not involve the presumption of normality. Rather, the presumption of normality is required only to establish the conceptual sampling distribution for the data-based values of Snedecor's central F test statistic under continual replication of the experiment test program of specific interest.

Now reconsider our Example One unreplicated RCBD experiment test program. Given the omnibus null hypothesis that the actual values for all of the $ctesc_j$'s are equal to zero (and presuming that no treatment-block inter-action has a physical basis), each of the four treatment datum values within each of its five blocks is equally likely to pertain to that treatment or to a different treatment. However, if the four treatment datum values within each of its five blocks were randomly reassigned, there would be $(4!)^5$ equally-likely outcomes for this Example One unreplicated RCBD experi-ment test program. Clearly, it is impractical to construct all of these equally-likely outcomes. Thus, we must adopt a randomization-based methodology to calculate the omnibus null hypothesis rejection probability. Microcomputer program *RRCBDONH* (randomization-based RCBD experiment test program given the omnibus null hypothesis) constructs n_{rbelo} equally-likely outcomes for our Example One unreplicated RCBD experi-ment test program by using uniform pseudorandom numbers to reorder its four treatment datum values within each of its five blocks. It then calculates the n_{rbelo} associated values of the distribution-free F test statistic and counts the number of these values that are equal to or greater than the data-based value for the experiment test program that was actually conducted. In turn, it calculates the randomization-based null hypothesis probability value. If this probability is sufficiently small, we must rationally reject the null hypothesis in favor of the composite alternative hypothesis that not all of the actual values for the $ctesc_j$'s are equal to zero.

Enumeration-based and randomization-based ANOVA's need not be restricted to omnibus null hypotheses. Rather, these ANOVA's can be pro-grammed for each specific null hypothesis of particular interest (although certain constraints may apply). Consider our Example Two unreplicated RCBD experiment test program with its four treatment combinations in a $(2)^2$ factorial arrangement. Suppose we wish to test the specific null hypoth-esis that the actual value for the *ct1esc* is equal to zero. If so, then we must construct all equally-likely outcomes of this experiment test program when only the treatment one datum values are rearranged within each block. Accordingly, we now discuss the rearrangement of only the treatment one datum values in a typical block. Recall that the orthogonal augment con-trast array for a typical block of this experiment test program is

|data|	|+1's|	|$ct1ec_i$'s|	|$ct2ec_i$'s|	|$ct1t2iec_i$'s|
$d1$	$+1$	-1	-1	$+1$
$d2$	$+1$	$+1$	-1	-1
$d3$	$+1$	-1	$+1$	-1
$d4$	$+1$	$+1$	$+1$	$+1$

```
C> COPY C6RARRY1 ARRAY

    1 file(s) copied

C> COPY C6RDATA1 DATA

    1 file(s) copied

C> RRCBDONH
```

The data-based value of the distribution-free F test statistic for the outcome of the unreplicated RCBD experiment test program that was actually conducted is equal to $0.112954D + 02$.

Given the omnibus null hypothesis that all of the actual values for the $ctesc_j$'s are equal to zero, this microcomputer program constructed 9999 equally-likely outcomes for this experiment test program by using uniform pseudorandom numbers to reorder its treatment datum values in each block. The number of these outcomes that had its distribution-free F test statistic value equal to or greater than $0.112954D + 02$ is equal to 23. Thus, given the omnibus null hypothesis that all of the actual values for the $ctesc_j$'s are equal to zero, the randomization-based probability that a randomly selected outcome of this experiment test program when continually replicated will have its distribution-free F test statistic value equal to or greater than $0.112954D + 02$ is equal to 0.0024. When this probability is sufficiently small, reject the null hypothesis in favor of the omnibus alternative hypothesis that not all of the actual values for the $ctesc_j$'s are equal to zero.

If the null hypothesis that the actual value of the $ct1esc$ is equal to zero is indeed correct, then $d1$ is interchangeable with $d2$, and $d3$ is interchangeable with $d4$ in each block. Moreover, when these datum values are interchanged in our typical block, the order of the respective datum values changes to 2143, viz.,

|data|	|+1's|	|$ct1ec_i$'s|	|$ct2ec_i$'s|	|$ct1t2iec_i$'s|
$d2$	+1	−1	−1	+1
$d1$	+1	+1	−1	−1
$d4$	+1	−1	+1	−1
$d3$	+1	+1	+1	+1

The calculation of est($ct1esc$) then includes a term that is equal to $[(d1 + d3) − (d2 + d4)]/4$ rather than $[(d2 + d4) − (d1 + d3)]/4$. Accordingly, the sign of this term changes when the treatment one datum values are interchanged in a block. In turn note that the calculation of est($ct1t2iesc$) now includes a term that is equal to $[(d2 + d3) − (d1 + d4)]/4$ rather than $[(d1 + d4) − (d2 + d3)]/4$. Thus, the sign of this term also changes when the treatment one datum values are interchanged in a block. Accordingly, each change in the calculated value of est($ct1esc$) is coupled with a corresponding change in the calculated value of est($ct1t2iesc$). The result of this coupling is that our enumeration-based ANOVA is not statistically credible unless it is reasonable to assert that the actual value of the $ct1t2iesc$ is equal to zero.

Next, consider the specific null hypothesis that the actual value of the $ct2esc$ is equal to zero. If this null hypothesis is indeed correct, then *d3 and d4* are interchangeable with *d1* and *d2*. Moreover, when these datum values are interchanged in our typical block, the order of the respective datum values changes to 3412, viz.,

|data|	|+1's|	|$ct1ec_i$'s|	|$ct2ec_i$'s|	|$ct1t2iec_i$'s|
$d3$	+1	−1	−1	+1
$d4$	+1	+1	−1	−1
$d1$	+1	−1	+1	−1
$d2$	+1	+1	+1	+1

The calculation of est($ct2esc$) then includes a term that is equal to $[(d1 + d2) − (d3 + d4)]/4$ rather than $[(d3 + d4) − (d1 + d2)]/4$. Accordingly, the sign of this term changes when the treatment two datum values are interchanged in a block. In turn, note that the calculation of est($ct1t2iesc$) now includes a term that is equal to $[(d2 + d3) − (d1 + d4)]/4$ rather than $[(d1 + d4) − (d2 + d3)]/4$. Thus, the sign of this term also changes when the treatment two datum values are interchanged in a block. Accordingly, each change in the calculated value of est($ct2esc$) is coupled with a corresponding change in the calculated value of est($ct1t2iesc$). The result of this coupling is that our enumeration-based ANOVA is not statistically credible unless it is reasonable to assert that the actual value for the *ct1t2iesc* is equal to zero.

Microcomputer programs *ERCBD2143* and *RRCBD2143* respectively generate enumeration-based and randomization-based ANOVA's that can be used to test the specific null hypothesis that the actual value for the *ct1esc* is equal to zero. Analogously, microcomputer programs *ERCBD3412* and *RRCBD3412* respectively generate enumeration-based and randomization-based ANOVA's that can be used to test the specific null hypothesis that the actual value for the *ct2esc* is equal to zero.

The respective enumeration-based microcomputer programs are actually randomization-based programs in which the respective n_{rbelo} equally-likely outcomes for this Example Two unreplicated RCBD experiment test program are first constructed and then grouped into n_{sbelo} equally-likely outcomes. It turns out that there are only $(2)^{nb-1}$ equally-likely enumeration-based values of the distribution-free F test statistic for this experiment test program. Thus, unless it is expanded to include at least 11 blocks, there is no enumeration-based ANOVA (or randomization-based ANOVA) available to test a specific null hypothesis if the selected value of null hypothesis rejection probability is as small as 0.001.

```
C> COPY C6RARRY2 ARRAY

     1 file(s) copied

C> COPY C6RDATA2 DATA

     1 file(s) copied

C> RRCB2143
```

The data-based value of the distribution-free F test statistic for the outcome of the unreplicated RCBD experiment test program with a 2^2 factorial arrangement for its four treatment combinations within blocks that was actually conducted is equal to 0.132211D + 02.

Given the specific null hypothesis that the actual value for the *ct1esc* is equal to zero and presuming that the actual value for the *ct1t2iesc* is equal to zero, this microcomputer program constructed 9999 equally-likely outcomes for this experiment test program by using uniform pseudorandom numbers to reorder its treatment one datum values in each block. The number of these outcomes that had its distribution-free F test statistic value equal to or greater than $0.132211D + 02$ is equal to 644. Thus, given the specific null hypothesis that the actual value for the *ct1esc* is equal to zero and presuming that the actual value for the *ct1t2iesc* is equal to zero, the randomization-based probability that a randomly selected outcome of this experiment test program when continually replicated will have its distribution-free F test statistic value equal to or greater than $0.132211D + 02$ is equal to 0.0645. When this probability is sufficiently small, reject the null hypothesis in favor of the specific composite (two-sided) alternative hypothesis that the actual value for the *ct1esc* is not equal to zero.

(Microcomputer program *ERBC2143* computes the corresponding enumeration-based probability as 0.0625)

```
C> COPY C6RARRY3 ARRAY

     1 file(s) copied

C> COPY C6RDATA3 DATA

     1 file(s) copied

C> RRCB3412
```

The data-based value of the distribution-free F test statistic for the outcome of the unreplicated RCBD experiment test program with a 2^2 factorial arrangement for its four treatment combinations within blocks that was actually conducted is equal to $0.271527D + 01$.

Given the specific null hypothesis that the actual value for the *ct2esc* is equal to zero and presuming that the actual value for the *ct1t2iesc* is equal to zero, this microcomputer program constructed 9999 equally-likely outcomes for this experiment test program by using uniform pseudorandom numbers to reorder its treatment two datum values in each block. The number of these outcomes that had its distribution-free F test statistic value equal to or greater than $0.271527D + 01$

is equal to 1896. Thus, given the specific null hypothesis that the actual value for the *ct2esc* is equal to zero and presuming that the actual value for the *ct1t2iesc* is equal to zero, the randomization-based probability that a randomly selected outcome of this experiment test program when continually replicated will have its distribution-free F test statistic value equal to or greater than $0.271527D+01$ is equal to 0.1897. When this probability is sufficiently small, reject the null hypothesis in favor of the specific composite (two-sided) alternative hypothesis that the actual value for the *ct2esc* is not equal to zero.

(Microcomputer program *ERBC3412* computes the corresponding enumeration-based probability as 0.1875.)

Reminder: The specific null hypothesis that the actual value for the *ct1t2iesc* must be tested in a classical ANOVA before running microcomputer programs *RRCB2143*, *ERCB2143*, *RRCB3412*, or *ERCB3412*.

Next, consider our Example Four unreplicated SPD experiment test program with its two main-plots within each of its five blocks and its two split-plots within each of its main-plots. Presuming that the actual value for the *cmptsptie* is equal to zero, the respective specific null hypotheses that the actual values for the *cmptsptie* and *csptesc* are equal to zero can be statistically tested by running microcomputer programs *RSPD2143* or *ESPD2143*, or, alternatively, microcomputer programs *RSPD3412* or *ESPD3412*.

Reminder: The specific null hypothesis that the actual value for the *cmptsptiesc* must be statistically tested in a classical ANOVA before running microcomputer programs *RSPD2143*, *ESPD2143*, *RSPD3412*, or *ESPD2143*.

6.7.1. Discussion

If the actual value for the *cmptsptiesc* is equal to zero for our Example Four unreplicated SPD experiment test program, then the two split-plot treatments within each of its 10 main-plots can legitimately be viewed as being *paired comparisons*. Accordingly, the specific null hypothesis that the actual value for the *csptesc* is equal to zero can also be statistically tested in this situation by performing the distribution-free analyses based on the test

```
C> COPY C6RARRY4 ARRAY

    1 file(s) copied

C> COPY C6RDATA4 DATA

    1 file(s) copied

C> RSPD2143
```

The data-based value of the distribution-free F test statistic for the outcome of the unreplicated split-plot experiment test program with two main-plots in each of its five blocks and two split-plots in each of its main-plots that was actually conducted is equal to 0.214040D + 02.

Given the specific null hypothesis that the actual value for the *csptesc* is equal to zero and presuming that the actual value for the *cmptsptiesc* is equal to zero, this microcomputer program constructed 9999 equally-likely outcomes for this experiment test program by using uniform pseudorandom numbers to reorder its split-plot treatment datum values in each main-plot. The number of these outcomes that had its distribution-free F test statistic value equal to or greater than 0.214040D + 02 is equal to 583. Thus, given the specific null hypothesis that the actual value for the *csptesc* is equal to zero and presuming that the actual value for the *cmptsptiesc* is equal to zero, the randomization-based probability that a randomly selected outcome of this experiment test program when continually replicated will have its distribution-free F test statistic value equal to or greater than 0.214040D + 02 is equal to 0.0584. When this probability is sufficiently small, reject the null hypothesis in favor of the specific composite (two-sided) alternative hypothesis that the actual value for the *csptesc* is not equal to zero.

(Microcomputer program *ESPD2143* computes the corresponding enumeration-based probability as 0.0625.)

```
C> COPY C6RARRY5 ARRAY

    1 file(s) copied

C> COPY C6RDATA5 DATA

    1 file(s) copied

C> RSPD3412
```

The data-based value of the distribution-free F test statistic for the outcome of the unreplicated split-plot experiment test program with two main-plots in each of its five blocks and two split-plots in each of its main-plots that was actually conducted is equal to $0.153873D+01$.

Given the specific null hypothesis that the actual value for the *cmptesc* is equal to zero and presuming that the actual value for the *cmptsptiesc* is equal to zero, this microcomputer program constructed 9999 equally-likely outcomes for this experiment test program by using uniform pseudorandom numbers to reorder its main-plot treatment datum values in each block. The number of these outcomes that had its distribution-free F test statistic value equal to or greater than $0.153873D+01$ is equal to 3094. Thus, given the specific null hypothesis that the actual value for the *cmptesc* is equal to zero and presuming that the actual value for the *cmptsptiesc* is equal to zero, the randomization-based probability that a randomly selected outcome of this experiment test program when continually replicated will have its distribution-free F test statistic value equal to or greater than $0.153873D+01$ is equal to 0.3095. When this probability is sufficiently small, reject the null hypothesis in favor of the specific composite (two-sided) alternative hypothesis that the actual value for the *cmptesc* is not equal to zero.

(Microcomputer program *ESPD3412* computes the corresponding enumeration-based probability as 0.3125.)

statistics employed in microcomputer programs *EBST*, *EBSRT*, *FEBT*, *FEBMPDT*, and *FRBT*.

6.8. CLOSURE

Fixed effects ANOVA is typically employed in mechanical reliability applications to analyze the outcome of comparative experiment test programs whose treatments are qualitative, e.g., cadmium plated or zinc plated, shot-peened or not shot-peened. However, fixed effects ANOVA is also conveniently employed to analyze the outcome of experiment test programs with only a few equally spaced quantitative levels of a single treatment, e.g., zinc plated 5, 10, and 15 mils thick. When the number of these quantitative levels is large or when these quantitative levels are not all equally spaced, then linear regression analysis (Chapter 7) can be viewed as being a natural extension of (adjunct to) classical fixed effects ANOVA. Both analyses are based on the fundamental presumptions that the respective experiment test program datum values are random, homoscedastic, and normally distributed. However, classical linear regression analyses pertain only to CRD experiment test programs, viz., to a single batch of experimental units. (This problem can be mitigated somewhat by structuring replicate linear regression experiment test programs within batches. See Supplemental Topic 6.C and Section 7.3.)

6.A. SUPPLEMENTAL TOPIC: EXACT STATISTICAL POWER CALCULATION EXAMPLE

Recall that the statistical power of the test of normality based on our modified Michael's *MDSPP* test statistic was examined empirically in Exercise Set 2(b), Chapter 5. Statistical power can also be calculated exactly when the respective conceptual sampling distributions pertaining to the null and alternative hypotheses are known. We now present an exact statistical power calculation example that is intended to reinforce the notion that the statistical power of a test of hypothesis depends on (a) the test statistic of specific interest, (b) the acceptable probability of committing a *Type I* error, (c) the magnitude of the difference between the specification values that respectively establish the null and alternative hypotheses, and (d) the size of the experiment test program.

6.A.1. Exact Statistical Power Calculation Example

Consider a quantitative CRD experiment test program. Let the respective *CRSIDV_i*'s each be denoted X_i. Presume that the respective X_i's are randomly selected from a conceptual (two-parameter) normal distribution whose variance is equal to 25. Next, consider the null hypothesis that mean(X) = 25 versus the alternative hypothesis that mean(X) = x^*, where (a) the test statistic of specific interest is the arithmetic average of four randomly selected realizations of X, subsequently denoted ave($4X$'s), and (b) x^* is greater than 25. Then, given an acceptable probability p of committing a *Type I* error equal to 0.05, we rationally reject the null hypothesis that mean(X) = 25 in favor of the alternative hypothesis that mean(X) = x^* whenever the data-based realization value for ave($4X$'s) exceeds $x_{critical}$, where $x_{critical}$ is such that

$$\text{Probability}\left[\text{ave}(4X\text{'s}) > x_{critical}, \text{given } H_n: \text{mean}(X) = 25\right] = 0.05$$

Thus,

$$x_{critical} = 25 + (1.6449) \cdot (25/4)^{1/2} = 29.11225$$

in which the standardized conceptual normal distribution variate $y(1 - p) = y > (0.95) = 1.6449$.

Figure 6.3 displays the conceptual sampling distribution PDF for the test statistic ave(48%) given the null hypothesis that mean(X) = 25 [and the stipulation that X is normally distributed with var(X) = 25]. The corresponding conceptual sampling distribution PDF given the alternative

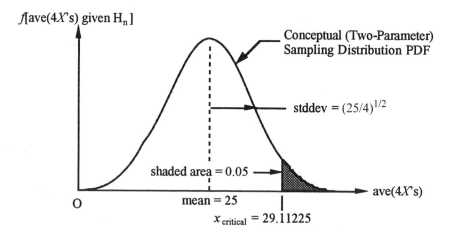

Figure 6.3 Plot of the conceptual sampling distribution PDF for the test statistic ave($4X$'s), given the null hypothesis that mean(X) = 25 [and the stipulation that X is normally distributed with var(X) = 25].

hypothesis that mean(X) = x^* is identical to this conceptual sampling distribution, except that it is translated to the right (because $x^* > 25$).

Now consider the alternative hypothesis that mean(X) = x^* = 26. When x^* = 26, the statistical power of this test of hypothesis is the probability of (correctly) rejecting the null hypothesis that mean(X) = 25 when the alternative hypothesis that mean(X) = 26 is correct. This statistical power is computed using the probability expression:

Probability[ave(4X's) > $x_{critical}$, given H_a: mean(X) = 26]

$$= \text{statistical power}$$

The value of the standardized conceptual normal distribution variate $y(p)$ when mean(X) = 26 is equal to $(29.11225 - 26)/(25/4)^{1/2} = 1.2449$. Thus, running microcomputer program PY with input $y(p) = 1.2449$, the statistical power is equal to 0.1066 given the alternative hypothesis that mean(X) = 26.

We now demonstrate that the statistical power for the test statistic ave(4X's) increases markedly as the alternative hypothesis specification value for mean(X) increases. Consider the following computed values of the statistical power (sp) given the alternative hypothesis that mean(X) = x^*, where x^* is successively set equal to 27 [$y(p) = 0.8449$, $sp = 0.1991$], 28 [$y(p) = 0.4449$, $sp = 0.3282$], 29 [$y(p) = 0.0449$, $sp = 0.4821$], 30 [$y(p) = -0.3551$, $sp = 0.6387$], 31 [$y(p) = 0.7551$, $sp = 0.7749$], 32 [$y(p) = -1.1551$, $sp = 0.8760$], 33 [$y(p) = 1.5551$, $sp = 0.9400$], 34 [$y(p) = -1.9551$, $sp = 0.9747$], and 35 [$y(p) = -2.3551$, $sp = 0.9907$]. These computed statistical power values are plotted in Figure 6.4.

Now suppose we extend this example by successively computing the exact statistical powers for the test statistics ave(16X's) and ave(64X's). Figure 6.5 displays the resulting *family* of statistical power curves. Figure 6.5 is intended to demonstrate that, for each candidate mean(X) given by the alternative hypothesis, the larger the experiment test program the greater the statistical power of the associated test of hypothesis, viz., the greater the probability of (correctly) rejecting the null hypothesis when the alternative hypothesis is correct.

The respective statistical power curves plotted in Figure 6.5 have two practicality problems because (a) the alternative hypothesis (parametric) value for [mean(X) given H_a] is stated in terms of an equality rather than an inequality, and (b) these statistical power calculations pertain only to var(X) = 25. Practicality problem (a) is easily solved by modifying the alternative hypothesis so that [mean(X) given H_a] is equal to or greater than (parametric) x^*. The actual statistical power is then equal to or greater than the statistical power pertaining to x^*. Practicality problem (b) is overcome

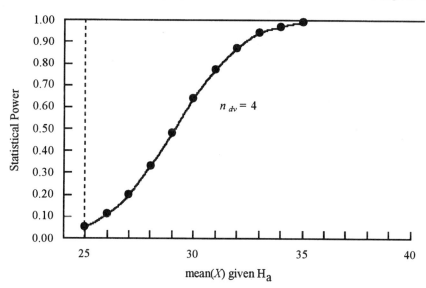

Figure 6.4 Plot of the statistical power of the statistical test of hypothesis, given the null hypothesis that mean$(X) = 25$ versus the alternative hypothesis that mean(X) is equal to the parametric value now denoted [mean(X) given H_a].

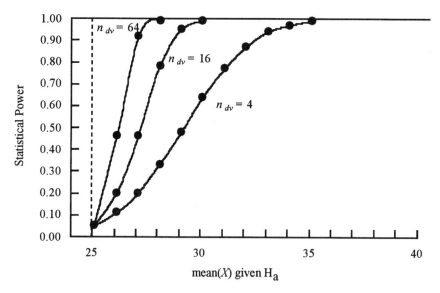

Figure 6.5 Plot of the family of statistical power curves pertaining to our exact statistical power calculation example. Note that the greater the size of the experiment test program, the greater the statistical power of the given test of hypothesis.

by working with a *standardized* minimum practical difference parameter, denoted *smpd*, where

$$smpd = \frac{\text{minimum practical difference } [\text{mean}(X) \text{ given } H_a - \text{mean}(X) \text{ given } H_n]}{\text{common } H_n \text{ and } H_a \text{ values for stddev}(X)}$$

and then replotting the statistical power curves in Figure 6.5 with a new abscissa stated in convenient values of *smpd* (Exercise Set 5, Exercise 3).

Exercise Set 5

These exercises are intended to verify the numerical results of our exact statistical power calculation example and to enhance your understanding of the concept of statistical power. Recall that statistical power is defined as the probability of (correctly) rejecting the null hypothesis when the alternative hypothesis is correct. It is the complement of the probability of committing a *Type II* error, where a *Type II* error is committed when the alternative hypothesis is correct and we (incorrectly) fail to reject the null hypothesis.

1. Compute the statistical power of our quantitative CRD experiment test program example with (a) $n_{dv} = 16$ or (b) $n_{dv} = 64$, when the alternative hypotheses that mean$(X) = x^*$, where x^* is successively set equal to 26, 27, 28, 29, and 30. Let the associated acceptable probability of committing a *Type I* error be equal to (c) 0.10, or (d) 0.05, or (e) 0.01.

2. Accurately sketch the respective conceptual sampling distribution PDF's for ave(4X's) given H_n: mean$(X) = 25$ and (a) H_a: mean$(X) = 29$ and (b) H_a: mean$(X) = 33$, when X is normally distributed with var$(X) = 25$. Then, using shading or cross-hatching, identify the respective areas under the appropriate PDF's that correspond to the probability of committing a *Type I* error and of committing a *Type II* error. In turn, identify the respective areas under the appropriate PDF's that correspond to the statistical power for the test of the null hypothesis that mean$(X) = 25$. Finally, compute the respective statistical powers pertaining to the alternative hypotheses in (a) and (b). Do your values agree with the text values?

3. Replot Figure 6.5 using *smpd* as the abscissa and *minimum statistical power* as the ordinate.

6.B.   SUPPLEMENTAL TOPIC: THE CONCEPT OF A WEIGHTED AVERAGE

Recall that one of the most intuitive perspectives in data analysis is that each datum point (datum value) has a mass (statistical weight). Homoscedastic datum values have equal statistical weights, whereas heteroscedastic datum values have different statistical weights. Accordingly, we now establish an expression for a *weighted average* pertaining to heteroscedastic datum values (as opposed to an arithmetic average pertaining to homoscedastic datum values).

We begin by constructing heteroscedastic datum values such that the actual values for the variances of their respective conceptual sampling distributions are known. Consider n_a mutually independent arithmetic averages respectively computed using n_{dv_i} realization values randomly selected from the conceptual statistical distribution that consists of all possible replicate realization values for the generic random variable X, where $i = 1, 2, \ldots, n_a$. Then, although each of these n_a arithmetic averages have mean(X) as its expected value, the actual values for the variances of their conceptual sampling distributions are respectively equal to var(X)]$/n_{dv_i}$.

Now suppose we wish to estimate mean(X) using the weighted average of our n_a constructed conceptual random heteroscedastic datum values. Let these constructed conceptual random heteroscedastic datum values be generically denoted $CRHeteroscedasticDV_i$'s, where for this example these are set equal to their corresponding [ave$_i(X$'s)]'s. Then, consider the intuitive linear estimator for mean(X):

$$
\begin{aligned}
\text{est}[\text{mean}(X)] &= \text{weighted average}\,(CRHeteroscedasticDV\text{'s}) \\
&= \text{weighted average}\,\big\{[\text{ave}_i(X\text{'s})]\text{'s}\big\} \\
&= \sum_{i=1}^{n_a} c_i \; \text{ave}_i(X\text{'s})
\end{aligned}
$$

in which the c_i's are coefficients that will be chosen to generate a minimum variance unbiased statistical estimator for mean(X). Recall that, by definition, this weighted average statistical estimator is unbiased when its expected value is equal to mean(X). Since the expected value of each of the respective arithmetic averages (weighted datum values) is equal to mean(X), the sum of the c_i's must clearly be set equal to one to create an unbiased statistical estimator for mean(X).

In turn, since the respective $CRHeteroscedasticDV_i$'s = [ave$_i(X$'s)]'s are mutually independent, the actual value for the variance of the conceptual sampling distribution that consists of all possible replicate estimated weighted averages is

$$\text{var}[\text{weighted average}(CRHeteroscedasticDV\text{'s})]$$

$$= \text{var}\left(\text{weighted average}\{[\text{ave}_i(X\text{'s})]\text{'s}\}\right) = \sum_{i=1}^{n_a}(c_i^2\,\text{var}[\text{ave}_i(X\text{'s})])$$

This variance takes on its minimum value, subject to the constraint that the sum of the c_i's is equal to one, when the c_i's are such that $c_j\text{var}[\text{ave}_j(X\text{'s})] = c_k\text{var}[\text{ave}_k(X\text{'s})]$, where dummy indices j and k are different.

Now let sw_i, the statistical weight pertaining to $CRHeteroscedasticDV_i$, be defined as the inverse of the actual value for the variance of the conceptual sampling distribution that consists of all possible replicate realization values for $CRHeteroscedasticDV_i$. Accordingly, $sw_i = n_{dv_i}/\text{var}(X)$. In turn,

$$c_i = \frac{sw_i}{\displaystyle\sum_{i=1}^{n_a} sw_i}$$

Then,

weighted average($CRHeteroscedasticDV$'s)

$$\text{weighted average}\{[\text{ave}_i(X\text{'s})]\text{'s}\} = \frac{\displaystyle\sum_{i=1}^{n_a} sw_i\,\text{ave}_i(X\text{'s})}{\displaystyle\sum_{i=1}^{n_a} sw_i}$$

and

$$\text{var}[\text{weighted average}(CRHeteroscedasticDV\text{'s})]$$

$$= \text{var}(\text{weighted average}\{[\text{ave}_i(X\text{'s})]\text{'s}\}) = \frac{1}{\displaystyle\sum_{i=1}^{n_a} sw_i}$$

6.B.1. Summary and Perspective

Given conceptual random heteroscedastic datum values randomly selected from conceptual statistical (sampling) distributions with the same actual values for their respective means, but difference actual values for the respective variances, the weighted average statistical estimator is not only unbiased, but the actual value for the variance of its conceptual sampling distribution is also smaller than the actual value for the variance of the conceptual sampling distribution pertaining to any alternative statistical estimator. Thus, a weighted average estimator is used to pool two or more statistical estimates of the same quantity that are based on different sample sizes (different numbers of statistical degrees of freedom). For example, it is used to pool the respective $[within(SS)]$'s associated with two or more unequally replicated treatments in a CRD experiment test program. (Note that the only reason that we have been able to avoid overt use of a weighted average estimator until now is that the experiment test programs presented in Chapter 2 each involved equal replication.)

Exercise Set 6

These exercises are intended to confirm that, given equal replication for the respective n_a arithmetic averages, the weighted average estimator is synonymous with an arithmetic average estimator.

1. Verify that, given equal replication for the respective n_a arithmetic averages in our weighted average example, the weighted average estimator for mean (X) can be re-expressed as

$$\text{Weighted average(all } X\text{'s)} = \text{ave(all } X\text{'s)} = \frac{\sum_{i=1}^{n_{dv_{total}}} X_i}{n_{dv_{total}}}$$

2. Verify that, given equal replication for the respective n_a arithmetic averages in our weighted average example, the variance expression for the weighted average estimator can be re-expressed as

$$\text{var[weighted average(all } X\text{'s)]} = \text{var[ave(all } X\text{'s)]} = \frac{\text{var}(X)}{n_{dv_{total}}}$$

Exercise Set 7

These exercises are intended to extend our weighted average example by generating additional insight regarding the statistical behaviors of weighted averages and arithmetic averages.

1. (a) Run microcomputer program $RANDOM2$ with $n_{elpri} = 8$ and $n_{digit} = 3$. Compute the arithmetic average of these eight pseudorandom numbers. Then, (b) compute the arithmetic average of the first five of these eight numbers. Next, (c) compute the arithmetic average of the last three of the eight numbers. In turn, (d) compute the arithmetic average of the arithmetic averages in (b)

and (c). Does this arithmetic average agree with the (correct) arithmetic average in (a)? Next, (e) compute the arithmetic average of the first four of the eight pseudorandom numbers. In turn, (f) compute the arithmetic average of the last four of the eight numbers. Then, (g) compute the arithmetic average of the arithmetic averages in (e) and (f). Does this arithmetic average agree with the (correct) arithmetic average in (a)? What is the principle that can be inferred by the outcome of this numerical example? Finally, (h) compute and compare the weighted averages of the respective arithmetic averages in (b) and (c) and in (e) and (f).

2. Given that the respective $\text{ave}_i(X$'s) have unequal variances and that their arithmetic average is used to estimate mean(X), is the arithmetic average an unbiased estimator of mean(X)? Discuss.

3. Given that the respective $\text{ave}_i(X$'s) have unequal variances and that their arithmetic average is used to estimate mean(X), is the arithmetic average a minimum variance estimator of mean(X)? Discuss.

6.C. SUPPLEMENTAL TOPIC: TESTING FOR BATCH-TO-BATCH EFFECTS

By far the most effective means to improve reliability is to decrease the batch-to-batch effects (b-t-be's) associated with materials and materials processing in the manufacture of components. For example, in fatigue testing of components, b-t-be's as large as 30 to 40% occur in terms of strength (resistance) and as large as 10,000% occur in terms of life (endurance). Yet b-t-be's are seldom ever included in a reliability-based experiment test program (or appropriately considered in establishing so-called design allowables).

We now present two illustrative example orthogonal augmented contrast arrays that can be used in ANOVA to test the null hypothesis that no b-t-be's exist versus the composite alternative hypothesis that b-t-be's exist. This statistical test is most effective when (a) the number of batches is small and the number of replicates is large, and (b) the respective batches are deliberately selected from diverse sources (or to represent processing extremes). These example arrays are also intended to provide insight regarding the inclusion of b-t-be's and/or replicate experimental units in the orthogonal augmented contrast array pertaining to the experiment test program of specific interest.

6.C.1. Example One

Suppose we have three batches, each of size two, viz., each batch has two replicate experiment units (test specimens). The associated orthogonal augmented contrast array appears below. The *between* batch-to-batch mean square is the sum of squares of the respective elements of the aggregated |est(b-t-be_i's)| column vector divided by two, the number of statistical degrees of freedom of the aggregated |est(b-t-be_i's)| column vector. In turn, the *within* batch replicates mean square is the sum of squares of the respective elements of the aggregated |est($CRHNDEE$'s)| column vector divided by three, the number of statistical degrees of freedom of the aggregated |est($CRHNDEE$'s)| column vector. Thus, microcomputer program *ANOVA* can be run to determine whether the null hypothesis that no b-t-be's exist can rationally be rejected. If the null hypothesis is not rejected, then (a) the *between* b-t-be's sums of squares can be aggregated with the *within* batch replicates sum of squares, (b) the *between* b-t-be's number of statistical degrees of freedom can be aggregated with the *within* batch replicates number of statistical degrees of freedom, and (c) the mean square pertaining to the column vector formed by aggregating the |est(b-t-be_i's)| and |est($CRHNDEE$'s)| column vectors can be used to estimate var($APRCRHNDEE$'s), which in turn is used to compute the classical (shortest) 100(scp)% (two-sided) confidence interval that allegedly includes the actual value for the *csdm*. However, if the null hypothesis is rejected, then clearly something must be done to reduce the b-t-be's. Note that the classical (shortest) 100(scp)% (two-sided) statistical confidence interval, if computed, would pertain only the specific batches employed in the experiment test program.

|experiment test program datum values|	|+1's|	|b-t-be's|		|batch replicate contrasts|	
Batch 1, replicate 1 datum value	+1	+1 −1	−1	0	0
Batch 1, replicate 2 datum value	+1	+1 −1	+1	0	0
Batch 2, replicate 1 datum value	+1	−1 −1	0	−1	0
Batch 2, replicate 2 datum value	+1	−1 −1	0	+1	0
Batch 3, replicate 1 datum value	+1	0 +2	0	0	−1
Batch 3, replicate 2 datum value	+1	0 +2	0	0	+1

6.C.2. Example Two

Suppose we have three batches, each of size three, viz., each batch has three replicate experiment units (test specimens). The associated orthogonal augmented contrast array appears below. The *between* b-t-be's mean square is the sum of squares of the respective elements of the aggregated |est(b-t-be_i's)|

column vector divided by two, the number of statistical degrees of freedom for the aggregated |est(b-t-be's)| column vector. In turn, the *within* batch replicates mean square is the sum of squares of the respective elements of the aggregated |est($CRHNDEE_i$'s)| column vector divided by six, the number of statistical degrees of freedom for the aggregated |est($CRHNDEE_i$'s)| column vector. Thus, as in Example One, microcomputer program *ANOVA* can be run to determine whether the null hypothesis that no batch-to-batch effects exist must be rationally rejected.

|experiment test program datum values|	|+1's|	|b-t-be's|		|*batch replicate contrasts*|				
Batch 1, replicate 1 datum value	+1	−1	−1	−1	−1	0	0	0	0
Batch 1, replicate 2 datum value	+1	−1	−1	+1	−1	0	0	0	0
Batch 1, replicate 3 datum value	+1	−1	−1	0	+2	0	0	0	0
Batch 2, replicate 1 datum value	+1	+1	−1	0	0	−1	−1	0	0
Batch 2, replicate 2 datum value	+1	+1	−1	0	0	+1	−1	0	0
Batch 2, replicate 3 datum value	+1	+1	−1	0	0	0	+2	0	0
Batch 3, replicate 1 datum value	+1	0	+2	0	0	0	0	−1	−1
Batch 3, replicate 2 datum value	+1	0	+2	0	0	0	0	+1	−1
Batch 3, replicate 3 datum value	+1	0	+2	0	0	0	0	0	+2

6.C.3. Discussion

These two arrays are also intended to indicate how the orthogonal augmented contrast arrays for unreplicated RCBD and SPD experiment test programs can be extended to include either equal replication or possible b-t-be's. Then, although this extension will double or triple the size of the experiment test program, we could either test the null hypothesis that no block, treatment interaction effects exist or (preferably) that no b-t-be's exist.

7

Linear Regression Analysis

7.1. INTRODUCTION

Classical fixed effects ANOVA can broadly be viewed as being concerned with statistically deciding whether differences among specifically selected independent variables (treatments, treatment levels, treatment combinations) affect the test response (outcome), whereas linear regression analysis can broadly be viewed as being concerned with establishing a statistically adequate analytical model for describing how specifically selected values for one or more independent variables affect the test response (outcome). Although this distinction is superficial, it is pedagogically convenient to present linear regression as a separate topic from classical fixed effects ANOVA, while at the same time emphasizing their common underlying statistical presumptions.

Linear regression analysis is based on a presumed conceptual statistical model that is constructed by replacing the conceptual location parameter (*clp*) of a conceptual (two-parameter) normal distribution (Chapter 5) by the analytical expression $clp0 + clp1 \cdot iv1v_i + clp2 \cdot iv2v_i + \ldots clpj \cdot ivjv_i$, where $ivjv_i$ is the ith specifically selected value for the jth independent variable. The distinction between the actual physical relationship and the presumed conceptual statistical model is illustrated schematically in Figure 7.1. The former is generally complex, whereas the latter is elementary.

308

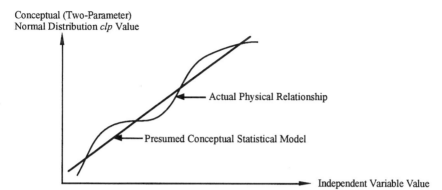

Figure 7.1 Schematic of the distinction between the actual physical relationship and the conceptual statistical model that is presumed in linear regression analysis. Unless the variability of replicate regression datum values is extremely small for all values of the independent variable(s) or unless the regression experiment test program has massive replication, it is unlikely that the actual physical model will ever be established experimentally.

Accordingly, the fundamental issue in linear regression analysis is whether the presumed conceptual statistical model is adequate to explain the observed experiment test program datum values (as judged by the outcome of an appropriate statistical test). If so, then the presumed conceptual statistical model can rationally be used to compute conceptual parameter estimates, statistical confidence intervals, etc.

The conceptual simple linear regression statistical model has a single independent variable. This model can be stated as

$$f(CRHNDRDV_i) \cdot$$

$$= \frac{1}{\sqrt{2\pi csp}} = \exp\left[-\frac{1}{2} \cdot \left(\frac{CRHNDRDV_i - clp0 - clp1 \cdot ivv_i}{csp} \right)^2 \right]$$

in which $CRHNDRDV_i$ is verbalized as $CRHNDRDV$ given the ivv (a parameter) is equal to the ivv_i (a specifically selected numerical value for the iv). Note that the homoscedasticity as well as the normality of this model is overt, viz., the actual value for the csp is not a function of the ivv. Note also that, by definition,

$$CRHNDRDV_i - clp0 - clp1 \cdot ivv_i = CRHNDREE_i$$

Thus, the conceptual simple linear regression model can also be expressed using our hybrid column vector notation as

$$|CRHNDRDV_i\text{'s}| = clp0 \cdot |+1\text{'s}| + clp1 \cdot |ivv_i\text{'s}| + |CRHNDREE_i\text{'s}|$$

Note, however, that the homosecdasticity of this simple linear regression model is evident only in the verbalization of the H in the $CRHNDRDV_i$'s and the $CRHNDREE_i$'s. Nevertheless, since this model is more succinct and tractable than the former, it is subsequently employed in simple linear regression analysis.

The analogous conceptual multiple linear regression statistical model has n_{clp} conceptual location parameters. It can be stated using our hybrid column vector notation as

$$|CRHNDRDV_i\text{'s}| = clp0 \cdot | + 1\text{'s}| + clp1 \cdot |iv1v_i\text{'s}| + clp2 \cdot |iv2v_i\text{'s}|$$
$$+ clp3|iv3v_i\text{'s}| + \ldots + |CRHNDREE_i\text{'s}|$$

This model can be reinterpreted as a polynomial model in a single independent variable, e.g.,

$$|CRHNDRDV_i\text{'s}| = clp0 \cdot |(ivv_i\text{'s})^0| + clp1 \cdot |(ivv_i\text{'s})^1 \cdot | + clp2 \cdot |(ivv_i\text{'s})^2|$$
$$+ \ldots + |CRHNDREE_i\text{'s}|$$

or as a polynomial in two or more independent variables, e.g.,

$$|CRHNDRDV_i\text{'s}| = clp00 \cdot |\left[(iv1v_i)^0 \cdot (iv2v_i)^0\right]\text{'s}|$$
$$+ clp10 \cdot |\left[(iv1v_i)^1 \cdot (iv2v_i)^0\right]\text{'s}| + clp20 \cdot |\left[(iv1v_i)^2 \cdot (iv2v_i)^0\right]\text{'s}| + \ldots +$$
$$+ clp01 \cdot |\left[(iv1v_i)^0 \cdot (iv2v_i)^1\right]\text{'s}| + clp02 \cdot |\left[(iv1v_i)^0 \cdot (iv2v_i)^2\right]\text{'s}| + \ldots +$$
$$+ clp11 \cdot |\left[(iv1v_i)^1 \cdot (iv2v_i)^1\right]\text{'s}| + \ldots + |CRHNDREE_i\text{'s}|$$

Thus, it should be clear that the terminology linear in linear regression refers to the conceptual location parameters rather than to the geometry of the conceptual statistical model. (Note that the power for each of the respective conceptual location parameters is equal to one.)

The conceptual linear regression model that employs the least number of terms to explain the observed experiment test program datum values in a statistically adequate manner is traditionally adopted in mechanical reliability (and other engineering) analyses.

Remember: The credibility of the geometry of the presumed conceptual linear regression model, and in particular the credibility of the associated homoscedasticity presumption, can severely limit the credibility of a linear regression analysis. Thus, the respective ranges of the regression experiment test program $ivjv$'s should be kept as short as practical and extrapolation should always be avoided.

7.2. SIMPLE LINEAR REGRESSION ANALYSIS

We now present simple linear regression analysis using a column vector perspective. However, since the conceptual simple linear regression statistical model only approximates the actual physical relationship, this model must be regarded as tentative until its credibility (adequacy) is subsequently established by appropriate statistical analysis. Given this caveat, we express the conventional simple linear regression statistical model in our hybrid column vector notation as

$$|CRHNDRDV_i\text{'s}| = clp0 \cdot | + 1\text{'s}| + clp1 \cdot |ivv_i\text{'s}| \ + \ |CRHNDREE_i\text{'s}|$$
$$= |[\text{mean}_i(APRCRHNDRDV\text{'s})]\text{'s}| \ + \ |CRHNDREE_i\text{'s}|$$

in which $\text{mean}_i(APRCRHNDRDV\text{'s})$ is technically verbalized as the actual value for the mean of the conceptual statistical distribution that consists of all possible replicate conceptual random homoscedastic normally distributed regression datum values given that $ivv = ivv_i$, where ivv_i is the ith ivv employed in the simple linear regression experiment test program. This conceptual simple linear regression statistical model is stated explicitly for $n_{rdv} = 3$ as

$$
\begin{vmatrix} CRHNDRDV_1 \\ CRHNDRDV_2 \\ CRHNDRDV_3 \end{vmatrix} = clp0 \cdot \begin{vmatrix} +1 \\ +1 \\ +1 \end{vmatrix} + clp1 \cdot \begin{vmatrix} ivv_1 \\ ivv_2 \\ ivv_3 \end{vmatrix} + \begin{vmatrix} CRHNDREE_1 \\ CRHNDREE_2 \\ CRHNDREE_3 \end{vmatrix}
$$
$$
= \begin{vmatrix} \text{mean}_1(APRCRHNDRDV\text{'s}) \\ \text{mean}_2(APRCRHNDRDV\text{'s}) \\ \text{mean}_3(APRCRHNDRDV\text{'s}) \end{vmatrix} + \begin{vmatrix} CRHNDREE_1 \\ CRHNDREE_2 \\ CRHNDREE_3 \end{vmatrix}
$$

It is schematically plotted in three-dimensional perspective in Figure 7.2 where the three mutually orthogonal co-ordinate axes are respectively denoted (1), (2), and (3). The $|CRHNDRDV_i\text{'s}|$ column vector is schematically plotted as the vector sum of its three components, with each component plotted along its associated co-ordinate axis. In turn, the $| + 1\text{'s}|$ identity column vector and the $|ivv\text{'s}|$ design column vector are schematically plotted as the vector sums of their three components, where the respective compo-

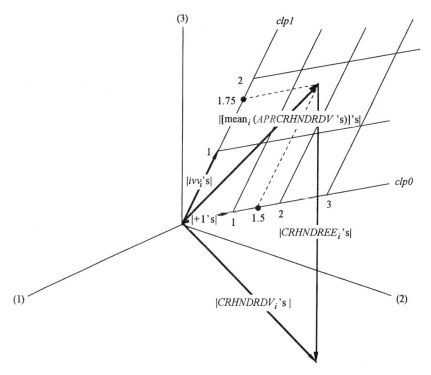

Figure 7.2   Three-dimensional schematic plot of the conventional conceptual simple linear regression statistical column vector model. The $|+1$'s$|$ identity column vector and the $|ivv_i$'s$|$ design column vector define the $clp0$, $clp1$ parameter plane in n_{rdv} space. Numerical values for the $clp0$, say 1.5, and for the $clp1$, say 1.75, as well as for the three components of the $|ivv_i$'s$|$ design column vector must be known to quantify the corresponding three components of the $|[mean_i(APRCRHNDRDV$'s$)]$'s$|$ column vector.

nents of each column vector are plotted along its associated co-ordinate axis. The latter two column vectors form a second set of co-ordinate axes that define the $clp0,clp1$ parameter plane. In turn, integer values for the $clp0$ and $clp1$ respectively generate tick-marks along these co-ordinate axes. A series of lines drawn through these tick-marks, each oriented parallel to the opposite co-ordinate axis, generate a grid in this $clp0,clp1$ parameter plane. In turn, given numerical values for the $clp0$ and the $clp1$, this grid is used to quantify the $|mean_i(APRCRHNDRDV$'s$)|$ column vector, whose components must respectively satisfy the scalar relationships:

$$\text{mean}_i(APRCRHNDRDV\text{'s}) = clp0 + clp1 \cdot ivv_i$$

for $i = 1$ to 3. Finally, the sum of three respective components of the $|CRHNDREE_i\text{'s}|$ column vector and the $|\text{mean}_i(APRCRHNDRDV\text{'s})|$ column vector must equal the corresponding three components of the $|CRHNDRDV_i\text{'s}|$ column vector.

The corresponding conventional estimated conceptual simple linear regression statistical model is written in our hybrid column vector notation as

$$
\begin{aligned}
|rdv_i\text{'s}| &= (clp0) \cdot |+1\text{'s}| + \text{est}(clp1) \cdot |ivv_i\text{'s}| + |[\text{est}(CRHNDREE_i)]\text{'s}| \\
&= |\{\text{est}[\text{mean}_i(APRCRHNDRDV\text{'s})]\}\text{'s} + |[\text{est}(CRHNDREE_i)]\text{'s}|
\end{aligned}
$$

in which $i = 1$ to n_{rdv}. Given that $n_{rdv} = 3$, this estimated conceptual simple linear regression model is explicitly stated as

$$
\begin{vmatrix} rdv_1 \\ rdv_2 \\ rdv_3 \end{vmatrix} = \text{est}(clp0) \cdot \begin{vmatrix} +1 \\ +1 \\ +1 \end{vmatrix} + \text{est}(clp1) \cdot \begin{vmatrix} ivv_1 \\ ivv_2 \\ ivv_3 \end{vmatrix} + \begin{vmatrix} \text{est}(CRHNDREE_1) \\ \text{est}(CRHNDREE_2) \\ \text{est}(CRHNDREE_3) \end{vmatrix}
$$

$$
= \begin{vmatrix} \text{est}[\text{mean}_1(APRCRHNDRDV\text{'s})] \\ \text{est}[\text{mean}_2(ARPCRHNDRDV\text{'s})] \\ \text{est}[\text{mean}_3(APRCRHNDRDV\text{'s})] \end{vmatrix} + \begin{vmatrix} \text{est}(CRHNDREE_1) \\ \text{est}(CRHNDREE_2) \\ \text{est}(CRHNDREE_3) \end{vmatrix}
$$

The associated three-dimensional schematic plot appears in Figure 7.3. The sum of squares of the est($CRHNDREE_i$s) takes on its minimum value when the $|[\text{est}(CRHNDREE_i)]\text{'s}|$ column vector is perpendicular to the plane formed by the $|+1\text{'s}|$ and $|ivv_i\text{'s}|$ column vectors. Thus, the $|[\text{est}(CRHNDREE_i)]\text{'s}|$ column vector is also perpendicular (normal) to both the $|+1\text{'s}|$ and $|ivv_i\text{'s}|$ column vectors. The resulting two normal equations, stated as dot products using our hybrid column vector notation, are

$$\left| [\text{est}(CRHNDREE_i)]\text{'s} \right| \bullet |+1\text{'s}| = 0$$

and

$$\left| [\text{est}(CRHNDREE_i)]\text{'s} \right| \bullet |ivv_i\text{'s}| = 0$$

Note that the orthogonality of the $|+1\text{'s}|$ and the $|[\text{est}(CRHNDREE_i)]\text{'s}|$ column vectors requires the respective est($CRHNDREE_i$s) to sum to zero.

Simultaneous solution of these two normal equations generates the conventional least-square plural squares expressions for est($clp0$) and est($clp1$). First, we evaluate the respective dot products to obtain two scalar equations in two unknowns, est($clp0$) and est($clp1$):

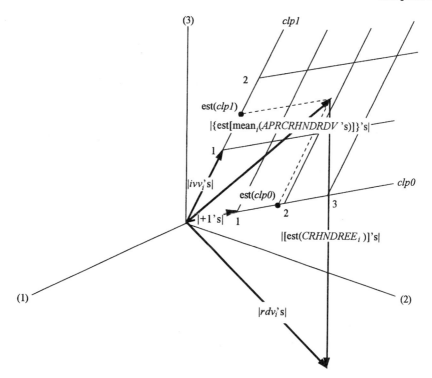

Figure 7.3 Three-dimensional schematic plot of the conventional estimated simple linear regression statistical vector model with est($clp0$) = 1.91 and est($clp1$) = 1.48. The sum of the squares of the components of the |[est($CRHNDREE_i$'s)]'s| column vector takes on its minimum value when this vector is perpendicular (normal) to the parameter plane formed by the |+1's| and the |ivv_i's| column vectors.

$$[|rdv_i\text{'s}| - \text{est}(clp0) \cdot |+1\text{'s}| - \text{est}(clp1)|ivv_i\text{'s}|] \bullet |+1\text{'s}| = 0$$

$$= \sum_{i=1}^{n_{rdv}} (rdv_i) \cdot (+1) - \text{est}(clp0) \cdot \sum_{i=1}^{n_{rdv}} (+1) \cdot (+1) - \text{est}(clp1) \cdot \sum_{i=1}^{n_{rdv}} (ivv_i) \cdot (+1)$$

and

$$\left[|rdv_i\text{'s}| - \text{est}(clp0) \cdot |+1\text{'s}| - \text{est}(clp1) \cdot |ivv_i\text{'s}|\right] \bullet |ivv_i| = 0$$

$$= \sum_{i=1}^{n_{rdv}} (rdv_i) \cdot (ivv_i) - \text{est}(clp0) \cdot \sum_{i=1}^{n_{rdv}} (+1) \cdot (ivv_i) - \text{est}(clp1) \cdot \sum_{i=1}^{n_{rdv}} (ivv_i) \cdot (ivv_i)$$

Then we solve these equations for est($clp0$) and est($clp1$), viz.,

$$\text{est}(clp0) = \frac{\sum_{i=1}^{n_{rdv}} ivv_i^2 \cdot \sum_{i=1}^{n_{rdv}} rdv_i - \sum_{i=1}^{n_{rdv}} ivv_i \cdot \sum_{i=1}^{n_{rdv}} ivv_i rdv_i}{n_{rdv} \cdot \sum_{i=1}^{n_{rdv}} ivv_i^2 - \left(\sum_{i=1}^{n_{rdv}} ivv_i\right)^2}$$

and

$$\text{est}(clp1) = \frac{n_{rdv} \cdot \sum_{i=1}^{n_{rdv}} ivv_i \cdot rdv_i - \sum_{i=1}^{n_{rdv}} ivv_i \cdot \sum_{i=1}^{n_{rdv}} rdv_i}{n_{rdv} \cdot \sum_{i=1}^{n_{rdv}} ivv_i^2 - \left(\sum_{i=1}^{n_{rdv}} ivv_i\right)^2}$$

Note that under continual replication of this simple linear regression experiment test program the conceptual sampling distributions for est($clp0$) and est($clp1$) are appropriately scaled conceptual (two-parameter) normal distributions (because the respective rdv_i's are presumed to be normally distributed). Note also that the respective ivv_i's must sum to zero for the $|+1$'s$|$ column vector and the $|ivv_i$'s$|$ column vector to be orthogonal. Since this orthogonality is unlikely in a simple linear regression experiment test program, it is unlikely that est($clp0$) and est($clp1$) will be statistically independent. Rather, est($clp0$) and est($clp1$) are almost always statistically correlated. Accordingly, we must now adopt an indirect procedure to establish the expressions for var[est($clp0$)] and var[est($clp1$)] that are required to compute the respective classical (shortest) $100(scp)\%$ (two-sided) statistical confidence intervals that allegedly include the actual values for the $clp0$ and the $clp1$ (based on the presumption that the conceptual simple linear regression model is correct).

Subsequent discussion of the conceptual simple regression model can be markedly simplified by appropriately re-expressing the estimated simple linear regression model to avoid the statistical problem that est($clp0$) and est($clp1$) are not independent. Accordingly, we now subtract the arithmetic average of all rdv_i's from each rdv_i and correspondingly subtract the arithmetic average of all ivv_i values from each ivv_i value. This transformation translates the co-ordinate origin so that the linear relationship between est[mean($APRCRHNDRDV$'s) given a parametric ivv] and this parametric ivv plots as a straight line that passes through the point [ave(rdv_i's),ave(ivv_i's)] and has a slope equal to est($clp1$). The associated regression variable values are subsequently denoted $trdv_i$ and $tivv_i$, where $trdv_i = rdv_i - \text{ave}(rdv_i\text{'s})$ and $tivv_i = ivv_i - \text{ave}(ivv_i\text{'s})$ Then, by definition, both the $trdv_i$'s and $tivv_i$'s sum to zero. Since the $|[\text{ave}(rdv_i\text{'s})]\text{'s}|$ column vector can be expressed as $\text{ave}(rdv_i\text{'s}) \cdot |+1\text{'s}|$, the estimated simple linear regression model can be re-stated in our hybrid column vector notation as

$$|rdv_i\text{'s}| = \text{ave}(rdv_i\text{'s}) \cdot |+1\text{'s}| + \text{est}(clp1) \cdot |tivv_i\text{'s}| + |[\text{est}(CRHNDREE_i)]\text{'s}|$$

$$= |\{\text{est}[\text{mean}_i(APRCRHNDRDV\text{'s})]\}\text{'s}| + |[\text{est}(CRHNDREE_i)]\text{'s}|$$

in which the $|+1\text{'s}|$ column vector and the $|trdv_i\text{'s}|$ column vector are orthogonal (because the $tivv_i$'s sum to zero). In turn, this estimated simple linear regression model can be schematically plotted in three-dimensional space as in Figure 7.4 when stated explicitly for $n_{rdv} = 3$ as

$$\begin{vmatrix} rdv_1 \\ rdv_2 \\ rdv_3 \end{vmatrix} = \text{ave}(rdv_i\text{'s}) \cdot \begin{vmatrix} +1 \\ +1 \\ +1 \end{vmatrix} + \text{est}(clp1) \cdot \begin{vmatrix} tivv_1 \\ tivv_2 \\ tivv_3 \end{vmatrix} + \begin{vmatrix} \text{est}(CRHNDREE_1) \\ \text{est}(CRHNDREE_2) \\ \text{est}(CRHNDREE_3) \end{vmatrix}$$

$$= \begin{vmatrix} \text{est}[\text{mean}_1(APRCRHNDRDV\text{'s})] \\ \text{est}[\text{mean}_2(APRCRHNDRDV\text{'s})] \\ \text{est}[\text{mean}_3(APRCRHNDRDV\text{'s})] \end{vmatrix} + \begin{vmatrix} \text{est}(CRHNDREE_1) \\ \text{est}(CRHNDREE_2) \\ \text{est}(CRHNDREE_3) \end{vmatrix}$$

The orthogonality of the $|+1\text{'s}|$ and $|tivv_i\text{'s}|$ column vectors is illustrated schematically in Figure 7.4 where it is evident that $\text{ave}(rdv_i\text{'s}) \cdot |+1\text{'s}| - \text{est}(clp1) \cdot \text{ave}(ivv_i\text{'s}) \cdot |+1\text{'s}| = \text{est}(clp0) \cdot |+1\text{'s}|$.

The generic least-squares plural expression for $\text{est}(clp1)$ is obtained by solving the (single) normal equation:

$$|[\text{est}(CRHNDREE_i)]\text{'s}| \bullet |tivv_i\text{'s}| = 0$$

$$= \{|trdv_i\text{'s}| - [\text{est}(clp1)] \cdot |tivv_i\text{'s}|\} \bullet |tivv_i\text{'s}|$$

viz.,

$$\text{est}(clp1) = \frac{\displaystyle\sum_{i=1}^{n_{dv}} tivv_i \cdot trdv_i}{\displaystyle\sum_{i=1}^{n_{dv}} tivv_i^2}$$

Clearly, this generic least-squares expression for $\text{est}(clp1)$ is algebraically much simpler than the conventional least-squares expression for $\text{est}(clp1)$ given previously.

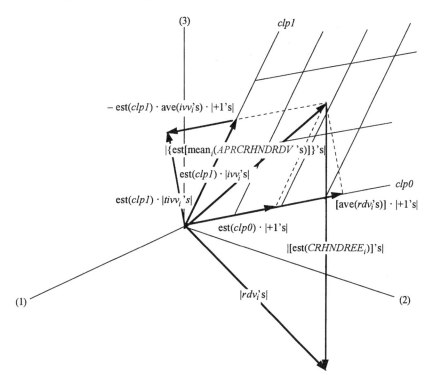

Figure 7.4 Three-dimensional schematic plot of the translated estimated conceptual simple linear regression statistical model, with the $|\{\text{est}[\text{mean}_i(APRCRHNDRDV\text{'s})]\text{'s}\}|$ column vector re-expressed in terms of the orthogonal $|\text{ave}(rdv_i\text{'s})|$ and $|tivv_i\text{'s}|$ column vectors. (Note that the $|rdv_i\text{'s}|$ and the $|[\text{est}(CRHNDREE_i)]\text{'s}|$ column vectors are the same for both the conventional and translated regression co-ordinate axes.)

Remark: Note that this generic least-squares expression for $\text{est}(clp1)$ is identical, except for notation, to the least-squares expression used to compute the respective $\text{est}(sc_j\text{'s})$ pertaining to the complete analytical models associated with the experiment test programs presented in Chapter 2 and statistically analyzed in Chapter 6.

Reminder: The orthogonality of the $|[\text{est}(CRHNDREE_i)]\text{'s}|$ and the $|+1\text{'s}|$ column vectors dictates that the $[\text{est}(CRHNDREE_i)]\text{'s}$ sum to zero.

Figure 7.4 is also intended to support the notion that, because the $|\text{ave}(rdv_i\text{'s})|$ and the $|tivv_i\text{'s}|$ column vectors are orthogonal, $\text{ave}(rdv_i\text{'s})$ and est(*clp1*) are statistically independent under continual replication of the simple linear regression experiment test program. The statistical independence of ave(*rdv$_i$*'s) and est(*clp1*) markedly simplifies the development of an expression for var{est[mean(*APRCRHNDRDV*'s) given the parametric *tivv* of specific interest, *tivv**}. First, we rewrite the estimated conceptual simple linear regression statistical model in scalar notation as

$$\text{est[mean}\,(APRCRHNDRDV\,\text{'s) given } ivv = ivv^*]$$
$$= \text{est[mean}\,(APRCRHNDRDV\,\text{'s) given } tivv = tivv^*]$$
$$= \text{ave}(rdv_i\text{'s}) + (tivv^*) \cdot \text{est}(clp1)$$

in which *ivv** and *tivv** are the corresponding regression metric values of specific interest. We then assert that

$$\text{var}\big\{\text{est[mean}(APRCRHNDRDV\,\text{'s) given } tivv = tivv^*]\big\}$$
$$= \text{var[ave}(rdv_i\text{'s})] + (tivv^*)^2 \cdot \text{var[est}(clp1)]$$

in which

$$\text{var[ave}(rdv_i\text{'s})] = \frac{\text{var}(APRCRHNDREE\,\text{'s})}{n_{rdv}}$$

where var(*APRCRHNDREE*'s) is unknown and must subsequently be estimated.

The expression for var[est(*clp1*)] can be deduced by noting that the generic least-squares expression for est(*clp1*) is merely the summation of a constant term, say *c*, times *trdv$_i$*. In turn, since each of the mutually independent *trdv$_i$*'s are presumed to be randomly selected from a conceptual (two-parameter) normal distribution with a (homoscedastic) variance equal to var(*APRCRHNDREE*'s), the variance of the conceptual sampling distribution that consists of all possible replicate realization values for est(*clp1*) is simply the summation of $c^2 \cdot$ var(*APRCRHNDREE*'s), where

$$c^2 = \frac{\displaystyle\sum_{i=1}^{n_{rdv}} tivv_i^2}{\left\{\displaystyle\sum_{i=1}^{n_{rdv}} tivv_i^2\right\}^2} = \frac{1}{\displaystyle\sum_{i=1}^{n_{rdv}} tivv_i^2}$$

Accordingly,

$$\text{var}[\text{est}(clp1)] = \frac{\text{var}(APRCRHNDREE\text{'s})}{\sum\limits_{i=1}^{n_{rdv}} tivv_i^2}$$

Note that this expression is identical, except for notation, to the expression for var[est(sc_j)] developed in Chapter 4 and used in the classical ANOVA examples in Chapter 6. In turn,

$$\text{var}\{(tivv^*) \cdot [\text{est}(clp1)]\} = \frac{(tivv^*)^2 \cdot \text{var}(APRCRHNDREE\text{'s})}{\sum\limits_{i=1}^{n_{rdv}} tivv_i^2}$$

Thus,

$$\text{var}\{\text{est}[\text{mean}\,(APRCRHNDRDV\text{'s}) \text{ given } tivv = tivv^*]\}$$
$$= \text{var}\{\text{est}[\text{mean}\,(APRCRHNDRDV\text{'s}) \text{ given } ivv = ivv^*]\}$$

$$= \left\{ \frac{1}{n_{rdv}} + \frac{(tivv^*)^2}{\sum\limits_{i=1}^{n_{rdv}} tivv_i^2} \right\} \cdot \text{var}(APRCRHNDREE\text{'s})$$

$$= \left\{ \frac{1}{n_{rdv}} + \frac{[ivv^* - \text{ave}(ivv_i\text{'s})]^2}{\sum\limits_{i=1}^{n_{rdv}} tivv_i^2} \right\} \cdot \text{var}(APRCRHNDREE\text{'s})$$

To estimate the actual value for the variance of the conceptual sampling distribution that consists of all possible replicate realization values for [est(mean($APRCRHNDRDV$'s) given the $tivv^*$ (or ivv^*) of specific interest], we must first compute est[var($APRCRHNDREE$'s)]. However, this calculation is not technically proper until we have first established the adequacy of the presumed conceptual simple linear regression model by conducting the appropriate statistical test (discussed later). Ignoring this problem for the present, we now proceed as if the null hypothesis that the presumed conceptual simple linear regression model is correct was not rejected. Then, est[var($APRCRHNDREE$'s)] is simply equal to the sum of the squares of the elements of the |[est($CRHNDREE_i$)]'s| column vector divided by the number of its statistical degrees of freedom, viz., by ($n_{rdv} - 2$).

Remark One: The square root of est[var($APRCRHNDREE$'s)] is numerically equal to est(csp) where the csp is the conceptual scale parameter (standard deviation) of the conceptual (two-parameter) normal distribution that consists of $APRCRHNDREE$'s.

Remark Two: To understand why the number of statistical degrees of freedom for the |[est($CRHNDREE_i$)]'s| column vector is equal to $(n_{rdv} - 2)$, recall that the $trdv_i$'s sum to zero by definition and therefore the |$trdv_i$'s| column vector has only $(n_{rdv} - 1)$ statistical degrees of freedom. The corresponding |{est[mean$_i$($APRCRH$ $NDTRDV$'s)]'s}| column vector has only a single statistical degree of freedom because its direction is established by the |$tivv_i$'s| column vector and only its scalar magnitude, viz., est($clp1$), requires specification. Accordingly, the |[est($CRHNDEE_i$)]'s| column vector has $(n_{rdv}-1) - 1 = (n_{rdv} - 2)$ statistical degrees of freedom. Another way to establish the number of statistical degrees of freedom for the |[est($CRHNDEE_i$)]'s| column vector is to assert that the |rdv_i's| column vector has n_{rdv} statistical degrees of freedom and that the numerical solution of the two normal equations used to compute est($clp0$) and est($clp1$) imposes two constraints on allowable values for the components of the |[est($CRHNDREE_i$)]'s| column vector, thereby reducing its number of statistical degrees of freedom to $(n_{rdv} - 2)$.

Exercise Set 1

These exercises are intended to familiarize you with the least-squares simple linear regression estimation expressions.

1. Verify that the two text least-squares expressions for est($clp1$) are algebraically equivalent.

2. (a) Write an expression for the sum of the squares of the [est($CRHNDREE_i$)]'s using the translated regression co-ordinate axes. Then, (b) take the derivative of this sum of squares with respect to est($clp1$). In turn, (c) equate this derivative expression to zero and solve for est($clp1$). Does the resulting estimation expression agree with the generic least-squares expression?

3. (a) Write an expression for the sum of the squares of the [est($CRHNDREE_i$)]'s using the conventional regression co-ordinate axes. Then, (b) take the derivatives of this sum with respect to est($clp0$) and est($clp1$). In turn, (c) equate these two derivative expressions to zero and solve for est($clp0$) and est($clp1$). Do the

resulting estimation expressions agree with the conventional least-squares estimation expressions?

Exercise Set 2

These exercises are intended to provide insight regarding (a) the classical (shortest) $100(scp)\%$ (two-sided) statistical confidence intervals that allegedly include the actual values for the $clp0$ and the $clp1$, and, in turn, (b) the associated classical (shortest) $100(scp)\%$ (two-sided) statistical confidence interval that allegedly includes [mean($APRCRHNDRDV$'s) given $ivv = ivv^*$], where the ivv^* of specific interest lies in the (shortest practical) ivv interval that was used in conducting the simple linear regression experiment test program.

1. Given the generic least-squares estimator est($clp1$)= $\sum tivv_i \cdot CRHNDTRDV_i$'s/ $\sum tivv_i^2$ and presuming that the conceptual simple linear regression model is correct, substitute the expected value of each $CRHNDTRDV_i$, viz., $clp1 \cdot tivv_i$, into this expression to demonstrate that the expected value of est($clp1$) is equal to $clp1$. Thus, est($clp1$) is statistically unbiased when the conceptual simple linear regression model is correct, viz., the actual value for the mean of the conceptual sampling distribution that consists of all possible replicate realization values for est($clp1$) is equal to $clp1$.

2. Given the least-squares estimator for est($clp0$) = ave($CRHNDTRDV_i$'s) − est($clp1$)·ave(ivv_i's) and presuming that the conceptual simple linear regression model is correct, extend Exercise 1 by substituting $clp0 + clp1 \cdot$ ave(ivv_i's) for the expected value of ave($CRHNDTRDV_i$'s) and $clp1$ for the expected value of est($clp1$) to demonstrate that the expected value of est($clp0$) is equal to $clp0$. Thus, est($clp0$) is also statistically unbiased when the conceptual simple linear regression model is correct, viz., the actual value for the mean of the conceptual sampling distribution that consists of all possible replicate realization values for est($clp0$) is equal to $clp0$.

3. Use the results of Exercises 1 and 2 to demonstrate that est[mean($APRCRHNDRDV$'s) given $ivv = ivv^*$] is statistically unbiased when the conceptual simple linear regression model is correct.

4. (a) Using the text expression for the var{est[mean($APRCRHNDRDV$'s) given $ivv = ivv^*$]}, let $ivv^* = 0$ and deduce the associated expression for var[est($clp0$)]. Then, (b) given the conventional simple linear regression expression:

est[mean $(APRCRHNDRDV\text{'s})$ given $ivv = ivv^*$]

$$= \text{est}(clp0) + \text{est}(clp1)ivv^*$$

demonstrate that the expression for var{est[mean $(APRCRHND\text{-}RDV\text{'s})$ given $ivv = ivv^*$]} cannot be derived by simply adding the expression for var[est$(clp0)$] to $(ivv^*)^2$ times the expression for var[est$(clp1)$]. In turn, (c) explain what presumption underlying simple arithmetic addition of variances must be violated here.

5. The expression for the actual value for the variance of conceptual sampling distribution that consists of all possible replicate realization values for est[mean$(APRCRHNDRDV\text{'s})$ given $ivv = ivv^*$] under continual replication of the experiment test program, when based on the conventional conceptual simple linear regression model, always includes a *covariance* term, viz.,

$$\text{var}\{\text{est}[\text{mean}\,(APRCRHNDRDV\text{'s})\text{ given } ivv = ivv^*]\}$$
$$= \text{var}[\text{est}(clp0)] + (ivv^*)^2 \cdot \text{var}[\text{est}(clp1)]$$
$$+ 2 \cdot ivv^* \cdot \text{covar}[\text{est}(clp0), \text{est}(clp1)]$$

(a) Accordingly, demonstrate that covar[est$(clp0)$,est$(clp1)$] must be given by the expression:

$$\text{covar}[\text{est}(clp0),\, \text{est}(clp1)] = -\frac{\text{ave}(ivv_i\text{'s})}{\displaystyle\sum_{i=1}^{n_{rdv}} tivv_i^2} \cdot \text{var}(APRCRHNDREE\text{'s})$$

Then, (b) verify that covar[est$(clp0)$,est$(clp1)$] is numerically equal to zero *only* when the sum of the ivv_i's is numerically equal to zero. (Recall that the sum of the $tivv_i$'s is always numerically equal to zero by the definition of the respective $tivv_i$'s.)

Discussion: Covariance is discussed in Supplemental Topic 7.A. For the present we merely note that when est[mean$(APRCRHNDRDV\text{'s})$ given $ivv = ivv^*$] is expressed in terms of est$(clp0)$ and est$(clp1)$, a covariance term is almost always required in the associated expression for var{est[mean $(APRCRHNDRDV\text{'s})$ given $ivv = ivv^*$]} *because* est$(clp0)$ and est$(clp1)$ are seldom statistically independent.

Recall that the $|ivv_i\text{'s}|$ column vector is not orthogonal to the $|+1\text{'s}|$ column vector unless the respective ivv_i's sum to zero. Thus, est$(clp0)$ and est$(clp1)$ are seldom statistically independent. On the other hand, because the respective $tivv_i$'s always sum to zero, the $|tivv_i$'s$|$ column vector is always orthogonal to the $|+1$'s$|$ column vector. Thus, ave(rdv's) and est($clp1$) are always statistically independent. Accordingly, covar[ave(rdv's), est($clp1$)] is always equal to zero. Recall also that var{est[mean($APRCRH$ $NDRDV$'s) given $tivv = tivv^*$]} is equal to var{est[mean (APR $CRHNDRDV$'s) given $ivv = ivv^*$]}, where $ivv^* = tivv^*$ + ave(ivv's).

6. (a) Can the respective least-squares expressions for est($clp0$), est($clp1$), and est[mean($APRCRHNDRDV$'s) given $ivv = ivv^*$] be rewritten as a sum of appropriate scalar constants times the associated mutually independent normally distributed datum values? If so, then the conceptual sampling distributions for est($clp1$), est($clp0$), and est[mean($APRCRHNDRDV$'s) given $ivv = ivv^*$] are appropriately scaled conceptual (two-parameter) normal distributions under continual replication of the regression experiment test program. (b) Write the analytical expression for the PDF pertaining to each these three conceptual sampling distributions.

7. Use the generic form of Student's central T test statistic, viz.,

$$\text{generic Student's central } T_{1,n_{dsdf}} \text{ statistic}$$

$$= \frac{\text{statistically unbiased normally distributed}}{\text{est[stddev (statistically unbiased normally}}$$
$$\frac{\text{estimator} - \text{its expected value}}{\text{distributed estimator)]}}$$

$$= \frac{[\text{est}(clp1)] - clp1}{\text{est}\{\text{stddev}[\text{est}(clp1)]\}} \quad \text{(for example)}$$

and the results of Exercises 1, 2, 3, and 6 to state probability expressions that can be reinterpreted to establish the classical (shortest) $100(scp)\%$ (two-sided) statistical confidence intervals that (a) allegedly include the actual values for the $clp0$ and the $clp1$, and (b) the classical (shortest) $100(scp)\%$ (two-sided) statistical confidence interval that allegedly includes [mean($APRCRH$-$NDRDV$'s) given $ivv = ivv^*$].

8. (a) Extend Exercise 7(b) by stating a probability expression that can be reinterpreted to establish the classical (shortest) $100(scp)\%$ (two-sided) statistical confidence interval that allegedly includes the actual value for the mean of n_f future $CRHNDRDV$'s given the ivv^* of specific interest that lies in the (shortest practical) ivv interval that was used in conducting the simple linear regression

experiment test program. Then, (b) explain what presumption underlying the addition of variances must be made to develop this statistical confidence interval expression. In turn, (c) state the caveat that is appropriate for this statistical prediction interval. (d) Explain how this interval could have practical application.

9. (a) Do the numerical values of the respective values of the ivv_i's, or the associated value of the ave(ivv_i's), change under continual replication of the simple linear regression experiment test program? If not, (b) do the respective expressions for var[est($clp0$)], var[est($clp1$), and var{est[mean($APRCRHNDRDV$'s) given ivv $= ivv*$]} numerically change under continual replication of the simple linear regression experiment test program? These answers underlie the use of the minimum variance strategy in selecting statistically effective ivv's when conducting simple linear regression experiment test programs.

10. (a) Plot hypothetical datum values generated by a regression experiment test program for which four different values (levels) of the independent variable are selected for replicate tests each involving five specimens. (b) Compare your plot to Figure 6.1 and demonstrate that the only thing to distinguish a CRD experiment test program (Chapter 6) from a regression experiment test program (Chapter 7) is the measurement metric for the independent variable and its associated terminology and notation. In turn, (c) based on the mass moment of inertia concept and using your own notation, state an analytical expression for the *within*(SS).

11. (a) Extend Exercise 10 by fairing (by eye) a straight line through your hypothetical regression experiment test program datum values and identify the resulting est($CRHNDREE_i$'s). Then, (b) based on the parallel axis theorem and using your own notation, state an analytical expressions for the *between*(SS). In turn, (c) explain how the *between*(MS) is related to the *within*(MS) in Exercise 10.

7.2.1. Testing the Statistical Adequacy of a Conceptual Simple (or Multiple) Linear Regression Model

Engineering texts typically present simple linear regression examples in which there is no replication and the ivv_i's are uniformly spaced. These examples are rational only in preliminary testing when the response versus independent variable relationship is unknown and the ivv_i's are selected in a "systematic" manner. Otherwise, replication is mandatory and the ivv_i's should be purposely selected to accomplish some statistical objective.

When some or all of the ivv_i's used in a (simple or multiple) linear regression experiment test program are replicated, the $|[\text{est}(CRHND\text{-}REE_i)]$'s$|$ column vector can be partitioned into two orthogonal components, viz., one component associated with the *within*(SS) and the other associated with the *between*(SS). Expressions for these two independent components of the sums of squares of the $[\text{est}(CRHNDREE_i)]$'s are easily established using the mass moment of inertia analogy and its associated parallel axis theorem. The resulting *within*(MS) and *between*(MS) expressions are then used to compute the data-based realization value of Snedecor's central F test statistic which in turn is employed to test the statistical adequacy of the proposed conceptual simple (or multiple) linear regression model.

First consider the component of the $|[\text{est}(CRHNDREE_i)]$'s$|$ column vector associated with the *within*(SS). Hypothetical linear regression experiment test program datum values are plotted schematically in Figure 7.5 where the different ivv's are denoted $divv$'s and the number of these $divv$'s is denoted n_{divv}. The *within*(SS) is conveniently computed as the mass moments of inertia of the replicated regression datum values $rdv_{k,kr}$ about their respective averages, $\text{ave}(rdv_k\text{'s})$, summed over each of the respective $divv$'s, viz.,

$$within(\text{SS}) = \sum_{k=1}^{n_{divv}} \sum_{kr=1}^{n_{rkdivv}} \left[rdv_{k,kr} - \text{ave}(rdv_k\text{'s}) \right]^2$$

in which all regression datum values have a unit (dimensionless) mass and the number of replicates at the kth $divv$ is denoted n_{rkdivv}. In turn, the *within*(SS) divided by its associated number of statistical degrees of freedom is the *within*(MS). The number of statistical degrees of freedom for the *within*(SS) is the sum over all n_{divv}'s of the respective $(n_{rkdivv}-1)$ statistical degrees of freedom. The expected value of the *within*(MS) is equal to var($APRCRHNDREE$'s). Note that this expected value does not depend on the assertion that the conceptual linear regression model is correct.

> *Remark*: This *within*(MS) in linear regression analysis is directly analogous to the *within*[est($CRHNDEE$'s)]MS for CRD experiment test programs in classical ANOVA. The credibility of its homoscedasticity presumption can be statistically tested by running either microcomputer program *RBBHT* or microcomputer program *BARTLETT* when the regression experiment test program has equal replication for all (or several) of its ivv_i's.

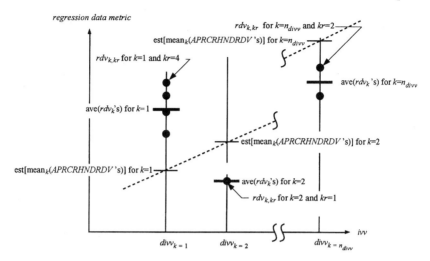

Figure 7.5 Plot of hypothetical linear regression experiment test program datum values and the associated estimated conceptual simple linear regression statistical model. Recall that nonuniform replication is not recommended (unless it accomplishes a well-defined statistical objective). It is depicted here only to illustrate how to establish (count) the proper number of statistical degrees of freedom for the *within*(MS). The *within* number of statistical degrees of freedom is equal to 3 $(4-1)$ for $k = 1$, *plus* 0 $(1-1)$ for $k = 2, \ldots$, *plus* 1 $(2-1)$ for $k = n_{divv}$.

Now consider the component of the $|[\text{est}(CRHNDREE_i)]\text{'s}|$ column vector associated with the *between*(SS). Refer again to Figure 7.5. The respective ave(rdv_k's) are normally distributed and their expected values under continual replication of the linear regression experiment test program are the corresponding mean$_k$($APRCRHNDRDV$'s)'s. However a similar statement can be made about the respective {est[mean$_k$($APRCRHNDRDV$'s)]}'s only when the conceptual linear regression model is correct. Nevertheless, the *between*(SS) component is conveniently computed using the parallel-axis theorem, where the coalesced unit masses at each *divv* are equal to n_{rkdivv} and the associated transfer distances are the {[ave(rdv_k's) − est[mean$_k$($APRCRHNDRDV$'s)]]}'s, viz.,

$$between(\text{SS}) = \sum_{k=1}^{n_{divv}} n_{rkdivv} \{ave(rdv_k\text{'s}) - est[mean_k(APRCRHNDRDV\text{'s})]\}^2$$

$$= \sum_{k=1}^{n_{divv}} \sum_{kr=1}^{n_{rkdivv}} \{ave(rdv_k\text{'s}) - est[mean_k(APRCRHNDRDV\text{'s})]\}^2$$

The *between*(SS) divided by its associated number of statistical degrees of freedom, viz., $(n_{divv} - n_{clp})$, is the *between*(MS). The expected value of the *between*(MS) is equal to var(*APRCRHNDREE*'s) only when the presumed (simple or multiple) linear regression model is correct.

Given the null hypothesis that the conceptual (simple or multiple) linear regression model is correct, the conceptual sampling distribution for the ratio of the *between*(MS) to the *within*(MS) under continual replication of the regression experiment test program is identical to Snedecor's central F conceptual sampling distribution with the corresponding numbers of statistical degrees of freedom. However, if the presumed conceptual linear regression model is not correct, then the *between*(SS) will be inflated by an amount proportional to the sum of squares associated with the discrepancies between the actual physical relationship and the presumed conceptual (simple or multiple) linear regression model. Recall Figure 7.1. Accordingly, when the observed linear regression experiment test program datum values generate a sufficiently large data-based value for Snedecor's central F test statistic, the null hypothesis that the presumed conceptual linear regression model is correct must rationally be rejected. In this event we should reanalyze the experiment test program datum values using a higher-order conceptual linear regression model. Then, in turn, we should test the statistical adequacy of the new higher-order conceptual linear regression model using the revised data-based value of Snedecor's central F test statistic.

A critical concept in establishing the range of the *divv*'s used in a regression experiment test program is to recognize that few if any physical relationships are actually linear over a relatively wide range of the *ivv*'s. Rather, the relevant issue is whether a linear approximation provides a statistically adequate explanation of the observed datum values for the range of the *ivv*'s employed in the given regression experiment test program. When the analytical objective is to examine whether or not a conceptual simple linear regression model is statistically adequate for some range of the *ivv*'s of specific interest, we recommend that one-third of the regression tests should be conducted at each of the two extreme *ivv*'s that establish this range, and the remaining one-third of the regression tests should be conducted at the midpoint of this *ivv* range. Recall that good test planning practice always favors (a) greater replication at fewer *ivv*'s (or treatments) and (b) testing at the practical extremes of the *ivv*'s (treatments) of specific interest. However, if there is not sufficient experience or information available to feel comfortable in planning any given regression experiment test program, then preliminary testing is clearly appropriate.

7.2.1.1. Simple Linear Regression Numerical Example

We now present a simple linear regression example to illustrate the numerical details of computing est(*clp0*), est(*clp1*), est{var[est(*clp0*)]}, and est{var[-est(*clp1*)]}, and in turn computing (a) the classical (shortest) 100(*scp*)% (two-sided) statistical confidence intervals that allegedly include the actual values of the *clp0* and the *clp1*, and (b) the associated classical (shortest) 100(*scp*)% (two-sided) statistical confidence interval that allegedly includes [mean(*APRCRHNDRDV*'s) given *ivv* = *ivv**]. Particular attention also is given to partitioning the *within*(SS) and the *between*(SS) components of the sum of squares for the respective [est(*CRHNDREE_i*)]'s.

Consider the following hypothetical regression datum values:

$ivv_i = ivv_{k,kr}$	$rdv_i = rdv_{k,kr}$
10	2
10	5
10	5
20	13
20	9
40	20
40	22
50	23
50	26
50	23

7.2.1.2. Preliminary Calculations

First, we use a column vector format to organize and tabulate the regression datum values. We then compute the sum of the ivv_i's (300) and the sum of the rdv_i's (148). Accordingly, ave(ivv_i's) = 300/10 = 30 and ave(rdv_i's) = 148/10 = 14.8. Also, $\sum[\text{ave}(ivv_i\text{'s})]^2 = 10[(30)^2] = 9000$ and $\sum[\text{ave}(rdv_i\text{'s})]^2 = 10[(14.8)^2] = 2190.4$. (These sums of squares will be of interest later.) In turn, we construct the following table to estimate the actual values for the *clp0* and the *clp1*.

ivv_i	rdv_i	$tivv_i$	$trdv_i$	ivv_i^2	rdv_i^2	$tivv_i^2$	$trdv_i^2$	$tivv_i \cdot trdv_i$
10	2	−20	−12.8	100	4	400	163.84	256
10	5	−20	−9.8	100	25	400	96.04	196
10	5	−20	−9.8	100	25	400	96.04	196
20	13	−10	−1.8	400	169	100	3.24	18
20	9	−10	−5.8	400	81	100	33.64	58
40	20	+10	+5.2	1600	400	100	27.04	52

ivv_i	rdv_i	$tivv_i$	$trdv_i$	ivv_i^2	rdv_i^2	$tivv_i^2$	$trdv_i^2$	$tivv_i \cdot trdv_i$
40	22	+10	+7.2	1600	484	100	51.84	72
50	23	+20	+8.2	2500	529	400	67.24	164
50	26	+20	+11.2	2500	676	400	125.44	224
50	23	+20	+8.2	2500	529	400	67.24	164
sum =		0	0	11,800	2922	2800	731.60	1400
		checks	checks					

Thus,

$$\text{est}(clp1) = \frac{\sum_{i=1}^{n_{rdv}} tivv_i \cdot trdv_i}{\sum_{i=1}^{n_{rdv}} tivv_i^2} = \frac{1400}{2800} = 0.5$$

Next, using the expression $\text{ave}(rdv_i\text{'s}) = \text{est}(clp0) + \text{est}(clp1) \cdot \text{ave}(ivv_i\text{'s})$:

$$\text{est}(clp0) = 14.8 - (0.5)(30) = -0.2$$

We then compute the respective $\{\text{est}[\text{mean}_k(APRCRHNDRDV\text{'s})]\}$'s and $[\text{ave}(rdv_k\text{'s})]$'s and extend these calculations by constructing the table on page 330.

We now use these tabular results to highlight certain orthogonality relationships of interest by demonstrating numerically that the corresponding partitioned sum-of-squares components add algebraically.

1. Our first example pertains to re-expressing the $|rdv_i\text{'s}|$ column vector in terms of two orthogonal components using our hybrid notation, viz.,

$$|rdv_i\text{'s}| = |[\text{ave}(rdv_i\text{'s})]\text{'s}| + |[rdv_i - \text{ave}(rdv_i\text{'s})]\text{'s}|$$

The associated partitioned sum of squares relationship is thus

$$\sum_{i=1}^{n_{rdv}} rdv_i^2 = \sum_{i=1}^{n_{rdv}} [\text{ave}(rdv_i\text{'s})]^2 + \sum_{i=1}^{n_{rdv}} [rdv_i - \text{ave}(rdv_i\text{'s})]^2$$
$$2922 = 2190.4 + 731.6$$

Similarly, for the $|ivv_i\text{'s}|$ column vector we write:

$$|ivv_i\text{'s}| = |[\text{ave}(ivv_i\text{'s})]\text{'s}| + |[ivv_i - \text{ave}(ivv_i\text{'s})]\text{'s}|$$

Its associated partitioned sums of squares relationship is

$$\sum_{i=1}^{n_{rdv}} ivv_i^2 = \sum_{i=1}^{n_{rdv}} [\text{ave}(ivv_i\text{'s})]^2 + \sum_{i=1}^{n_{rdv}} [ivv_i - \text{ave}(ivv_i\text{'s})]^2$$
$$11,800 = 9000 + 2800$$

$rdv_{k,kr}$	est[mean$_k$ ($APRCRHNDRDV$'s)]	ave(rdv_k's)	$rdv_{k,kr} -$ est[mean$_k$($APRCRHNDRDV$'s)]		$rdv_{k,kr} -$ ave(rdv_k's)		ave(rdv_k's) $-$ est[mean$_k$($APRCRHNDRDV$'s)]	
2	4.8	4.0	-2.8	$(7.84)^{(a)}$	-2.0	(4.0)	-0.8	(0.64)
5	4.8	4.0	$+0.2$	(0.04)	$+1.0$	(1.0)	-0.8	(0.64)
5	4.8	4.0	$+0.2$	(0.04)	$+1.0$	(1.0)	-0.8	(0.64)
13	9.8	11.0	$+3.2$	(10.24)	$+2.0$	(4.0)	$+1.2$	(1.44)
9	9.8	11.0	-0.8	(0.64)	-2.0	(4.0)	$+1.2$	(1.44)
20	19.8	21.0	$+0.2$	(0.04)	-1.0	(1.0)	$+1.2$	(1.44)
22	19.8	21.0	$+2.2$	(4.84)	$+1.0$	(1.0)	$+1.2$	(1.44)
23	24.8	24.0	-1.8	(3.24)	-1.0	(1.0)	-0.8	(0.64)
26	24.8	24.0	$+1.2$	(1.44)	$+2.0$	(4.0)	-0.8	(0.64)
23	24.8	24.0	-1.8	(3.24)	-1.0	(1.0)	-0.8	(0.64)
			0.0	(31.60)	0.0	(22.0)	0.0	(9.60)
			$\sum\{[\text{est}(CRHNDREE_i)]\text{'s}\}^2$		$within(SS)$		$between(SS)$	

(a) $(-2.8)^2 = 7.84$

2. Next, we partition the sums of squares pertaining to the $|trdv_i$'s$|$ column vector by first writing

$$|trdv_i\text{'s}| = |[rdv_i - ave(rdv_i\text{'s})]\text{'s}|$$

$$= |\{rdv_i - \text{est}[\text{mean}_i(APRCRHNDRDV\text{'s})]\}\text{'s}|$$

$$+ |\{\text{est}[\text{mean}_i(APRCRHNDRDV\text{'s})] - ave(rdv_i\text{'s})\}\text{'s}|$$

and then substitute

$$|\{\text{est}[\text{mean}_i(APRCRHNDRDV\text{'s})] - ave(rdv_i\text{'s})\}\text{'s}| = \text{est}(clp1) \cdot |tivv_i\text{'s}|$$

to generate a partitioned sums of squares expression:

$$\sum_{i=1}^{n_{rdv}} trdv_i^2 = \sum_{i=1}^{n_{rdv}} \{rdv_i - \text{est}[\text{mean}_i(APRCRHNDRDV\text{'s})]\}^2$$

$$+ [\text{est}(clp1)]^2 \cdot \sum_{i=1}^{n_{rdv}} tivv_i^2$$

$$731.6 \quad = \qquad\qquad 31.6 \qquad\qquad + (0.5)^2 \times 2800 = 700$$

that can be used to test the null hypothesis that the actual value for the $clp1$ is equal to zero versus the alternative hypothesis that the actual value for the $clp1$ is not equal to zero. (See Exercise Set 4, Exercise 2.)

3. Finally, we partition the $|[\text{est}(CRHNDREE_i)]\text{'s}|$ column vector into two orthogonal components respectively associated with the $within$(SS) and the $between$(SS) in Snedecor's central F test for the statistical adequacy of the presumed conceptual linear regression model. First, we note that

$$|\{rdv_i - \text{est}[\text{mean}_i(APRCRHNDRDV\text{'s})]\}\text{'s}|$$

$$= |\{rdv_{k,kr} - \text{est}[\text{mean}_k(APRCRHNDRDV\text{'s})]\}\text{'s}|$$

and then we restate the $|\{rdv_{k,kr} - \text{est}[\text{mean}_k(APRCRHNDRDV\text{'s})]\}\text{'s}|$ column vector as

$$|\{rdv_{k,kr} - \text{est}[\text{mean}_k(APRCRHNDRDV\text{'s})]\}\text{'s}|$$

$$= |[rdv_{k,kr} - ave(rdv_k\text{'s})]\text{'s}|$$

$$+ |\{ave(rdv_k\text{'s}) - \text{est}[\text{mean}_k(APRCRHNDRDV\text{'s})]\}\text{'s}|$$

The associated partitioned sum of squares expression is

$$\sum_{k=1}^{n_{divv}} \sum_{kr=1}^{n_{rkdivv}} \left\{ rdv_{k,kr} - \text{est}[\text{mean}_k(APRCRHNDRDV\text{'s})] \right\}^2$$

(31.6)

$$= \sum_{k=1}^{n_{divv}} \sum_{kr=1}^{n_{rkdivv}} \left[rdv_{k,kr} - \text{ave}(rdv_k\text{'s}) \right]^2$$

(22)

$$+ \sum_{k=1}^{n_{divv}} \sum_{kr=1}^{n_{rkdivv}} \left\{ \text{ave}(rdv_k\text{'s}) - \text{est}[\text{mean}_k(APRCRHNDRDV\text{'s})] \right\}^2$$

(9.6)

Exercise Set 3

These exercises are intended to verify various orthogonality relationships associated with simple linear regression by verifying that the corresponding dot product are equal to zero when the example simple linear repression experiment test program datum values and associated calculated values are substituted appropriately.

1. Verify numerically that the dot product of the $|\{\text{est}[\text{mean}_i(APRCRHNDRDV\text{'s})]\}\text{'s}|$ and the $|[\text{est}(CRHNDR\text{-}EE_i)]\text{'s}|$ column vectors is equal to zero.

2. (a) Verify numerically that the $|\{\text{est}[\text{mean}_k(APRCRHNDRDV\text{'s})] - \text{ave}(rdv_i\text{'s})\}\text{'s}|$ column vector is identical to the $\text{est}(clp1) \cdot |tivv_k\text{'s}|$ column vector. Explain this result using the appropriate sketch that includes both the conventional ivv and rdv co-ordinate axes and the translated $tivv$ and $trdv$ co-ordinate axes. Then, (b) verify numerically that the dot product of each of these two column vectors with the $|\{rdv_{k,kr} - \text{est}[\text{mean}_k(APRCRHNDRDV\text{'s})]\}\text{'s}|$ column vector is equal to zero.

3. State a relationship that involves both the $|\{\text{ave}(rdv_k\text{'s}) - \text{est}[\text{mean}_k(APRCRHNDRDV\text{'s})]\}\text{'s}|$ and $|[rdv_{k,kr} - \text{ave}(rdv_k\text{'s})]\text{'s}|$ column vectors, then restate these vectors in explicit numerical form and demonstrate that the dot product of the two vectors is equal to zero.

4. (a) Demonstrate numerically that the sum of the number of statistical degrees of freedom pertaining to the *within*(SS) and the *between*(SS) is equal to the number of statistical degrees of freedom pertaining to the [est($CRHNDREE_i$)]'s, viz., is equal to ($n_{rdv} - 2$). Then, (b) account for the two remaining statistical degrees of freedom in this simple linear regression numerical example.

7.2.1.3. Simple Linear Regression Numerical Example (Continued)

We now have the numerical information needed to test the statistical adequacy of the presumed conceptual simple linear regression model. Using the same notation associated with our prior expressions for the *within*(MS) and *between*(MS), we write:

data-based value for Snedecor's central F *within* n_{sdf}, *between* n_{sdf} test statistic

$$
= \frac{\dfrac{\displaystyle\sum_{k=1}^{n_{divv}} \sum_{kr=1}^{n_{rkdivv}} \left\{ \text{ave}(rdv_k\text{'s}) - \text{est}[\text{mean}_k(APRCRHNDRDV\text{'s})] \right\}^2}{between\ n_{sdf}}}{\dfrac{\displaystyle\sum_{k=1}^{n_{divv}} \sum_{kr=1}^{n_{rkdivv}} \left[rdv_{k,kr} - \text{ave}(rdv_k\text{'s}) \right]^2}{within\ n_{sdf}}}
$$

$$
= \frac{9.6/2}{22.0/6} = 1.3091
$$

Microcomputer program *PF* indicates that the probability that a randomly selected value of Snedecor's central F test statistic is equal to or greater than 1.3091 equals 0.3374. Thus we opt not to reject the null hypothesis that the presumed conceptual linear regression model is correct. Accordingly, this decision provides a rational basis for computing the classical (shortest) 100(*scp*)% (two-sided) statistical confidence intervals that allegedly include the actual values for the *clp0* and the *clp1*, and the associated classical (shortest) 100(*scp*)% (two-sided) statistical confidence interval that allegedly includes [mean(*APRCRHNDRDV*'s) given $ivv = ivv^*$].

Caveat: Numerous low-cycle (controlled-strain) fatigue tests have been conducted with replicated tests at two or more ivv_i's. The *majority* of these tests generated datum values that fail Snedecor's central F test for the statistical adequacy of a bi-linear $\log_e$(strain component)–$\log_e$(number of strain reversals to failure) model that is almost universally used in low-cycle (controlled-strain) fatigue testing. Thus, the widespread use of a mechanical behavior model does not assure that it is either statistically adequate or physically credible.

Remark: When a conceptual simple linear regression mechanical behavior model fails Snedecor's central F test for its statistical adequacy, it is rational to propose a conceptual mechanical behavior model whose parametric mean(*APRCRHNDRDV*'s) versus *ivv* relationship is parabolic (second-order) in terms of *ivv*. However, the sign of the resulting est(*clp2*) must be physically credible.

We now extend our simple linear regression numerical example by computing the respective classical (shortest) 100(*scp*)% (two-sided) statistical confidence intervals that allegedly include the actual values of the *clp0* and the *clp1*, and in turn the associated classical (shortest) 100(*scp*)% (two-sided) statistical confidence interval that allegedly includes [mean(*APRCRHNDRDV*'s) given *ivv* = *ivv**]. Recall that this extension is valid because when we performed Snedecor's central F test for the statistical adequacy of the conceptual simple linear regression model, we did not reject the null hypothesis that the presumed model is correct. Accordingly, we assert that it is statistically rational to estimate var(*APRCRHNDREE*'s) using the sum of the *between*(SS) and the *within*(SS), divided by the sum of the number of *between*(SS) and *within*(SS) statistical degrees of freedom. However, the following equivalent estimation expression is much more commonly used to estimate var(*APRCRHNDREE*'s):

est[var(*APRCRHNDREE*'s)]

$$
= \sum_{i=1}^{n_{rdv}} \text{est}(CRHNDREE_i\text{'s})^2 \Big/ (n_{rdv} - 2)
$$

$$
= \sum_{i=1}^{n_{rdv}} \{rdv_i - \text{est}[\text{mean}(APRCRHNDRDV_i\text{'s})]\}^2 \Big/ (10 - 2)
$$

$$
= 31.6/8 = 3.95 \text{ (with 8 statistical degrees of freedom)}
$$

The associated est[stddev(*APRCRHNDREE*'s)] is 1.9875 (with eight statistical degrees of freedom). Accordingly, est(*csp*) = 1.9875.

Remark: The details of computing a 100(*scp*)% (two-sided) statistical confidence interval that allegedly includes the actual value for the *csp* are found in Section 5.9.1. However, make sure to use the proper number of statistical degrees of freedom for est[var(*APRCRHNDREE*'s), viz., replace $(n_{dv} - 1)$ by $(n_{rdv} - 2)$ for simple linear regression, or by $(n_{rdv} - n_{clp})$ for multiple linear regression. (Recall that this statistical confidence interval does not have its minimum possible width.)

Now consider the classical (shortest) 100(*scp*)% (two-sided) statistical confidence interval that allegedly includes the actual value for the *clp1*. Recall that

$$\text{var[est}(clp1)] = \frac{1}{\sum\limits_{i=1}^{n_{rdv}} tivv_i^2} \cdot \text{var}(APRCRHNDREE\text{'s})$$

Thus,

$$\text{var[est}(clp1)] = \frac{1}{2800} \cdot \text{var}(APRCRHNDREE\text{'s})$$

In turn, est{var[est(*clp1*)]} is computed using the expression:

$$\text{est}\{\text{var[est}(clp1)]\} = \frac{1}{2800} \cdot \text{est[var}(APRCRHNDREE\text{'s})]$$

$$= \frac{1}{2800} \cdot 3.95 = 0.0014$$

Recall also that the classical (shortest) 95% (two-sided) statistical confidence interval that allegedly includes the actual value for the *clp1* is computed using complementary percentiles of Student's central t conceptual sampling distribution. Accordingly, we assert that

$$\text{Probability}[\text{est}(clp1) - t_{1,8;0.975}\text{est}\{\text{stddev[est}(clp1)]\} < clp1$$
$$< \text{est}(clp1) - t_{1,8;0.025}\text{est}\{\text{stddev[est}(clp1)]\}] = 0.95$$

in which $- t_{1,8;0.025} = t_{1,8;0.975} = 2.3060$. Accordingly, the corresponding classical (shortest) 95% (two-sided) statistical confidence interval that allegedly includes the actual value for the *clp1* is

$$\left[0.5 - (2.3060)(0.0014)^{1/2}, 0.5 + (2.3060)(0.0014)^{1/2}\right] = [0.4134, 0.5866]$$

Next, consider the classical (shortest) 100(*scp*)% (two-sided) statistical confidence interval that allegedly includes the actual value for the *clp0*. Recall that the *clp0* is the alias of the [mean(*APRCRHNDRDV*'s) given *ivv* = *ivv** = 0] and that

$$\text{var}\{\text{est}[\text{mean}\,(APRCRHNDRDV\text{'s})\text{ given }ivv=ivv^*]\}$$

$$= \left\{ \frac{1}{n_{rdv}} + \frac{[ivv^* - \text{ave}(ivv_i\text{'s})]^2}{\sum\limits_{i=1}^{n_{rdv}} tivv_i^2} \right\} \cdot [\text{var}(APRCRHNDREE\text{'s})]$$

Accordingly,

$$\text{var}[\text{est}(clp0)] = \left\{ \frac{1}{n_{rdv}} + \frac{[0 - \text{ave}(ivv_i\text{'s})]^2}{\sum\limits_{i=1}^{n_{rdv}} tivv_i^2} \right\} \cdot [\text{var}(APRCRHNDREE\text{'s})]$$

Thus,

$$\text{var}[\text{est}(clp0)] = \left\{ \frac{1}{10} + \frac{[-30]^2}{2800} \right\} \cdot [\text{var}(APRCRHNDREE\text{'s})]$$

$$= \{0.1 + 0.3214\} \cdot [\text{var}(APRCRHNDREE\text{'s})]$$

$$= 0.4214 \cdot [\text{var}(APRCRHNDREE\text{'s})]$$

Substituting est[var($APRCRHNDREE$'s)] for var($APRCRHNDREE$'s) gives est{var[est($clp0$)]} = (0.4214)(3.95) = 1.6646. In turn, the classical (shortest) 95% (two-sided) statistical confidence interval that allegedly includes the actual value for the $clp0$ is based on the probability expression:

$$\text{Probability}\big[\text{est}(clp0) - t_{1,8;0.975} \cdot \text{est}\{\text{stddev}[\text{est}(clp0)]\} < clp0$$
$$< \text{est}(clp0) - t_{1,8;0.025} \cdot \text{est}\{\text{stddev}[\text{est}(clp0)]\}\big] = 0.95$$

Accordingly, the corresponding classical (shortest) 95% (two-sided) statistical confidence interval that allegedly includes the actual value for the $clp0$ is

$$\big[-0.2 - (2.3060) \cdot (1.6646)^{1/2}, -0.2 + (2.3060) \cdot (1.6646)^{1/2}\big]$$
$$= [-3.1752, +2.7752]$$

Finally, we compute the associated 95% (two-sided) statistical confidence interval that allegedly includes [mean($APRCRHNDRDV$'s) given $ivv = ivv^*$], where the ivv^* of specific interest lies in the (shortest practical) ivv interval that was used in conducting the regression experiment test program. Suppose $ivv^* = 15$, then

est[mean $(APRCRHNDRDV$'s) given $ivv = ivv^* = 15$]
$$= \text{est}(clp0) + \text{est}(clp1) \cdot ivv^* = -0.2 + (0.5)(15) = 7.3$$

and

var{est[mean $(APRCRHNDRDV$'s) given $ivv = ivv^* = 15$]}

$$= \left\{ \frac{1}{n_{rdv}} + \frac{[ivv^* - \text{ave}(ivv_i\text{'s})]^2}{\sum_{i=1}^{n_{rdv}} tivv_i^2} \right\} \cdot [\text{var}(APRCRHNDREE\text{'s})]$$

$$= \left\{ \frac{1}{10} + \frac{[15 - 30]^2}{2800} \right\} \cdot [\text{var}(APRCRHNDREE\text{'s})]$$

$$= \{0.1 + 0.0804\} \cdot [\text{var}(APRCRHNDREE\text{'s})]$$

$$= 0.1804 \cdot [\text{var}(APRCRHNDREE\text{'s})]$$

In turn, since est[var$(APRCRHNDREE$'s)] $= 3.95$:

est$\big($var{est[mean $(APRCRHNDRDV$'s) given $ivv = ivv^* = 15$]}$\big)$
$$= 0.1804 \cdot 3.95 = 0.7124$$

Thus,

est$\big($stddev{est[mean $(APRCRHNDRDV$'s) given $ivv = ivv^* = 15$]}$\big)$
$$= (0.1124)^{1/2} = 0.8440$$

Accordingly, the classical (shortest) 95% (two-sided) statistical confidence interval that allegedly includes [mean$(APRCRHNDRDV$'s) given $ivv = ivv^* = 15$] is

$$[7.3 - (2.3060)(0.8440), 7.3 + (2.3060)(0.8440)] = [5.3536, 9.2464]$$

Microcomputer program $ATCSLRM$ (adequacy test conceptual simple linear regression model) tests the null hypotheses that (a) the conceptual simple linear regression model is correct and (b) the actual value for the $clp1$ is equal to zero. Then, if the first null hypothesis is not rejected and the second, null hypothesis is rejected, it computes est[mean($APRCRHN-DRDV$'s) given a parametric ivv] and the classical (shortest) $100(scp)\%$ (two-sided) statistical confidence interval that allegedly includes [mean($APRCRHNDRDV$'s) given $ivv = ivv^*$)]. It also computes the analogous Wald–Wolfowitz approximate (two-sided) statistical tolerance interval that allegedly includes $(p)\%$ of [$APRCRHNDDV$'s given $ivv = ivv^*$].

```
COPY EXSLRDTA DATA

   1 file(s) copied

C>ATCSLRM
```

Given an acceptable probability of committing a *Type I* error equal to 0.050, the null hypothesis that the conceptual simple linear regression model is correct is not rejected in favor of the omnibus alternative hypothesis that the conceptual simple linear regression model is not correct.

Given an acceptable probability of committing a *Type I* error equal to 0.050, the null hypothesis that the actual value for the *clp1* is equal to zero is rejected in favor of the composite (two-sided) alternative hypothesis that the actual value for the *clp1* is not equal to zero.

The estimated conceptual simple linear regression model is

$$\text{est[mean}(APRCRHNDRDV\text{'s) given } ivv] = 0.2000 + 0.5000 * ivv$$

The classical (shortest) 95% (two-sided) statistical confidence interval that allegedly includes [mean($APRCRHNDRDV$'s) given $ivv = ivv^* = 15.00$] is [5.3536, 9.2464].

The analogous Wald–Wolfowitz approximate 95% (two-sided) statistical tolerance interval that allegedly includes 90% of [$APRCRHNDRDV$'s) given $ivv = ivv^* = 15.00$] is [1.2224, 13.3776].

7.2.1.4. Discussion

Common sense and engineering judgment is required in the selection of appropriate values for the *scp* and the proportion *p* of the conceptual statistical distribution that consists of [$APRCRHNDRDV$'s given $ivv = ivv^*$]. The selection of very large values for the *scp* and/or the proportion *p* requires an unwarranted reliance on the accuracy of the presumed analytical expression for the linear regression model and on the presumptions of random, homoscedastic, normally distributed datum values and no batch-to-batch effects. Accordingly, in mechanical reliability applications, the *scp* value should not exceed 0.95 and proportion p should not exceed 0.90.

Exercise Set 3 (Extended)

5. Given the expression:

    $$\text{covar}[\text{est}(clp0), \text{est}(clp1)] =$$

    $$-\left[\frac{\text{ave}(ivv_i\text{'s})}{\sum\limits_{i=1}^{n_{rdv}} tivv_i^2}\right] \cdot [\text{var}(APRCRHNDREE\text{'s})]$$

    verify that

    $$\text{covar}[\text{est}(clp0), \text{est}(clp1)] = -\left[\frac{30}{2800}\right] \cdot [\text{var}(APRCRHNDRDV\text{'s})]$$
    $$= -0.0107 \cdot [\text{var}(APRCRHNDRDV\text{'s})]$$

6. Given the expression:

    $$\text{var}\{\text{est}[\text{mean}\,(APRCRHNDRDV\text{'s}) \text{ given } ivv = ivv^*]\}$$
    $$= \text{var}[\text{est}(clp0)] + (ivv^*)^2 \cdot \text{var}[\text{est}(clp1)]$$
    $$+ 2 \cdot ivv^* \cdot \text{covar}[\text{est}(clp0), \text{est}(clp1)]$$

    verify that when $ivv = ivv^* = 15$:

    $$\text{var}\{\text{est}[\text{mean}\,(APRCRHNDRDV\text{'s}) \text{ given } ivv = ivv^* = 15]\}$$
    $$= (0.4214 + 0.0804 - 0.3214) \cdot [\text{var}(APRCRHNDRDV\text{'s})]$$
    $$= (0.1804) \cdot [\text{var}(APRCRHNDRDV\text{'s})]$$

    Then verify that est(var$\{$est[mean($APRCRHNDRDV$'s) given $ivv = ivv^* = 15]\}$) = 0.7124 and that est(stddev$\{$est[mean $(APRCRHNDRDV$'s) given $ivv = ivv^* = 15]\}$) = 0.8440.

7. Note that in Exercise 6 the difference $(0.4214 - 0.3214) = 0.1000 = 1/n_{rdv}$. Is this result general, or does it merely (fortuitously) apply to $ivv^* = 15$ in our numerical example? Support your answer both numerically and algebraically.

8. We demonstrate in Supplemental Topic 8.A that given the expression:

    $$\text{ave}(rdv_i\text{'s}) = \text{est}(clp0) + \text{est}(clp1) \cdot \text{ave}(ivv_i\text{'s})$$

the (exact) propagation of variability expression for var[ave(rdv_i's)] is

$$\text{var[ave}(rdv_i\text{'s})] = \text{var[est}(clp0)] + [\text{ave}(ivv_i\text{'s})]^2 \cdot \text{var[est}(clp1)]$$
$$+ 2 \cdot \text{ave}(ivv_i\text{'s}) \cdot \text{covar[est}(clp0), \text{est}(clp1)]$$

Substitute the respective variance and covariance expressions into this propagation of variability expression to confirm that

$$\text{var[ave}(rdv_i\text{'s})] = \frac{\text{var}(APRCRHNDREE\text{'s})}{n_{rdv}}$$

7.2.2. Minimum Variance Strategy

Suppose that experience indicates that the simple linear regression model is statistically adequate, or, as in low-cycle (strain-controlled) fatigue testing, the bi-linear model continues to be used despite its shortcomings. Then, the only issue is to select the ivv_i's such that the resulting estimates of interest are most (or more) precise. For example, suppose that the test objective is to obtain the most precise estimate of the actual value for the $clp1$ in the conceptual simple linear regression model. Accordingly, we first write the associated variance expression, viz.,

$$\text{var[est}(clp1)] = \frac{\text{var}(APRCRHNDREE\text{'s})}{\sum\limits_{i=1}^{n_{rdv}} [ivv_i - \text{ave}(ivv_i\text{'s})]^2}$$

and then we systematically examine all practical ivv's relative to minimizing the magnitude of this expression. Note that the ivv_i's in the denominator of this variance expression are under our control and thus can be advantageously selected. Moreover, only those ivv_i's that are displaced as far as practical from their arithmetic average make a substantial contribution to the denominator of this variance expression. Accordingly, it is intuitively obvious that (a) only two ivv's should be used in this simple linear regression experiment test program, (b) these two ivv's should be as widely spaced as practical and reasonable, viz., such that one pertains to the largest ivv of practical interest and the other to the smallest ivv of practical interest, and (c) one-half of the experiment test program specimens should be tested at each of these two extreme ivv's. If the numerical value for the variance resulting from this intuitively optimal allocation scheme is compared to the numerical value for the variance associated with equal spacing of the ivv_i's, this ratio approaches $1/3$ as n_{dv} increases—because only a minority of the equally spaced ivv_i's make a substantial contribution to the denominator sum of squares. A similar result is obtained for var{est[mean($APRCRHNDRDV$'s) given $ivv = ivv^*$]} when ivv^* is remote to ave(ivv_i's). The analogous ratio approaches 1/4 for extreme cases of extrapolation. (Remember, however, that extrapolation is not recommended.)

The best way to make a simple linear regression experiment test program statistically effective and efficient is to select all of its ivv's before the experiment test program is begun. However, the optimal allocation strategy can be adopted at any point during the experiment test program. Suppose that the unplanned simple linear regression experiment test program is either about 90% completed and only a few specimens remain, or preferably that a few specimens have been deliberately reserved to verify the predictions made on the basis of a statistically planned simple linear regression experiment test program. The analytical procedure is still the same: (a) state the estimate or estimates of specific interest; (b) write the corresponding variance expressions; (c) substitute alternative (candidate) ivv_i's into these expressions and evaluate the respective variance expressions pertaining to these alternative ivv_i's; and (d) select those ivv_i's that generate the most precise estimates of specific interest.

Figure 7.6 is an example taken from Little and Jebe (1975) pertaining to Wöhler's s_a–$\log_e(fnc)$ fatigue tests on wrought iron axles subjected to eight equally spaced alternating stress amplitudes (levels): 320, 300, 280, 260, 240, 220, 200, and 180 centners/zoll^2. The respective s_a's are the ivv_i's, whereas the natural logarithms of the number of alternating stress cycles to fatigue failure, denoted [$\log_e(fnc)$]'s, are the rdv's. Suppose that Wöhler's objective was to estimate the actual value for the median (mean) of the presumed normally distributed conceptual fatigue life distribution that consists of all possible replicate $\log_e(fnc)$ datum values given $s_a = s_a^*$. Suppose further that, after running these eight fatigue tests, Wöhler still had one specimen remaining. The ivv (s_a) at which this ninth specimen should be tested is plotted in Figure 7.6 versus the normalized variance of the conceptual sampling distribution pertaining to five example ivv's (s_a's) of potential interest. Clearly, when the ivv_{opi} ($s_{a_{opi}}$) differs markedly from the arithmetic average of all ivv's (s_a's), the selection of ivv_9 (s_{a_9}) markedly affects (a) the precision of the statistical estimate of the actual value for the median (mean) of the presumed normally distributed conceptual fatigue life distribution given the s_a^* value of specific interest, and (b) the associated width of the classical (shortest) $100(scp)$% (two-sided) statistical confidence interval that allegedly includes the actual value for this median (mean), given $s_a = s_a^*$. Accordingly, a poor choice for ivv_9 (s_{a9}) can effectively waste this specimen and its associated test time and cost. On the other hand, when the ivv_{opi} (s_{aopi}) does not differ markedly from the arithmetic

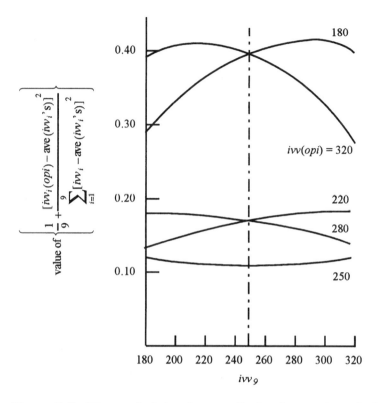

Figure 7.6 Diagram depicting the normalized variance ratio var{est[mean(APR $CRHNDRDV$'s) given $ivv = ivv_{opi}$]}/var($APRCRHNDREE$'s), plotted versus ivv_9, where ivv_{opi} pertains to five example ivv's of potential interest at which the most precise estimate of the actual value for the median of the conceptual normal distribution that consists of ($APRCRHNDRDV$'s) given $ivv = ivv_{opi}$ is desired. The ivv's pertain to Wöhler's s_a's and the rdv's pertain to Wöhler's [$\log_e(fnc)$]'s for his eight actual s_a–$\log_e(fnc)$ fatigue tests and our hypothetical ninth test. (From Little and Jebe, 1975.)

average of all ivv's (s_a's), then the selection of ivv_9 (s_{a9}) does not markedly affect either (a) or (b) above.

The importance of rational (purposeful) selection of the respective ivv's in linear regression experiment test program planning cannot be over emphasized. Unfortunately, almost all simple linear regression experiment test program datum values found in the mechanical metallurgy literature pertain to unreplicated ivv's that are more or less uniformly spaced. These experiment test programs are statistically inept because the majority of their experimental units (and the associated test time and cost) are not effectively allocated relative to improving the precision of the statistical estimates of potential or of specific interest.

7.2.3. Inverse Simple Linear Regression (One-Sided) Statistical Confidence Limits

Consider the case where the conceptual simple linear regression model is presumed to be correct and that the actual value for the *clp1* is presumed to be markedly positive. Suppose that we compute the maximum (single-valued) *ivv** such that the upper 100(*scp**)% (one-sided) statistical confidence limit that allegedly bounds [mean(*APRCRHNDRDV*'s) given *ivv* = *ivv**] is less than the specification value of specific interest for the regression data metric. This calculation is called inverse regression. Microcomputer programs *ISLRCLPS* and *ISLRCLNS* cover the four cases of specific interest.

```
COPY IRPSDATA DATA

    1 file(s) copied


C>ISLRCLPS
```

Given that the presumed conceptual simple linear regression model is correct and that the actual value for the *clp1* is markedly positive, the upper 95% (one-sided) statistical confidence limit that allegedly bounds the actual value for [mean(*APRCRHNDRDV*'s) given *ivv* = *ivv**] is less than 7.3000 when *ivv** is less than 11.5173.

Given that the presumed conceptual simple linear regression model is correct and that the actual value for the *clp1* is markedly positive, the lower 95% (one-sided) statistical confidence limit that allegedly bounds the actual value for [mean(*APRCRHNDRDV*'s) given *ivv* = *ivv**] is greater than 7.3000 when *ivv** is greater than 17.8857.

```
COPYIRNSDATA DATA

    1 file(s) copied

C>ISLRCLNS
```

Given that the presumed conceptual simple linear regression model is correct and that the actual value for the *clp1* is markedly negative, the upper 95% (one-sided) statistical confidence limit that allegedly bounds the actual value for [mean($APRCRHNDRDV$'s) given $ivv = ivv^*$] is less than 7.3000 when ivv^* is greater than 48.4827.

Given that the presumed conceptual simple linear regression model is correct and that the actual value for the *clp1* is markedly negative, the lower 95% (one-sided) statistical confidence limit that allegedly bounds the actual value for [mean($APRCRHNDRDV$'s) given $ivv = ivv^*$] is greater than 7.3000 when ivv^* is at less than 42.1143.

7.2.4. Inverse Simple Linear Regression (One-Sided) Statistical Tolerance Limits

In most mechanical reliability applications it is more practical to compute inverse simple linear regression statistical tolerance limits than to compute inverse simple linear regression statistical confidence limits. Microcomputer programs *ISLRTLPS* and *ISLRTLNS* are analogous to microcomputer programs *ISLRCLPS* and *ISLRCLNS*, except that the respective lower and upper statistical tolerance limits allegedly bound the actual value for the $(1 - p)^{th}$ percentile of the conceptual statistical distribution that consists of [$APRCRHNDRDV$'s) given $ivv = ivv^*$].

7.2.5. Classical Hyperbolic Lower 100(*scp*)% (One-Sided) Statistical Confidence Band

We now present the classical hyperbolic lower 100(*scp*)% (one-sided) statistical confidence band that allegedly bounds mean($APRCRHNDRDV$'s) for all *ivv*'s (simultaneously) that lie in the (shortest practical) *ivv* interval that was used in conducting the regression experiment test program.

```
COPY IRPSDATA DATA

    1 file(s) copied

C>ISLRTLPS
```

Given that the presumed conceptual simple linear regression model is correct and that the actual value for the *clp1* is markedly positive, the upper 95% (one-sided) statistical tolerance limit that allegedly bounds 90% of [*APRCRHNDRDV*'s given $ivv = ivv^*$] is less than 7.3000 when ivv^* is less than 3.7183.

Given that the presumed conceptual simple linear regression model is correct and that the actual value for the *clp1* is markedly positive, the lower 95% (one-sided) statistical tolerance limit that allegedly bounds 90% of [*APRCRHNDRDV*'s given $ivv = ivv^*$] is greater than 7.3000 when ivv^* is greater than 24.7436.

```
COPY IRNSDATA DATA

    1 file(s) copied

C>ISLRTLNS
```

Given that the presumed conceptual simple linear regression model is correct and that the actual value for the *clp1* is markedly negative, the upper 95% (one-sided) statistical tolerance limit that allegedly bounds 90% of [*APRCRHNDRDV*'s given $ivv = ivv^*$] is less than 7.3000 when ivv^* is greater than 56.2817.

Given that the presumed conceptual simple linear regression model is correct and that the actual value for the *clp1* is markedly negative, the lower 95% (one-sided) statistical tolerance limit that allegedly bounds 90% of [*APRCRHNDRDV*'s given $ivv = ivv^*$] is greater than 7.3000 when ivv^* is less than 35.2564.

classical lower hyperbolic $100(scp)$ % (one-sided) statistical confidence band

$$= \text{est[mean}(APRCRHNDRDV\text{'s) given } ivv] - k_{chlosseb} \cdot \text{stddevterm}$$

in which

$$k_{chlosseb} = [2 \cdot \text{Snedecor's central } F_{2,n_{rdv}-2;(1+scp)/2}]^{1/2}$$

and

$$\text{stddevterm} = \text{est(stddev}\{\text{est[mean}(APRCRHNDRDV\text{'s) given } ivv]\})$$

$$= \sqrt{\frac{1}{n_{rdv}} + \frac{[ivv - \text{ave}(ivv_i\text{'s})]^2}{\sum_{i=1}^{n_{dv}} tivv_i^2}} \cdot \{\text{est[stddev}(APRCRHNDRDV\text{'s})]\}$$

See Supplemental Topic 7.C for background regarding the development of the associated classical hyperbolic $100(scp)$% (two-sided) statistical confidence band that allegedly bounds mean($APRCRHNDRDV$'s) for all ivv's (simultaneously) that lie in the (shortest practical) ivv interval that was used in conducting the regression experiment test program.

7.2.6. Uniform Width Lower 100(scp)% (One-Sided) Statistical Confidence Band

The shape of the classical lower hyperbolic $100(scp)$% (one-sided) statistical confidence band in Figure 7.7(a) does not agree with the intuitive engineering notion of scatter bands that are faired (by eye) on a data plot to form an envelope for the majority of datum points. Rather, intuition dictates either (a) uniform width $100(scp)$% (two-sided) statistical confidence bands or (b) a uniform width lower (or upper) $100(scp)$% (one-sided) statistical confidence band. Given that the presumed conceptual simple linear regression model is correct, microcomputer program $UWLOSSCB$ (uniform width lower one-sided statistical confidence band) computes the (parallel straight-line) uniform width lower $100(scp)$% (one-sided) statistical confidence band in Figure 7.7(b) that allegedly bounds [mean($APRCRHNDRDV$'s) given ivv] for all ivv's that lie in the interval from ivv_{low} to ivv_{high}, provided that this ivv interval is either equal to or lies completely within the (shortest practical) ivv interval that was used in conducting the simple linear regression experiment test program. (See Little and Jebe, 1975.)

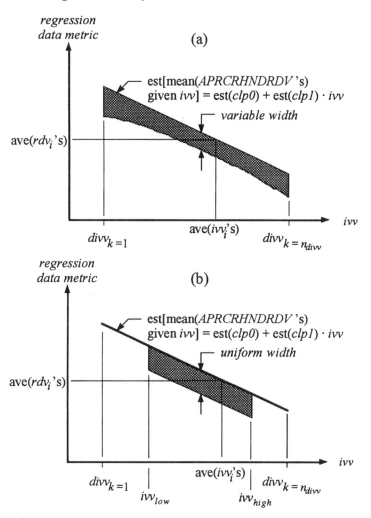

Figure 7.7 Schematic (a) depicts the classical hyperbolic lower $100(scp)\%$ (one-sided) statistical confidence band that allegedly bounds [mean($APRCRHNDRDV$'s) given ivv] for all ivv's that lie in the (shortest practical) ivv interval that was used in conducting the simple linear regression experiment test program. Schematic (b) depicts either (i) a uniform width lower $100(scp)\%$ (one-sided) statistical confidence band that allegedly bounds [mean($APRCRHNDRDV$'s) given ivv] for all ivv's that lie in the interval from ivv_{low} to ivv_{high}, or (ii) a uniform width lower $100(scp)\%$ (one-sided) statistical tolerance band that allegedly bounds $100(p)\%$ of [$APRCRHNDRDV$'s given ivv] for all ivv's that lie in the interval from ivv_{low} to ivv_{high}.

```
COPY UWLCBDTA DATA

    1 file(s) copied


C>UWLOSSCB
```

Given that the presumed conceptual simple linear regression model is correct, the uniform width of the lower 95% (one-sided) statistical confidence band that allegedly bounds [mean($APRCRHNDRDV$'s) given ivv] for all ivv's that lie in the interval from 4.2000 to 4.5000 is equal to 0.1797.

7.2.7. Uniform Width Lower 100(scp)% (One-Sided) Statistical Tolerance Band

Microcomputer program $UWLOSSTB$ (uniform width lower one-sided statistical tolerance band) computes the analogous (parallel straight-line) uniform width lower 100(scp)% (one-sided) statistical tolerance band that allegedly bounds 100(p)% of [$APRCRHNDRDV$'s given ivv] for all ivv's that lie in the interval from ivv_{low} to ivv_{high}, provided that this ivv interval is either equal to or lies completely within the (shortest practical) ivv interval that was used in conducting the simple linear regression experiment test program. (See Little and Jebe, 1975.)

```
COPY UWLTBDTA DATA

    1 file(s) copied


C>UWLOSSTB
```

Given that the presumed conceptual simple linear regression model is correct, the uniform width of the lower 95% (one-sided) statistical tolerance band that allegedly bounds 90% of [$APRCRHNDRDV$'s given ivv] for all ivv's that lie in the interval from 4.2000 to 4.5000 is equal to 0.4825.

7.2.8. Linearizing Transformations

Sometimes a simple transformation of either the independent variable or the random variable, or both, is necessary to establish linearity for the statistical model subsequently employed in simple linear regression analysis. For example, Ludwig's true-stress, true-strain model for plastic deformation in a tension test is stated (except for notation) as

$$(\text{true-stress}) = c1 \cdot (\text{true-strain})c2$$

Taking the natural logarithms of the terms on both sides of Ludwig's model generates an equivalent simple linear regression model, viz.,

$$\log_e(\text{true-stress}) = \log_e c1) + c2 \cdot \log_e(\text{true-strain})$$

In turn, we must add a normally distributed homoscedastic experimental error term to this linearized model to make it amenable to simple linear regression analysis. Then, re-expressing the linearized model in text notation, we write:

$$CRHND\left[\log_e(\text{true-stress})\right]DV_i\text{'s}$$
$$= \log_e(clp0) + clp1 \cdot \{\left[\log_e(\text{true-strain})\right]ivv_i\text{'s}\}$$

Following statistical analysis for the linearized simple linear regression model, the exponentials (antilogs) of the conceptual location parameter estimates and the respective upper and lower limits of any classical (shortest) $100(scp)\%$ (two-sided) statistical confidence interval of specific interest are taken to conform again to the original measurement metric.

> *Remark*: Note that we have arbitrarily presumed that Ludwig's model is properly expressed in terms of its independent and dependent variables. Recall, however, that the first step in simple linear regression analysis is to decide which variable is properly viewed as the independent variable and which is properly viewed as the dependent variable.

7.2.9. Checking the Presumptions Underlying Linear Regression Analysis

Always statistically examine the est($CRHNDREE_i$'s) relative to normality, homoscedasticity, and randomness (independence, viz., lack of association).

7.2.9.1. Normality

As in ANOVA, given that the presumed conceptual statistical model is correct, the value of the generalized modified Michael's $MDSPP$ test statistic does not depend on the actual values for any of the parameters in the conceptual statistical model. Accordingly, the generalized modified Michael's $MDSPP$ test statistic is employed in microcomputer programs $NTCSLRM$ and $NTCMLRM$ to test the null hypotheses of normality for the $CRHNDREE_i$'s in the presumed conceptual simple and multiple linear regression models.

7.2.9.2. Homoscedasticity

Given equal replication at two or more $divv_k$'s, run either microcomputer program $RBBHT$ or $BARTLETT$ to test the null hypothesis of homoscedasticity for the respective sets of $CRHNDREE_{k,kr}$'s. If there is no replication, microcomputer program $RBKTAU$ provides a very weak surrogate test for the homoscedasticity of the $CRHNDREE_i$'s when the respective [est($CRHNDREE_i$)]'s are paired with the magnitudes of their associated ivv_i's.)

7.2.9.3. Randomness (Independence, viz., Lack of Association)

Run microcomputer program $RBKTAU$ to examine the respective [est($CRHNDREE_i$)]'s relative to a lack of a monotonic time-order-of-testing association and relative to all other monotonic associations that may be plausible.

Exercise Set 4

These exercises are intended to provide perspective regarding certain details of simple regression analysis.

1. Consider the following relative wear resistance data for normalized plain carbon steels. What is the independent variable and what is the test outcome? Can the ivv's also have experimental errors? Discuss the issues involved in selecting the ivv's in this example. In particular, indicate why it is important to make independent attempts to attain the specification target value for

each *divv* (or each different treatment, treatment level, or treatment combination).

Carbon content	Relative wear resistance
0.05	1.00
0.20	1.13
0.38	1.28
0.55	1.39
0.80	1.65
1.02	1.68
1.17	1.87

2. Given the following hybrid column vector expression for the translated estimated conceptual simple linear regression model:

$$|trdv_i\text{'s}| = \text{est}(clp1) \cdot |tivv_i\text{'s}| + |[\text{est}(CRHNDREE_i)]\text{'s}|$$

and the associated expression for Snedecor's central $F_{1,n_{dv}-2}$ test statistic.

Snedecor's central $F_{1,n_{dv}-2}$ test statistic

$$= \frac{[\text{est}(clp1)]^2 \cdot \sum\limits_{i=1}^{n_{rdv}} tivv_i^2 / 1}{\sum\limits_{i=1}^{n_{rdv}} [\text{est}(CRHNDREE_i)]^2 / (n_{rdv} - 2)}$$

(a) explain why the data-based realization value for Snedecor's central $F_{1,n_{dv}-2}$ test statistic can be used to test the specific null hypothesis that the actual value for the *clp1* is equal to zero versus the specific composite (two-sided) alternative hypothesis that the actual value for the *clp1* is not equal to zero. Then, (b) extend the text simple linear regression numerical example by testing the specific null hypothesis that the actual value for the *clp1* is equal to zero versus the specific composite (two-sided) alternative hypothesis that the actual value for the *clp1* is not equal to zero. Let the acceptable probability of committing a *Type I* error be equal to 0.05. Is your null hypothesis rejection decision consistent with the classical (shortest) 95% (two-sided) statistical confidence interval that allegedly includes the actual value for the *clp1*, viz., [0.4134, 0.5866]?

7.3. MULTIPLE LINEAR REGRESSION ANALYSIS

Multiple linear regression analysis pertains to the situation where two or more independent variables are employed in the regression experiment test program. The conceptual multiple linear regression statistical model can be written in hybrid matrix notation as

$$|CRHNDRDV_i\text{'s}| = |ivjv_i\text{'s}||clpj\text{'s}| + |CRHNDREE_i\text{'s}|$$
$$= |[\text{mean}_i(APRCRHNDRDV\text{'s}) \text{ given } |(ivjv_i)\text{'s}|]\text{'s}|$$
$$+ |CRHNDREE_i\text{'s}|$$

in which $|CRHNDRDV_i\text{'s}|$ is a $(n_{rdv} \times 1)$ column vector, $|ivjv_i\text{'s}|$ is a $(n_{rdv} \times n_{clp})$ array with its first $(j = 0)$ column consisting of plus ones and its $(n_{clp} - 1)$ remaining j columns consisting of the specific $ivjv_i$ values selected for the $(n_{clp} - 1)$ independent variables, $|clpj\text{'s}|$ is a $(n_{clp} \times 1)$ column vector whose transpose is $[clp0, clp1, \ldots, clp(n_{clp} - 1)]$, and $|CRHNDREE_i\text{'s}|$ is a $(n_{rdv} \times 1)$ column vector.

The $clpj$'s in this conceptual multiple linear regression matrix statistical model can be estimated by first writing the normal equations (Draper and Smith, 1966):

$$|ivjv_i\text{'s}|^t|ivjv_i\text{'s}| \, |\text{est}(clpj\text{'s})| = |ivjv_i\text{'s}|^t|rdv_i\text{'s}|$$

and then premultiplying both sides of these normal equations by $[\,|ivjv_i\text{'s}|^t\,|ivjv_i\text{'s}|]^{-1}$ to obtain

$$|[\text{est}(clpj)]\text{'s}| = [|ivjv_i\text{'s}|^t|ivjv_i\text{'s}|]^{-1}[|ivjv_i\text{'s}|^t|rdv_i\text{'s}|]$$

In turn, the n_{clp} by n_{clp} covariance matrix for these $[\text{est}(clpj)]$'s is computed using the expression:

$$|\text{covar}[\text{est}(clpm), \text{est}(clpn)]| = \text{var}(APRCRHNDREE\text{'s})[ivjv_i\text{'s}|^t|ivjv_i\text{'s}|]^{-1}$$

in which m and n are dummy indices, each 0 to $(n_{clp} - 1)$. The actual values for the elements of this (symmetrical) covariance matrix are estimated by substituting est[var($APRCRHNDREE$'s)] with $(n_{divjv\text{'s}} - n_{clp})$ statistical degrees of freedom for var($APRCRHNDREE$'s). In turn, given the elements of the estimated covariance matrix, the propagation of variability methodology (Supplemental Topic 7.A) could be used to compute est(var{est[mean($APRCRHNDRDV$'s) given $(ivjv)$'s equal to any single arbitrary but complete set of the $(ivjv^*)$'s of specific interest]}). However, it is much more convenient to use the matrix expression:

$$\text{est}\big(\text{var}\{\text{est}[\text{mean}\,(APRCRHNDRDV\,\text{'s})\text{ given }|(ivjv)\text{'s}| = |(ivjv^*)\text{'s}|]\}\big)$$

$$= \text{est}[\text{var}(APRCRHNDREE\,\text{'s})] \cdot \left[|(ivjv^*)\text{'s}|^t \Big(|(ivjv_i)\text{'s}|^t |(ivjv_i)\text{'s}|\Big)^{-1} |(ivjv^*)\text{'s}|\right]$$

in which $|(ivjv^*)\text{'s}|^t = [1,\ iv1v^*,\ iv2v^*, \ldots,\ iv(n_{clp}-1)v^*]$.

Microcomputer program $ATCMLRM$ (adequacy test conceptual multiple linear regression model) tests the null hypotheses that the conceptual multiple linear regression model is correct and if this null hypothesis is not rejected, the conceptual parameters of this model are estimated and the classical (shortest) $100(scp)\%$ (two-sided) statistical confidence interval that allegedly includes [mean($APRCRHNDRDV$'s) given $|(ivjv)\text{'s}| = |(ivjv^*)\text{'s}|$] is computed.

```
COPY EXMLRDTA MLRDATA

    1 file(s) copied

COPY EXDMX DESIGNMX

    1 file(s) copied

COPY EXIVJVAS XIVJVAS

    1 file(s) copied

C>ATCMLRM
```

Given an acceptable probability of committing a *Type I* error equal to 0.050, the null hypothesis that the conceptual multiple linear regression model is correct is not rejected in favor of the omnibus alternative hypothesis that the conceptual multiple linear regression model is not correct.

The elements of the est($clpj$) column vector are, respectively,

$clp0 = 12.738095$
$clp1 = 0.342857$
$clp2 = 0.021905$

Given the following $ivjv^*$ values of specific interest

$$iv0v^* = \quad 1.000000$$
$$iv1v^* = \quad 16.750000$$
$$iv2v^* = 280.562500$$

est[mean($APRCRHNDRDV$'s)] is equal to 24.626607

The corresponding classical (shortest) 90% (two-sided) statistical confidence interval that allegedly includes mean($APRCRHNDRDV$'s) is [22.5224, 26.7308].

Exercise Set 5

These exercises are intended to verify that multiple linear regression matrix expressions can be algebraically expanded to establish corresponding simple linear regression expressions. Hopefully, this verification process will generate greater familiarity with the matrix expressions.

1. Symbolically state the elements in each of the following arrays and then perform the relevant respective matrix operations. Compare the result obtained in (d) to the corresponding algebraic expressions in Section 7.2 and comment appropriately. State the statistical interpretation for each element of the array obtained in (b).

    (a) $|ivjv_i\text{'s}|^t|ivjv_i\text{'s}|$ (b) $[|ivjv_i\text{'s}|^t|ivjv_i\text{'s}|]^{-1}$

    (c) $|ivjv_i\text{'s}|^t|tdv_i\text{'s}|$ (d) $[|ivjv_i\text{'s}|^t|ivjv_i\text{'s}|]^{-1}[|ivjv_i\text{'s}|^t|rdv_i\text{'s}|]$

2. Symbolically state the elements in each of the following arrays and then perform the relevant respective matrix operations.

    $$[|ivjv^*\text{'s}|^t[|ivjv_i\text{'s}|]^{-1}|ivjv^*\text{'s}|]$$

    In turn, manipulate your algebraic expression appropriately to verify that

$$\text{var}\{\text{est}[\text{mean}\,(APRCRHNDRDV\text{'s}) \text{ given } ivv=ivv^*]\}$$

$$= \left\{ \frac{1}{n_{rdv}} + \frac{[ivv^* - \text{ave}(ivv_i\text{'s})]^2}{\sum\limits_{i=1}^{n_{rdv}} tivv_i^2} \right\} \text{var}(APRCRHNDREE\text{'s})$$

7.3.1. Indicator Variables

Indicator variables can be used in multiple linear regression analysis to couple quantitative and qualitative effects. The following elementary examples introduce this concept.

7.3.1.1. Example One

Consider the conceptual multiple linear regression statistical model:

$$|CRHNDRDV_i\text{'s}| = clp0 \cdot | + 1\text{'s}| + clp1 \cdot |iv1v_i\text{'s}| + clp2 \cdot |iv2v_i\text{'s}|$$
$$+ |CRHNDREE_i\text{'s}|$$

in which $iv2v_i$ takes on the value zero when treatment A pertains, but takes on the value one when treatment B pertains. Then,

If treatment A:

$$|CRHNDRDV_i\text{'s}| = clp0 \cdot | + 1\text{'s}| + clp1 \cdot |iv1v_i\text{'s}| + |CRHNDREE_i\text{'s}|$$

If treatment B:

$$|CRHNDRDV_i\text{'s}| = (clp0 + clp2) \cdot | + 1\text{'s}| + clp1 \cdot |iv1v_i\text{'s}| + |CRHNDEE_i\text{'s}|$$

This conceptual multiple linear regression model thus generates two parallel conceptual simple linear regression models. One pertains strictly to treatment A and the other pertains strictly to treatment B.

7.3.1.2. Example Two

Consider the conceptual multiple linear regression statistical model:

$$|CRHNDRDV_i\text{'s}| = clp0 \cdot | + 1\text{'s}| + clp1 \cdot |iv1v_i\text{'s}| + clp2 \cdot |iv2v_i\text{'s}|$$
$$+ clp12 \cdot |[(iv1v_i) \cdot (iv2v_i)]\text{'s}| + |CRHNDEE_i\text{'s}|$$

in which $iv2v_i$ takes on the value zero when treatment A pertains, but takes on the value plus one when treatment B pertains. Then,

If treatment A:

$$|CRHNDRDV_i\text{'s}| = clp0 \cdot | + 1\text{'s}| + clp1 \cdot |iv1v_i\text{'s}|; +|CRHNDEE_i\text{'s}|$$

If treatment B:

$$|CRHNDRDV_i\text{'s}| = (clp0 + clp12) \cdot | + 1\text{'s}| + (clp1 + clp2) \cdot |iv1v_i\text{'s}|$$
$$+ |CRHNDEE_i\text{'s}|$$

This conceptual multiple linear regression model generates two nonparallel conceptual simple linear models. Again, one pertains strictly to treatment A and the other pertains strictly to treatment B.

> *Remark*: Note that the conceptual simple linear regression models for treatments A and B share a common variance, viz., var($APRCHRHNDREE$'s) associated with the underlying conceptual multiple linear regression model.

7.4. BALANCING SPURIOUS EFFECTS OF NUISANCE VARIABLES IN LINEAR REGRESSION EXPERIMENT TEST PROGRAMS

Two statistical abstractions that were tacitly presumed in the preceding discussion of linear regression analyses are (a) all experimental units are randomly selected from an infinite population of nominally identical experimental units and (b) all nuisance variable are negligible. Otherwise, the concepts of statistical confidence and tolerance intervals have limited statistical credibility. However, experimental units are typically produced in small batches in mechanical reliability applications. Moreover, all experiment test programs involve actual or potential nuisance variables. Thus, special planning is required to balance (mitigate) the spurious effects of batch-to-batch variability and nuisance variables on the estimated parameters of the conceptual regression model.

First, suppose that a specific batch of experimental units is large enough to conduct the entire CRD experiment test program. If there are no batch,*ivjv* interaction effects, the specific batch of experimental units used in the experiment test program will have the same incremental effect on each of the observed regression datum values—but it will have no effect on the corresponding *trdv_i*'s. Thus, the est(*clpj*'s) for $j > 0$ will not be pragmatically biased by the use of a specific batch of experimental units (provided that there are no batch,*ivjv* interaction effects). Note that a similar conclusion pertains to any collection of nuisance variables.

Next, suppose that the experimental units pertain to several batches conveniently structured in replicates of size n_{divv}. Again, the respective experimental unit batch-to-batch effects will have no effect on the *trdv_i*'s. Thus, by balancing these effects, we can still generate est(*clpj*'s) for $j > 0$ that are statistically unbiased provided that the conceptual statistical model is correct and that no batch,*ivjv* interaction effects exist. If the actual magnitudes of the estimated batch-to-batch effects are of specific interest, then a combination of linear regression analysis and ANOVA, termed analysis of covariance, is required (see Little and Jebe, 1975).

Finally, we note that it is unwise to undertake any CRD experiment test program without proper regard for potential test equipment breakdown and other inadvertencies. Thus, a well-planned linear regression experiment test program should always include time blocks (even when all other nuisance variables can reasonably be ignored). Then, if no equipment breakdown or other inadvertences occur, the time blocks can be ignored after first examining the [est(*CRHNDREE_i*)]'s relative to time trends within blocks.

7.5. CLOSURE

We do not recommend the use of linear regression analysis for modes of failure with long lives (endurances) because the presumption of homoscedasticity is seldom if ever statistically credible. Rather, we recommend the use of maximum likelihood (ML) analysis because the experiment test program datum values can be presumed to be either homoscedastic or heteroscedastic. Moreover, the respective datum values can be presumed to have been randomly selected from any of several alternative conceptual statistical distributions.

We discuss and illustrate ML analysis in Chapter 8. First, however, we present the propagation of variability methodology in Supplemental Topic 7.A to provide the analytical tools needed to compute the statistical confidence intervals, bands, and limits that are of specific interest in ML analyses.

7.A. SUPPLEMENTAL TOPIC: PROPAGATION OF VARIABILITY

7.A.1. Introduction

Recall that, although the simple linear regression model is traditionally written as

$$\text{est}[\text{mean}(APRCRHNDRDV\text{'s}) \text{ given } ivv = ivv^*] = \text{est}(clp0) + \text{est}(clp1) \cdot ivv^*$$

est($clp0$) and est($clp1$) are seldom independent—because the $|+1\text{'s}|$ column vector and the $|ivv_i\text{'s}|$ column vector are seldom orthogonal. Next, suppose that we continually replicate the associated simple linear regression experiment test program and plot the resulting collection of *paired* data-based values for est($clp0$) and est($clp1$) on orthogonal $clp0,clp1$ co-ordinates. The limiting form of the resulting three-dimensional histogram for proportions would be a *bivariate* conceptual four- or five-parameter normal distribution. The replicate realizations for both est($clp0$) and est($clp1$), *individually*, generate conceptual (two-parameter) normal sampling distributions. In addition to these four parameters, a fifth parameter is required when est($clp0$) and est($clp1$) are not independent. This fifth conceptual parameter is called the conceptual correlation coefficient (Section 7.A.5.2). However, it is analytically more convenient to work with an associated conceptual parameter called covariance in presenting and illustrating the methodology called propagation of variability.

7.A.2. Basic Concept

Consider any collection of n_{rvos} random variables (or statistics), H_1, H_2, $H_3, \ldots, H_{n_{rvos}}$, that are jointly normally distributed under continual replication of the experiment test program—where jointly normally distributed connotes that given a specific realization value for each of these random variables (or statistics) except one, the conceptual statistical (or sampling) distribution for the remaining random variable (or statistic) is a conceptual (two-parameter) normal distribution. In turn, consider a random variable (or statistic) Z whose realization value z_k depends only on the respective realization values for the H_1, H_2, $H_3, \ldots, H_{n_{rvos}}$ random variables (or statistics). Let $z = h(h_1, h_2, h_3, \ldots, h_{n_{rvos}})$, where h is the (deterministic) functional relationship of specific interest. When h is a smooth continuous function along each of its $h_1, h_2, h_3, \ldots, h_{n_{rvos}}$ metrics, the actual value for the variance of the (presumed) conceptual normal sampling distribution that consists of all possible replicate realization values for the random variable (or statistic)

Z can be computed using the approximate expression (Hahn and Shapiro, 1967):

$$\mathrm{var}(Z) \approx \sum_{i=1}^{n_{rvos}} \sum_{j=1}^{n_{rvos}} \left[\left(\frac{\partial h}{\partial h_i} \right) \cdot \left(\frac{\partial h}{\partial h_j} \right) \cdot \mathrm{covar}(H_i, H_j) \right]$$

in which the respective partial derivatives are evaluated in theory at the (unknown) point [mean(H_1), mean(H_2), mean(H_3), . . . , mean($H_{n_{rvos}}$)]—because this approximate expression is based on a multivariate Taylor's series expansion of z about the point [mean(H_1), mean(H_2), mean(H_3), . . . , mean($H_{n_{rvos}}$)], retaining only the first-order terms. However, in practice, the respective partial derivatives must be evaluated at the point {est[mean(H_1)], est[mean(H_2)], est[mean(H_3)], . . . , est[mean($H_{n_{rvos}}$)]}. In turn, by definition, covar(H_i, H_i) equals var(H_i) and covar(H_i, H_j) equals covar(H_j, H_i). Thus, the propagation of variability expression is conventionally rewritten as

$$\mathrm{var}(Z) \approx \sum_{i=1}^{n_{rvos}} \left[\left(\frac{\partial h}{\partial h_i} \right)^2 \cdot \mathrm{var}(H_i) \right] + 2 \cdot \sum_{i=1}^{n_{rvos}} \sum_{j>i}^{n_{rvos}} \left[\left(\frac{\partial h}{\partial h_i} \right) \cdot \left(\frac{\partial h}{\partial h_j} \right) \cdot \mathrm{covar}(H_i, H_j) \right]$$

Note that a covariance term is required for each pair of H_i and H_j that are not independent. However, when certain pairs of H_i's and H_j's are statistically independent, the associated covariance terms are equal to zero. In particular, when all H_i's and H_j's are (mutually) statistically independent, then all covariance terms are equal to zero. In a broader perspective, covariances pertain to the off-diagonal elements of the symmetrical $n_{clp} \times n_{clp}$ covariance matrix, whereas variances pertain to the associated diagonal elements. Numerical calculation of all the elements of this covariance matrix is required to compute the classical (shortest) $100(scp)\%$ (two-sided) statistical confidence interval that allegedly includes mean(Z). (These numerical calculations are discussed in Chapter 8.)

7.A.3. Use of Propagation of Variability in Maximum Likelihood Analyses

The primary application of the propagation of variability methodology in maximum likelihood (ML) analysis (Chapter 8) pertains to the computation of asymptotic statistical confidence intervals and limits. We now present two examples to illustrate this primary application.

7.A.3.1. Example 1

Consider the conceptual two-parameter Weibull distribution, written as

$$F(x) = 1 - \exp - \left(\frac{x}{cdp1}\right)^{cdp2}$$

in which *cdp1* and *cdp2* are (generic) conceptual distribution parameters. The inverse CDF expression for this conceptual two-parameter Weibull distribution is

$$y = \log_e\{-\log_e[1 - F(x)]\} = cdp2 \cdot \left[\log_e(x) - \log_e(cdp1)\right] = a + b \cdot \log_e(x)$$

This inverse CDF expression is used to construct Weibull probability paper, viz., when y is plotted along the (linear) ordinate and $\log_e(x)$ is plotted along the (linear) abscissa, the conceptual two-parameter Weibull CDF is represented by a straight line with intercept $a = -cdp2 \cdot \log_e(cdp1)$ and slope $b = cdp2$.

> *Caveat*: Do not plot x along an abscissa with a (nonlinear) logarithmic scale (even though it is the traditional way to construct Weibull probability paper). It is always more rational to plot $\log_e(x)$ along a linear metric than to plot x along a logarithmic metric.

The est(CDF) for this parameterization of the conceptual two-parameter Weibull distribution is more explicitly expressed as

$$\left[\text{est}(y) \text{ given } \log_e(x)\right] = \text{est}(cdp2) \cdot \left\{\log_e(x) - \log_e[\text{est}(cdp1)]\right\}$$

Then, employing the conventional form of the propagation of variability expression, est{var[est(y) given $\log_e(x)$]} can be computed using the approximation:

est$\left\{\text{var}\left[\text{est}(y) \text{ given } \log_e(x)\right]\right\}$

$\approx \left[\dfrac{-\text{est}(cdp2)}{\text{est}(cdp1)}\right]^2 \cdot \text{est}\left\{\text{var}[\text{est}(cdp1)]\right\}$

$+ \left\{\log_e(x) - \log_e[\text{est}(cdp1)]\right\}^2 \cdot \text{est}\left\{\text{var}[\text{est}(cdp2)]\right\}$

$+ 2 \cdot \left[\dfrac{-\text{est}(cdp2)}{\text{est}(cdp1)}\right] \cdot \left\{\log_e(x) - \log_e[\text{est}(cdp1)]\right\} \cdot \text{est}\left\{\text{covar}[\text{est}(cdp1), \text{est}(cdp2)]\right\}$

We assert in Chapter 8 that asymptotic sampling distributions for ML estimates of the actual values for *cdp1* and *cdp2* are conceptual (two-parameter) normal distributions and that ML est(*cdp1*) and ML est(*cdp2*) are asymptotically unbiased. Accordingly, given the outcome of an appropriate ML analysis, a lower 95% (one-sided) asymptotic statistical confidence limit that allegedly bounds the actual value for [y given $\log_e(x)$] is, for example, approximately equal to [ML est(y) given $\log_e(x)$] $- 1.6499 \cdot$ (est{var[ML est(y) given $\log_e(x)$]})$^{1/2}$.

Remark: We discuss empirical and pragmatic statistical bias corrections for ML analyses of specific interest in mechanical reliability in Supplemental Topic 8.D.

7.A.3.2. Example 2

Consider the conceptual (two-parameter) smallest-extreme-value distribution, written as

$$F\left[\log_e(x)\right] = 1 - \exp^{-\exp\left[\frac{\log_e(x)-clp}{csp}\right]}$$

in which, akin to the conceptual (two-parameter) normal distribution, *clp* and *csp* are conceptual location and scale parameters. Note, however, that $\log_e(x)$ rather than x is chosen as the continuous measurement metric for this conceptual (two-parameter) smallest-extreme-value distribution so that it is directly analogous to the corresponding conceptual two-parameter Weibull distribution in Example 1. Its inverse CDF expression, viz.,

$$y = \log_e\left(-\log_e\left\{1 - F\left[\log_e(x)\right]\right\}\right) = \frac{\log_e(x) - clp}{csp} = a + b \cdot \log_e(x)$$

is used to construct smallest-extreme-value probability paper by plotting y along the (linear) ordinate and $\log_e(x)$ along the (linear) abscissa, so that the sigmoidal conceptual (two-parameter) smallest-extreme-value distribution CDF is transformed into a straight-line plot.

Remark: Corresponding smallest-extreme-value and Weibull probability papers are identical when the latter is properly constructed. These probability papers differ only when x is plotted along a logarithmic abscissa for the conceptual two-parameter Weibull distribution and $\log_e(x)$ is plotted along a linear abscissa for the corresponding conceptual (two-parameter) smallest-extreme-value distribution. A linear–linear plot is always geometrically preferable.

The est(CDF) for this parameterization of the conceptual (two-parameter) smallest-extreme-value distribution is more explicitly expressed as

$$\left[\text{est}(y) \text{ given } \log_e(x)\right] = \frac{\log_e(x) - \text{est}(clp)}{\text{est}(csp)}$$

Then, employing the conventional form of the propagation of variability expression, est$\{$var[est(y) given $\log_e(x)$]$\}$ can be computed using the approximation:

$$
\begin{aligned}
&\text{est}\{\text{var}[\text{est}(y) \text{ given } \log_e(x)]\} \\
&\approx \left[\frac{-1}{\text{est}(csp)}\right]^2 \cdot \text{est}\{\text{var}[\text{est}(clp)]\} \\
&\quad + \left\{\frac{\log_e(x) - \text{est}(clp)}{-[\text{est}(csp)]^2}\right\}^2 \cdot \text{est}\{\text{var}[\text{est}(csp)]\} \\
&\quad + 2\left[\frac{-1}{\text{est}(csp)}\right] \cdot \left\{\frac{\log_e(x) - \text{est}(clp)}{-[\text{est}(csp)]^2}\right\} \cdot \text{est}\{\text{covar}[\text{est}(clp), \text{est}(csp)]\}
\end{aligned}
$$

which in turn can be re-written as

$$
\begin{aligned}
&\text{est}\{\text{var}[\text{est}(y) \text{ given } \log_e(x)]\} \\
&\approx \left[\frac{1}{\text{est}(csp)}\right]^2 \cdot \left(\begin{array}{c} \text{est}\{\text{var}[\text{est}(clp)]\} + [\text{est}(y)]^2 \cdot \text{est}\{\text{var}[\text{est}(csp)]\} \\ +2 \cdot \text{est}(y) \cdot \text{est}\{\text{covar}[\text{est}(clp), \text{est}(csp)]\} \end{array}\right)
\end{aligned}
$$

Accordingly, as in Example 1, given the outcome of an appropriate ML analysis, a lower 95% (one-sided) asymptotic statistical confidence limit that allegedly bounds the actual value for [y given $\log_e(x)$] is, for example, approximately equal to [ML est(y) given $\log_e(x)$] $-1.6499 \cdot$ (est$\{$var[ML est(y) given $\log_e(x)$]$\})^{1/2}$. It is numerically demonstrated in Chapter 8 that this lower 95% (one-sided) asymptotic statistical confidence limit is equal to the lower 95% (one-sided) asymptotic statistical confidence limit in Example 1.

7.A.3.3. Discussion

The equivalence of the conceptual two-parameter Weibull distribution in Example 1 and its corresponding conceptual (two-parameter) smallest-extreme-value distribution in Example 2 is the direct consequence of the invariance of ML estimates under transformation (reparameterization). We now restate the fundamental propagation of variability expression in a form more amenable to relating the asymptotic variances and covariances for the respective ML estimates of the actual values for the parameters of equivalent conceptual two-parameter distributions.

Given any two equivalent conceptual two-parameter statistical distributions, generically denote their conceptual parameters $cp1$ and $cp2$ (distribution one) and $cp3$ and $cp4$ (distribution two). Let these conceptual parameters be related as follows: $cp3 = g(cp1, cp2)$ and $cp4 = h(cp1, cp2)$, where g and h are continuous functions that are established by comparing the respective inverse CDF expressions for the two equivalent conceptual two-parameter statistical distributions. Next, suppose that ML est($cp1$), ML est($cp2$), ML est{var[est($cp1$)]}, ML est{var[est($cp2$)]}, and ML est{covar[-est($cp1$),est($cp2$)]} have been established numerically by ML analysis (Chapter 8). Then, ML est($cp3$) = g[ML est($cp1$),ML est($cp2$)] and ML est($cp4$) = h[ML est($cp1$),ML est($cp2$)]. Moreover,

$$est\{var[ML\ est(cp3)]\}$$

$$= \left(\frac{\partial g}{\partial cp1}\right)^2 \cdot est\{var[ML\ est(cp1)]\}$$

$$+ \left(\frac{\partial g}{\partial cp2}\right)^2 \cdot est\{var[ML\ est(cp2)]\}$$

$$+ 2 \cdot \left(\frac{\partial g}{\partial cp1}\right) \cdot \left(\frac{\partial g}{\partial cp2}\right) \cdot est\{covar[ML\ est(cp1),\ ML\ est(cp2)]\}$$

$$est\{var[ML\ est(cp4)]\}$$

$$= \left(\frac{\partial h}{\partial cp1}\right)^2 \cdot est\{var[ML\ est(cp1)]\}$$

$$+ \left(\frac{\partial h}{\partial cp2}\right)^2 \cdot est\{var[ML\ est(cp2)]\}$$

$$+ 2 \cdot \left(\frac{\partial h}{\partial cp1}\right) \cdot \left(\frac{\partial h}{\partial cp2}\right) \cdot est\{covar[ML\ est(cp1),\ ML\ est(cp2)]\}$$

and

$$est\{covar[ML\ est(cp3),\ ML\ est(cp4)]\}$$

$$= \left(\frac{\partial g}{\partial cp1}\right) \cdot \left(\frac{\partial h}{\partial cp1}\right) \cdot est\{var[ML\ est(cp1)]\}$$

$$+ \left(\frac{\partial g}{\partial cp2}\right) \cdot \left(\frac{\partial h}{\partial cp2}\right) \cdot est\{var[ML\ est(cp2)]\}$$

$$+ \left(\frac{\partial g}{\partial cp1}\right) \cdot \left(\frac{\partial h}{\partial cp2}\right) \cdot est\{covar[ML\ est(cp1),\ ML\ est(cp2)]\}$$

$$+ \left(\frac{\partial g}{\partial cp2}\right) \cdot \left(\frac{\partial h}{\partial cp1}\right) \cdot est\{covar[ML\ est(cp2),\ ML\ est(cp1)]\}$$

in which the respective partial derivatives are evaluated at (and pertain strictly to) the point [ML est($cp1$), ML est($cp2$)]. These expressions have direct application in establishing the *algebraic* equivalence of the respective expressions for est{var[ML est(y) given $\log_e(x)$]}in Examples 1 and 2.

7.A.3.4. Example 3

Consider the respective inverse CDF expressions for the conceptual two-parameter Weibull distribution considered in Example 1 and its equivalent conceptual (two-parameter) $\log_e$ smallest-extreme-value distribution considered in Example 2, viz.,

$$y = a + b \cdot \log_e(x) = cdp2[\log_e(x) - \log_e(cdp1)] = \frac{\log_e(x) - clp}{csp}$$

In turn, suppose that a ML analysis has been conducted for the conceptual two-parameter Weibull distribution considered in Example 1. The equivalent outcome for a ML analysis pertaining to the conceptual (two-parameter) smallest-extreme-value distribution considered in Example 2 is established by substituting the following expressions for g and h into our three restated propagation of variability expressions:

$$clp = g(cdp1, cdp2) = \log_e(cdp1) \quad \text{and} \quad csp = h(cdp1, cdp2) = \frac{1}{cdp2}$$

This substitution leads to ML est(clp) = $\log_e$[ML est($cdp1$)], ML est(csp) = 1/[ML est($cdp2$)], and the following estimated asymptotic variance and covariance expressions for ML est(clp) and ML est(csp):

$$\text{est}\{\text{var}[\text{ML est}(clp)]\} = \left[\frac{1}{\text{ML est}(cdp1)}\right]^2 \cdot \text{est}\{\text{var}[\text{ML est}(cdp1)]\}$$

$$\text{est}\{\text{var}[\text{ML est}(csp)]\} = \left\{\frac{-1}{[\text{ML est}(cdp2)]^2}\right\}^2 \cdot \text{est}\{\text{var}[\text{ML est}(cdp2)]\}$$

and

$$\text{est}\{\text{covar}[\text{ML est}(clp), \text{ML est}(csp)]\}$$
$$= 2 \cdot \frac{1}{\text{ML est}(cdp1)} \cdot \frac{-1}{[\text{ML est}(cdp2)]^2} \cdot \text{est}\{\text{covar}[\text{ML est}(cdp1), \text{ML est}(cdp2)]\}$$

On the other hand, suppose that a ML analysis has been conducted for the conceptual (two-parameter) smallest-extreme-value distribution considered in Example 2. Then, the equivalent outcome for a ML analysis pertaining to the conceptual two-parameter Weibull distribution considered in Example 1 is established by substituting the following expressions for g and h into our three restated propagation of variability expressions:

$$cdp1 = g(clp, csp) = \exp^{clp} \quad \text{and} \quad cdp2 = h(clp, csp) = \frac{1}{csp}$$

This substitution leads to ML est($cdp1$) = exp[ML est(clp)], ML est($cdp2$) = 1/[ML est(csp)], and the following estimated asymptotic variance and covariance expressions for ML est($cdp1$) and ML est($cdp2$):

$$\text{est}\{\text{var}[\text{ML est}(cdp1)]\} = \left[\exp^{\text{ML est}(clp)}\right]^2 \cdot \text{est}\{\text{var}[\text{ML est}(clp)]\}$$

$$\text{est}\{\text{var}[\text{ML est}(cdp2)]\} = \left\{\frac{-1}{[\text{ML est}(csp)]^2}\right\}^2 \cdot \text{est}\{\text{var}[\text{ML est}(csp)]\}$$

and

$$\text{est}\{\text{covar}[\text{ML est}(cdp1), \text{ML est}(cdp2)]\}$$

$$= 2 \cdot \exp^{\text{ML est}(clp)} \cdot \frac{-1}{[\text{ML est}(csp)]^2} \cdot \text{est}\{\text{covar}[\text{ML est}(clp), \text{ML est}(csp)]\}$$

The equivalence of the est{var[ML est(y) given $\log_e(x)$]} expressions in Examples 1 and 2 can now be demonstrated by algebraically substituting the respective est{var[ML est(clp)]}, est{var[ML est(csp)]}, and est{covar[ML est(clp), ML est(csp)]} propagation of variability expressions into the respective est{var[ML est($cdp1$)]}, est{ML var[est($cdp2$)]}, and est{covar[ML est($cdp1$), ML est($cdp2$)]} propagation of variability expressions, and vice versa.

7.A.4. Use of Propagation of Variability in Linear Regression Analyses

The simple linear regression expression for var{{est[mean-($APRCRH$ $NDRDV$'s) given $ivv = ivv^*$]} was derived in Section 7.2 without introducing the concept of covariance. We now employ the propagation of variability methodology to generate an algebraically equivalent expression that includes a covariance term (as in Exercise Set 2, Exercise 5).

The estimated parameters in simple and multiple linear regression are jointly normally distributed when the conceptual regression datum values are all presumed to be mutually independent and normally distributed. Recall that for simple linear regression, exact expressions for the elements of the 2 by 2 covariance matrix for est($clp0$) and est($clp1$) are, respectively,

$$\text{est}\{\text{var}[\text{est}(clp0)]\} = \left\{ \frac{1}{n_{rdv}} + \frac{[\text{ave}(ivv_i\text{'s})]^2}{\sum\limits_{i=1}^{n_{rdv}} tivv_i^2} \right\} \cdot \text{est}[\text{var}(APRCRHNDREE\text{'s})]$$

$$\text{est}\{\text{covar}[\text{est}(clp0), \text{est}(clp1)]\} = \text{est}\{\text{covar}[\text{est}(clp1), \text{est}(clp0)]\}$$

$$= -\left[\frac{\text{ave}(ivv_i\text{'s})}{\sum\limits_{i=1}^{n_{rdv}} tivv_i^2} \right] \cdot \text{est}[\text{var}(APRCRHNDREE\text{'s})]$$

$$\text{est}\{\text{var}[\text{est}(clp1)]\} = \frac{1}{\sum\limits_{i=1}^{n_{rdv}} tivv_i^2} \cdot \text{est}[\text{var}(APRCRHNDREE\text{'s})]$$

We now use the propagation of variability methodology in conjunction with these variance and covariance expressions to state an exact expression for $\text{var}\{\text{est}[\text{mean}(APRCRHNDRDV\text{'s}) \text{ given } ivv = ivv^*]\}$.

In ML analysis the partial derivatives must be evaluated at the estimated conceptual parameter values because the expected values of the respective conceptual location parameters are not known, but for linear regression, two differences occur. First, the statistical estimators of the actual values for the conceptual location parameters are unbiased; thus, the associated change to expected values are known. Second, the partial derivatives of specific interest do not involve these conceptual location parameters. Note that, given the simple linear regression expression [mean($APRCRHNDRDV$'s) given $ivv = ivv^*$] = $clp0$ + $clp1 \cdot ivv^*$, the partial derivative of [mean($APRCRHNDRDV$'s) given $ivv = ivv^*$] with respect to $clp0$ is equal to one and the partial derivative of [mean($APRCRHNDRDV$'s) given $ivv = ivv^*$] with respect to $clp1$ is equal to ivv^*. Therefore, based on the propagation of variability methodology:

$$\text{var}\{\text{est}[\text{mean}(APRCRHNDRDV\text{'s}) \text{ given } ivv = ivv^*]\}$$

$$\approx 1^2 \cdot \text{var}[\text{est}(clp0)] + (ivv^*)^2 \cdot \text{var}[\text{est}(clp1)]$$
$$+ 2 \cdot 1 \cdot ivv^* \cdot \text{covar}[\text{est}(clp0), \text{est}(clp1)]$$

Note that this approximate expression is algebraically identical to the exact expression that appears in Exercise Set 2, Exercise 5. In turn, appropriate substitution of the exact expressions for the elements of the 2 by 2 simple linear regression covariance matrix gives

$$\text{var}\{\text{est}[\text{mean}\,(APRCRHNDRDV\text{'s})\text{ given } ivv = ivv^*]\}$$

$$\approx \left\{ \frac{1}{n_{rdv}} + \frac{[ivv^* - \text{ave}(ivv_i\text{'s})]^2}{\displaystyle\sum_{i=1}^{n_{rdv}} tivv_i^2} \right\} \cdot \text{var}(APRCRHNDREE\text{'s})$$

Next, suppose that the Taylor's series expansion underlying the conventional propagation of variability methodology had included both first- and second-order terms. The following approximation would then pertain (Hahn and Shapiro, 1967):

$$\text{mean}(Z) \approx h[\text{mean}(H_1), \text{mean}(H_2), \text{mean}(H_3), \ldots]$$

$$+ \frac{1}{2} \cdot \sum_{i=1}^{n_{rvos}} \left[\frac{\partial^2 h}{\partial h_i^2} \cdot \text{var}(H_i) \right] + \sum_{i=1}^{n_{rv}} \sum_{j>i}^{n_{rv}} \left[\frac{\partial^2 h}{\partial h_i \partial h_j} \cdot \text{covar}(H_i, H_j) \right]$$

However, all second and higher partial derivatives are equal to zero for linear functions. Thus, for linear functions (such as our simple linear regression expressions), Z is exactly normally distributed. Moreover, est[mean(Z)] is statistically unbiased, viz., the actual value for the mean of the conceptual normal statistical distribution that consists of all possible replicate realization values for est[mean(Z)] is equal to mean(Z). On the other hand, Z is not normally distributed for nonlinear functions. Then, the critical issue is whether the h function surface can reasonably be approximated as a hyperplane in the vicinity of the probable realization values for (H_1, H_2, H_3, $\ldots$, $H_{n_{rvos}}$). The greater the curvature(s) of the h function surface, the greater the non-normality of the actual conceptual statistical distribution that consists of all possible replicate realization values for the random variable Z and the poorer the resulting propagation of variability approximations. When the coefficient of variation for H_i, defined as

$$\text{coefficient of variation}\,(H_i) = \frac{\text{stddev}\,(H_i)}{\text{mean}\,(H_i)}$$

is very small for all H_i's, then the approximations underlying the conventional propagation of variability equations can cautiously be presumed to be reasonable accurate. However, if there is concern regarding the accuracy of the standard propagation of variability equations, additional terms in the multivariate Taylor's series expansion are appropriate and Appendix 7B in Hahn and Shapiro (1967) should be consulted.

7.A.5. Use of Propagation of Variability in Establishing Cognate Conceptual Sampling Distributions

The propagation of methodology has extensive application in establishing cognate conceptual sampling distributions pertaining to physical quantities that cannot be measured directly, e.g., the elastic modulus (*em*) of a material. However, before illustrating its use relative to this application, we first discuss two background topics. The first topic is the propagation of variability expression pertaining to the simple relationship $Z = X^a \cdot Y^b$. The second topic pertains to how to estimate the covariance of X and Y.

7.A.5.1. Background Topic One

Suppose that X and Y are generic random variables that are jointly normally distributed. We now develop the standard propagation of variability approximate expressions for mean(Z) and var(Z) given the relationship $Z = X^a \cdot Y^b$. The approximate expression for mean(Z) is obtained by direct substitution, viz.,

$$\text{mean}(Z) \approx [\text{mean}(X)]^a \cdot [\text{mean}(Y)]^b$$

The associated approximate expression for var(Z) is developed as follows. First, we take the required partial derivatives to obtain the expression:

$$\text{var}(Z) \approx \left(a \cdot x^{a-1} \cdot y^b\right)^2 \cdot \text{var}(X) + \left(b \cdot x^a \cdot y^{b-1}\right)^2 \cdot \text{var}(Y)$$
$$+ 2 \cdot \left(a \cdot x^{a-1} \cdot y^b\right) \cdot \left(b \cdot x^a \cdot y^{b-1}\right) \cdot \text{covar}(X, Y)$$

and then we evaluate these partial derivatives by substituting mean(X) for x and mean(Y) for y. In turn, we divide both sides of the resulting expression by $[\text{mean}(Z)]^2 = \{[\text{mean}(X)]^a[\text{mean}(Y)]^b\}^2$ to obtain the desired result, viz.,

$$\frac{\text{var}(Z)}{[\text{mean}(Z)]^2} \approx a^2 \cdot \frac{\text{var}(X)}{[\text{mean}(X)]^2} + b^2 \cdot \frac{\text{var}(Y)}{[\text{mean}(Y)]^2}$$
$$+ 2 \cdot a \cdot b \cdot \frac{\text{covar}(X, Y)}{[\text{mean}(X)] \cdot [\text{mean}(Y)]}$$

Note that each of the terms in this expression are dimensionless. Note also that, when X and Y are independent, a and b can be either positive or negative and still yield the same numerical value for var(Z). This apparently anomalous result is due to the use of only linear terms in the multivariate

Taylor's series expansion underlying the development of the standard propagation of variability variance expression.

> *Remark*: When paired measurement values are indeed independent, their estimated covariance value will generally be small compared to their associated estimated variance values. Accordingly, when the associated covariance term is (incorrectly) included in a propagation of variability expression, it will seldom markedly affect the numerical limits of the resulting approximate (shortest) $100(scp)\%$ (two-sided) statistical confidence interval that allegedly includes the deterministic (actual) value of specific interest.

7.A.5.2. Background Topic Two

We now discuss how to estimate the covariance of generic random variables X and Y that are jointly normally distributed. Consider replicate *paired* measurement values (realization values) for these two generic random variables, where the number of replicate paired measurement values (realization values) is denoted n_{rpmv}. Under continual replication of the experiment test program the respective replicate paired measurement values (realization values) for generic random variables X and Y generate a conceptual bivariate (five-parameter) normal distribution. The actual values for these five parameters are estimated as follows:

(1) $\text{est}[\text{mean}(X)] = \text{ave}(X_i\text{'s})$

(2) $\text{est}[\text{mean}(Y)] = \text{ave}(Y_i\text{'s})$

(3) $\text{est}[\text{var}(X)] = \dfrac{\left\{[X_1 - \text{ave}(X\text{'s})]^2 + [X_2 - \text{ave}(X\text{'s})]^2 + [X_3 - \text{ave}(X\text{'s})]^2 + \ldots\right\}}{n_{rpmv} - 1}$

(4) $\text{est}[\text{var}(Y)] = \dfrac{\left\{[Y_1 - \text{ave}(Y\text{'s})]^2 + [Y_2 - \text{ave}(Y\text{'s})]^2 + [Y_3 - \text{ave}(Y\text{'s})]^2 + \ldots\right\}}{n_{rpmv} - 1}$

$$(7.120)$$

and

(5) $[\text{covar}(X, Y)] =$
$$\dfrac{\left\{[X_1 - \text{ave}(X\text{'s})][Y_1 - \text{ave}(Y\text{'s})] + [X_2 - \text{ave}(X\text{'s})][Y_2 - \text{ave}(Y\text{'s})] + \ldots\right\}}{n_{rpmv} - 1}$$

However, the conceptual correlation coefficient, *ccc*, is almost always employed as an algebraic surrogate for $\text{covar}(X, Y)$ in the analytical expression for the conceptual bivariate five-parameter normal distribution, viz., the *ccc* is its fifth parameter. It is defined as $\text{covar}(X, Y)/[\text{var}(X) \cdot \text{var}(Y)]^{1/2}$.

(The actual value for the *ccc* is dimensionless and lies in the interval from -1 to $+1$.)

It is particularly important to remember that we cannot compute est[covar(X, Y)] unless the associated respective measurement values are paired. Thus, we should always group experiment test program measurement values into n_{rpmv} time blocks whenever statistical correlation is plausible.

7.A.5.3. Cognate Conceptual Sampling Distribution Example

Material behavior parameters are regarded as deterministic (invariant) values in the mechanics and mechanical metallurgy literature. Nevertheless, any value that must be measured can be modeled as having a deterministic physically based component and a random statistically based component. Then, the deterministic physically based component is statistically viewed as the mean of the conceptual sampling distribution that consists of all possible measurement values that would occur if the experiment test program were continually replicated—provided that the measurement process and the associated statistical estimation process are both unbiased.

We now develop the cognate conceptual sampling distribution that pertains to "measuring" the elastic modulus (*em*) of a material by conducting a three-point bending test. Suppose that, to begin this development, replicate three-point bending tests are conducted in which the same specimen is repeatedly tested. Suppose also that (a) this test specimen has a perfectly uniform rectangular cross-section with (specification) width w and (specification) thickness t, (b) a dead load p is applied by a calibrated 5 lb weight and acts at the midspan of the test fixture, and (c) the test fixture span s is invariant with a nominal dimension of 6 in. Then, the actual value for the deflection d at the midspan of the three-point bending specimen is theoretically computed using the expression:

$$d = \frac{p \cdot s^3}{4 \cdot e \cdot m \cdot w \cdot t^3} = \frac{270}{e \cdot m \cdot w \cdot t^3}$$

Solving for *em* gives

$$em = \frac{270}{w \cdot t^3 \cdot d}$$

Then, denoting the respective replicate measurement values for w, t, and d as w_i, d_i, and t_i gives

$$em_i = \frac{270}{w_i \cdot t_i^3 \cdot d_i}$$

Next, under continual replication of this hypothetical experiment test program, we assert that the respective three sets of paired measurement values for w_i, t_i, and d_i respectively generate three jointly normally distributed conceptual sampling distributions. If so, then (a) the means, variances, and covariances of these three conceptual sampling distributions can be estimated using the expressions given in Section 7.A.5.2, and (b) the conceptual sampling distribution that consists of the corresponding (computed) em_i values can be approximated by a conceptual (two-parameter) normal distribution with

$$\text{mean}\,(em_i\text{'s}) \approx \frac{270}{[\text{mean}\,(w_i\text{'s})] \cdot [\text{mean}\,(t_i\text{'s})]^3 \cdot [\text{mean}\,(d_i\text{'s})]}$$

and

$$\frac{\text{var}(em_i\text{'s})}{[\text{mean}\,(em_i\text{'s})]^2} \approx (-1)^2 \cdot \frac{\text{var}(w_i\text{'s})}{[\text{mean}\,(w_i\text{'s})]^2}$$

$$+ (-3)^2 \cdot \frac{\text{var}(t_i\text{'s})}{[\text{mean}(t_i\text{'s})]^2} + (-1)^2 \cdot \frac{\text{var}(d_i\text{'s})}{[\text{mean}\,(d_i\text{'s})]^2}$$

$$+ 2 \cdot (-1) \cdot (-3) \cdot \frac{\text{covar}[(w_i\text{'s}), (t_i\text{'s})]}{[\text{mean}(w_i\text{'s})] \cdot [\text{mean}\,(t_i\text{'s})]}$$

$$+ 2 \cdot (-1) \cdot (-1) \cdot \frac{\text{covar}[(w_i\text{'s}), (d_i\text{'s})]}{[\text{mean}(w_i\text{'s})] \cdot [\text{mean}(d_i\text{'s})]}$$

$$+ 2 \cdot (-1) \cdot (-1) \cdot \frac{\text{covar}[(t_i\text{'s}), (d_i\text{'s})]}{[\text{mean}\,(t_i\text{'s})] \cdot [\text{mean}(d_i\text{'s})]}$$

Thus, the (shortest) 95% (two-sided) statistical confidence interval that allegedly includes the deterministic (actual) value for the elastic modulus is approximately equal to $\text{mean}(em_i\text{'s}) \mp 1.9600[\text{var}(em_i\text{'s})]^{1/2}$.

7.A.5.4. Discussion

The experiment test program underlying this example application was deliberately contrived so that the variabilities in the w_i's, t_i's, and d_i's pertained only to their respective measurement errors. However, the variability that must be properly assessed in this application is the "intrinsic" variability in physical behavior (elastic modulus) from specimen to specimen. Accordingly, suppose we now modify the contrived experiment test program, viz., suppose that the respective w_i, t_i, and d_i measurement values pertain to a single measurement of w, t, and d in replicate three-point bending tests performed on nominally identical specimens taken from different sheets or plates. Then, (a) the variability of the respective w_i's and t_i's would be the confounded sum of the measurement variability and the dimensional variability from specimen to specimen, and (b) the variability of the respective d_i's would be the confounded sum of these two variabilities and the "intrinsic" variability in physical behavior (elastic modulus) from specimen to specimen. However, in practice, because neither specimen width nor thickness is perfectly uniform along the entire span, it is never prudent to make only a single measurement of either w or t for any test specimen. Rather, the "measured" values of w and t should be the arithmetic average of at least three sets of n_{rep} replicate measurements, one set taken at each end of the span and one set taken at its center. The use of arithmetic averages not only makes each resulting w_i and t_i value more credible relative to its modeling of the actual geometry of the test specimen, but it also (almost surely) reduces the magnitude of the resulting var(em_i's).

It is always good statistical practice to plan an experiment test program to minimize nuisance variabilities, viz., measurement variability and dimensional variability in this example application. Thus, it is always good statistical practice to make the respective w_i's and t_i's the arithmetic average of replicate measurements. In turn, thickness variability could be reduced by taking all specimens from adjacent locations in the same sheet or plate. However, this procedure cannot be recommended because it will not provide a proper assessment of the "intrinsic" variability in physical behavior from specimen to specimen. On the other hand, it is very effective statistically to machine all (or large groups of) test specimens in a sandwich configuration to reduce the resulting width variability.

7.A.6. Use of Propagation of Variability in Dimensioning Components

Consider a component with a series of uniformly spaced holes. It used to be standard practice to dimension only one of the spaces between adjacent holes and to write "typical" following that dimension. This out-dated practice resulted in the variability between the first and the last hole being excessively large (presuming independent manufacturing errors in establishing each respective distance):

$$\text{var(first to last)} = \text{var(first to second)} + \text{var(second to third)} + \ldots$$
$$+ \text{var(next-to-last to last)}$$

On the other hand, when each hole location is dimensioned relative to a common reference line, the actual value for the variance of the conceptual sampling distribution pertaining to the spacing between the first and the last hole is the same as the actual value for the variance of the conceptual sampling distribution pertaining to the spacing between any two holes of specific interest (presuming independent manufacturing errors in establishing each respective distance). It is merely twice the actual value for the variance of the conceptual sampling distribution pertaining to the location of any given hole relative to the common reference line.

Another out-dated standard practice was the so-called limit-stack analysis in which a component was drawn 10 times its actual size with all of its dimensions simultaneously taken at their highest tolerances to determine if this (unreasonably) extreme geometry would interfere with the opposite extreme geometry for a mating component in an assembled configuration. Potential interference problems can now be investigated using pseudorandom numbers to simulate manufacturing errors for the mating components of specific interest. This simulation-based process is based on the creation of a population of hypothetical manufactured components by coupling pseudorandom manufacturing errors with the nominal component dimensions. Then, by randomly selecting mating hypothetical manufactured components from these populations, the proportion of assemblies that exhibit an interference problem can be simulated.

Unfortunately, this simulation process is not properly exploited. Alternative designs (or alternative dimensioning) are seldom examined unless a serious interference problem is likely to occur. However, in situations where a serious interference problem is unlikely to occur there is still the strong possibility that various other manufacturing tolerances can be increased. If so, the associated cost savings can either be used to increase profits or in a trade-off with another component to increase its reliability.

> *Remark*: The major advantage of the propagation of variability methodology over simulation is that the former permits the examination of its expressions to identify the major sources (causes) of the variability of specific interest. Unfortunately, the major advantage of the propagation of variability methodology over simulation is also its major disadvantage. We cannot always write an analytical expression for the variability of specific interest.

Exercise Set 6

(These exercises are intended to familiarize you with the sources of and certain terms used to describe variability components.)

1. (a) Respectively define measurement variability, dimensional variability, and "intrinsic" variability. (b) Does the variability of actual experiment test program datum values (almost surely) involve all three variability components? (c) If so, are these variability components mutually independent?

2. How small does one variability component have to be relative to the other two variability components so that it is negligible for practical purposes? State your answer in terms of the ratio of their respective variances.

3. State the American Society for Testing and Materials (ASTM) definitions for repeatability and reproducibility as used in their standards.

4. (a) Which is typically the largest: (i) measurement variability, (ii) dimensional variability, (iii) "intrinsic" variability, (iv) laboratory-to-laboratory variability, or (v) batch-to-batch variability? (b) Is batch-to-batch variability considered in either of the ASTM definitions for repeatability or reproducibility? (c) Is batch-to-batch variability considered in propagation of variability? (d) Is laboratory-to-laboratory variability considered in propagation of variability?

7.B. SUPPLEMENTAL TOPIC: WEIGHTED SIMPLE LINEAR REGRESSION ANALYSIS

When the variances of the conceptual statistical distributions for the respective simple linear regression experiment test program datum values are not (presumed to be) homoscedastic (for all ivv's of potential interest), then weighted simple linear regression analysis is mandatory. The additional complexity associated with weighted simple linear regression provides important perspective regarding the effect of the presumption of homoscedasticity in simple linear regression. Moreover, weighted simple linear regression analysis also provides useful background relative to understanding ML analysis for strength test datum values (Supplemental Topic 8.D). Accordingly, we now present a weighted simple linear regression example that is directly analogous to our text simple linear regression example.

Suppose we take the arithmetic average of our simple linear regression example datum values at each ivv as our weighted simple linear regression example datum values. We then have the following constructed heteroscedastic datum values for our weighted simple linear regression example:

independent variable ivv_i	weighted regression datum value $wrdv_i = \text{ave}(rdv_i\text{'s})$	conceptual variance of these weighted regression datum value
10	4	var($APRCRHNDREE$'s)/3
20	11	var($APRCRHNDREE$'s)/2
40	21	var($APRCRHNDREE$'s)/2
50	24	var($APRCRHNDREE$'s)/3

7.B.1. Weighted Simple Linear Regression Example

The conceptual weighted simple linear regression model is expressed in our hybrid column vector notation as

$$|CR\,Heteroscedastic NDRDV_i's| = clp0 \cdot |+1's|$$
$$+ clp1 \cdot |ivv_i's| + |CR\,Heteroscedastic NDREE_i's|$$

Clearly, the only difference between a conceptual simple linear regression statistical model and a conceptual weighted simple linear regression statistical model is the *heteroscedasticity* of the weighted regression datum values. The following matrix estimation expression for the $clpj$'s ($j = 1,2$) pertains when the respective weighted regression datum values, $WRDV_i$, are all presumed to be mutually independent:

$$|[est(clpj)]'s| = \left(|ivv_i's|^t |covar(WRDV_i, WRDV_i)|^{-1}(ivv_i's)|\right)^{-1}$$
$$\underset{2\times1}{} \quad \underset{2\times n_{wrdv}}{} \quad \underset{n_{wrdv}\times n_{wrdv}}{} \quad \underset{n_{wrdv}\times 2}{}$$

$$\left(|ivv_i's|^t |covar(WRDV_i, WRDV_i)|^{-1} |wrdv_i's|\right)$$
$$\underset{2\times n_{wrdv}}{} \quad \underset{n_{wrdv}\times n_{wrdv}}{} \quad \underset{n_{wrdv}\times 1}{}$$

in which $|covar(WRDV_i, WRDV_i)|$ is a diagonal matrix whose elements are the variances of the respective conceptual sampling distributions consisting of all possible replicate $WRDV_i$'s. (The off-diagonal elements are all equal to zero because the respective $WRDV_i$'s are presumed to be mutually independent.) Thus,

$$|covar(WRDV_i, WRDV_i)|$$

$$= var(APRCRHNDREE's) \cdot \begin{vmatrix} 1/3 & 0 & 0 & 0 \\ 0 & 1/2 & 0 & 0 \\ 0 & 0 & 1/2 & 0 \\ 0 & 0 & 0 & 1/3 \end{vmatrix}$$

Next, recall that statistical weights are inversely related to variances. Moreover, because the $|covar(WRDV_i, WRDV_i)|$ matrix is diagonal, its inverse, the statistical weight matrix $|sw_i's|$, is also diagonal. Moreover, its diagonal elements are the inverses of the diagonal elements of the $|covar(WRDV_i, WRDV_i)|$ matrix, viz.,

$$|sw_i\text{'s}| = |\text{covar}(WRDV_i, WRDV_i)|^{-1}$$

$$= [\text{var}(APRCRHNDREE\text{'s})]^{-1} \cdot \begin{vmatrix} 3 & 0 & 0 & 0 \\ 0 & 2 & 0 & 0 \\ 0 & 0 & 2 & 0 \\ 0 & 0 & 0 & 3 \end{vmatrix}$$

We now substitute $|sw_i\text{'s}|$ for $|\text{covar}(WRDV_i,WRDV_i)|^{-1}$ in the prior matrix-based expression for $|[\text{est}(clpj)]\text{'s}|$ to obtain

$$\left|[\text{est}(clpj)]\text{'s}\right| = \left(\left|ivv_i\text{'s}\right|^t\left|sw_i\text{'s}\right| \left|ivv_i\text{'s}\right|\right)^{-1}\left(\left|ivv_i\text{'s}\right| \left|sw_i\text{'s}\right| \left|wrdv_i\text{'s}\right|\right)$$

in which, for our constructed weighted simple linear regression example, the *relative* statistical weights replace the actual statistical weights because the scalar multiplier var($APRCRHNDREE$'s) cancels out. (Nevertheless, we shall retain the notation sw, rather than use the notation rsw, because one of our objectives in this constructed weighted simple linear regression example is to develop estimation expressions that can be compared to the ML estimation expressions that are subsequently developed in Supplemental Topic 8.D.) We next expand the right-hand side of this revised matrix estimation expression in incremental steps. First,

$$\left(\left|ivv_i\text{'s}\right|^t\left|sw_i\text{'s}\right|\left|ivv_i\text{'s}\right|\right)$$

$$= \left[\frac{1}{\text{var}(APRCRHNDREE\text{'s})}\right] \cdot \begin{vmatrix} \sum_{i=1}^{n_{wrdv}} sw_i & \sum_{i=1}^{n_{wrdv}} sw_i \cdot ivv_i \\ \sum_{i=1}^{n_{wrdv}} sw_i \cdot ivv_i & \sum_{i=1}^{n_{wrdv}} sw_i \cdot ivv_i^2 \end{vmatrix}$$

Then,

$$\left(\left|ivv_i\text{'s}\right|^t\left|sw_i\text{'s}\right|\left|ivv_i\text{'s}\right|\right)^{-1}$$

$$= \left[\frac{\text{var}(APRCRHNDREE\text{'s})}{(\text{determinant})}\right] \cdot \begin{vmatrix} \sum_{i=1}^{n_{wrdv}} sw_i \cdot ivv_i^2 & -\sum_{i=1}^{n_{wrdv}} sw_i \cdot ivv_i \\ -\sum_{i=1}^{n_{wrdv}} sw_i \cdot ivv_i & \sum_{i=1}^{n_{wrdv}} sw_i \end{vmatrix}$$

in which

$$\text{determinant} = \sum_{i=1}^{n_{wrdv}} sw_i \sum_{i=1}^{n_{wrdv}} sw_i ivv_i^2 - \left(\sum_{i=1}^{n_{wrdv}} sw_i ivv_i \right)^2$$

In turn,

$$(|ivv_i\text{'s}||sw_i\text{'s}||wrdv_i\text{'s}|)$$

$$= \{1/[\text{var}(APRCRHNDREE\text{'s})]\} \cdot \begin{vmatrix} \sum_{i=1}^{n_{wrdv}} sw_i \cdot wrdv_i \\ \sum_{i=1}^{n_{wrdv}} sw_i \cdot ivv_i \cdot wrdv_i \end{vmatrix}$$

Finally,

$$\begin{vmatrix} \text{est}(clp0) \\ \text{est}(clp1) \end{vmatrix}$$

$$= \begin{vmatrix} \dfrac{\sum_{i=1}^{n_{wrdv}} sw_i \cdot ivv_i^2 \cdot \sum_{i=1}^{n_{wrdv}} sw_i \cdot wrdv_i}{\text{determinant}} - \dfrac{\sum_{i=1}^{n_{wrdv}} sw_i \cdot ivv_i \cdot \sum_{i=1}^{n_{wrdv}} sw_i \cdot ivv_i \cdot wrdv_i}{\text{determinant}} \\[4ex] \dfrac{\sum_{i=1}^{n_{wrdv}} sw_i \cdot \sum_{i=1}^{n_{wrdv}} sw_i \cdot ivv_i \cdot wrdv_i}{\text{determinant}} - \dfrac{\sum_{i=1}^{n_{wrdv}} sw_i \cdot ivv_i \cdot \sum_{i=1}^{n_{wrdv}} sw_i \cdot wrdv_i}{\text{determinant}} \end{vmatrix}$$

However, before we can use these estimation expressions generate to compute est($clp0$) and est($clp1$), we must first compute the *weighted averages* (wt ave) of the ivv_i's and $wrdv_i$'s. Accordingly, we now develop the following table:

ivv_i	sw_i	$sw_i ivv_i$	$wrdv_i$	$sw_i \cdot wrdv_i$
10	3	30	4	12
20	2	40	11	22
40	2	80	21	42
50	3	150	24	72
(sum)	10	300		148

Observe that (a) wt ave(ivv_i's) $= 300/10 = 30$, which agrees (checks) with the arithmetic average of the ivv_i's in the text simple linear regression example, and (b) wt ave($wrdv_i$'s) $= 148/10 = 14.8$, which agrees (checks) with the arithmetic average of the rdv_i's in the text simple linear regression example. Now we can augment our table to compute est($clp0$) and est($clp1$):

ivv_i	sw_i	$sw_i \cdot ivv_i$	$sw_i \cdot ivv_i^2$	$wrdv_i$	$sw_i \cdot ivv_i \cdot wrdv_i$
10	3	30	300	4	120
20	2	40	800	11	440
40	2	80	3200	21	1680
50	3	150	7500	24	3600
(sum)		300	11,800		5840

Thus,

$$
\begin{aligned}
\text{est}(clp0) &= [(11{,}800) \cdot (148) - (300) \cdot (5840)]/[(10) \cdot (11{,}800) - (300) \cdot (300)] \\
&= [1{,}746{,}400 - 1{,}752{,}000]/[118{,}000 - 90{,}000] \\
&= [-5600]/[28{,}000] \\
&= -0.2 \quad \text{(checks)}
\end{aligned}
$$

and

$$
\begin{aligned}
\text{est}(clp1) &= [(10) \cdot (5840) - (300) \cdot (148)]/[28{,}000] \\
&= [58{,}400 - 44{,}400]/[28{,}000] \\
&= [14{,}000]/[28{,}000] \\
&= 0.5 \quad \text{(checks)}
\end{aligned}
$$

Next, the covariance matrix for est($clp0$) and est($clp1$) is expressed as

$$
\left| \text{covar}[\text{est}(clpm), \text{est}(clpn)] \right|
$$

$$
= \text{var}(APRCRHNDREE\text{'s}) \cdot \left(\left| ivv_i\text{'s} \right|^t \left| sw_i\text{'s} \right| \left| ivv_i\text{'s} \right| \right)^{-1}
$$

$$
= \left| \begin{array}{cc} \text{var}[\text{est}(clp0)] & \text{covar}[\text{est}(clp0), \text{est}(clp1)] \\ \text{covar}[\text{est}(clp0), \text{est}(clp1)] & \text{var}[\text{est}(clp1)] \end{array} \right|
$$

in which m and n are dummy indices, $m = 0,1$ and $n = 0,1$. Thus,

$$\frac{\text{var}[\text{est}(clp0)]}{\text{var}(APRCRHNDREE\text{'s})} = \frac{\sum\limits_{i=1}^{n_{wrdv}} sw_i \cdot ivv_i^2}{\sum\limits_{i=1}^{n_{wrdv}} sw_i \cdot \sum\limits_{i=1}^{n_{wrdv}} sw_i \cdot ivv_i^2 - \left(\sum\limits_{i=1}^{n_{wrdv}} sw_i \cdot ivv_i\right)^2}$$

$$\frac{\text{var}[\text{est}(clp1)]}{\text{var}(APRCRHNDREE\text{'s})} = \frac{\sum\limits_{i=1}^{n_{wrdv}} sw_i}{\sum\limits_{i=1}^{n_{wrdv}} sw_i \cdot \sum\limits_{i=1}^{n_{wrdv}} sw_i \cdot ivv_i^2 - \left(\sum\limits_{i=1}^{n_{wrdv}} sw_i \cdot ivv_i\right)^2}$$

$$\frac{\text{covar}[\text{est}(clp0), \text{est}(clp1)]}{\text{var}(APRCRHNDREE\text{'s})} = \frac{-\sum\limits_{i=1}^{n_{wrdv}} sw_i ivv_i}{\sum\limits_{i=1}^{n_{wrdv}} sw_i \cdot \sum\limits_{i=1}^{n_{wrdv}} sw_i \cdot ivv_i^2 - \left(\sum\limits_{i=1}^{n_{wrdv}} sw_i \cdot ivv_i\right)^2}$$

Solving numerically gives

$$\text{var}[\text{est}(clp0)] = \frac{11,800}{28,000} \cdot \text{var}(APRCRHNDREE\text{'s})$$
$$= (0.4214) \cdot \text{var}(APRCRHNDREE\text{'s}) \quad \text{(checks)}$$

$$\text{var}[\text{est}(clp1)] = \frac{10}{28,000} \cdot \text{var}(APRCRHNDREE\text{'s})$$
$$= (0.000357) \cdot \text{var}(APRCRHNDREE\text{'s}) \quad \text{(checks)}$$

and

$$\text{covar}[\text{est}(clp0), \text{est}(clp1)] = -\frac{300}{28,000} \cdot \text{var}(APRCRHNDREE\text{'s})$$
$$= (0.0107) \cdot \text{var}(APRCRHNDREE\text{'s}) \quad \text{(checks)}$$

Finally, algebraic substitution of the respective expressions for var[est($clp0$)], var[est($clp1$)], and covar[est($clp0$),est($clp1$)] into the expression:

$$\text{var}\{\text{est}[\text{mean}(APRCRHeteroscedasticNDRDV\text{'s}) \text{ given } ivv = ivv^*]\}$$
$$= \text{var}[\text{est}(clp0)] + (ivv^*)^2 \cdot \text{var}[\text{est}(clp1)]$$
$$+ 2 \cdot ivv^* \cdot \text{covar}[\text{est}(clp0), \text{est}(clp1)]$$

yields

$$\frac{\mathrm{var}\left\{\mathrm{est}[\mathrm{mean}\,(APRCRHeteroscedasticNDRDV\text{'s})\text{ given }ivv = ivv^*]\right\}}{\mathrm{var}(APRCRHNDREE\text{'s})}$$

$$= \frac{\displaystyle\sum_{i=1}^{n_{wrdv}} sw_i \cdot ivv_i^2}{\displaystyle\sum_{i=1}^{n_{wrdv}} sw_i \cdot \sum_{i=1}^{n_{wrdv}} sw_i \cdot ivv_i^2 - \left(\sum_{i=1}^{n_{wrdv}} sw_i \cdot ivv_i\right)^2}$$

$$+ \frac{(ivv^*)^2 \cdot \displaystyle\sum_{i=1}^{n_{wrdv}} sw_i}{\displaystyle\sum_{i=1}^{n_{wrdv}} sw_i \cdot \sum_{i=1}^{n_{wrdv}} sw_i \cdot ivv_i^2 - \left(\sum_{i=1}^{n_{wrdv}} sw_i \cdot ivv_i\right)^2}$$

$$+ \frac{-2ivv^* \cdot \displaystyle\sum_{i=1}^{n_{wrdv}} sw_i \cdot ivv_i}{\displaystyle\sum_{i=1}^{n_{wrdv}} sw_i \cdot \sum_{i=1}^{n_{wrdv}} sw_i \cdot ivv_i^2 - \left(\sum_{i=1}^{n_{wrdv}} sw_i \cdot ivv_i\right)^2}$$

which, when evaluated numerically, gives

$$\mathrm{var}\left\{\mathrm{est}[\mathrm{mean}\,(APRCRHeteroscedasticNDRDV\text{'s})\text{ given }ivv = ivv^* = 15]\right\}$$
$$= (0.4214 + 0.0804 - 0.3214) \cdot \mathrm{var}(APRCRHNDREE\text{'s})$$
$$= 0.1808 \cdot \mathrm{var}(APRCRHNDREE\text{'s})$$

Clearly, this expression for $\mathrm{var}\{\mathrm{est}[\mathrm{mean}(APRCRHeteroscedastic\ NDRDV\text{'s})$ given $ivv = ivv^* = 15]\}$ in our constructed weighted simple linear regression example checks with the corresponding expression for $\mathrm{var}\{\mathrm{est}[\mathrm{mean}(APRCRHomoscedasticNDRDV\text{'s})$ given $ivv = ivv^* = 15]\}$ pertaining to the underlying text simple linear regression example.

Exercise Set 7

These exercises are intended to verify that the derived expressions for our the constructed weighted simple linear regression example agree (check) with the corresponding expressions for the text simple linear regression example.

    1. (a) Verify algebraically that, for our constructed weighted simple linear regression example, $\mathrm{var}[\mathrm{est}(clp1)]$ can be expressed as

$$\text{var}[\text{est}(clp1)] = \frac{\text{var}(APRCRHNDREE\text{'s})}{\sum_{i=1}^{n_{wrdv}} sw_i \cdot \{ivv_i - [\text{wt ave}(ivv_i\text{'s})]\}^2}$$

in which

$$\text{wt ave}(ivv_i\text{'s}) = \frac{\sum_{i=1}^{n_{wrdv}} sw_i \cdot ivv_i}{\sum_{i=1}^{n_{wrdv}} sw_i}$$

Then, (b) verify that this expression reduces to the simple linear regression expression for est(clp1) when all of the relative statistical weights are set equal to one and n_{wrdv} is set equal to n_{rdv}.

2. (a) Verify algebraically that, for our constructed weighted simple linear regression example, var[est(clp0)] can be expressed as

var[est(clp0)]

$$= \left(\frac{1}{\sum_{i=1}^{n_{wrdv}} sw_i} + \frac{[\text{wt ave}(ivv_i\text{'s})]^2}{\sum_{i=1}^{n_{wrdv}} sw_i \cdot \{ivv_i - [\text{wt ave}(ivv_i\text{'s})]\}^2} \right) \cdot$$

$$\text{var}(APRCRHNDREE\text{'s})$$

in which

$$\text{wt ave}(ivv_i\text{'s}) = \frac{\sum_{i=1}^{n_{wrdv}} sw_i \cdot ivv_i}{\sum_{i=1}^{n_{wrdv}} sw_i}$$

Then, (b) verify that this expression reduces to the simple linear regression expression for est(clp0) when all of the relative statistical weights are set equal to one and n_{wrdv} is set equal to n_{rdv}.

3. (a) Given our constructed weighed simple linear regression example and the propagation of variability expression:

$$\text{var}\{\text{est}[\text{mean}(APRCRHeteroscedasticNDRDv\text{'s}) \text{ given } ivv = ivv^*]\}$$

$$= \text{var}[\text{est}(clp0)] + (ivv^*)^2 \cdot \text{var}[\text{est}(clp1)]$$

$$+ 2 \cdot ivv^* \cdot \text{covar}[\text{est}(clp0), \text{est}(clp1)]$$

substitute expressions for the respective elements of the |covar[est $(clp0)$,est$(clp1)$]| covariance matrix to develop an expression for var{mean[($APRCRHeteroscedasticNDRDV$'s) given $ivv = ivv^*$]}. Then, (b) verify that this expression can be algebraically manipulated to read

$$\frac{\text{var}\{\text{est}[\text{mean}(APRCRHeteroscedasticNDRDV\text{'s}) \text{ given } ivv = ivv^*]\}}{\text{var}(APRCRHNDREE\text{'s})}$$

$$= \frac{1}{\displaystyle\sum_{i=1}^{n_{wrdv}} sw_i} + \frac{[\text{wt ave}(ivv_i\text{'s})]^2}{\displaystyle\sum_{i=1}^{n_{wrdv}} sw_i \cdot \left\{ ivv_i - [\text{wt ave}(ivv_i\text{'s})] \right\}^2}$$

In turn, (c) verify that this expression reduces to the simple linear regression expression for var{est[mean($APRCRHomoscedastic$ $NDRDV$'s) given $ivv = ivv^*$]} when the relative statistical weights sw_i's are set equal to one and n_{wrdv} is set equal to n_{rdv}.

Exercise Set 8

These exercises are intended to demonstrate that, given our constructed weighted simple linear regression example, the sum of the *weighted* [est($CRHeteroscedasticNDREE_i$)]'s is equal to zero and that the weighted sum of squares of the [est($CRHeteroscedasticNDREE_i$)]'s is identical to the *between*(SS) for its underlying text simple linear regression example.

1. Given the following tabulation for our constructed weighted simple linear regression example:

ivv_i	$wrdv_i$	=	est$(clp0)$	+	est$(clp1)ivv_i$	+	est($CHeteroscedastic$ $NDREE_i$)	relative statistical weight sw_i
10	4		−0.2		5		−0.8	3
20	11		−0.2		10		+1.2	2
30	21		−0.2		20		+1.2	2
40	24		−0.2		25		−0.8	3

verify that the values given for the respective [est($CRHeteroscedasticNDREE_i$)]'s are (a) algebraically correct, but (b) do not sum to zero. Then, (c) demonstrate that the *weighted* sum of the [est($CRHeteroscedasticNDREE_i$)]'s is equal to zero. In turn, (d) plot the estimated weighted simple linear regression model with its associated weighted regression datum values, using datum points with different diameters to reflect their respective magnitudes of the relative weights.

2. (a) Verify that $\sum sw_i \cdot \{[\text{est}(CRHeteroscedasticNDREE_i)]\text{'s}\}^2 = 9.6$, viz., equals the *between*(SS) for the text simple linear regression example. Then, (b) explain why the relative statistical weights can be used instead of the actual statistical weights to compute this weighted sum of squares. [Hint: Were the actual or the relative weights used in computing the *between*(SS) for the simple linear regression example?]

7.B.2. Discussion

Replication is not required (nor used) in weighted linear regression—because var($APRCRHeterocedasticNDRDV$'s) must be known (or can be estimated) at each $divv_i$ used in the regression experiment test program so that the associated statistical weights sw_i's are known (or can be estimated) at each $divv_i$. Thus, instead of using the experiment test program value of Snedecor's central F test statistic in linear regression analysis to test the adequacy of the presumed conceptual statistical model, we must use the associated experiment test program value of Pearson's central χ^2 test statistic to judge the statistical adequacy for the presumed conceptual statistical model in weighted linear regression analysis. In effect, this adequacy test examines whether the estimated deviations of the weighed normally distributed datum values from their respective est$\{[\text{mean}_i(APRCRHeteroscedastic NDRDV$'s)]$\}$'s are statistically consistent with the presumed known (or estimated) values of their associated $\{\text{est}[\text{var}_i(APRCRHeteroscedastic NDRDV$'s)]$\}$'s.

7.B.3. Perspective

Because the actual values of the statistical weights are almost never known in mechanical testing and mechanical reliability applications, our constructed weighted simple linear regression example is primarily intended to provide perspective and background. Hopefully it accomplishes two objectives: (a) to demonstrate how the presumption of homoscedasticity simplifies linear regression analyses; and (b) to establish a background that provides perspective regarding maximum likelihood analysis for strength experiment test program datum values (Supplemental Topic 8.D).

The presumption of homoscedasticity so markedly simplifies linear regression analysis that it is often rationalized as being physically credible. Do not be deceived. The universal experience of humankind is that small measurement values exhibit less variability than large measurement values. Thus, the presumption of homoscedasticity should always be suspect when the experiment test program range selected for the *ivv*'s is large. Then, whatever the application, the homoscedasticity versus heteroscedasticity issue is of paramount importance in establishing a statistically adequate (statistically credible) mechanical behavior model. We recommend the use of maximum likelihood analysis for all mechanical behavior models (Chapter 8)—because the associated likelihood ratio methodology provides a simple intuitive way to examine the homoscedasticity versus heteroscedasticity issue statistically (Supplemental Topic 8.E).

7.C. SUPPLEMENTAL TOPIC: CLASSICAL HYPERBOLIC 100(scp)% (TWO-SIDED) STATISTICAL CONFIDENCE BANDS IN SIMPLE LINEAR REGRESSION

Recall that the classical (shortest) $100(scp)\%$ (two-sided) statistical confidence interval in simple linear regression that allegedly includes [mean($APRCRHNDRDV$'s) given $ivv = ivv^*$] was presented in the text, where ivv^* lies in the (shortest practical) ivv interval that was used in conducting the experiment test program. We now present the corresponding classical hyperbolic $100(scp)\%$ (two-sided) statistical confidence bands that allegedly include mean($APRCRHNDRDV$'s) for all ivv's *simultaneously* that lie in this ivv interval.

When the conceptual simple linear regression statistical model:

$$|CRHNDRDV_i\text{'s}| = clp0|+1\text{'s}| + clp1|ivv_i\text{'s}| + |CRHNDREE_i\text{'s}|$$

is correct, the joint $100(scp)\%$ statistical confidence region for the actual values of the $clp0$ and $clp1$ can be obtained by appropriate simplification of an analogous multiple linear regression (matrix) expression found in Draper and Smith (1966). This simplification generates an explicit equation for the elliptical boundary of this joint $100(scp)\%$ statistical confidence region, viz.,

$$2 \cdot \text{Snedecor's central } F_{2,n_{rdv}-2;scp}$$

$$= \frac{\sum_{m=0}^{1}\sum_{n=0}^{1}\{[clpm - \text{est}(clpm)] \cdot [\text{term}_{m,n}] \cdot [clpn - \text{est}(clpn)]\}}{\text{est}[\text{var}(APRCRHNDRDV\text{'s})]}$$

in which (a) coupled values of $clpm$ and $clpn$ define points lying on this elliptical boundary, and (b) term$_{m,n}$, with dummy indices m and n, respectively, refers to the four elements of the $[\,|ivv_i\text{'s}|^t|ivv_i\text{'s}|\,]$ array, where

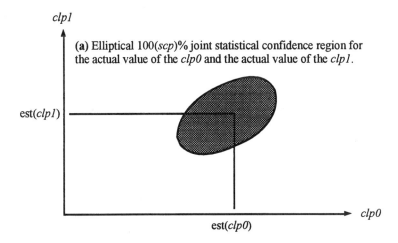

clp1

est(*clp1*)

(a) Elliptical 100(*scp*)% joint statistical confidence region for the actual value of the *clp0* and the actual value of the *clp1*.

est(*clp0*)

clp0

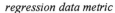

regression data metric

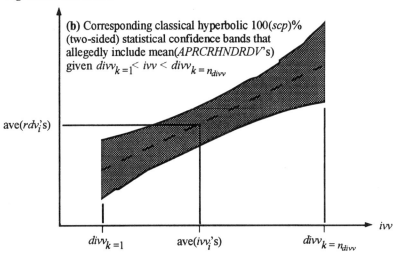

ave(*rdv*$_i'$s)

(b) Corresponding classical hyperbolic 100(*scp*)% (two-sided) statistical confidence bands that allegedly include mean(*APRCRHNDRDV*'s) given *div*$_{k=1}$ < *ivv* < *div*$_{k=n_{div}}$

div$_{k=1}$ ave(*ivv*$_i'$s)   *div*$_{k=n_{div}}$

ivv

Figure 7.8 Given the conceptual simple linear regression model expressed as the [mean(*APRCRHNDRDV*'s) given *ivv*] = *clp0* + *clp1* · *ivv*, the collection of all coupled *clp0* and *clp1* values that lie inside the boundary of the elliptical joint 100(*scp*)% statistical confidence region for the actual values of the *clp0* and the *clp1* that is plotted schematically in (a) correspond one-to-one to the collection of conceptual simple linear regression models that lie inside of the classical hyperbolic 100(*scp*)% (two-sided) statistical confidence bands schematically plotted in (b). These bands allegedly include [mean(*APRCRHNDRDV*'s)] for all *ivv*'s that lie in the (shortest practical) *ivv* interval that was used in conducting the simple linear regression experiment test program.

$$\left| ivv_i\text{'s} \right|^t \left| ivv_i\text{'s} \right| = \begin{vmatrix} n_{rdv} & \sum\limits_{i=1}^{n_{rdv}} ivv_i \\ \sum\limits_{i=1}^{n_{rdv}} ivv_i & \sum\limits_{i=1}^{n_{rdv}} ivv_i^2 \end{vmatrix}$$

Figure 7.8(a) schematically displays this elliptical joint $100(scp)\%$ statistical confidence region. Its center is located at the point [est($clp0$), est($clp1$)], but its major and minor areas are parallel and perpendicular to the $clp0$ and $clp1$ axes only when the $|+1\text{'s}|$ column vector is orthogonal to the $|ivv_i\text{'s}|$ column vector. Every $clp0,clp1$ point lying *inside* of the boundary of this elliptical joint $100(scp)\%$ statistical confidence region corresponds to coupled $clp0$ and $clp1$ values that could be taken as the actual $clp0$ and $clp1$ values and the resulting data-based value for Snedecor's central F test statistic would not lead to rejection of this null hypothesis—provided that the associated acceptable probability of committing a *Type I* error is equal to $(1 - scp)$. Accordingly, each of these coupled $clp0$ and $clp1$ values establishes a hypothetical conceptual simple linear regression model. In turn, given the collection of all of these hypothetical models, the respective maximum and minimum values for the {mean($APRCRHNDRDV$'s)] given $divv_{k=1} < ivv < divv_{k=n_{divv}}$}'s generate the classical hyperbolic $100(scp)\%$ (two-sided) statistical confidence bands illustrated in Figure 7.8(b). These bands allegedly include [mean($APRCRHNDRDV$'s) given ivv] for all ivv's that lie in the (shortest practical) ivv interval that was used in conducting the simple linear regression experiment test program, viz.,

classical hyperbolic $100(scp)\%$ (two-sided) statistical confidence bands

$$= \text{est[mean}\,(APRCRHNDRDV\text{'s) given } ivv] \pm k_{chscb} \cdot \text{stddevterm}$$

in which

$$k_{chscb} = [2 \cdot \text{Snedecor]'s central } F_{2,n_{rdv}-2;scp}]^{1/2}$$

and

stddevterm = est(stddev{est[mean $(APRCRHNDRDV\text{s})$ given ivv]})

$$= \sqrt{\frac{1}{n_{rdv}} + \frac{[ivv - \text{ave}(ivv_i\text{'s})]^2}{\sum\limits_{i=1}^{n_{dv}} tivv_i^2}} \cdot \{\text{est[stddev}(APRCRHNDREE\text{'s})]\}$$

8

Mechanical Reliability Fundamentals and Example Analyses

8.1. INTRODUCTION

The first step in mechanical design for a new product is to synthesize (configure) the product such that it performs its desired function. Design synthesis is markedly enhanced by first recognizing functional analogies among existing designs, then comparing alternative feasible designs, and in turn proposing the design that appears to have the greatest overall advantage. The second step in mechanical design for a new product is to try to assure that the proposed design will reliably perform its function. Tentative assurance of adequate reliability requires a new set of comparisons interwoven in a combination of design analyses and laboratory tests whose goal is to extrapolate service-based performance data for similar products to the proposed product. However, it is imperative to understand that adequate reliability is established only by actual service performance. It cannot be established by a combination of design analyses based on analytical bogies (design allowables, factors of safety) and laboratory tests based on experimental bogies (extreme load and environment histories). Nevertheless, a combination of design analysis and laboratory testing can be effectively employed to maintain or to improve the reliability of a product.

When the mechanical design objective is to maintain the service-proven reliability of a product following its redesign, the redesign is required to meet the same set of analytical and laboratory test bogies that were met by the original design. However, when the mechanical design objective is to improve product reliability, the performance of the redesign must excel the performance of the original design. This improved performance must be evident in laboratory tests before it can be presumed that it will also be evident in service operation. Accordingly, reliability improvement laboratory tests should always be conducted using load and environment histories that are nominally identical to service load and environment histories. In particular, all failures in these laboratory tests should be identical in location, mode of failure, and appearance to corresponding service failures.

This chapter covers reliability concepts and reliability applications of statistical analyses that can be used to improve the reliability of a product. The more credible the performance comparison of the redesign to the original design, the more likely the predicted reliability improvement will actually be realized in service operation.

8.2. MECHANICAL RELIABILITY TERMINOLOGY

All solid materials resist failure. When the stimulus for failure is stated in terms of stress, the corresponding resistance is technically termed strength. However, the terms stress and strength are widely used in a generic sense to connote stimulus and resistance, respectively.

Strengths and resistances are always established experimentally by strictly following a standardized laboratory test method that has been severely constrained and simplified so that the resulting datum values will be repeatable within the given laboratory and reproducible between different laboratories. However, from a mechanical reliability perspective, all such standardized laboratory test methods are self-defeating—because there is no dependable way to use the resulting datum values to predict service behavior. There is no direct relationship between the environmental and loading histories pertaining to standardized laboratory test methods and actual service operation for any mechanical mode of failure.

The test duration in a reliability experiment test program is defined in terms of the environmental and loading histories imposed prior to failure, where failure connotes the discontinuation of satisfactory performance. When these environmental and loading histories exhibit recognizable increments such as cycles or repeated sequences, then life (endurance) is stated in terms of the number of these increments endured prior to failure. Similar recognizable increments seldom, if ever, occur in service. Thus, there is seldom, if ever, a direct relationship between a well-defined laboratory test life (endurance) and the ill-defined service operation life (endurance).

Mechanical reliability is technically defined as the probability that a given device, randomly selected from the population of all such devices, will perform satisfactorily under certain specified service operation environmental and loading histories for at least an arbitrarily specified life (duration). However, this technical definition for reliability is seldom practical (for reasons that will be evident later). Rather, it is customary to compute a lower $100(scp)\%$ (one-sided) statistical confidence limit that allegedly bounds the actual value for the metric pertaining to the pth percentile of the presumed conceptual two-parameter life (endurance) or strength (resistance) statistical distribution—where p is usually selected to be either 0.01 or 0.10 and the associated scp is usually selected to be 0.95 (see Supplemental Topic 8.A).

8.3. CONCEPTUAL STATISTICAL MODEL FOR FATIGUE FAILURE AND ASSOCIATED LIFE (ENDURANCE) AND STRENGTH (RESISTANCE) EXPERIMENT TEST PROGRAMS

We now present a conceptual statistical model for fatigue failure that ostensibly pertains to conventional laboratory data, but also applies in concept to all duration-dependent mechanical modes of failure when the appropriate analogies are made. Figure 8.1 depicts the PDF of the conceptual statistical distribution that consists of all possible replicate fatigue life datum values that could be generated by continually replicating the conduct of a quantitative CRD experiment test program (Chapter 2) with a fixed value s_a^* for its alternating stress amplitude s_a. Note that the metric for this PDF is $\log_e(fnc)$, where fnc connotes failure number of cycles. Note in addition that the traditional s_a–$\log_e(fnc)$ curve pertains to the median fatigue life. The corresponding CDF is depicted in Figure 8.2. It also establishes the values for $\log_e[fnc(pf)]$ that pertain to other probabilities of failure (pf) of specific interest, e.g., 0.01 and 0.99. In turn, a conceptual s_a–$\log_e[fnc(pf)]$ fatigue failure model, Figure 8.3, is developed when the respective $\log_e[fnc(pf)]$ values of specific interest are augmented with corresponding $\log_e[fnc(pf)]$ values pertaining to other values for s_a^*. However, extensive fatigue test experience clearly indicates that the actual value for the variance of the conceptual statistical distribution that consists of all possible replicate $\log_e(fnc)$ datum values increases markedly as s_a^* decreases, especially at long fatigue lives. Thus, the presumption of homoscedasticity for a conceptual s_a–$\log_e[fnc(pf)]$ fatigue failure model is statistically invalid—as it is for all other mechanical modes of failure that are time dependent (duration dependent). In contrast, this experience indicates that empirical scatter

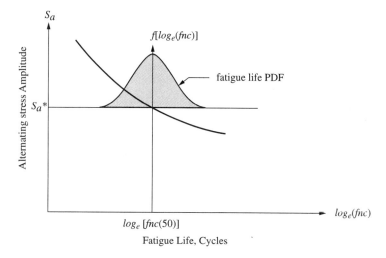

Figure 8.1 Example life (endurance) experiment test program that involves test-ing replicate fatigue specimens at a specific alternating stress amplitude s_a^*. Each replicate test is ideally conducted until failure occurs (which may not be practical). Note the improper reversal of independent and dependent (random) variables for the traditional method of plotting the median S–N curve for which we use the uncon-ventional notation, s_a–$\log_e[fnc(50)]$ model, where $fnc(50)$ connotes median failure number of cycles.

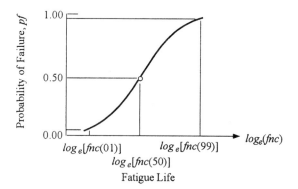

Figure 8.2 Schematic of the conceptual fatigue life CDF associated with Figure 8.1. Note that the conceptual fatigue life CDF pertains to the probability of failure, pf, whereas reliability is stated in terms of the probability of survival, ps.

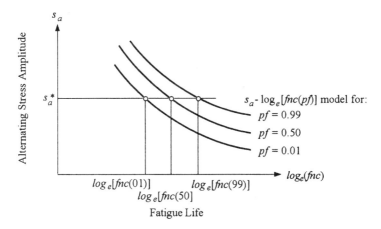

Figure 8.3 Conceptual s_a–$\log_e[fnc(pf)]$ model that is based on a series of life (endurance) experiment test programs conducted at several values for the alternating stress amplitude s_a^* (Figure 8.1). All of the associated laboratory fatigue tests must pertain to the same set of nominally identical test specimens and test conditions (mean stress, notch configuration, environmental history, etc.).

bands faired to datum values that pertain to various time-dependent (duration-dependent) modes of failure (including fatigue) are almost always approximately parallel when their width is stated in terms of stimulus levels (alternating stress amplitudes) as in Figure 8.4. Accordingly, the notion that the associated conceptual fatigue strength statistical distribution, Figure 8.5, has a homoscedastic variance is physically credible. Finally, the conceptual s_a–$\log_e[fnc(pf)]$ fatigue failure model can be visualized as a surface in s_a (stimulus level), $\log_e(fnc)$ (duration to failure), and pf (probability of failure) space. See Figure 8.6.

8.4. LIFE (ENDURANCE) EXPERIMENT TEST PROGRAMS

Given a mechanical mode of failure whose stimulus, duration-to-failure model is illustrated in Figure 8.1, the conceptual statistical distribution that consists of all possible replicate realization values for the random variable (duration to failure) will exhibit (a) a heteroscedastic variance that markedly increases as the stimulus level decreases and approaches a failure initiation threshold, and (b) a PDF that is markedly skewed to the right. Moreover, run-outs (viz., suspended tests) will occur when the test stimulus level is (deliberately or inadvertently) selected to lie below the associated

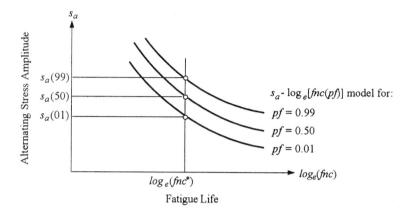

Figure 8.4   Experience demonstrates that s_a–$\log_e(fnc)$ fatigue experiment test programs almost always generate datum values whose faired scatter bands are approximately parallel when stated in terms of s_a. Accordingly, we subsequently employ a homoscedastic conceptual strength distribution to model fatigue behavior for all $\log_e(fnc^*)$ values of practical interest (Figure 8.5).

failure initiation threshold. A logarithmic scale for the test duration tends to mitigate the PDF skewness problem somewhat (and thus is almost always employed). Nevertheless, a proper statistical analysis must account for heteroscedastic datum values and for run-outs.

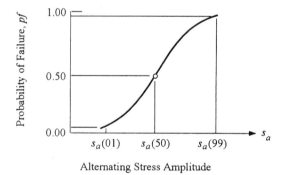

Figure 8.5   Schematic of the CRD pertaining to the conceptual fatigue strength distribution associated with Figure 8.4. The actual values for percentiles of this conceptual fatigue strength distribution, given the value for $\log_e(fnc^*)$ of specific interest, can be estimated by conducting a strength (resistance) experiment test program (Section 8.5).

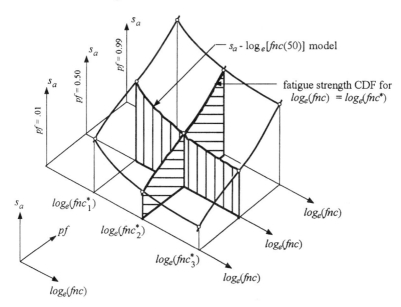

Figure 8.6 Schematic of the conceptual s_a–$\log_e[fnc(pf)]$ fatigue model in three-dimensional perspective. Massive fatigue testing is required to generate the estimated s_a–$\log_e[fnc(pf)]$ model that pertains to each specific set of test conditions that might conceivably be regarded as practical.

8.4.1. Types of Life (Endurance) Data

8.4.1.1. Complete Data

All test items fail (no test suspensions), and the respective life (endurance) datum values are known (schematically depicted by an X below). Complete data occur *only* when the experiment test program stimulus level for each test item exceeds its corresponding failure initiation level (threshold level):

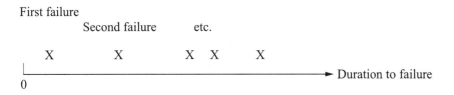

8.4.1.2. Complete Grouped Data

All test items fail (no test suspensions), but the respective life (endurance) datum values are grouped in a histogram format so that the individual life (endurance) datum values are unknown at the time of analysis. Complete grouped data are most likely to occur when a standardized report form is scanned to compile service failure data:

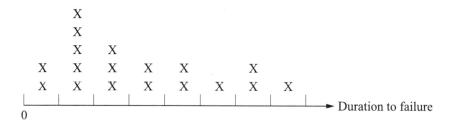

8.4.1.3. Censored Data with *Type I* Suspended Tests

All test items with *Type I* censoring either fail prior to a predetermined test duration d^* (*fnc**), or the test is suspended at this predetermined duration d^* (*fnc**). Note that *Type I* suspended tests occur only in statistically planned experiment test programs:

8.4.1.4. Censored Data with Arbitrarily Suspended Tests

One or more tests in the experiment test program are arbitrarily suspended (perhaps due to a test equipment problem or the need to run a test with a higher priority):

> *Remark*: Test equipment breakdown during very long experiment test programs is surely foreseeable. Equipment breakdown problems can be mitigated in comparative test programs by employing time blocks (and subsequently ignoring these time blocks if no equipment breakdown occurs).

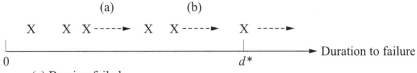

(a) Bearing failed
(b) Machine required for a more important test

8.4.1.5. Censored Data: The Four General Cases

(a) It is only known that the life (endurance) of the test item is less than some specific value; (b) it is only known that the life (endurance) of the test item is bounded by two specific values; (c) it is only known that the life (endurance) of the test item is greater than some specific value; (d) the life (endurance) of the test item is known.

```
        (a)            (b)          (c)         (d)
   ◄----- X      |◄-----X----►|    X-----►      X
   └                                              ► Duration to failure
   0
```

8.4.1.6. Discussion

Type II (so-called item) censoring has deliberately been excluded from this listing of types of life (endurance) data because it is very seldom practical in mechanical testing. It requires the *simultaneous* testing of a single item in each of n_i test machines that are alleged to be nominally identical, or using a single test fixture that allegedly exposes each of n_i items to nominally identical stimulus levels, whereas *Type I* censoring typically pertains to *sequential* testing of n_i items using a single test machine (and presuming invariant test conditions during the entire experiment test program).

8.5. STRENGTH (RESISTANCE) EXPERIMENT TEST PROGRAMS

Strength (resistance) experiment test programs are generally more practical than life (endurance) experiment test programs for long-life reliability applications, but strength (resistance) experiment test programs typically require considerably more test time to conduct. Accordingly, more attention must be given to the statistical planning of a strength (resistance) experiment test program so that the specimen allocation strategies employed in testing generate a more precise statistical estimate of the actual value for the median of the conceptual strength (resistance) distribution.

> *Remark*: The CDF's for all conceptual statistical distributions used in the analysis of life (endurance) and strength (resistance) data are similar near their respective medians and differ markedly only at their extreme percentiles. Moreover, there is no practical way to distinguish one CDF statistically from another at these extreme percentiles. Accordingly, strength (resistance) estimates have traditionally pertained to medians. On the other hand, life (endurance) estimates have traditionally pertained to (relatively) extreme percentiles, thus requiring the presumption that the actual conceptual statistical distribution is known. But this presumption, if believed, is either naive or foolish.

8.5.1. Strength (Resistance) Experiment Test Program Strategies

8.5.1.1. Conventional Up-and-Down Test Method

The first test in the conventional up-and-down test method is conducted at the "guestimated" stimulus level (e.g., alternating stress amplitude) that hopefully corresponds to the actual value for the median of the presumed normally distributed conceptual strength (resistance) distribution. Then, if failure occurs before some preselected test duration d^*, the second test is conducted at a decreased stimulus level. However, if the first test item survives to this preselected test duration d^*, then the second test is conducted using an increased stimulus level. This strategy continues until all available items have been tested. For purposes of test conduct simplification and the corresponding use of statistical formulas and tables, the increment used to decrease or increase the stimulus level is fixed and ideally selected to be equal to the actual value for the standard deviation of the presumed normally distributed conceptual strength (resistance) distribution (which can only be guestimated on the basis of preliminary testing or prior experience.)

The conventional up-and-down test method is illustrated below for the example data given in ASTM D-3029-90, "Standard Test Methods for Impact Resistance of Flat, Rigid Plastic Specimens by Means of a Tup (Falling Weight)," in which the weight of the tup (dart) is decreased or increased depending on the prior test outcome. (Predetermined test duration d^* is not relevant in this example.) An analogous test method pertains to a tup (dart) of fixed weight dropped from a height that is either increased or decreased depending on the prior test outcome.

```
tup (dart)
weight (kg)            test outcome code: X = failure; O = nonfailure
   9                        X
   8                  O   X   X           X   X
   7          X   O          O   X   O   O   X   X   O
   6      O   O                   O               O   O
          1   2   3   4   5   6   7   8   9  10  11  12  13  14  15  16  17  18  19  20
```

8.5.1.2. The Two-Point Strategy

Maximum likelihood analysis of the ASTM D-3029-90 example data indicates that the five tests conducted with a dart weight of 6 kg have a negligible effect on the statistical estimate of the actual value for the median of the conceptual impact resistance distribution and its precision. Accordingly, the two-point strategy was proposed to alleviate this statistically ineffective test item allocation problem. In the two-point strategy, the conventional up-and-down test method strategy is terminated as soon as both O's and X's have been observed at two distinct stimulus levels. The test program then continues with all future tests being conducted at only these two levels. Considering the ASTM D-3029-90 up-and-down test method example data, the two-point strategy gives:

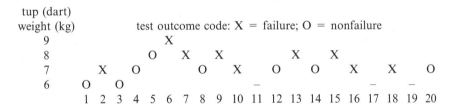

```
tup (dart)
weight (kg)            test outcome code: X = failure; O = nonfailure
   9                        X
   8                  O   X   X           X   X
   7          X   O          O   X   O   O   X   X   O
   6      O   O                  —               —   —
          1   2   3   4   5   6   7   8   9  10  11  12  13  14  15  16  17  18  19  20
```

in which the three specimens not tested at 6 kg (indicated by dashes) are much more effectively tested at either 7 or 8 kg. [In fact, as discussed later, an even more statistically effective tup (dart) weight can be calculated using the minimum variance strategy.]

8.5.1.3. Conventional Small Sample Up-and-Down Test Method

An experiment test program that is conducted following the conventional up-and-down test method usually involves at least 20 to 40 specimens, whereas an experiment test program that is conducted following the conventional small sample up-and-down test method usually involves only five to 10 specimens. However, the conceptual strength (resistance) distribution and its standard deviation must be presumed to be known to accomplish this sample size reduction. Accordingly, only the actual value for the median of the presumed conceptual strength (resistance) distribution is estimated in statistical analysis. The spacing between successive stimulus levels is also fixed for the conventional small sample up-and-down test method. This spacing is set equal to the presumed known (guestimated) value for the standard deviation of the presumed conceptual strength (resistance) statistical distribution. The nominal sample size for a conventional small sample up-and-down test method program is established by counting only the last response of the beginning sequence of like responses, and all of the remaining responses. For example, given a conventional small sample up-and-down experiment test program outcome of O O O O X O X O, the nominal sample size is 5. Although, when employed to conduct fatigue tests, the conventional small sample up-and-down test method strictly pertains to estimating the actual value for the median of the conceptual endurance limit distribution, it can also be used to estimate the actual value for the median of the conceptual fatigue strength distribution at very long fatigue lives, e.g., at 10^7 to 10^8 alternating stress cycles.

8.5.1.4. Minimum Variance Strategy

The spacing (increment) between successive stimulus levels is fixed in the conventional small sample up-and-down test method. However, the minimum variance methodology indicates that the stimulus level that most effectively increases the statistical precision of the resulting estimate of the actual value for the median of the presumed conceptual strength (resistance) distribution changes from test to test. Thus, as soon as a reasonably precise estimate of the actual value for the median of the presumed conceptual strength (resistance) distribution has been established, all subsequent stimulus levels should be selected to maximize the statistical precision of the resulting estimate of the actual value for the median of the presumed conceptual strength (resistance) distribution. This strategy is also effective when a conventional small sample up-and-down test method program is augmented by testing a few additional items to improve the statistical precision of the final estimate of the actual value for the median of the presumed conceptual strength (resistance) distribution.

8.5.2. Preliminary Strength (Resistance) Experiment Test Program Strategies (Little, 1990)

When the number of items available for testing is quite limited or costly, or when the time available for testing is very limited, and when there is little if any information available to guestimate the initial stimulus level to begin up-and-down testing, the strategies used in preliminary testing should be particularly efficient and effective. These strategies should either markedly reduce the number of test items or the amount of test time required to home in on the actual value for the median of the presumed conceptual strength (resistance) distribution, thereby permitting the remaining test items to be allocated to stimulus levels that are statistically more effective in estimating the actual value for the median of the presumed conceptual strength (resistance) distribution. Two examples follow which pertain to an (unpublished) unique bolt pull-through fatigue experiment test program on a composite material that is markedly stronger in the longitudinal direction than in the transverse direction. The respective stimulus levels are stated in terms of force (load) rather than stress because one of the main objectives of the experiment test program was to establish the mode of failure and the associated governing stress expression(s) for each different composite material tested. The third example strategy is intended to reduce the number of tests required to obtain a change in response from X to O or vice versa.

8.5.2.1. Example One: Run-Up Preliminary Test Strategy (Table 8.1)

When the number of test items is severely limited or are extremely costly, the run-up preliminary test strategy is appropriate in which a single test item is used to begin to home in on the actual value for the median of the presumed conceptual strength (resistance) distribution. The first test is conducted with its stimulus level guestimated to be well below the actual value for the median of the presumed conceptual strength (resistance) distribution. If the test item does not fail before predetermined test duration d^*, it is thoroughly examined and then retested at an increased stimulus level. This strategy continues until the test item eventually fails. (If sample size, test time, and cost constraints permit, this failure stimulus level could be replicated and the data for the run-up item relegated to ancillary information status because of potential damage or potential coaxing effects.) The fixed increment between the successive stimulus levels for these run-up tests should be relatively large, so that relatively few run-up test increments are required to cause the first failure. Accordingly, this run-up increment must be markedly reduced before beginning to employ the up-and-down test method strategy. The half-interval strategy is conveniently used to reduce this increment while continuing to home in on the actual value for the median of the presumed conceptual strength (resistance) distribution, viz., the increment between successive tests is reduced by one-half of the difference in adjacent stimulus levels with X and O outcomes. Once the stimulus level increment has been reduced to its preselected minimum level using the half-interval strategy, the up-and-down test method strategy is adopted and

Table 8.1 Example One: Run-Up Preliminary Test Strategy

Preliminary test planning choices:
  Initial alternating force: 50 lbs
  Run-up increment: 25 lbs
  Minimum practical interval value: 5% of est[$f_a(50)$]

Test data:
  X = failure before 10^7 cycles; O = run-out (did not fail, DNF)

Test no.	Alternating force (f_a), lbs	Test outcome	Test strategy
1a	50	O	Test specimen did not fail. Thus, the unfailed specimen is retested with $f_a = 50 + 25$
1b	75	O	Test specimen did not fail. Thus, this unfailed specimen is retested with $f_a = 75 + 25$
1c	100	O	Test specimen did not fail. Thus, this unfailed specimen is retested with $f_a = 100 + 25$
1d	125	X	Test specimen failed. Thus, the half-interval strategy begins. Accordingly, the next specimen is tested with $f_a = 125 - (25/2)$
2	112.5	O	Test specimen did not fail. Thus, the next specimen is tested with $f_a = 112.5 + (12.5/2)$. Note that this new increment was approximately equal to its preselected minimum practical value. Accordingly, the up-and-down test method strategy begins
3	118.75	X	Test specimen failed. Thus, the next specimen is tested with $f_a = 118.75 - 6.25$

continued until a preselected nominal sample size is reached (or until some other criterion is satisfied). The minimum variance strategy is then used to allocate all subsequent specimens to their statistically most effective stimulus levels.

8.5.2.2. Example Two: Run-Down Preliminary Test Strategy (Table 8.2)

The run-down preliminary test strategy is recommended to begin to home in on the actual value for the median of the presumed conceptual strength (resistance) distribution when ample test items are available and it is desired to limit the test time as much as practical. The failure time (endurance) datum values can also be used to decide when the change to the half-interval strategy is appropriate. For example, the fatigue test specimen tested at $f_a = 75$ lbs in Table 8.2 failed just before the preselected test duration $d^* = fnc^* = 10^7$ load cycles was reached. Hence, the third specimen could have been tested with an alternating force f_a equal to 62.5 lbs rather than strictly following the run-down strategy.

This experiment test program followed the Example One test program and employed the same composite material and test fixture, but the test specimen was oriented in the transverse direction rather than in the longitudinal direction. Thus, the initial alternating force of 100 lbs was expected to be much larger than the actual value for the median of its conceptual failure load distribution.

8.5.2.3. Example Three: Wide-Spacing Preliminary Test Strategy (Little and Thomas, 1993)

When sufficient prior testing has been conducted to assure a reasonably precise estimate of the actual value for the standard deviation of the presumed conceptual strength (resistance) distribution, an effective preliminary test strategy is to use a stimulus level spacing equal to twice the estimated actual value for the standard deviation in the run-up or run-down tests used to begin the experiment test program, and then to change to the conventional small sample up-and-down test method strategy with a stimulus level spacing equal to the estimated actual value for the standard deviation as soon as the test outcome changes character from X to O, or vice versa. Then, after a few up-and-down tests have been conducted, the statistical precision of the final estimate of the actual value for the median of the presumed conceptual strength (resistance) distribution can be improved by subsequently employing the minimum variance strategy to allocate the remaining specimens to their statistically most effective stimulus levels.

Table 8.2 Example Two: Run-Down Preliminary Test Strategy

Preliminary test conduct choices:
   Initial alternating force: 100 lbs
   Run-down increment: 25 lbs
   Minimum practical interval value: 5% of est[$f_a(50)$]

Test data:
   X = failure before 10^7 cycles; O = run-out (did not fail, DNF)

Test no.	Alternating force (f_a), lbs	Test outcome	Test strategy
1	100	X	Test specimen failed. Thus, the next specimen is tested with $f_a = 100 - 25$
2	75	X	Test specimen failed. Thus, the next specimen is tested with $f_a = 75 - 25$
3	50	O	Test specimen did not fail. Thus, the half-interval strategy begins. Accordingly, the next specimen is tested with $f_a = 50 + (25/2)$
4	62.5	X	Test specimen failed. Thus, the next specimen is tested with $f_a = 62.5 - (12.5/2)$
5	56.25	O	Test specimen did not fail. Thus, the next specimen is tested with $f_a = 56.25 + (6.25/2)$. Note that this new increment was approximately equal to its preselected minimum practical value. Accordingly, the up-and-down test method strategy begins
6	59.375	X	Test specimen failed. Thus, the next specimen is tested with $f_a = 59.375 - 3.125$

8.6. CONCEPTUAL STATISTICAL DISTRIBUTIONS FOR MODELING OUTCOMES OF LIFE (ENDURANCE) AND STRENGTH (RESISTANCE) EXPERIMENT TEST PROGRAMS

The outcomes of life (endurance) and strength (resistance) experiment test programs are modeled by conceptual statistical distributions that are actually mathematical abstractions. The conceptual statistical distribution that is properly employed in modeling, if any, is unknown for mechanical modes of failure. Moreover, even when some conceptual statistical distribution is widely alleged to model a given mechanical mode of failure, alternative conceptual statistical distributions should always be employed in analysis for comparative purposes. It is, therefore, important to understand the similarities and the differences among alternative conceptual life and strength statistical distributions.

8.7. CONCEPTUAL LIFE (ENDURANCE) DISTRIBUTIONS

The two-parameter conceptual $\log_e$–normal and Weibull distributions are most commonly used to model life (endurance) datum values, but $\log_e$–normal datum values are seldom analyzed directly. Rather, the natural logarithms of the respective life (endurance) datum values are first calculated and then these transformed datum values are presumed to have been randomly selected from a (two-parameter) conceptual normal distribution. In turn, following appropriate analysis, the exponentials (antilogarithms) of the particular estimated values of specific interest are evaluated. Weibull datum values should be analyzed similarly, viz., the natural logarithms of the respective life (endurance) datum values should be calculated and then these transformed datum values should be presumed to have been randomly selected from a conceptual (two-parameter) smallest-extreme-value distribution. In turn, following appropriate analysis, the exponentials (antilogarithms) of the particular estimated values of specific interest should be evaluated.

8.7.1. Conceptual Two-Parameter Weibull Distribution

The CDF for the conceptual two-parameter Weibull can be written as

$$F(fnc) = 1 - \exp^{-\left(\frac{fnc}{cdp1}\right)^{cdp2}}$$

in which *fnc*, the failure number of cycles, is the life (endurance) metric. The inverse CDF expression is

$$y = \log_e\{-\log_e[1 - F(fnc)]\} = cdp2 \cdot [\log_e(fnc) - \log_e(cdp1)]$$
$$= a + b \cdot \log_e(fnc)$$

Accordingly, the conceptual two-parameter Weibull CDF plots as a straight line on Weibull probability paper with ordinate *y* and abscissa $\log_e(fnc)$, where *y* is a function of $pf = F(fnc)$ given by the inverse CDF expression.

8.7.2. Corresponding Conceptual (Two-Parameter) Smallest-Extreme-Value Distribution

The CDF for the corresponding conceptual (two-parameter) smallest-extreme-value distribution is written as

$$F[\log_e(fnc)] = 1 - \exp^{-\exp\left[\dfrac{\log_e(fnc) - clp}{csp}\right]}$$

in which $\log_e(fnc)$ is the life (endurance) metric rather than *fnc*. The corresponding inverse CDF expression is

$$y = \log_e\left(-\log_e\{1 - F[\log_e(fnc)]\}\right) = \frac{\log_e(fnc) - clp}{csp}$$
$$= a + b \cdot \log_e(fnc)$$

Accordingly, when $csp = (1/cdp2)$ and $clp = \log_e(cdp1)$ the conceptual (two-parameter) smallest-extreme-value distribution pertaining to the (transformed) random variable $\log_e(FNC)$ is identical to the conceptual two-parameter Weibull distribution pertaining to the random variable *FNC*. Observe, however, that the *clp* and the *csp* for the conceptual (two-parameter) smallest-extreme-value distribution have intuitive geometric interpretation because its life (endurance) metric $\log_e(fnc)$ is plotted along a linear abscissa.

8.7.3. Conceptual Three-Parameter $\log_e$–Normal and Weibull Distributions

The addition of a third conceptual parameter in the conceptual two-parameter $\log_e$–normal and Weibull distributions provides their respective CDF's more curve-fitting versatility relative to describing life (endurance) datum values and thus has led to the unwarranted popularity of three-parameter distributions. The third conceptual parameter, the so-called conceptual minimum life parameter, *cmlp*, is absolutely fictitious relative to its physical interpretation. Accordingly, three-parameter distribution cannot be recommended for use in mechanical reliability applications.

> *Perspective*: It is important to understand that no conceptual statistical distribution, regardless of the number or the nature of its parameters, ever exactly models life (endurance) or strength (resistance) datum values for any mechanical mode of failure. All conceptual statistical distributions are mathematical abstractions.

8.8. CONCEPTUAL LIFE (ENDURANCE) DISTRIBUTIONS WITH A CONCOMITANT INDEPENDENT VARIABLE

Any conceptual life (endurance) distribution that employs the concomitant independent variable s_a to augment its *clp* can be used to construct a conceptual s_a–$\log_e[fnc(pf)]$ model. For example, the conceptual (two-parameter) smallest-extreme-value distribution can be stated as

$$F[\log_e(fnc)] = 1 - \exp^{-\exp\left[\dfrac{\log_e(fnc) - clp0 - clp1 \cdot s_a - clp2 \cdot s_a^2}{csp}\right]}$$

in which its *clp* is replaced by the quadratic (modal) expression:

$$clp = clp0 + clp1 \cdot s_a + clp2 \cdot s_a^2$$

However, this example conceptual s_a–$\log_e[fnc(pf)]$ model is not physically credible for long-life fatigue applications because it presumes a homoscedastic fatigue life distribution, viz., its *csp* does not increase appropriately as $\log_e[fnc(50)]$ increases. Although modifications of analogous conceptual s_a–$\log_e[fnc(pf)]$ models have been proposed to account for the heteroscedasticity of long-life fatigue datum values, none can be recommended.

A conceptual s_a–$\log_e[fnc(pf)]$ model that employs a homoscedastic strength distribution, which is physically much more credible, is based on the preponderance of s_a–$\log_e(fnc)$ data with extensive replication. For example, the conceptual (two-parameter) smallest-extreme-value distribution can be stated as

$$F(s_a) = 1 - \exp^{-\exp\left\{\dfrac{s_a - clp0 - clp1 \cdot \log_e(fnc) - clp2 \cdot [\log_e(fnc)]^2}{csp}\right\}}$$

An analogous alternative s_a–$\log_e[fnc(pf)]$ model can be constructed using the conceptual (two-parameter) $\log_e$–normal distribution. Moreover, each of these alternative s_a–$\log_e[fnc(pf)]$ models can (a) be either linear or quadratic, and (b) employ either a linear or a logarithmic alternating stress amplitude. Thus, eight different alternative s_a–$\log_e[fnc(pf)]$ models should be examined in mechanical reliability analyses.

8.9. CONCEPTUAL STRENGTH (RESISTANCE) DISTRIBUTIONS

The conceptual (two-parameter) logistic distribution is sometimes used to analyze strength (resistance) datum values. Its symmetrical PDF is very similar to the conceptual (two-parameter) normal distribution PDF, but its longer tails have greater probability content. The conceptual (two-parameter) logistic distribution CDF can be written as

$$F(s) = \left[1 + \exp^{-\left(\frac{s - clp}{csp}\right)} \right]^{-1}$$

Because of its longer tails, it serves as an excellent adjunct to the conceptual (two-parameter) normal distribution for purposes of comparing the respective lower $100(scp)\%$ (one-sided) statistical confidence limits based on alternative statistical models, just as does the conceptual (two-parameter) smallest-extreme-value distribution, which is skewed to the left and its mirror image, the conceptual (two-parameter) largest-extreme-value distribution, which is skewed to the right. This latter distribution can be written as

$$F(s) = \exp^{-\exp^{-\left(\frac{s - clp}{csp}\right)}}$$

These four CDF's, with both linear and logarithmic abscissa scales, generate eight alternative analyses for estimating the actual value for the median (or other percentiles) of the presumed conceptual strength (resistance) distribution of specific interest.

Exercise Set 1

These exercises are intended to familiarize you with the PDF's and the corresponding CDFs for the alternative conceptual (two-parameter) life (endurance) and strength (resistance) distributions presented in this section.

First, given the analytical expression for each of the CDF's in Exercises 1 through 6 below, develop analytical expressions for the PDF, mean, median, mode, and variance. (See Supplemental Topic 8.D to check results.) Then tabulate x, $f(x)$, and $y(p)$ for $F(x) = p = 0.001$, 0.005, 0.01, 0.025, 0.05, 0.1, 0.2, 0.3, 0.4, 0.5, 0.6, 0.7, 0.8, 0.9, 0.95, 0.975, 0.99, 0.995, and 0.999. Finally, using your tabulation of p and $y(p)$, construct a sheet of probability paper for each of these conceptual (two-parameter) life (endurance) and strength (resistance) distributions with both p and $y(p)$ plotted along its respective nonlinear and linear ordinates and with either x or $\log_e(x)$ (as appropriate) plotted along its linear abscissa.

1. Given the conceptual (two-parameter) normal distribution CDF,

$$F(x) = \frac{1}{\sqrt{2\pi} \cdot csp} \int_{-\infty}^{x} \exp\left[-\frac{1}{2}\left(\frac{u - clp}{csp}\right)^2\right] du$$

   in which u is the dummy variable of integration. Plot its PDF when the actual values for its mean and variance are both equal to 100. In turn, plot its corresponding CDF both on normal probability paper and on rectilinear graph paper (with a linear p scale along its ordinate and a linear x scale along its abscissa).

2. Given the conceptual (two-parameter) smallest-extreme-value distribution CDF:

$$F(x) = 1 - \exp^{-\exp\left(\frac{x - clp}{csp}\right)}$$

   Plot its PDF when the actual values for its mean and variance are both equal to 100. In turn, plot its corresponding CDF both on smallest-extreme-value probability paper and on rectilinear graph paper (with a linear p scale along its ordinate and a linear x scale along its abscissa).

3. Given the conceptual (two-parameter) largest-extreme-value distribution CDF:

$$F(x) = \exp^{-\exp^{-\left(\frac{x - clp}{csp}\right)}}$$

   Plot its PDF when the actual values for its mean and variance are both equal to 100. In turn, plot its corresponding CDF both on largest-extreme-value probability paper and on rectilinear graph paper (with a linear p scale along its ordinate and a linear x scale along its abscissa).

4. Given the conceptual (two-parameter) logistic distribution CDF:

$$F(x) = \left[1 + \exp^{-\left(\frac{x - clp}{csp}\right)} \right]^{-1}$$

Plot its PDF when the actual values for its mean and variance are both equal to 100. In turn, plot its corresponding CDF both on logistic probability paper and on rectilinear graph paper (with a linear p scale along its ordinate and a linear x scale along its abscissa).

5. Given the conceptual two-parameter Weibull distribution CDF:

$$F(x) = 1 - \exp^{-\left(\frac{x}{cdp1}\right)^{cdp2}}$$

Plot its PDF when the actual values for its mean and variance are both equal to 100 ($cdp1 = 104.30376808$ and $cdp2 = 12.15343419$). In turn, plot its corresponding CDF on smallest-extreme-value probability paper (with a linear x scale along its abscissa), on Weibull probability paper [with a linear $\log_e(x)$ scale along its abscissa], and on rectilinear graph paper (with a linear p scale along its ordinate and a linear x scale along its abscissa).

6. Given the conceptual two-parameter log–normal distribution CDF:

$$F(x) = \frac{1}{\sqrt{2\pi} \cdot cdp2} \int_0^x \frac{1}{u} \exp \left\{ -\frac{1}{2} \left[\frac{\log_e(u) - \log_e(cdp1)}{cdp2} \right]^2 \right\} du$$

in which u is the dummy variable of integration. Plot its PDF when the actual values for its mean and variance are both equal to 100 ($cdp1 = 99.50371902$ and $cdp2 = 0.099751345$). In turn, plot its corresponding CDF both on normal probability paper (with a linear x scale along its abscissa), on $\log_e$–normal probability paper [with a linear $\log_e(x)$ scale along its abscissa], and on rectilinear graph paper (with a linear p scale along its ordinate and a linear x scale along its abscissa).

Exercise Set 2

These exercises supplement Exercise Set One and are intended to make specific comparisons between alternative conceptual life (endurance) and strength (resistance) CDF's to enhance your perspective. For each of the required plots, let p cover the range from 0.001 to 0.999 using the values enumerated in Exercise Set One.

1. Given a conceptual (two-parameter) smallest-extreme-value distribution with the actual values for both its mean and variance equal to 100, plot its CDF on normal probability paper. Then, given a conceptual (two-parameter) normal distribution with the actual values for both its mean and variance equal to 100, plot its CDF on smallest-extreme-value probability paper.

2. Given a conceptual (two-parameter) largest-extreme-value distribution with the actual values for both its mean and variance equal to 100, plot its CDF on normal probability paper. Then, given a conceptual (two-parameter) normal distribution with the actual values for both its mean and variance equal to 100, plot its CDF on largest-extreme-value probability paper.

3. Given a conceptual (two-parameter) logistic distribution with the actual values for both its mean and variance equal to 100, plot its CDF on normal probability paper. Then, given a conceptual (two-parameter) normal distribution with the actual values for both its mean and variance equal to 100, plot its CDF on logistic probability paper.

4. Given a conceptual two-parameter Weibull distribution with the actual values for both its mean and variance equal to 100, plot its CDF on $\log_e$–normal probability paper [with a linear $\log_e(x)$ scale along its abscissa]. Then, given a conceptual two-parameter $\log_e$–normal distribution with both its mean and variance equal to 100, plot its CDF on Weibull probability paper [with a linear $\log_e(x)$ scale along its abscissa].

5. Given a conceptual two-parameter Weibull distribution with the actual values for both its mean and variance equal to 100, plot its CDF on normal probability paper (with a linear x scale along its abscissa). Then, given a conceptual two-parameter $\log_e$–normal distribution with both its mean and variance equal to 100, plot its CDF on smallest-extreme-value probability paper (with a linear x scale along its abscissa).

8.10. QUANTITATIVE ANALYSES FOR THE OUTCOMES OF LIFE (ENDURANCE) AND STRENGTH (RESISTANCE) EXPERIMENT TEST PROGRAMS

Only the maximum likelihood (ML) methodology is sufficiently versatile to deal with the outcomes of both statistically planned and ad hoc experiment test programs. This methodology, coupled with the associated likelihood ratio test, provides the backbone of mechanical reliability analysis. We present several examples of quantitative ML analyses in this section. (Recall, however, that the fundamental statistical abstraction of a continually replicated experiment test program is obscure for ad hoc experiment test programs.)

8.11. QUANTITATIVE MAXIMUM LIKELIHOOD ANALYSIS

Likelihood is analogous to probability. Consider a random variable X whose PDF is properly stated as $f(x$ given the respective cdp's). Then, for any collection of n_{dv} values of specific interest for x, the joint PDF g for these x_i's is stated as

$g(x_1, x_2, \ldots, x_{n_{dv}}$ given the respective cdp's)

$$= \prod_{i=1}^{n_{dv}} f(x_i \text{ given the respective } cdp\text{'s})$$

Note that the x_i's are analytical parameters and the actual values for the respective cdp's must be known. In contrast, for the algebraically identical likelihood expression, the x_i's are known (because the experiment test program has been conducted and the x_i's are the resulting datum values), whereas the actual values of the respective cdp's are unknown and are, therefore, regarded as analytical parameters. Accordingly, in ML analysis, the actual values for the respective cdp's are estimated by (analytically or numerically) establishing the particular set of cdp's that maximizes the likelihood (probability) of obtaining (observing) the experiment test program datum values.

Given datum values randomly selected from a conceptual two-parameter statistical distribution whose metric range does not depend on the actual values of its cdp's, the ML estimator is *asymptotically unexcelled*, viz., ML estimators (a) are asymptotically unbiased, (b) have minimum variance, and (c) are normally distributed. However, for experiment test programs with practical sizes, these extraordinary statistical behaviors seldom prevail. Accordingly, simulation-based empirical sampling distributions and simulation-based statistical bias corrections must be generated for each experiment test program of specific interest. (Simulation-based empirical sampling distributions and simulation-based statistical bias corrections for ML analyses are discussed and illustrated in Supplemental Topic 8.D.)

Because it is analytically expedient to work with a sum rather than a product, it is traditional in ML analysis to maximize $\log_e$(likelihood) rather than likelihood. Thus, ML estimation for statistical models with n_{cdp} conceptual distribution parameters traditionally involves a search over n_{cdp}-dimensional conceptual parameter space for the global maximum of the $\log_e$(likelihood). Given the ML estimates of the actual values for the respective conceptual distribution parameters, the second partial derivatives of the $\log_e$(likelihood) expression, evaluated at the respective est(cdp) values, can be used to compute the associated n_{cdp} by n_{cdp} estimated asymptotic covariance matrix. In turn, this matrix can be used to compute, for example, a classical lower $100(scp)\%$ (one-sided) asymptotic statistical confidence limit that allegedly bounds the actual value for the metric that pertains to the pf^{th} percentile of the presumed conceptual life (endurance) distribution of specific interest. [However, it should be remembered that a statistical estimate of the actual value for the metric pertaining to pf equal to 01 (or smaller) is markedly dependent on the presumed CDF, whereas a statistical estimate of the actual value for the metric pertaining to pf equal to 50 (the median) is relatively insensitive to the presumed CDF.]

Convergence problems for numerical maximization procedures increase markedly as n_{cdp} increases. Moreover, the $\log_e$(likelihood) surface in n_{cdp}-dimensional conceptual parameter space is often so flat in the vicinity of its apparent maximum that classical $100(scp)\%$ (two-sided) asymptotic statistical confidence intervals are too wide to be practical. Thus, while ML analysis is technically unbounded relative to the number of parameters that can be included in the conceptual statistical distribution (model), common sense dictates that the number of est(cdp's) be kept to a minimum.

Bartlett's likelihood ratio (LR) test statistic should be used to examine the statistical adequacy of each proposed conceptual statistical model, just as Snedecor's central F test statistic should be used to test the statistical adequacy of a conceptual linear regression model. Suppose the adequacy of a conceptual two-parameter statistical model is to be tested versus an alternative conceptual three-parameter statistical model. Then, the null hypothesis is that the conceptual two-parameter statistical model is correct, whereas the alternative hypothesis is that the conceptual three-parameter statistical model is correct. When ML analyses are conducted for these alternative models, the respective two estimated values for the maximized $[\log_e$(likelihood)]'s establish the experiment test program data-based realization value for Bartlett's LR test statistic, viz.,

$$\text{Bartlett's LR test statistic} = -2 \cdot \log_e \left(\frac{\text{est(ML)}_{n_{cp}=2}}{\text{est(ML)}_{n_{cp}=3}} \right)$$

$$= 2 \cdot \left\{ \log_e \left[\text{est(ML)}_{n_{cp}=3} \right] - \log_e \left[\text{est(ML)}_{n_{cp}=2} \right] \right\}$$

Then, under continual replication of the experiment test program the respective data-based realization values for Bartlett's LR test statistic asymptotically generate Pearson's central $\chi^2_{n_{sdf}=3-2=1}$ conceptual sampling distribution. However, in general, Bartlett's LR test statistic is expressed as

$$\text{Bartlett's LR test statistic} = -2 \cdot \log_e\left(\frac{\text{est(ML)}_{n_{cpH_n}}}{\text{est(ML)}_{n_{cpH_a}}}\right)$$

$$= 2 \cdot \left\{\log_e\left[\text{est(ML)}_{n_{cpH_a}}\right] - \log_e\left[\text{est(ML)}_{n_{cpH_n}}\right]\right\}$$

Then, under continual replication of the experiment test program the respective data-based realization values for Bartlett's LR test statistic asymptotically generate Pearson's central $\chi^2_{n_{sdf}=n_{cpH_a}-n_{cpH_n}}$ conceptual sampling distribution. (A numerical example of testing the adequacy of a proposed statistical model using Bartlett's LR test is found in Supplemental Topic 8.F.)

Both the engineering and statistics literature is sadly deficient in terms of proposing alternative conceptual reliability models and then comparing the respective analyses. Nevertheless, all reasonable alternative models should be employed in statistical analysis and an engineering decision should be made only after comparing the respective analyses.

Bartlett's LR test statistic can also be used to compute asymptotic statistical confidence intervals that are viable alternatives to classical asymptotic statistical confidence intervals. Accordingly, Bartlett's LR test is the most versatile and useful statistical tool available in mechanical reliability analysis.

8.12. QUANTITATIVE MAXIMUM LIKELIHOOD EXAMPLES

We now present four examples of quantitative ML analyses. The first example pertains to ML analysis of datum values generated by a series of independent strength (resistance) tests that were conducted with the same stress (stimulus). The second example pertains to ML analysis of datum values generated in a series of independent life (endurance) tests with *Type I* censoring and presumes that these datum values were randomly selected from a conceptual two-parameter Weibull distribution. These first two examples are intended to illustrate the fundamentals of ML analysis. In contrast, the last two examples are intended to illustrate typical applications. The third example pertains to the outcome of a s_u–$\log_e(fnc)$ experiment test program with *Type I* censoring, whereas the fourth example pertains to the outcome of a strength (resistance) experiment test program that was conducted using the up-and-down test strategy.

Comparative ML analyses are presented in Section 8.13. These analyses can be used either to establish statistically whether or not treatment effects (or batch-to-batch effects) exist or whether or not the presumed conceptual model is statistically adequate to explain the observed experiment test program datum values. (Recall that in contrast to quantitative analyses, statistical bias corrections are not of paramount importance in comparative analyses.)

8.12.1. Example One: ML Analysis for Outcome of a Strength (Resistance) Experiment Test Program

Consider the outcome of a strength (resistance) experiment test program that consists of n_{st} independent strength (resistance) tests conducted at the same stress (stimulus) level. Since each test item has a priori the same (unknown) probability ps of surviving a predetermined duration d^*, each of the respective strength (resistance) tests is statistically viewed as being a binomial trial. Next, let $y = 1$ connote that the test item survived some predetermined duration d^* and let $y = 0$ connote that the test item failed prior to duration d^*. Then, the conceptual joint PDF for the respective n_{st} independent strength (resistance) tests can be written as

$$\text{conceptual joint PDF} = \frac{n_{st}!}{(n_{is})!(n_{st} - n_{is})!} \prod_{i=1}^{n_{st}} ps^{y_i}(1 - ps)^{(1 - y_i)}$$

$$= \frac{n_{st}!}{(n_{is})!(n_{st} - n_{is})!} ps^{n_{is}}(1 - ps)^{(n_{st} - n_{is})}$$

where n_{is} is the number of test items that survived the predetermined duration d^*, viz., the number of y_i's that are equal to 1, and ps is the fixed probability of surviving for at least the predetermined duration d^* in each respective reliability test. This conceptual joint PDF is subsequently reinterpreted, without analytical change, as the likelihood expression, viz.,

$$\text{likelihood} = \frac{n_{st}!}{(n_{is})!(n_{st} - n_{is})!} ps^{n_{is}}(1 - ps)^{(n_{st} - n_{is})}$$

Although this product expression for likelihood is directly amenable to numerical maximization, analytical maximization procedures have traditionally dominated. Accordingly, the product expression for likelihood is converted into a summation expression for $\log_e(\text{likelihood})$, viz.,

$$\log_e(\text{likelihood}) = n_{is} \log_e(ps) + (n_{st} - n_{is})[\log_e(1 - ps)]$$

in which the nonessential factorial expression has been ignored. [Recall that the natural logarithm is a monotonic transformation. Thus, the likelihood and the $\log_e$(likelihood) reach their respective maximums simultaneously.]

The maximum value of the $\log_e$(likelihood) is obtained by equating the derivative of the $\log_e$(likelihood) expression with respect to ps equal to zero, viz.,

$$\frac{d\left[\log_e(\text{likelihood})\right]}{d(ps)} = 0 = \frac{n_{is}}{ps} - \frac{n_{st} - n_{is}}{1 - ps}$$

which, when solved analytically for ps, yields the ML estimate:

$$\text{ML est}(ps) = \frac{n_{is}}{n_{st}}$$

This ML estimate clearly agrees with our intuition. Nevertheless, it is criticized for allowing estimates of the actual value for ps to be equal to either zero or one.

We now establish an intuitive $100(scp)\%$ (two-sided) statistical confidence interval that allegedly includes the actual value for the fixed probability of survival, ps, in each respective reliability test. First, however, we revise our notation to conform to a generic $100(scp)\%$ (two-sided) statistical confidence interval pertaining to a series of binomial trials. Accordingly, n_{st} now becomes n_{bt}, the number of binomial trials; n_{is} now becomes n_{fo}, the number of favorable outcomes; and ps now becomes pfo, the fixed probability of a favorable outcome in each independent binomial trial. The intuitive way to compute a $100(scp)\%$ (two-sided) statistical confidence interval that allegedly includes the actual value for the pfo is to compute two related probability values, one so low that the probability of observing n_{fo} or more favorable outcomes in n_{bt} binomial trials is equal to $[(1 - scp)/2]$, and one so high that the probability of observing n_{fo} or fewer favorable outcomes in n_{bt} binomial trials is also equal to $[(1 - scp)/2]$. Then, the statistical confidence probability that the actual value of conceptual binomial probability pfo lies in the interval from pfo_{low} to pfo_{high} is equal to $\{1 - [(1 - scp)/2] - [(1 - scp)/2]\} = scp$. Microcomputer program *IBPSCI* (intuitive binomial probability statistical confidence interval) calculates this intuitive $100(scp)\%$ (two-sided) statistical confidence interval as $[0.4439, 0.9748]$ when the scp is selected to be equal to 0.95. For comparison, the corresponding LR-based 95% (two-sided) statistical confidence interval (Figure 8.7) is $[0.5006, 0.9636]$.

> *Remark*: See Natrella (1963) for an analogous $100(scp)\%$ (two-sided) statistical confidence interval that is shorter than our intuitive $100(scp)\%$ (two-sided) statistical confidence interval.

```
C> IBPSCI

Input the number of binomial trials of specific interest

10

Input the number of favorable outcomes of specific interest

8

Input the statistical confidence probability of specific interest in per
cent (integer value)

95
```

The intuitive 95% (two-sided) statistical confidence interval that allegedly includes the actual value for the *pfo* is [0.4439, 0.9748].

Classical $100(scp)\%$ (two-sided) statistical confidence intervals are
computed using the estimated asymptotic covariance matrix. The individual
elements of the estimated asymptotic covariance matrix prior to its inversion
are negatives of the second and mixed partial derivatives of the $\log_e$(likelihood) expression with respect to the conceptual distribution parameters.
For the discrete (one-parameter) conceptual binomial distribution, this estimated asymptotic covariance matrix has a single element, viz., the variance
of the asymptotic conceptual sampling distribution that consists of all possible replicate realization values for the statistic [est(*pfo*)]. Thus, est{var[-
est(*pfo*)]} is the negative of the reciprocal of the second derivative of the
$\log_e$(likelihood) with respect to *pfo*. Accordingly, we first take the second
derivative of the generic $\log_e$likelihood expression with respect to *pfo*, viz.,

$$\frac{d^2[\log_e(\text{likelihood})]}{d(pfo)^2} = -\frac{n_{fo}}{pfo^2} - \frac{n_{bt} - n_{fo}}{(1 - pfo)^2}$$

and then we evaluate this second derivative expression in theory at *pfo*.
However, because *pfo* is unknown, this second derivative expression must

```
C> LRBBPSCI

Input the number of binomial trials of specific interest

10

Input the number of favorable outcomes of specific interest

8

Input the statistical confidence probability of specific interest in per
cent (integer value)

95

The LR 95% (two-sided) asymptotic statistical confidence interval
that allegedly includes the actual value for the pfo is [0.5006, 0.9636].
```

be evaluated by substituting $\text{est}(pfo) = n_{fo}/n_{bt}$ for pfo. This substitution gives

$$\frac{d^2[\log_e(\text{likelihood})]}{d(pfo)^2} = -\frac{n_{bt}}{\text{est}(pfo)[1 - \text{est}(pfo)]}$$

Finally, we take the inverse of the negative of this second derivative expression to generate the desired estimated variance expression, viz.,

$$\text{ML est}\{\text{var}[\text{est}(pfo)]\} = \frac{\text{est}(pfo)[1 - \text{est}(pfo)]}{n_{bt}}$$

Comparing this estimated asymptotic variance expression to the corresponding exact variance expression developed in Supplemental Topic 3.B, we see that it becomes exact as n_{bt} increases without bound, viz., when $\text{est}(pfo) = pfo$. In turn, using the conceptual (two-parameter) normal distribution asymptotic approximation to the conceptual (one-parameter) binomial distribution (Figure 3B.2), we compute the classical 95% (two-sided) asymptotic statistical confidence interval that allegedly includes the actual value for the *pfo* as

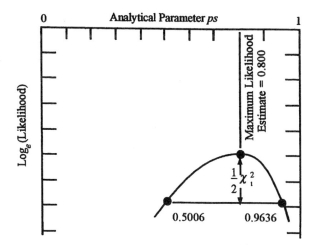

Figure 8.7 The $\log_e$(likelihood) for $n_{st} = 10$ independent strength (resistance) tests with $n_{is} = 8$ test items surviving these n_{st} strength (resistance) tests, plotted versus all alternative possible values for the probability of surviving, ps. Microcomputer program *LRBBPSCI* employs the numerical likelihood ratio (LR) method to establish the respective limits of the 95% (two-sided) asymptotic statistical confidence interval that allegedly includes the actual value for ps. Recall that under continual replication of the experiment test program the respective realizations values for the maximized $\log_e$(likelihood) statistic asymptotically generate Pearson's central $\chi^2_{n_{sdf}=1}$ conceptual sampling distribution.

$$\left[0.8 - 1.96\left(\frac{(0.8)(0.2)}{10}\right)^{1/2}, \; 0.8 + 1.96\left(\frac{(0.8)(0.2)}{10}\right)^{1/2} \right] = [0.5521, 1.0479]$$

Since the maximum actual value of the *pfo* is 1.0 by definition, the conceptual (two-parameter) normal distribution asymptotic approximation to the (one-parameter) binomial distribution is clearly unacceptable for the small number of replicate tests in our numerical example.

Remark: This elementary example clearly illustrates that alternative $100(scp)$% (two-sided) statistical confidence intervals can differ markedly. Accordingly, good statistical practice requires that alternative intervals be computed and compared.

8.12.2. Example Two: ML Analysis for Outcome of a Life (Endurance) Experiment Test Program with *Type I* Censoring, Presuming Weibull Datum Values

Consider the outcome of a life (endurance) experiment test program such that n_f test items failed and n_s test items were *Type I* censored (suspended) at $snc = snc^*$. Suppose it is presumed that the respective datum values were randomly selected from a conceptual two-parameter Weibull distribution whose CDF is expressed as

$$F(fnc) = 1 - \exp^{-\left(\frac{fnc}{cdp1}\right)^{cdp2}}$$

The corresponding conceptual PDF is then expressed as

$$f(fnc) = \frac{cdp2}{cdp1^{cdp2}} \cdot fnc^{(cdp2-1)} \cdot \exp^{-\left(\frac{fnc}{cdp1}\right)^{cdp2}}$$

The likelihood expression is the product of two likelihood expressions when the respective life (endurance) tests are all mutually independent: (a) the likelihood expression that pertains to the respective fnc_i's, and (b) the likelihood expression that pertains to the n_s *Type I* censored tests. Accordingly,

$$\text{likelihood} = \frac{(n_f + n_s)!}{n_f! n_s!} \prod_{i=1}^{n_f} f(fnc_i) \cdot \prod_{j=1}^{n_s} [1 - F(snc^*)]$$

in which the nonessential factorial expression enumerates the equally-likely time orders of generating n_f failed test items failed and n_s *Type I* test items. This likelihood expression can be rewritten as

$$\text{likelihood} = \frac{cdp2^{n_f}}{cdp1^{n_f \cdot cdp2}} \cdot \prod_{i=1}^{n_f} (fnc_i)^{cdp2-1} \cdot \prod_{i=1}^{n_f} \exp^{-\left(\frac{fnc_i}{cdp1}\right)^{cdp2}}$$
$$\cdot \prod_{j=1}^{n_s} \exp^{-\left(\frac{snc^*}{cdp1}\right)^{cdp2}}$$

The corresponding $\log_e$(likelihood) expression is

$$\log_e(\text{likelihood}) = n_f \cdot \log_e(cdp2) - n_f \cdot cdp2 \log_e(cdp1)$$

$$+ (cdp2 - 1) \cdot \sum_{i=1}^{n_f} \log_e(fnc_i) - \sum_{i=1}^{n_f} \left(\frac{fnc_i}{cdp1}\right)^{cdp2} - \sum_{j=1}^{n_s} \left(\frac{snc^*}{cdp1}\right)^{cdp2}$$

The last two terms in this $\log_e$ likelihood expression can be combined under a summation that pertains to all experiment test program datum values. Accordingly, we subsequently use the index k for all test items, where $n_{lt} = n_f + n_s$ and $k = 1$ to n_{lt}, the number of individual life (endurance) tests in the experiment test program. Correspondingly, we include both fnc_i and snc^* in anc_k.

The conceptual distribution parameters $cdp1$ and $cdp2$ can then be estimated by simultaneously solving the partial derivative expressions:

$$\frac{\partial \log_e(\text{likelihood})}{\partial cdp1} = 0 = -n_f \cdot \frac{cdp2}{cdp1} + \frac{cdp2}{cdp1} \cdot \sum_{k=1}^{n_{rlt}} \left(\frac{anc_k}{cdp1}\right)^{cdp2}$$

and

$$\frac{\partial \log_e(\text{likelihood})}{\partial cdp2} = 0 = \frac{n_f}{cdp2} - n_f \cdot \log_e(cdp1) + \sum_{i=1}^{n_f} \log_e(fnc_i)$$

$$- \sum_{k=1}^{n_{rlt}} \left[\left(\frac{anc_k}{cdp1}\right)^{cdp2} \cdot \log_e\left(\frac{anc_k}{cdp1}\right) \right]$$

or, if so desired, the second and third terms in the latter partial derivative expression can be combined under a single summation, viz.,

$$-n_f \log_e(cdp1) + \sum_{i=1}^{n_f} \log_e(fnc_i) = \sum_{i=1}^{n_f} \log_e\left(\frac{fnc_i}{cdp1}\right)$$

The two resulting nonlinear partial derivative equations, whatever their algebraic form, are conveniently solved numerically by using the iterative Newton–Raphson (N–R) methodology in which these two equations are simultaneously expanded in a Taylor's series and higher order terms are ignored, viz.,

$$0 = \frac{\partial \log_e(\text{likelihood})}{\partial cdp1} + \frac{\partial^2 \log_e(\text{likelihood})}{\partial cdp1^2} \cdot \Delta(cdp1)$$

$$+ \frac{\partial^2 \log_e(\text{likelihood})}{\partial cdp1 \partial cdp2} \cdot \Delta(cdp2)$$

and

$$0 = \frac{\partial \log_e(\text{likelihood})}{\partial cdp2} + \frac{\partial^2 \log_e(\text{likelihood})}{\partial cdp2\, \partial cdp1} \cdot \Delta(cdp1)$$

$$+ \frac{\partial^2 \log_e(\text{likelihood})}{\partial cdp2^2} \cdot \Delta(cdp2)$$

which are subsequently viewed simply as two equations in two unknowns, $\Delta(cdp1)$ and $\Delta(cdp2)$. Given initial numerical estimates for the $cdp1$ and $cdp2$, these analytical equations become two numerical equations that are easily solved for $\Delta(cdp1)$ and $\Delta(cdp2)$. These two Δ's are first-order corrections to the initial estimates for the $cdp1$ and $cdp2$. In turn, these improved (corrected) estimates of the actual values for the $cdp1$ and $cdp2$ can be further improved (corrected) using the iterative N–R solution methodology, viz.,

$cdp1$ (first iteration) $= cdp1$ (initial estimate)

$+$ N–R $\Delta(cdp1)$ numerical correction

and

$cdp2$ (first iteration) $= cdp2$ (initial estimate)

$+$ N–R $\Delta(cdp2)$ numerical correction

In turn,

$cdp1$ (second iteration) $= cdp1$ (first iteration)

$+$ new N–R $\Delta(cdp1)$ numerical correction

and

$cdp2$ (second iteration) $= cdp2$ (first iteration)

$+$ new N–R $\Delta(cdp2)$ numerical correction

etc.

The iterative N–R estimation algorithm should continue until successive iterations produce absolute values for both $\Delta(cdp$'s) numerical corrections less than about 10^{12}, at which time, the numerical values for $cdp1$ and $cdp2$ are regarded as the respective maximum likelihood estimates, viz., as $est(cdp1)$ and $est(cdp2)$

The primary advantage of the iterative N–R methodology in ML analysis is that numerical values for the second partial derivatives are directly available for subsequent use in computing the estimated asymptotic covariance matrix, viz., the negatives of the numerical values of the respective second partial derivatives form elements of a 2 by 2 symmetrical array

```
C> TYPE WBLDATA
```

5   Number of Items Failed, Followed by the Respective Failure
           Number of Cycles (kilocycles)
277
310
374
402
456
1   Number of *Type I* Censored Tests, Followed by the *Type I*
           Censored Number of Cycles (kilocycles)
500
95   Statistical Confidence Probability of Specific Interest in Per
           Cent (Integer Value, 95 Maximum)
01   Conceptual CDF Percentile of Specific Interest in Per Cent
           (Integer Value)

whose inverse is the estimated asymptotic covariance matrix. In turn, given the estimated asymptotic covariance matrix, the propagation of variability methodology can be used to compute the classical lower $100(scp)$% (one-sided) asymptotic statistical confidence limit that allegedly bounds the actual value for the Weibull CDF percentile of specific interest.

Microcomputer program *WEIBULL* presumes a conceptual two-parameter Weibull distribution. It computes two alternative ML-based classical lower $100(scp)$% (one-sided) asymptotic statistical confidence limits (*closascl*'s) that allegedly bound the actual value for *fnc(pf)*. The first is based on the propagation of the variability expression for est[var(est{$\log_e$[*fnc (pf)*]})]. It computes the value for *fnc* such that the asymptotic probability is equal to *scp* that the actual value for *fnc(pf)* is greater than this value. The second is based on the propagation of variability expression for est{var[est(*y* given *fnc* = *closascl*)]}. It is akin to inverse regression, viz., it computes the value for *y* given *fnc* = *closascl* such that the asymptotic probability is equal to *scp* that the actual value for *y(pf)* is less than this value. Although these two alternative statistical confidence limits are asymptotically identical, the second is always smaller and more accurate for finite n_{dv}. However, even this smaller more accurate statistical confidence limit requires a statistical bias correction for experiment test programs of practical sizes. (See Supplemental Topic 8.D.)

Remark: If the life (endurance) experiment test program includes arbitrarily suspended tests, then the respective (identical) $snc*$'s are merely replaced by the respective (different) snc_j's in microcomputer file *WBLDATA*. Recall, however, that the fundamental statistical concept of a continually replicated experiment test program is obscure when arbitrarily suspended tests occur.

```
C> COPY WBLDATA DATA

    1 files(s) copied

C> WEIBULL
```

$$\text{Given } F(fnc) = 1 - \exp^{-\left(\frac{fnc}{cdp1}\right)^{cdp2}}$$

est($cdp1$) = 0.4289595976D + 03
est($cdp2$) = 0.4731943406D + 01
est{var[est($cdp1$)]} = 0.1666932095D + 04
est{var[est($cdp2$)]} = 0.3013638227D + 01
est{covar[est($cdp1$),est($cdp2$)]} = 0.8393830857D + 01
est(*conceptual correlation coefficient*) = 0.1184283562D + 00

fnc	est(y)	est(pf)
277.000	−2.0694929	0.1186053
310.000	−1.5368900	0.1934980
374.000	−0.6487823	0.4070717
402.000	−0.3071535	0.5207523
456.000	0.2892640	0.7369587

snc	est(y)	est(pf)
500.000	0.7251484	0.8731865

Classical Lower 95% (One-Sided) Asymptotic Statistical Confidence Limits that Allegedly Bound the Actual Value for $fnc(01)$
86.868 – Computed Using the Propagation of Variability Expression
for est[var(est{$\log_e[fnc(01)]$})]
34.584 – Computed Using the Propagation of Variability Expression
for est{var[est(y given $fnc = closascl$)]}

8.12.2.1. Discussion

The Weibull distribution CDF can be expressed in eight different parameterizations. We present the four most common parameterizations below:

(1) $\quad F(fnc) = 1 - \exp^{-\left(\frac{fnc}{cdp1}\right)^{cdp2}}$

(2) $\quad F(fnc) = 1 - \exp^{-cdp1 \cdot (fnc)^{cdp2}}$

(3) $\quad F(fnc) = 1 - \exp^{-\left(\frac{fnc}{cdp1}\right)^{\frac{1}{cdp2}}}$

(4) $\quad F(fnc) = 1 - \exp^{-\left(\frac{1}{cdp1}\right) \cdot (fnc)^{\frac{1}{cdp2}}}$

The corresponding four parameterizations for the conceptual (two-parameter) $\log_e$ smallest-extreme-distributions are employed in ML analyses given by microcomputer programs *LSEV1A*, *LSEV2A*, *LSEV3A*, and *LSEV4A* (pages 424–427). The respective outputs for these four programs demonstrate that (a) ML estimates are the same regardless of the parameterization selected for the CDF, and (b) the associated classical lower $100(scp)\%$ (one-sided) asymptotic statistical confidence limits that allegedly bound the actual value for any $fnc(pf)$ of specific interest are the same when these intervals are computed using propagation of variance methodology.

8.12.2.2. Perspective

Supplemental Topic 8.B presents the analytical details of the analogous ML analysis that pertains to a conceptual two-parameter $\log_e$–normal distribution. Given the same set of life (endurance) datum values, the respective $\{est[fnc(pf)]\}$'s for small values of pf pertaining to the analogous conceptual two-parameter $\log_e$–normal distribution are larger than for the conceptual two-parameter Weibull distribution. Thus, quantitative life (endurance) estimates based on the conceptual two-parameter Weibull distribution are safer than corresponding quantitative reliability estimates based on the conceptual two-parameter $\log_e$–normal distribution—which is the only rational reason for preferring the Weibull distribution over the $\log_e$–normal distribution.

From a statistical point of view, the conceptual two-parameter Weibull and $\log_e$–normal distributions are interchangeable for experiment test program sizes that are practical in mechanical reliability applications.

```
C> COPY WBLDATA DATA

    1 files(s) copied

C> LSEV1A
```

Given the Standardized Conceptual Smallest-Extreme-Value Distribution Variate $y = csp[\log_e(fnc) - clp]$

est$(clp) = 0.6061362736D + 01$
est$(csp) = 0.4731943406D + 01$
est$\{$var[est$(clp)]\} = 0.9059101593D–02$
est$\{$var[est$(csp)]\} = 0.3013638227D + 01$
est$\{$covar[est(clp),est$(csp)]\} = 0.1956788216D–01$
est$(conceptual\ correlation\ coefficient) = 0.1184283562D + 00$

fnc	est(y)	est(pf)
277.000	−2.0694929	0.1186053
310.000	−1.5368900	0.1934980
374.000	−0.6487823	0.4070717
402.000	−0.3071535	0.5207523
456.000	0.2892640	0.7369587

snc	est(y)	est(pf)
500.000	0.7251484	0.8731865

Classical Lower 95% (One-Sided) Asymptotic Statistical Confidence Limits that Allegedly Bound the Actual Value for $fnc(01)$

86.868 – Computed Using the Propagation of Variability Expression for est[var(est$\{\log_e[fnc(01)]\})$)]
34.584 – Computed Using the Propagation of Variability Expression for est$\{$var[est$(y$ given $fnc = closascl)]\}$

The ratio of the respective estimated ML's was used as a test statistic by Dumonceaux and Antle (1973) in an attempt to discern between these two distributions statistically. Their simulation-based study demonstrated that, even given an acceptable probability of committing a *Type I* error as large as 0.10, a sample size of almost 50 is required to have a probability equal to 0.90 of correctly rejecting the wrong conceptual statistical distribution when it is incorrectly taken as the null hypothesis.

C> COPY WBLDATA DATA

    1 files(s) copied

C> LSEV2A

Given the Standardized Conceptual Smallest-Extreme-Value Distribution Variate $y = clp + csp\log_e(fnc)$

est(clp) = 0.2868202543D + 02
est(csp) = 0.4731943406D + 01
est{var[est(clp)]} = 0.1120467627D + 03
est{var[est(csp)]} = 0.3013638227D + 01
est{covar[est(clp),est(csp)]} = −0.1835934856D + 02
est(*conceptual correlation coefficient*) = 0.9991071169D + 00

fnc	est(y)	est(pf)
277.000	−2.0694929	0.1186053
310.000	−1.5368900	0.1934980
374.000	−0.6487823	0.4070717
402.000	−0.3071535	0.5207523
456.000	0.2892640	0.7369587

snc	est(y)	est(pf)
500.000	0.7251484	0.8731865

Classical Lower 95% (One-Sided) Asymptotic Statistical Confidence Limits that Allegedly Bound the Actual Value for $fnc(01)$

86.868 - Computed Using the Propagation of Variability Expression for est[var(est{$\log_e[fnc(01)]$})]
34.584 - Computed Using the Propagation of Variability Expression for est{var[est(y given $fnc = closascl$)]}

```
C> COPY WBLDATA DATA

    1 files(s) copied

C> LSEV3A
```

Given Standardized Conceptual Smallest-Extreme-Value Distribution Variate $y = [\log_e(fnc) - clp]/csp$

$\text{est}(clp) = 0.6061362736\text{D}+01$ $\text{est}(csp) = 0.2113296619\text{D}+00$
$\text{est}\{\text{var}[\text{est}(clp)]\} = 0.9059101593\text{D}{-}02$
$\text{est}\{\text{var}[\text{est}(csp)]\} = 0.6010809287\text{D}{-}02$
$\text{est}\{\text{covar}[\text{est}(clp),\text{est}(csp)]\} = -0.8739060394\text{D}{-}03$
$\text{est}(conceptual\ correlation\ coefficient) = -0.1184283562\text{D}\mid00$

fnc	$\text{est}(y)$	$\text{est}(pf)$
277.000	−2.0694929	0.1186053
310.000	−1.5368900	0.1934980
374.000	−0.6487823	0.4070717
402.000	−0.3071535	0.5207523
456.000	0.2892640	0.7369587

snc	$\text{est}(y)$	$\text{est}(pf)$
500.000	0.7251484	0.8731865

Classical Lower 95% (One-Sided) Asymptotic Statistical Confidence Limits that Allegedly Bound the Actual Value for $fnc(01)$

86.868 – Computed Using the Propagation of Variability Expression for $\text{est}[\text{var}(\text{est}\{\log_e[fnc(01)]\})]$

34.584 – Computed Using the Propagation of Variability Expression for $\text{est}\{\text{var}[\text{est}(y\ \text{given}\ fnc = closascl)]\}$

Exercise Set 3

These exercises are intended to use propagation of variability expressions (Supplemental Topic 7.A) to verify numerically the analytical relationships between (a) the respective $\text{est}(cdp\text{'s})$ for the conceptual two-parameter Weibull distribution and the corresponding conceptual (two-parameter) smallest-extreme-value distributions with a logarithmic metric, and (b) the elements of their respective estimated asymptotic covariance matrices. Use at least eight digits in your calculations to avoid so-called round-off errors.

C> COPY WBLDATA DATA

    1 files(s) copied

C> LSEV4A

Given the Standardized Conceptual Normal Distribution Variate
$y = \{[\log_e(fnc)]/csp\} - clp$

$\text{est}(clp) = 0.2868202543D+02$ $\text{est}(csp) = 0.2113296619D+00$
$\text{est}\{\text{var}[\text{est}(clp)]\} = 0.1120467627D+03$
$\text{est}\{\text{var}[\text{est}(csp)]\} = 0.6010809287D{-}02$
$\text{est}\{\text{covar}[\text{est}(clp),\text{est}(csp)]\} = -0.8199326557D+00$
$\text{est}(\textit{conceptual correlation coefficient}) = -0.9991071169D+00$

fnc	$\text{est}(y)$	$\text{est}(pf)$
277.000	−2.0694929	0.1186053
310.000	−1.5368900	0.1934980
374.000	−0.6487823	0.4070717
402.000	−0.3071535	0.5207523
456.000	0.2892640	0.7369587

snc	$\text{est}(y)$	$\text{est}(pf)$
500.000	0.7251484	0.8731865

Classical Lower 95% (One-Sided) Asymptotic Statistical Confidence
Limits that Allegedly Bound the Actual Value for $fnc(01)$

86.868 – Computed Using the Propagation of Variability Expression
for $\text{est}[\text{var}(\text{est}\{\log_e[fnc(01)]\})]$
34.584 – Computed Using the Propagation of Variability Expression
for $\text{est}\{\text{var}[\text{est}(y \text{ given } fnc = closascl)]\}$

1. Given the numerical values for the respective est(cdp)'s in the
   example microcomputer program *WEIBULL* output, verify the
   corresponding numerical values for the respective est(cdp)'s
   given in the example output for microcomputer program (a)

LSEV1A, (b) *LSEV2A*, (c) *LSEV3A*, or (d) *LSEV4A*. Then, given the numerical values for the respective elements of the estimated asymptotic covariance matrix in the *WEIBULL* output, verify the corresponding numerical values for the respective elements of the estimated asymptotic covariance matrix given in the (a) *LSEV1A*, (b) *LSEV2A*, (c) *LSEV3A*, or (d) *LSEV4A* output.

2. Given the numerical values for the respective est(cdp)'s in the example microcomputer program *LSEV1A* output, verify the corresponding numerical values for the est(cdp)'s given in the example output for microcomputer program (a) *WEIBULL*, (b) *LSEV2A*, (c) *LSEV3A*, or (d) *LSEV4A*. Then, given the numerical values for the respective elements of the estimated asymptotic covariance matrix in the *LSEV1A* output, verify the corresponding numerical values for the respective elements of the estimated asymptotic covariance matrix given in the (a) *WEIBULL*, (b) *LSEV2A*, (c) *LSEV3A*, or (d) *LSEV4A* output.

3. Given the numerical values for the respective est(cdp)'s in the example microcomputer program *LSEV2A* output, verify the corresponding numerical values for the est(cdp)'s given in the example output for microcomputer program (a) *WEIBULL*, (b) *LSEV1A*, (c) *LSEV3A*, or (d) *LSEV4A*. Then, given the numerical values for the respective elements of the estimated asymptotic covariance matrix in the *LSEV2A* output, verify the corresponding numerical values for the respective elements of the estimated asymptotic covariance matrix given in the (a) *WEIBULL*, (b) *LSEV1A*, (c) *LSEV3A*, or (d) *LSEV4A* output.

4. Given the numerical values for the respective est(cdp)'s in the example microcomputer program *LSEV3A* output, verify the corresponding numerical values for the est(cdp)'s given in the example output for microcomputer program (a) *WEIBULL*, (b) *LSEV1A*, (c) *LSEV2A*, or (d) *LSEV4A*. Then, given the numerical values for the respective elements of the estimated asymptotic covariance matrix in the *LSEV3A* output, verify the corresponding numerical values for the respective elements of the estimated asymptotic covariance matrix given in the (a) *WEIBULL*, (b) *LSEV1A*, (c) *LSEV2A*, or (d) *LSEV4A* output.

5. Given the numerical values for the respective est(cdp)'s in the example microcomputer program *LSEV4A* output, verify the corresponding numerical values for the est(cdp)'s given in the example output for microcomputer program (a) *WEIBULL*, (b) *LSEV1A*, (c) *LSEV2A*, or (d) *LSEV3A*. Then, given the numer-

ical values for the respective elements of the estimated asymptotic covariance matrix in the *LSEV4A* output, verify the corresponding numerical values for the respective elements of the estimated asymptotic covariance matrix given in the (a) *WEIBULL*, (b) *LSEV1A*, (c) *LSEV2A*, or (d) *LSEV3A* output.

8.12.3. Example Two (Extended): ML Analysis for Outcome of a Life (Endurance) Experiment Test Program with Competing Modes of Failure

Mechanical devices (and sometimes even their components) can exhibit competing modes of failure. Then, for each life (endurance) failure datum value pertaining to a specific mode of failure, a suspended test datum value is created at that failure life (endurance) value for each of its competing modes of failure. In turn, if all competing modes of failure are presumed to be mutually independent, the desired likelihood expression is simply the product of the respective individual likelihood expressions. However, the presumption that all competing modes of failure are mutually independent is usually doubtful. Moreover, the notion of a continually replicated experiment test program suffers markedly, because the respective suspension times are not statistically planned. Thus, ML analysis for mutually independent competing models of failure has both practical and statistical problems. Nevertheless, it is statistically more credible than an analysis that lumps all competing modes of failure into a single failure model.

Suppose that an alternator can fail to operate properly only by the failure of its front bearing, or its near bearing, or its electrical diode. Presuming that the respective service failures are all mutually independent, the likelihood expression for alternator failure is the product of the respective individual likelihood expressions, viz.,

likelihood (alternator failure) = [likelihood(front bearing failures)]

· [likelihood(rear bearing failures)]

· [likelihood(electrical diode failures)]

· [likelihood(front bearing suspensions)]

· [likelihood(rear bearing suspensions)]

· [likelihood(electrical diode suspensions)]

in which, for example,

[likelihood(electrical diode failures)] · [likelihood(electrical diode suspensions)]

$$= \prod_{i=1}^{n_{edf}} f[fnc(\text{electrical diode}_i)] \cdot \prod_{j=1}^{n_{eds}} \left(1 - F[snc(\text{electrical diode}_j)]\right)$$

where n_{edf} is the number of alternators with electrical diode failures and n_{eds} is the number of alternators whose electrical diode did not fail (because one or the other of the two bearings failed first or because the alternator itself did not fail during the experiment test program). Numerically n_{eds} is equal to the sum of the number of alternators with front bearing failures, n_{fbf}, plus the number of alternators with rear bearing failures, n_{rbf}, plus the number of alternators still operating satisfactorily when the experiment test program was suspended, n_{as}. The associated electrical diode snc's are the fnc's for the front and rear bearings and the snc's for the alternators still operating satisfactorily when the experiment test program was suspended.

Recall that all competing modes of failure for the mechanical device of specific interest are presumed to be mutually independent. Accordingly, the respective partial derivatives of the $\log_e$(likelihood) expression with respect to the parameters of each conceptual failure mode distribution are nonzero *only* for that failure mode. Thus, ML analysis for mutually independent modes of failure is, for practical purposes, merely the amalgamation of ML analyses similar to the ML Weibull analysis presented in Section 8.12.1. In fact, if the duration to failure, say fnc, for n_m mutually independent competing modes of failure were modeled using conceptual two-parameter Weibull distributions, then microcomputer program *WEIBULL* could be run n_m times consecutively, each time with the appropriate failure and aggregate suspension durations, and the resulting estimated CDF would be given by the expression:

$$\text{est}[F(fnc)] = 1 - (\{1 - \text{est}[F(fnc)_{\text{mode }1}]\} \cdot \{1 - \text{est}[F(fnc)_{\text{mode }2}]\} \cdots$$
$$\cdot \{1 - \text{est}[F(fnc)_{\text{mode }n_m}]\})$$

Unfortunately, classical lower $100(scp)\%$ (one-sided) asymptotic statistical confidence limits analogous to those given in microcomputer program *WEIBULL* cannot be computed using propagation of variability. However, given the value for pf of specific interest, a lower $100(scp)\%$ (one-sided) asymptotic statistical confidence limit that allegedly bounds the actual value for $fnc(pf)$ can be approximated numerically.

The failure duration for each mode of failure can, of course, be modeled using a different conceptual statistical distribution. Then, ML analysis for mutually independent competing modes of failure merely requires having a library of alternative ML analyses available for each mode of failure of potential interest. (But care must be taken to assure that the respective duration metrics are coherent.)

> *Remark One*: Antifriction bearing failures are typically a competing modes of failure problem. Pitting fatigue failure usually initiates at the surface of the rolling elements (balls and rollers), but it can initiate at the corresponding contact surface of either the inner or the outer race. Moreover, even the rolling elements experience competing modes of failure, viz., their pitting fatigue life depends on the specific type of nonmetallic inclusion that initiates the fatigue crack.

> *Remark Two*: Even low-cycle (strain-controlled) fatigue is typically a competing modes of failure problem, especially at elevated temperatures. Microscopic examination usually indicates the existence of (at least) two distinct types of fatigue crack initiation processes.

8.12.4. Example Three: ML Analysis for Outcome of an s_a–fnc Experiment Test Program, Presuming Homoscedastic Fatigue Strength Distribution

Consider a quadratic s_a–$\log_e[fnc(pf)]$ model with a homoscedastic conceptual fatigue strength distribution. It can be expressed so that its standardized homoscedastic strength variate y pertains to any conceptual (two-parameter) statistical distribution of specific interest. Its concomitant variate can be either s_a or $\log_e(s_a)$. The s_a–$\log_e[fnc(pf)]$ model that is quadratic in terms of $\log_e(fnc)$ is physically more credible, both at long life and at shorter lives, than the s_a–$\log_e[fnc(pf)]$ model that is quadratic in s_a: (a) its slope approaches zero at a very long lives and its curvature increases as its slope aproaches zero, and (b) its slope becomes very steep at short lives and its curvature decreases as its slope increases.

Given a quadratic s_a–$\log_e[fnc(pf)]$ model with a homoscedastic conceptual (two-parameter) normal fatigue strength distribution, microcomputer programs *SAFNCM3A* and *SAFNCM3B* (s_a–$\log_e[fnc(pf)]$ model – versions 3A and 3B) compute (a) ML estimates for the actual values for $s_{fs}(50)$, conditional on (given the) respective experiment test program failure and suspension number of cycles datum values, and (b) the respective lower $100(scp)\%$ (one-sided) asymptotic statistical confidence limits that allegedly bound the actual values for $s_{fs}(50)$ and $s_{fs}(pf)$, given the fnc value of specific interest. Recall that ML estimates are typically biased and therefore require appropriate statistical bias corrections, especially for

```
C> COPY SAFNCDTA DATA (see page 436)

    1 file(s) copied

C> SAFNCM3A
```

Presumed Quadratic s_a–$\log_e[fnc(pf)]$ Model:

standardized normal distribution variate $y = (s_a - clp)/csp$, where
$$clp = clp0 + clp1[\log_e(fnc)] + clp2[\log_e(fnc)]^2$$

fnc	s_a	est[$s_{fs}(50)$]
56430	320	317.6
99000	300	300.9
183140	280	283.5
479490	260	258.0
909810	240	242.3
3632590	220	211.6
4917990	200	205.5
19186790	180	180.7

snc	s_a	est[$s_{fs}(50)$]
32250000	160	172.4

Lower 95% (One-Sided) Asymptotic Statistical Confidence Limits that Allegedly Bound the Actual Value for $s_{fs}(50)$ at $fnc = 25000000$ Cycles

169.4 – Computed Using the Propagation of Variability Expression for est(var{est[$s_{fs}(50)$]})
168.8 – Computed Using the Propagation of Variability Expression for est{var[est(y given $s_{fs} = closascl$)]}

Lower 95% (One-Sided) Asymptotic Statistical Confidence Limits that Allegedly Bound the Actual Value for $s_{fs}(10)$ at $fnc = 25000000$ Cycles

164.0 – Computed Using the Propagation of Variability Expression for est(var{est[$s_{fs}(10)$]})
162.3 – Computed Using the Propagation of Variability Expression for est{var[est(y given $s_{fs} = closascl$)]}

Employing Ad Hoc Statistical Bias Corrections (Version A)

Lower 95% (One-Sided) Asymptotic Statistical Confidence Limits that Allegedly Bound the Actual Value for $s_{fs}(50)$ at $fnc = 25000000$ Cycles

167.6 – Computed Using the Propagation of Variability Expression
for est(var$\{$est$[s_{fs}(50)]\}$)
166.2 – Computed Using the Propagation of Variability Expression
for est$\{$var[est(y given $s_{fs} = closascl$)]$\}$

Lower 95% (One-Sided) Asymptotic Statistical Confidence Limits that Allegedly Bound the Actual Value for $s_{fs}(10)$ at $fnc = 25000000$ Cycles

160.7 – Computed Using the Propagation of Variability Expression
for est(var$\{$est$[s_{fs}(10)]\}$)
156.4 – Computed Using the Propagation of Variability Expression
for est$\{$var[est(y given $s_{fs} = closascl$)]$\}$

---

small quantitative CRD experiment test programs. However, until the advent of the present generation of microcomputers, ML analyses either included *ad hoc* statistical bias corrections (occasionally) or none at all (typically). For example, microcomputer program *SAFNCM3A* includes an *ad hoc* multiplicative statistical bias correction factor equal to $[(n_f)/(n_f - n_{clp})]^{1/2}$ for est(csp) in the computation of the respective statistical confidence limits. Microcomputer program *SAFNCM3B* additionally substitutes a *scp*-based value for Student's central $t_{1,n_{dsdf}=n_f - n_{clp}}$ variate for the classical *scp*-based value of the standardized conceptual normal distribution variate y in the calculation of the respective lower $100(scp)\%$ (one-sided) asymptotic statistical confidence limits. This additional *ad hoc* substitution makes the resulting ML statistical confidence limits identical to linear regression statistical confidence limits (Chapter 7) when there are no *Type I* suspended fatigue tests. However, it should be clear that bias corrections based on simulation-based empirical sampling distributions are markedly preferable to *ad hoc* bias corrections.

We discuss simulation-based generation of pragmatic statistical bias corrections for s_a–$\log_e[fnc(pf)]$ models with a homoscedastic conceptual fatigue strength distribution in Supplemental Topic 8.D. The resulting ML analyses are without peer.

```
C> COPY SAFNCDTA DATA (see page 436)

    1 file(s) copied

C> SAFNCM3B
```

Presumed Quadratic s_a–$\log_e[fnc(pf)]$ Model:

standardized normal distribution variate $y = (s_a - clp)/csp$, where
$$clp = clp0 + clp1[\log_e(fnc)] + clp2[\log_e(fnc)]^2$$

fnc	s_a	est$[s_{fs}(50)]$
56430	320	317.6
99000	300	300.9
183140	280	283.5
479490	260	258.0
909810	240	242.3
3632590	220	211.6
4917990	200	205.5
19186790	180	180.7

snc	s_a	est$[s_{fs}(50)]$
32250000	160	172.4

Lower 95% (One-Sided) Asymptotic Statistical Confidence Limits that Allegedly Bound the Actual Value for $s_{fs}(50)$ at $fnc = 25000000$ Cycles

169.4 – Computed Using the Propagation of Variability Expression
for est(var$\{$est$[s_{fs}(50)]\}$)
168.8 – Computed Using the Propagation of Variability Expression
for est$\{$var[est(y given $fnc = closascl$)]$\}$

Lower 95% (One-Sided) Asymptotic Statistical Confidence Limits that Allegedly Bound the Actual Value for $s_{fs}(10)$ at $fnc = 25000000$ Cycles

164.0 – Computed Using the Propagation of Variability Expression
for est(var$\{$est$[s_{fs}(10)]\}$)
162.3 – Computed Using the Propagation of Variability Expression
for est$\{$var[est(y given $fnc = closascl$)]$\}$

Employing Ad Hoc Statistical Bias Corrections (Version B)

Lower 95% (One-Sided) Asymptotic Statistical Confidence Limits that Allegedly Bound the Actual Value for $s_{fs}(50)$ at *fnc* = 25000000 Cycles

165.6 – Computed Using the Propagation of Variability Expression
for est(var{est[$s_{fs}(50)$]})
162.6 – Computed Using the Propagation of Variability Expression
for est{var[est(y given *fnc* = *closascl*)]}

Lower 95% (One-Sided) Asymptotic Statistical Confidence Limits that Allegedly Bound the Actual Value for $s_{fs}(10)$ at *fnc* = 25000000 Cycles

158.6 – Computed Using the Propagation of Variability Expression
for est(var{est[$s_{fs}(10)$]})
150.3 – Computed Using the Propagation of Variability Expression
for est{var[est(y given *fnc* = *closascl*)]}

8.12.4.1. Discussion

Although linear and quadratic s_a–log$_e$[*fnc*(*pf*)] models are theoretically valid for all materials that do not exhibit a endurance limit (viz., a threshold fatigue crack initiation alternating stress amplitude), problems can arise when only the strength at a very long fatigue life is of specific interest. When the estimated segment of the s_a–log$_e$[*fnc*(*pf*)] model pertains only to long-life tests and thus is relatively short, the resulting ML estimated median s_a–log$_e$(*fnc*) curve may not be physically credible. Nevertheless, it is good practice to allocate all test specimens to alternating stress amplitudes that generate the range of fatigue lives of specific interest. Accordingly, the up-and-down strategy is recommended for use in allocating specimens to alternating stress amplitudes that pertain to very long fatigue lives. Then, if the ML-estimated median s_a–log$_e$(*fnc*) curve is not physically credible, it can be replaced by an experience-based median s_a–log$_e$(*fnc*) curve and the alternating stress amplitudes associated with the fatigue failure datum values can be pragmatically translated to pertain to the *Type I* censored duration of specific interest. In turn, the actual value for the median fatigue strength at this duration can be estimated and the associated lower 100(*scp*)% (one-sided) asymptotic statistical confidence limit can be computed.

```
C> COPY SAFNCDTA DATA

    1 file(s) copied

C> TYPE DATA
```

8	Number of Alternating Stress Amplitudes With One or More Fatigue Failures
1	Number of Replicate Tests at the Highest Alternating Stress Amplitude with Fatigue Failures
320 56430	Highest Alternating Stress Amplitude (space) Corresponding Failure Number of Cycles
1	Number of Replicate Tests at the Second Highest Alternating Stress Amplitude
300 99000	etc.
1	
280 183140	
1	
260 479490	
1	
240 909810	
1	
220 3632590	
1	
200 4917990	
1	
180 19186790	
1	Number of Alternating Stress Amplitudes With One or More Suspended Tests
1	Number of Suspended Tests at the Highest Alternating Stress Amplitude with Suspended Tests
160 32250000	Highest Alternating Stress Amplitude (space) Corresponding Suspension Number of Cycles, etc.
95	Statistical Confidence Probability of Specific Interest in Per Cent (Integer Value, Maximum = 95)
25000000	Fatigue Life of Specific Interest
10	Conceptual Strength CDF Percentile of Specific Interest in Per Cent (Integer Value, Minimum = 10)

8.12.5   Example Four: ML Analysis for Outcome of a Strength (Resistance) Experiment Test Program Conducted Using the Up-and-Down Strategy

Although the up-and-down test method strategy is strictly valid only for threshold phenomena, this strategy nevertheless can be used in a very effective manner to estimate the actual value for the median of the presumed conceptual one-parameter strength (resistance) distribution at very long durations for those modes of failure whose presumed stimulus-duration models asymptotically reach a limiting (threshold) value, e.g., fatigue, stress rupture, and stress-corrosion cracking. This sequential strategy is particularly effective when only a few (nominally identical) test items are available for testing because it efficiently allocates these items to stress (stimulus) levels in the vicinity of the actual value for the median of the presumed conceptual one-parameter strength (resistance) distribution. Little (1981) presents tabulated quantities for use in estimating the actual value for the median of the presumed conceptual one-parameter strength (resistance) distribution given the outcome of a conventional small-sample up-and-down experiment test program with a fixed increment between successive alternating stress amplitudes. These tabulated quantities pertain to four alternative conceptual one-parameter strength distributions: normal (symmetrical with short tails), logistic (symmetrical with long tails), smallest-extreme-value (nonsymmetrical, skewed toward low values), and the largest-extreme value (nonsymmetrical, skewed toward high values), and include enumeration-based (exact) statistical bias corrections. (In Supplemental Topic 8.C we explain why up-and-down test programs with fixed increments do not generate minimum variance estimates.)

The example data set in microcomputer file *UADDATA* pertains to a fatigue experiment test program that employed the conventional small-sample up-and-down test method strategy for the first six test specimens, followed by allocation of the next four test specimens to their respective optimal alternating stress amplitudes successively computed by running microcomputer program *N1A* (normal distribution, one-parameter–version A). Because program *N1A* has the actual values for the respective alternating stress amplitudes as input (rather than a beginning value and a fixed increment), it is conveniently used at any time during the experiment test program to allocate the next fatigue specimen to its statistically most effective alternating stress amplitude regardless of the strategy (strategies) employed for testing prior fatigue specimens.

Microcomputer program *N1A* has the presumed known (guestimated) value for the standard deviation of the presumed conceptual one-parameter normal strength (resistance) distribution as input. However, this presumed

```
C> TYPE UADDATA
```

10	Number of Alternating Stress Amplitudes	Fatigue Life, cycles	Temperature Increase (°F)
3000 1 1	Alternating Stress Amplitude (space)	3,133,000	18
2750 1 1	Number of Items Tested (space)	9,932,000	9
2500 1 0	Number of Items Failed	10^7 DNF	7
2750 1 1		4,900,000	10
2500 1 0		10^7 DNF	5
2750 1 0		10^7 DNF	16
2710 1 1		4,548,000	6
2655 1 0		10^7 DNF	11
2700 1 0		10^7 DNF	9
2740 1 0		10^7 DNF	6

250   Presumed Standard Deviation of the Presumed Conceptual One-Parameter Strength (Resistance) Distribution

95   Statistical Confidence Probability of Specific Interest in Per Cent (Integer Value, 95 Maximum)

Note: DNF Denotes a Run-Out (Did Not Fail) Under *Type I* Censoring

---

```
C> COPY UADDATA DATA

    1 file(s) copied


C> N1A
```

Presuming a Conceptual One-Parameter Normal Distribution

$$\text{est}[s(50)] = 2778.0$$

Lower 95% (One-Sided) Asymptotic Statistical Confidence Limits that Allegedly Bound the Actual Value for $s(50)$

2604.7 – Computed Using the Propagation of Variability Expression for $\text{est}\{\text{var}[\text{est}(y \text{ given } s = closascl)]\}$

2602.9 – Computed Using the Likelihood Ratio Method

C>L1A

Presuming a Conceptual One-Parameter Logistic Distribution

est[$s(50)$] = 2769.8

Lower 95% (One-Sided) Asymptotic Statistical Confidence Limits that Allegedly Bound the Actual Value for $s(50)$

2610.6 – Computed Using the Propagation of Variability Expression for est{var[est(y given $s = closascl$)]}
2612.7 – Computed Using the Likelihood Ratio Method

C> SEV1A

Presuming a Conceptual One-Parameter Smallest-Extreme-Value Distribution

est[$s(50)$] = 2768.1

Lower 95% (One-Sided) Asymptotic Statistical Confidence Limits that Allegedly Bound the Actual Value for $s(50)$

2603.6 – Computed Using the Propagation of Variability Expression for est{var[est(y given $s = closascl$)]}
2615.9 – Computed Using the Likelihood Ratio Method

C> LEV1A

Presuming a Conceptual One-Parameter Largest-Extreme-Value Distribution

est[$s(50)$] = 2773.2

Lower 95% (One-Sided) Asymptotic Statistical Confidence Limits that Allegedly Bound the Actual Value for $s(50)$

2630.8 – Computed Using the Propagation of Variability Expression for est{var[est(y given $s = closascl$)]}
2613.3 – Computed Using the Likelihood Ratio Method

known (guestimated) value can be changed later (a) to obtain a revised estimate of the actual value for the median (mean) of the presumed conceptual one-parameter normal strength (resistance) distribution if a more accurate known (guestimated) value for its standard deviation becomes available, or (b) to study the sensitivity of the ML estimate for the actual value for the median (mean) of the presumed conceptual one-parameter normal strength (resistance) distribution to alternative presumed known (guestimated) values for its standard deviation.

Microcomputer programs *L1A* (logistic distribution, one-parameter–version A), *SEV1A* (smallest-extreme-value distribution, one-parameter–version A), and *LEV1A* (largest-extreme-value distribution, one-parameter–version A) on page 439 demonstrates that the ML estimate of the actual value for the metric pertaining to the median of the presumed conceptual one-parameter strength (resistance) distribution is relatively insensitive to the presumed CDF.

8.12.5.1. Discussion

The statistical literature does not exploit the versatility of ML analysis for strength (resistance) experiment test programs. The individual test outcomes do not have to be limited to realization values that are equal to zero or one. Rather, we can establish a damage index that lies in the interval from zero (no damage whatsoever) to one (complete failure). Each test item (specimen) pertaining to each suspended test can then be examined for signs of physical damage. For example, each of the automotive composite fatigue specimens pertaining to a suspended test in microcomputer file *UADDATA* was visually examined relative to the presence of small fatigue cracks and/or noticeable local delamination. The damage index for two of these specimens was subjectively assessed as 0.50 as indicated in microcomputer file *MUADDATA*. Accordingly, based on the damage index assessment for the re-interpreted experiment test program outcomes, microcomputer program *N1A* computes a revised estimate of the actual value for the median (mean) of the presumed conceptual one-parameter normal fatigue strength (fatigue limit) distribution at 10^7 stress cycles.

Little and Kosikowski (unpublished) conducted a modified up-and-down experiment test program to establish a nontraditional measure for the fatigue notch sensitivity of an automotive composite material. The size of a small central hole was decreased or increased depending on whether the fatigue crack that led to failure did or did not originate at the central hole. Predictably there were several failures where the origin could not be stated with certainty. Accordingly, although the test strategy merely required replicating the test with the same size hole, a subjective probability index was required in ML analysis to estimate the diameter of the hole for which fatigue failure is equally likely or not to originate at the hole.

Finally, Little (unpublished) conducted several up-and-down long-life fatigue experiment test programs with an automotive composite specimen attached to a steel grip at each of its ends with nominally identical bolted lap joints. Accordingly, fatigue failure was equally likely to occur at either the top or the bottom lap joint. The conceptual strength distribution for this example thus pertains to the smallest of two independent realizations randomly selected from the conceptual strength distribution presumed for each independent realization, viz.,

$$F_1(s) = 1 - [1 - F(s)]^2$$

in which

$$F(s) = \frac{1}{(2\pi)^{1/2} \cdot \text{stddev}(S)} \int_{-\infty}^{s} \exp\{-[(u - \text{median}(S)]/\text{stddev}(S)\}^2 du$$

```
C> TYPE MUADDATA
```

		Fatigue Life	Temperature
10	Number of Alternating Stress Amplitudes		Increase (°F)
3000 1 1	Alternating Stress Amplitude (space)	3,133,000	18
2750 1 1	Number of Item Tested (space)	9,932,000	9
2500 1 0	Number of Items Failed	10^7 DNF	7
2750 1 1		4,900,000	10
2500 1 0		10^7 DNF	5
2750 1 0.5	(*Subjective Damage Index*)	10^7 DNF	16
2710 1 1		4,548,000	6
2655 1 0		10^7 DNF	11
2700 1 0		10^7 DNF	9
2740 1 0.5	(*Subjective Damage Index*)	10^7 DNF	6
250	Presumed Standard Deviation of the Presumed Conceptual One-Parameter Strength (Resistance) Distribution		
95	Statistical Confidence Probability of Specific Interest, Stated in Per cent (Integer Value)		

Note: DNF Denotes a Run-Out (Did Not Fail) Under *Type I* Censoring

```
C> COPY MUADDATA DATA

    1 file(s) copied

C> N1A
```

Presuming a Conceptual One-Parameter Normal Distribution

$$\text{est}[s(50)] = 2704.9$$

Lower 95.0% (One-Sided) Asymptotic Statistical Confidence Limits
that Allegedly Bound the Actual Value for $s(50)$

2534.0 – Computed Using the Propagation of Variability Expression
for $\text{est}\{\text{var}[\text{est}(y \text{ given } s = closascl)]\}$

for the normal distribution;

$$F(s) = \left(1 + \exp\{-\pi \cdot [s - \text{median}(S)]/[3^{1/2} \cdot \text{stddev}(S)]\}\right)^{-1}$$

for the logistic distribution;

$$F(s) = 1 - \exp(-\exp\{+\pi \cdot [s - \text{median}(S) - 0.28577 \cdot \text{stddev}(S)]/$$
$$[6^{1/2} \cdot \text{stddev}(S)]\})$$

for the smallest-extreme-value distribution; and

$$F(s) = \exp(-\exp\{-\pi \cdot \{s - \text{median}(S) + 0.28577 \cdot \text{stddev}(S)]/$$
$$[6^{1/2} \cdot \text{stddev}(S)]\})$$

for the largest-extreme-value distribution. Note that these four respective conceptual strength (resistance) distributions are scaled to have the same (presumed known) standard deviation (variance). This scaling is also done in microcomputer programs *N1A*, *L1A*, *SEV1A*, and *LEV1A* so that the standard deviation (rather than the associated *csp*) is the appropriate input information to each respective microcomputer program. Otherwise, the presumed value for the *csp* depends on which conceptual strength (resistance) distribution is presumed in analysis.

Little (1975) presents tabulated quantities pertaining to each of the four alternative strength distributions for use in estimating (a) the actual value for median(*S*) with two specimens "in series," and (b) the associated enumeration-based (exact) statistical bias corrections, given the outcome of a conventional small-sample up-and-down experiment test program with a nominal sample size from 2 to 6.

8.13. COMPARATIVE ANALYSES FOR OUTCOMES OF LIFE (ENDURANCE) AND STRENGTH (RESISTANCE) EXPERIMENT TEST PROGRAMS

The ML analyses presented in Sections 8.10–8.12 all pertain to quantitative CRD experiment test programs. Recall that, for a quantitative estimate to have credible statistical inference, it must be presumed that (a) all possible batch-to-batch effects are negligible, (b) all nuisance test variables are negligible, and (c) all calculations are accurate. Fortunately, the likelihood ratio (LR) method provides a statistical means to make comparative (rather than quantitative) mechanical reliability analyses. This feature of the LR method is illustrated in Section 8.14. A distribution-free comparative mechanical reliability analysis is presented in Section 8.15. It may be useful in certain mechanical reliability applications.

8.14. COMPARATIVE MAXIMUM LIKELIHOOD ANALYSES

Suppose that two s_a–fnc experiment test programs pertaining to different batches of the same material have been conducted and the issue is whether or not the respective data sets can legitimately be combined. Microcomputer program *C2SFNCM7* (compare $2s_a$–$log_e[fnc(pf)]$ models – version 7) performs a LR-based analysis, presuming that the respective s_a–$log_e[fnc(pf)]$ models that are quadratic in $log_e(fnc)$, have linear s_a metrics, and employ homoscedastic conceptual (two-parameter) smallest-extreme-value strength distributions. The example input s_a–fnc data for this microcomputer program are constructed using the s_a–fnc values (data set one) that appear in microcomputer file *SAFNCDTA*, plus additional s_a–fnc values (data set two) whose s_a's are computed by subtracting 20 from each of the corresponding s_a's in data set one. (See microcomputer file C25NDATA.) A linear metric for the alternating stress amplitude was deliberately used in constructing these example input s_a–fnc data to generate parallel s_a–$log_e[fnc(pf)]$ models for the respective s_a–fnc data sets. Accordingly, the respective [est(*clp1*)]'s and [est(*clp2*)]'s pertaining to our two example s_a–fnc data sets will be identical and that the respective [est(*clp0*)]'s will differ by exactly 20 units. However, regardless of the alternating stress amplitude metric that is presumed for the conceptual s_a–$log_e[fnc(pf)]$

model, the fundamental issue is that the respective data-based values for the asymptotic LR test statistic will establish null hypothesis rejection probability values for the two statistical tests of hypothesis that are of specific interest: (*i*) the null hypothesis that a single s_a–$\log_e[fnc(pf)]$ model suffices statistically to explain the two aggregated s_a–$\log_e(fnc)$ data sets versus the alternative hypothesis that two distinct s_a–$\log_e[fnc(pf)]$ models are statistically required to explain these two data sets; and (*ii*) the null hypothesis that two parallel s_a–$\log_e[fnc(pf)]$ models suffice statistically versus the alternative hypothesis that two distinct s_a–$\log_e[fnc(pf)]$ models are statistically required to explain these two data sets. Given our two example s_a–$\log_e(fnc)$ data sets, we must rationally opt to reject null hypothesis (*i*) in favor of alternative hypothesis (*i*), but we cannot rationally reject null hypothesis (*ii*) in favor of alternative hypothesis (*ii*). Accordingly, we must rationally conclude that a batch-to-batch effect exists and that this effect is not affected by the actual value of the concomitant variable s_a.

```
C> COPY C2SNDATA DATA


     1 file(s) copied


C> C2SFNCM7
```

Presumed Quadratic s_a–$\log_e[fnc(pf)]$ Model with a Homoscedastic Smallest-Extreme-Value Fatigue Strength Distribution

Presuming a Distinct s_a–$\log_e[fnc(pf)]$ Model for Data Set One
est(*clp0*) = 0.751737D + 03
est(*clp1*) = −0.505715D + 02
est(*clp2*) = 0.995199D + 00
est(*csp*) = 0.404219D + 01
Estimated Maximum $\log_e$(Likelihood) = −0.23451179D + 02

Presuming a Distinct s_a–$\log_e[fnc(pf)]$ Model for Data Set Two
est(*clp0*) = 0.731737D + 03
est(*clp1*) = −0.505715D + 02
est(*clp2*) = 0.995199D + 00
est(*csp*) = 0.404219D + 01
Estimated Maximum $\log_e$(Likelihood) = −0.23451179D + 02

Presuming Parallel s_a–$\log_e[fnc(pf)]$ Models with a Common csp
$\text{est}(clp01) = 0.751737\text{D}+03$
$\text{est}(clp02) = 0.731737\text{D}+03$
$\text{est}(clp1) = -0.505715\text{D}+02$
$\text{est}(clp2) = 0.995199\text{D}+00$
$\text{est}(csp) = 0.404219\text{D}+01$
Estimated Maximum $\log_e$(Likelihood) $= -0.46902359\text{D}+02$

Presuming a Single s_a–$\log_e[fnc(pf)]$ Model for the Aggregated Data
$\text{est}(clp0) = 0.800413\text{D}+03$
$\text{est}(clp1) = -0.587405\text{D}+02$
$\text{est}(clp2) = 0.129182\text{D}+01$
$\text{est}(csp) = 0.942785\text{D}+01$
Estimated Maximum $\log_e$(Likelihood): $-0.61570256\text{D}+02$

Likelihood Ratio Tests

(*i*) H_a: Two Distinct s_a–$\log_e[fnc(pf)]$ Models *versus* H_n: A Single
s_a–$\log_e[fnc(pf)]$ Model (8–4 n_{sdf})
H_n Rejection Probability $= 0.6681\text{D-05}$

(*ii*) H_a: Two Distinct s_a–$\log_e[fnc(pf)]$ Models *versus* H_n: Two Parallel
s_a–$\log_e[fnc(pf)]$ Models (8–5 n_{sdf})
H_n Rejection Probability $= 0.1000\text{D}+01$

Also (*iii*) H_a: Two Parallel s_a–$\log_e[fnc(pf)]$ Models *versus* H_n: A Single
s_a–$\log_e[fnc(pf)]$ Model (5–4 n_{sdf})
H_n Rejection Probability $= 0.6086\text{D}-07$

```
C> TYPE C2SNDATA


8            Data Set One (Same as given in Microcomputer File
             SAFNCDTA)
1
320    56430
1
300    99000
1
280    183140
```

1
260   479490
1
240   909810
1
220   3632590
1
200   4917990
1
180   19186790
1
1
160   32250000
8   Data Set Two (Constructed by subtracting 20 from
                           each alternating stress
1   amplitude in Data Set One)
300   56430
1
280   99000
1
260   183140
1
240   479490
1
220   909810
1
200   3632590
1
180   4917990
1
160   19186790
1
1
140   32250000

```
C> COPY C2SDDATA DATA

   1 file(s) copied

C> C2NSDDS
```

Presuming a Strength Model with a Distinct Conceptual (Two-Parameter) Normal Distribution for Data Set One

$$\text{est}[s(50)] = 86.1 \quad \text{est}(csp) = 6.8$$
$$\text{Estimated Maximum } \log_e(\text{Likelihood}) = -6.2841$$

Presuming a Strength Model with a Distinct Conceptual (Two-Parameter) Normal Distribution for Data Set Two
$$\text{est}[s(50)] = 80.2 \quad \text{est}(csp) = 14.5$$
$$\text{Estimated Maximum } \log_e(\text{Likelihood}) = -10.5833$$

Presuming a Strength Model with Two Conceptual (Two-Parameter) Normal Distributions that Have a Common csp

$$\text{est}[s(50)] \text{ for Data Set One} = 86.0$$
$$\text{est}[s(50)] \text{ for Data Set Two} = 81.0$$
$$\text{est}(\text{Common } csp) = 11.1$$
$$\text{Estimated Maximum } \log_e(\text{Likelihood}) = -17.7893$$

Presuming a Strength Model with a Single Conceptual (Two-Parameter) Normal Distribution for the Aggregated Data
$$\text{est}[s(50)] = 83.3$$
$$\text{est}(csp) = 11.2$$
$$\text{Estimated Maximum } \log_e(\text{Likelihood}) = -18.2358$$

Likelihood Ratio Tests

(*i*) H_a: Two Distinct Normal Distributions *versus* H_n: A Single Normal Distribution (42 n_{sdf})
H_n Rejection Probability $= 0.2545$

(*ii*) H_a: Two Distinct Normal Distributions *versus* H_n: Two Normal Distributions with a Common csp (4–3 n_{sdf})
H_n Rejection Probability $= 0.1745$

Also (*iii*) H_a: Two Normal Distributions with a Common csp *versus* H_n: A Single Normal Distribution (3–2 n_{sdf})
H_n Rejection Probability $= 0.3447$

```
C> TYPE C2SDDATA

4          Number of Data Set One Program Alternating Stress
           Amplitudes
100 4 4    Alternating Stress Amplitude (space) Number of Items Tested
           (space) Number of Items Failed
90 6 4
80 4 1
70 5 0
3          Number of Data Set Two Alternating Stress Amplitudes
95 8 7     Alternating Stress Amplitude (space) Number of Items Tested
           (space) Number of Items Failed
85 7 4
65 6 1
```

Analogously, suppose that two strength (endurance) experiment test programs pertaining to different batches of the same material have been conducted at two or more stress (resistance) levels and the issue is whether the respective data sets can be legitimately combined. Microcomputer program *C2NSDDS* (compare 2 normal strength distribution data sets) performs a LR-based analysis, presuming that the four respective strength (resistance) models of potential interest each employ an appropriate conceptual (two-parameter) normal distribution. It first computes the [est(clp)]'s and the [est(csp)]'s for each of these strength (resistance) models. It then computes the data-based values for the two asymptotic LR test statistics of specific interest and establishes the associated null rejection probabilities for (*i*) the null hypothesis that a strength (resistance) model that employs a single conceptual (two-parameter) normal distribution suffices statistically to explain the aggregated experiment test program datum sets versus the alternative hypothesis that a strength (resistance) model that employs two distinct conceptual (two-parameter) normal distributions is statistically required to explain these two data sets, and (*ii*) the null hypothesis that a strength (resistance) model that employs two conceptual (two-parameter) normal distributions with a common *csp* suffices statistically to explain these two data sets versus the alternative hypothesis that a strength (resistance) model that employs two distinct conceptual (two-parameter) normal distributions are statistically required to explain these two data sets. Given the two example data sets that appear in microcomputer file *C2SNDATA*, LR-based analysis indicates that we cannot rationally opt to reject null hypothesis (*i*) in favor of alternative hypothesis (*i*). Accordingly, a strength (resistance) model that employs a single conceptual (two-parameter) normal distribution suffices statistically to explain the aggregated example data sets. Thus, we conclude that the batch-to-batch effect, if it exists, is evidently relatively small compared to the actual value for the standard deviation of the presumed conceptual (two-parameter) normal strength (resistance) distribution.

Remark: Note that if the experiment test program objective is to compare the respective (mode) s_a–$\log_e(fnc)$ curves pertaining to treatments B and A, then data set one would pertain to treatment B (instead of batch one) and data set two would pertain to treatment A (instead of batch two).

8.15. COMPARATIVE ANALYSIS FOR OUTCOME OF A LIFE (ENDURANCE) EXPERIMENT TEST PROGRAM, BASED ON A GENERALIZED SAVAGE DISTRIBUTION-FREE (NON-PARAMETRIC) TEST STATISTIC

We now reconsider a CRD experiment test program that is conducted to test the null hypothesis that $B = A$ statistically versus the alternative hypothesis that $B > A$ statistically. Suppose that certain of the tests in this experiment test program are *Type I* censored. The log-rank algorithm (Mantel, 1981) can be used to assign ranks to the aggregated A and B experiment test program datum values with *Type I* censoring. Given a CRD experiment test program with complete data and two treatments, Savage's test statistic is asymptotically unexcelled. Since this test statistic can be generalized by using the log-rank algorithm to include suspended tests, it is employed in microcomputer program *C2DSWST* (compare 2 data sets with suspended tests) to test the null hypothesis that $(B-mpd) = A$ statistically versus the alternative hypothesis that $(B-mpd) > A$ statistically. In turn, a lower $100(scp)\%$ (one-sided) confidence limit that allegedly bounds the actual value for the *mpd* can be computed by iteratively running microcomputer program *C2DSWST* with appropriate input values for the *mpd* of specific interest.

Recall that *Type I* censoring occurs only in laboratory testing, whereas arbitrarily suspended tests are statistically analogous to products that are still operating satisfactorily in service. Thus, arbitrarily suspended tests tend to be more practical than *Type I* censoring in a mechanical reliability perspective. Accordingly, microcomputer program *C2DSWST* was written to accommodate arbitrarily suspended tests. (Observe that each individual suspended duration appears in microcomputer file *C2STDATA*.)

Remember, however, that the fundamental statistical concept of a continually replicated experiment test program is obscure when arbitrarily suspended tests occur.

```
C> TYPE C2STDATA

200      Minimum Practical Difference of Specific Interest
10       Total Number of A Datum Values Pertaining to Failures,
         Followed by A Failure Lives (Endurances)
112
143
151
177
231
345
378
401
498
512
2        Total Number of A Suspended Tests, Followed by A
         Suspended Lives (Durations)
1000.01
1000.01
2        Total Number of B Datum Values Pertaining to Failures,
         Followed by B Failure Lives (Endurances)
592
712
5        Total Number of B Suspended Tests, Followed by B
         Suspended Lives (Durations)
1000.01
1000.01
1000.01
1000.01
1000.01
```

Note: An increment equal to 0.01 is added to each datum value pertaining to a test suspension to avoid the possibility of a tie between a suspended test duration and a failure life (endurance).

When a paired-comparison test program is clearly appropriate, arbitrarily suspended tests within blocks can pose problems in establishing the relevant $(b-a)$ difference, e.g., when the suspended test pertains to a shorter duration than the corresponding failure life (endurance). However, given *Type I* censoring, the durations of all suspended tests are greater than their corresponding observed failure lives. Then, all paired-comparison $(b - a)$ differences are either known or can be treated as arbitrarily suspended tests. When the paired-comparison $(b-a)$ difference is positive, it is treated as *B* data in a CRD test program. On the other hand, when the paired-comparison $(b-a)$ difference is negative, it is treated as *A* data in this CRD test program. [If both specimens fail at the same time or if both tests are suspended at the same time, the resulting paired-comparison $(b - a)$ difference is equal to zero and this paired-comparison outcome can be treated as a tie.] Accordingly, microcomputer program *C2DSWST* can be used to test the null hypothesis $(B-mpd) = A$ statistically versus the alternative hypothesis that $(B-mpd) > A$ statistically for any *mpd* value of specific interest, even when a paired-comparison experiment test program is conducted.

8.16. ESTIMATING SUB SYSTEM RELIABILITY

A system is comprised of subsystems. In turn, subsystems are comprised of subsubsystems, and subsubsystems are comprised of subsubsubsystems, etc. Because systems, subsystems, subsubsystems, and subsubsubsystems can be parsed in various ways, their distinction is often a matter of semantics. We choose to use the terminology subsystem in our presentation.

The proper method to estimate the reliability of a mechanical (electromechanical) subsystem is to conduct a life (endurance) experiment test program using the subsystem itself as the test specimen. However, if this experiment-based methodology is not practical, then the reliability of a mechanical (electromechanical) subsystem can be analytically guestimated using estimates of the reliabilities of its components. In theory, the reliability of a mechanical (electromechanical) subsystem can be analytically expressed as a function of duration *d* when (a) all of its components are presumed to operate independently and (b) the reliabilities of these components are expressed as a function of, or in terms of, the *same* stimulus and duration metrics. However, in practice, the respective component reliabilities must be estimated by the appropriate life (endurance) experiment test programs whose stimulus and duration metrics are unlikely to be identical and may not even be coherent (compatible). Accordingly, it is extremely unlikely that the necessary component reliability estimates exist—unless these estimates were deliberately generated for use in estimating the subsystem reliability of specific interest. Even then, the only way to test the credibility of the pre-

```
C> COPY C2STDATA DATA

   1 file(s) copied

C> C2DSWST
```

For a Minimum Practical Difference (*mpd*) = 200.0
Aggregated *A* and *B* Datum Values with Suspended Tests,
Ranked Using the Log-Rank Algorithm

datum value	code	rank
0.1120000000D + 03	1	0.1000000000D + 01
0.1430000000D + 03	1	0.2000000000D + 01
0.1510000000D + 03	1	0.3000000000D + 01
0.1770000000D + 03	1	0.4000000000D + 01
0.2310000000D + 03	1	0.5000000000D + 01
0.3450000000D + 03	1	0.6000000000D + 01
0.3780000000D + 03	1	0.7000000000D + 01
0.3920000000D + 03	3	0.8000000000D + 01
0.4010000000D + 03	1	0.9000000000D + 01
0.4980000000D + 03	1	0.1000000000D + 02
0.5120000000D + 03	1	0.1150000000D + 02
0.5120000000D + 03	3	0.1150000000D + 02
0.1000010000D + 04	0	0.1600000000D + 02
0.1000010000D + 04	0	0.1600000000D + 02
0.8000100000D + 03	2	0.1600000000D + 02
0.8000100000D + 03	2	0.1600000000D + 02
0.8000100000D + 03	2	0.1600000000D + 02
0.8000100000D + 03	2	0.1600000000D + 02
0.8000100000D + 03	2	0.1600000000D + 02

The data-based value of the generalized Savage test statistic for the CRD experiment test program that was actually conducted is equal to 11.195.

Given the null hypothesis that $(B - mpd) = A$ statistically, this microcomputer program constructed exactly 50,388 equally-likely outcomes for this experiment test program by using Ehrlich's method to reassign its a and $(b-mpd)$ datum values to treatments A and B. The number of these outcomes that had its generalized Savage test statistic value equal to or greater than 11.195 is equal to 232. Thus, given the simple null hypothesis that $(B - mpd) = A$ statistically, the enumeration-based probability that a randomly selected outcome of this experiment test program when continually replicated will have its generalized Savage test statistic value equal to or greater than 11.195 is equal to 0.0046. When this probability is sufficiently small, reject the null hypothesis in favor of the simple (one-sided) alternative hypothesis that $(B - mpd) > A$ statistically.

Given the null hypothesis that $(B - mpd) = A$ statistically, the asymptotic probability that a randomly selected outcome of this experiment test program when continually replicated will have its generalized Savage test statistic value equal to or greater than 11.195 is equal to 0.0054.

Note: Code 0 = suspended A test duration,
Code 1 = A datum value
Code 2 = suspended B test duration $- mpd$
Code 3 = B datum value $- mpd$

sumption that the respective components operate independently within this subsystem is to test this subsystem itself. Despite these practical limitations, subsystem reliability estimates based on questionable presumptions and analyses can nevertheless have useful application in comparing alternative designs and redesigns in an iterative reliability improvement process.

We now define a mechanical (electromechanical) subsystem as being comprised of components in a specific configuration, where a component by definition has no redundancy. Accordingly, at any given duration, each of the components in a subsystem is either operating satisfactorily or it is not. In contrast, depending on the configuration of its components, a subsystem can (should) have redundancy, viz., certain components can fail and the subsystem will continue to operate satisfactorily.

The physical understanding of a subsystem with redundancy is enhanced by limiting the subsystem to having a maximum of six to eight components. This can be accomplished by parsing larger subsystems into smaller subsystems (that could be called subsubsystems or subsubsubsystems if so desired.) Consider a subsystem that is comprised of n_c components. Its reliability can be established by examining the operational state for each of these n_c components. This examination requires that each of the 2^{n_c} distinct sets of component operational states be enumerated in terms of whether each individual component is operating satisfactorily or not.

Consider a component denoted *A*. Let (capital) *A* connote that it is operating satisfactorily and let (lower case) *a* denote that it is not operating satisfactorily. Next, consider a subsystem consisting of four components: *A*, *B*, *C*, and *D*. The reliability of this subsystem can be estimated by enumerating all 2^4 ($= 16$) of its distinct sets of component operational states. The required enumeration is conveniently accomplished using Yate's enumeration algorithm in which, e.g., *a* corresponds to -1 and *A* corresponds to $+1$ in column one, *b* corresponds to -1 and *B* corresponds to $+1$ in column two, etc. The resulting 16 distinct sets of component operational states for this subsystem example are:

$$
\begin{array}{llll}
a & b & c & d \\
A & b & c & d \\
a & B & c & d \\
A & B & c & d \\
a & b & C & d \\
A & b & C & d \\
a & B & C & d \\
A & B & C & d \\
a & b & c & D \\
A & b & c & D \\
a & B & c & D \\
A & B & c & D \\
a & b & C & D \\
A & b & C & D \\
a & B & C & D \\
A & B & C & D \\
\end{array}
$$

To simplify subsequent notation, let *A* also denote the estimated probability that component *A* is operating satisfactorily (*A* = estimated reliability) and let *a* also denote the complementary probability that component *A* is not operating satisfactorily (*a* = estimated unreliability).

8.16.1. Reliability of a New Subsystem

Now consider a new subsystem with all of its components operating satisfactorily at duration $d = 0$. For simplicity, let this subsystem have only two components, say *A* and *B*. The four distinct sets of component operational states for this subsystem are:

component operational states	associated probability
a b	a · b
A b	A · b
a B	a · B
A B	A · B

The estimated reliability of this subsystem can be established by summing the respective estimated probabilities pertaining to each distinct set of component operational states such that the subsystem continues to operate satisfactorily. When the estimated reliabilities for components A and B have been established by either appropriate life (endurance) or strength (resistance) experiment test programs, the reliability of the subsystem can be tabulated up to the duration used in *Type I* censoring for the life (endurance) experiment test programs, or at the specific duration d^* employed in the strength experiment test programs to suspend each individual test.

8.16.1.1. Elementary Example One

Suppose that a subsystem is configured to have no redundancy, viz., all of its components must operate satisfactorily for the subsystem to operate satisfactorily. Then, only the distinct component operational state set A B pertains. Accordingly, the estimated reliability of this subsystem is the product $A \cdot B$. Note that if this Example One subsystem had been comprised of numerous components, then its estimated reliability would have been the product of each of the individual estimated component reliabilities. Clearly, this estimated reliability becomes ever smaller as the number of components increases. Therefore, this type of subsystem configuration should be avoided whenever practical.

8.16.1.2. Elementary Example Two

Suppose that a subsystem is configured to have maximum redundancy, viz., only one of its components must operate satisfactorily for the subsystem to operate satisfactorily. Then, only distinct component operational state set a b fails the condition for satisfactory operation. Accordingly, the estimated unreliability of this subsystem with maximum redundancy is the product $a \cdot b$. Note that if this Example Two subsystem had been comprised of numerous components, then its estimated unreliability would have been the product of the respective estimated component unreliabilities. Clearly, this estimated unreliability becomes smaller and smaller as the number of its redundant components increases.

> *Message*: Redundant components increase subsystem reliability and should, therefore, be used whenever practical.

Elementary Examples One and Two are intended to demonstrate that the reliability of a subsystem decreases with an increase in the number of its components, unless the subsystem is appropriately configured with redundant components. Unfortunately, mechanical (electromechanical) subsystems are never quite as simple as either of these two elementary examples. Nevertheless, we now present three additional elementary examples that are intended to demonstrate the computational simplicity of estimating subsystem reliability when (a) all distinct sets of component operational states have been enumerated, (b) the component operational states that pertain to satisfactory subsystem operation can be unambiguously defined (or the distinct sets of component operational states that pertain to unsatisfactory subsystem operation can be unambiguously defined), and (c) all components are presumed to operate independently.

8.16.1.3. Elementary Example Three

Suppose satisfactory subsystem operation is defined such that at least two of its four components must operate satisfactorily. Then, by inspection, the estimated reliability of this subsystem is the sum of the products of the respective estimated probabilities in the second main column of the following example array:

component operational states	Elementary Example Three (sum these estimated probability products)	Elementary Example Four (sum these estimated probability products)	Elementary Example Five (sum these estimated probability products)
a b c d	—	—	—
A b c d	—	—	—
a B c d	—	—	—
A B c d	A B c d	—	A B c d
a b C d	—	—	—
A b C d	A b C d	—	—
a B C d	a B C d	—	—
A B C d	A B C d	A B C d	A B C d
a b c D	—	—	—
A b c D	A b c D	—	—
a B c D	a B c D	—	—
A B c D	A B c D	A B c D	A B c D
a b C D	a b C D	—	a b C D
A b C D	A b C D	A b C D	A b C D
a B C D	a B C D	a B C D	a B C D
A B C D	A B C D	A B C D	A B C D

8.16.1.4. Elementary Example Four

Suppose that at least three components must operate satisfactorily for this subsystem to operate satisfactorily. Then, by inspection, the estimated reliability for this subsystem is the sum of the products of the respective estimated probabilities given by the entries in the third main column of our example array.

8.16.1.5. Elementary Example Five

Suppose that this subsystem operates satisfactorily only when both A and B *or* when both C and D operate satisfactorily. Accordingly, by inspection, the estimated reliability of this subsystem is the sum of the products of the respective estimated probabilities given by the entries in the fourth main column of our example array.

8.16.1.6. Discussion

These five elementary examples are intended to demonstrate that the estimated reliability of a subsystem can be established by enumerating all distinct sets of component operational states. This estimation methodology has three basic steps: (a) enumerate all component operational states using Yate's enumeration algorithm; (b) identify those distinct sets of component operational states that pertain either to satisfactory or unsatisfactory subsystem operation; and (c) sum the products of the associated estimated probabilities. Step (b) requires a special microcomputer program for each different subsystem—because the specifications defining satisfactory or unsatisfactory operation for each different subsystem involve a distinct set of computer logic statements. This special microcomputer program is easily obtained by inserting the logic statements appropriate to the subsystem into the code for a generic enumeration microcomputer program.

When the generic enumeration microcomputer program pertains to more than six to eight components, the number of component operational states may be so large that physical understanding is compromised. Accordingly, each subsystem should be established (defined) such that its completely enumerated operational states are all easily comprehensible.

8.16.2. Reliability of a Repaired Subsystem

We now presume that when a component fails, the failed component is eventually replaced (say at a routinely scheduled maintenance) even if the subsystem continues to operate satisfactorily. Then, the respective durations endured by different components in the repaired subsystem can (will) differ. Satisfactory replacement of each failed component depends on (a) its availability and (b) the duration required for its satisfactory replacement, where (a) must be stated in terms of an (estimated probability)–(duration to obtain) expression and (b) must be expressed using a (estimated probability)–(duration to its satisfactory replacement) expression. Accordingly, reliability estimation for a repaired subsystem is best handled by simulation (with a minimum of 1000 trials).

To begin our discussion of a simulation-based estimate of subsystem reliability, presume that all components operate satisfactorily at duration $d = 0$. Then, given estimated reliabilities for each of the respective components generated by appropriate life (endurance) experiment test programs, let duration d be successively incremented by an appropriately small duration interval di during each simulation trail. In turn, for each such appropriately small duration interval, let a uniform pseudorandom number generator, zero to one, be used to establish whether each respective component fails during that interval. However, a component cannot fail in the interval of specific interest unless it has survived to the beginning of that interval. Accordingly, for each component in the system, the enumeration-based probability of failure in the interval of specific interest is given by the expression

enumeration-based probability of failure in the interval of
specific interest

$$= \frac{\int_{d}^{d+di} f(u)du}{\int_{d}^{\infty} f(u)du} = \frac{f(d)(di)}{\int_{d}^{\infty} f(u)du}$$

in which d is the duration to the beginning of the appropriately small duration interval di of specific interest. This probability expression is typically parsed into two terms: (a) the instantaneous failure rate function IFRF and (b) the appropriately small duration increment di of specific interest, where, by definition:

$$\text{IFRF} = \frac{f(d)}{1 - F(d)}$$

Clearly, the instantaneous failure rate function IFRF is analogous to a probability density function PDF. The IFRF is also commonly called the hazard rate function HRF, which is sometimes re-expressed as

$$\text{HRF} = \frac{f(d)}{\text{Reliability}(d)}$$

The replacement strategy for redundant components that fail during a given duration interval di must be explicitly defined so that the duration endured by each replacement component is known. (Remember that the duration required for satisfactory replacement of a component can extend over many appropriately short duration intervals.)

Each respective simulation trial is ended when the duration is reached that pertains to the *Type I* censoring duration used in conducting the respective component life (endurance) experiment test programs. Once all of the (mutually independent) simulation trials have been conducted, numerous statistics will be available for analysis, e.g., (a) the simulation-based estimate of the reliability at duration d for a new subsystem (which can be compared to its analytically estimated value), (b) the average duration to failure for a new subsystem, or (c) the simulation-based estimate of the reliability at duration d for a repaired subsystem, (d) the average number of failures for each component in the subsystem (which can be compared to its expected value), or (e) the average number of components that had to be repaired before any duration d of specific interest, etc. A comparison of the relevant statistics for various alternative designs should aid in selecting a more reliable configuration of components in the subsystem of specific interest.

Exercise Set 4

These exercises are intended to provide further insight into the sum of probabilities associated with enumerating all possible mutually exclusive and exhaustive outcomes for two or more random variables whose realizations are dichotomous.

1. Given the array of all distinct sets of component operational states for two components A and B, arbitrarily select estimated reliability values for A and B and then demonstrate that the sum of all four estimated probability products is one.
2. Given the array of all distinct sets of component operational states for three components A, B, and C, arbitrarily select estimated reliability values for A, B, and C and then demonstrate that the sum of all eight estimated probability products is one.
3. Given the array of all distinct sets of component operational states for four components A, B, C, and D, arbitrarily select estimated reliability values for A, B, C, and D and then demonstrate that the sum of all 16 estimated probability products is one.

8.17. CLOSURE

The dominance of quantitative CRD experiment test programs presented in this chapter should cause substantial concern regarding the applicability of these analyses in estimating the reliability of a subsystem (or even a component) in service operation. All CRD experiment test programs presume nominally identical test specimens and test conditions, whereas mechanical components are always produced in batches. Accordingly, it is seldom if ever reasonable to presume that a quantitative CRD test program will generate an unbiased estimate of any conceptual statistical model parameter of specific interest. Moreover, it is likely that vendor-supplied prototype components will markedly excel subsequent vendor-supplied production components. In estimating mechanical reliability, skepticism and investigation are always preferable to credulity and supinity.

All competent mechanical design methodologies attempt to establish adequate subsystem (component) reliability by extrapolating well-documented service-proven reliability performance for similar subsystems (components) to the subsystem (component) design of specific interest. The statistical and test planning tools presented herein can be used to enhance these mechanical design methodologies. However, regardless of the amount and type of design analysis and prototype testing involved, and regardless of the associated statistical methodology, the actual value for the reliability of a product can only be demonstrated by its service performance.

8.A. SUPPLEMENTAL TOPIC: EXACT (UNBIASED) A-BASIS AND B-BASIS STATISTICAL TOLERANCE LIMITS BASED ON UNCENSORED REPLICATE DATUM VALUES

Quantitative materials behavior data and the associated design allowables values are often stated in terms of A-basis and B-basis statistical tolerance limits. An A-basis statistical tolerance limit is the metric value that is allegedly exceeded by 99% of all possible replicate datum values with 0.95 probability, whereas a B-basis statistical tolerance limit is the metric value that is allegedly exceeded by 90% of all possible replicate datum values with 0.95 probability.

This supplemental topic presents microcomputer programs to compute A-basis and B-basis statistical tolerance limits based on uncensored replicate datum values that are randomly selected from (a) a conceptual (two-parameter) normal distribution, (b) a conceptual two-parameter $\log_e$-normal distribution, or (c) a conceptual two-parameter Weibull distribution. These A-basis and B-basis statistical tolerance limits are termed exact (unbiased) because the proportion of their associated probability statements (assertions) that are actually correct asymptotically approaches

0.95 under continual replication of the experiment test program, where 0.95 is the *scp* value of specific interest. However, it is important to understand that exact (unbiased) does not connote precise. When replicate experiment test programs are conducted, e.g., as in a simulation study, *A*-basis and *B*-basis statistical tolerance limit (realization) values typically exhibit marked variability. Thus, it is not sufficient merely to compute *A*-basis and *B*-basis statistical tolerance limits, we must also examine their precision.

Microcomputer programs and methodologies to compute *A*-basis and *B*-basis statistical tolerance limits based on replicate life (endurance) datum values with *Type I* censoring are presented in Supplemental Topic 8.D.

8.A.1. Exact (Unbiased) *A*-basis and *B*-basis Statistical Tolerance Limits Based on Uncensored Replicate Normally Distributed Datum Values

These statistical tolerance limits are computed using theoretical factors that are based on Student's noncentral t conceptual sampling distribution. Microcomputer programs *ABNSTL* and *BBNSTL* first calculate this tolerance limit factor and then calculate the associated *A*-basis and *B*-basis statistical tolerance limits.

Programs *ABNSTL* and *BBNSTL* pertain to quantitative CRD experiment test programs that generate six to 32 replicate datum values. Even the smallest mechanical test (or mechanical reliability) experiment test program should include *at least* six replicate datum values. On the other hand, it is unlikely that a mechanical test (or mechanical reliability) experiment test program will pertain to a batch of more than 32 nominally identical test specimens.

```
C> COPY STLDATA DATA

    1 file(s) copied


C> ABNSTL
```

Based on the presumption that the replicate datum values of specific interest were randomly selected from a conceptual (two-parameter) normal distribution, the *A*-basis statistical tolerance limit is equal to 0.113809D + 00.

```
C> COPY STLDATA DATA

     1 file(s) copied

C> BBNSTL
```

Based on the presumption that the replicate datum values of specific interest were randomly selected from a conceptual (two-parameter) normal distribution, the *B*-basis statistical tolerance limit is equal to 0.118762D + 00.

```
C> TYPE STLDATA

   21    Number of Replicate Datum Values
  0.120  Replicate Datum Value One, etc.
  0.132
  0.123
  0.128
  0.123
  0.124
  0.126
  0.129
  0.120
  0.132
  0.123
  0.126
  0.129
  0.128
  0.123
  0.124
  0.126
  0.129
  0.120
  0.126
  0.129
```

8.A.1.1. Discussion

Consider a reference set of replicate normally distributed pseudorandom datum values and their associated A-basis and B-basis statistical tolerance limits. If each of these datum values were multiplied by a positive constant c, the resulting A-basis and B-basis statistical tolerance limits would be multiplied by c. However, to generate this new set of pseudorandom datum values we would have to multiply both the mean and standard deviation of the reference normal distribution by c, or, equivalently, we could specify the value for the coefficient of variation (the ratio of the standard deviation to the mean) and then multiply the normal distribution mean by c. Accordingly, given any value for the coefficient of variation of specific interest, A-basis and B-basis statistical tolerance limits can be stated in terms of a (positive) proportion of the normal distribution mean. Thus, we can use this value for the coefficient of variation to generate n_{rep} replicate normally distributed data sets and thereby construct the empirical sampling distributions for the A-basis and B-basis statistical tolerance limits stated in terms of this unitless metric.

Given the actual value for the coefficient of variation of specific interest, microcomputer programs *SNABSTL* and *SNBBSTL* generate 30,000 replicate normally distributed data sets, each of size n_{dv}, and then delimit certain percentiles of the empirical sampling distributions for the respective A-basis and B-basis statistical tolerance limits in terms of their proportion of the conceptual distribution mean. These programs were run to generate the empirical sampling distribution data summarized in Tables 8.3 and 8.4. Note that these exact normal statistical limits are not precise for small numbers of replicate datum values when the coefficient of variation is relatively large.

The conventional estimate of the actual value for the coefficient of variation should be modified by an appropriate empirical or pragmatic statistical bias correction before running either microcomputer program *SNABSTL* or *SNBBSTL*. This methodology is presented in Supplemental Topic 8.D.

> *Remark*: When an estimated value is substituted for the actual value in the generation of a sampling distribution, we term the resulting sampling distribution pragmatic. We then assert (conjecture) that the pragmatic sampling distribution is a reasonably accurate approximation to its corresponding empirical sampling distribution.

Table 8.3 Proportions of the Conceptual (Two-Parameter) Normal Distribution Mean that Delimit the Central $p\%$ of the Empirical Sampling Distribution for A-Basis Statistical Tolerance Limits Based on 30,000 Replicate Normal Pseudorandom Data Sets, Each of Size n_{dv} [a]

n_{dv}	(stddev)/(mean) = 0.0125			(stddev)/(mean) = 0.025			(stddev)/(mean) = 0.05			(stddev)/(mean) = 0.1		
	90%	95%	99%	90%	95%	99%	90%	95%	99%	90%	95%	99%
6	**0.91–0.97**	**0.90–0.98**	**0.88–0.98**	**0.81–0.94**	**0.79–0.95**	0.76–0.97	0.62–0.88	0.59–0.90	0.53–0.94	0.24–0.77	0.18–0.81	0.06–0.87
8	**0.92–0.97**	**0.92–0.97**	**0.91–0.98**	**0.84–0.94**	**0.83–0.95**	0.81–0.96	0.69–0.88	0.67–0.90	0.62–0.93	0.37–0.77	0.33–0.80	0.25–0.86
10	**0.93–0.97**	**0.93–0.97**	**0.92–0.98**	**0.86–0.94**	0.85–0.95	0.84–0.96	0.72–0.88	0.71–0.90	0.67–0.92	0.45–0.77	0.41–0.79	0.34–0.84
12	**0.94–0.97**	**0.93–0.97**	**0.93–0.98**	**0.87–0.94**	**0.87–0.95**	0.85–0.96	0.75–0.88	0.73–0.90	0.70–0.92	0.49–0.77	0.46–0.79	0.41–0.83
16	0.94–0.97	**0.94–0.97**	**0.93–0.98**	**0.89–0.94**	**0.88–0.95**	**0.87–0.96**	0.77–0.88	0.67–0.89	0.74–0.91	0.54–0.77	0.52–0.78	0.47–0.82
24	**0.95–0.97**	**0.95–0.97**	**0.94–0.98**	**0.90–0.94**	**0.90–0.95**	**0.89–0.95**	**0.80–0.88**	0.79–0.89	0.77–0.91	0.60–0.77	0.58–0.78	0.55–0.81
32	**0.95–0.97**	**0.95–0.97**	**0.95–0.98**	**0.91–0.94**	**0.90–0.95**	**0.90–0.95**	**0.81–0.88**	**0.81–0.89**	0.79–0.90	0.63–0.77	0.61–0.78	0.59–0.80

[a] Bold entries denote that $p\%$ of replicate A-basis statistical tolerance limits lie within $\pm 5\%$ of the estimated mean of their empirical sampling distribution.

Table 8.4 Proportions of the Conceptual (Two-Parameter) Normal Distribution Mean that Delimit the Central $p\%$ of the Empirical Sampling Distribution for B-Basis Statistical Tolerance Limits Based on 30,000 Replicate Normal Pseudorandom Data Sets, Each of Size n_{dv}[a]

n_{dv}	(stddev)/(mean) = 0.0125			(stddev)/(mean) = 0.025			(stddev)/(mean) = 0.05			(stddev)/(mean) = 0.1		
	90%	95%	99%	90%	95%	99%	90%	95%	99%	90%	95%	99%
6	**0.94–0.98**	**0.94–0.99**	**0.93–0.99**	**0.89–0.97**	**0.88–0.97**	0.86–0.99	0.77–0.94	0.75–0.95	0.71–0.97	0.54–0.87	0.50–0.90	0.44–0.95
8	**0.95–0.98**	**0.95–0.99**	**0.94–0.99**	**0.91–0.97**	**0.90–0.97**	**0.89–0.98**	0.81–0.94	0.80–0.95	0.77–0.96	0.62–0.87	0.59–0.89	0.54–0.93
10	**0.96–0.98**	**0.96–0.99**	**0.95–0.99**	**0.92–0.97**	**0.91–0.97**	**0.90–0.98**	0.83–0.94	0.82–0.95	0.80–0.96	0.66–0.87	0.64–0.89	0.60–0.93
12	**0.96–0.98**	**0.96–0.99**	**0.95–0.99**	**0.92–0.97**	**0.92–0.97**	**0.91–0.98**	0.85–0.94	0.84–0.94	0.82–0.96	0.69–0.87	0.67–0.89	0.64–0.92
16	**0.97–0.98**	**0.96–0.99**	**0.96–0.99**	**0.93–0.97**	**0.93–0.97**	**0.92–0.98**	0.86–0.94	**0.86–0.94**	0.84–0.95	0.72–0.87	0.71–0.88	0.68–0.91
24	**0.97–0.98**	**0.97–0.99**	**0.97–0.99**	**0.94–0.97**	**0.94–0.97**	**0.93–0.98**	0.88–0.94	0.87–0.94	**0.86–0.95**	0.76–0.87	0.75–0.88	0.73–0.90
32	**0.97–0.98**	**0.97–0.99**	**0.97–0.99**	**0.94–0.97**	**0.94–0.97**	**0.94–0.97**	0.89–0.94	0.88–0.94	**0.88–0.95**	0.78–0.87	0.77–0.88	0.75–0.90

[a]Bold entries denote that $p\%$ of replicate B-basis statistical tolerance limits lie within $\pm5\%$ of the estimated mean of their empirical sampling distribution.

8.A.2. Exact (Unbiased) *A*-basis and *B*-basis Statistical Tolerance Limits Based on Uncensored Replicate $\log_e$–Normal Life (Endurance) Datum Values

Exact (unbiased) *A*-basis and *B*-basis statistical tolerance limits based on uncensored replicate $\log_e$–normal life (endurance) datum values are computed using the same theoretical factors that are used to compute exact (unbiased) *A*-basis and *B*-basis statistical tolerance limits for replicate uncensored normal datum values. Thus, microcomputer programs *ABLNSTL* and *BBLNSTL* are analogous to microcomputer programs *ABNSTL* and *BBNSTL*.

```
C> COPY STLDATA DATA

    1 file(s) copied

C> ABLNSTL
```

Based on the presumption that the replicate datum values of specific interest were randomly selected from a conceptual two-parameter $\log_e$–normal distribution, the *A*-basis statistical tolerance limit is equal to 0.114304D + 00.

```
C> COPY STLDATA DATA

    1 file(s) copied

C> BBLNSTL
```

Based on the presumption that the replicate datum values of specific interest were randomly selected from a conceptual two-parameter $\log_e$–normal distribution, the *B*-basis statistical tolerance limit is equal to 0.118900D + 00.

Microcomputer programs *SLNABSTL* and *SLNBBSTL* are analogous to microcomputer programs *SWABSTL* and *SWBBSTL*, which follow. These programs were run to generate the empirical sampling distribution data summarized in Tables 8.5 and 8.6. Note that *A*-basis and *B*-basis statistical tolerance limits exhibit such marked variability for practical experiment test program sizes that very few of these statistical tolerance limits can be regarded as being reasonably precise.

8.A.3. Exact (Unbiased) *A*-basis and *B*-basis Statistical Tolerance Limits Based on Uncensored Replicate Weibull Life (Endurance) Datum Values

Exact (unbiased) *A*-basis and *B*-basis statistical tolerance limits based on uncensored replicate Weibull life (endurance) datum values can be computed using simulation-based factors (based on best linear invariant estimates) tabulated by Mann and Fertig (1973). Microcomputer program *BLISTL* employs these factors to compute exact (unbiased) *A*-basis and *B*-basis statistical tolerance limits given the uncensored replicate Weibull life (endurance) datum values of specific interest. The variability of these statistical tolerance limits can be examined by running microcomputer programs *SWABSTL* and *SWBBSTL* with any combination of the Weibull distribution mean and standard deviation that is of specific interest, e.g., the estimated mean and standard deviation pertaining to the datum values for the experiment test program that was actually conducted. These programs were run to generate the empirical sampling distribution data summarized in Tables 8.7 and 8.8. Note that exact (unbiased) *A*-basis and *B*-basis statistical tolerance limit values for Weibull life (endurance) datum values are slightly less precise (exhibit slightly more variability) than the corresponding statistical tolerance limit values for $\log_e$–normal life (endurance) datum values.

8.A.3.1. Summary and Recommendations

Tables 8.3 through 8.8 demonstrate that, depending on the values for n_{dv} and the coefficient of variation, exact (unbiased) *A*-basis and *B*-basis statistical tolerance limits computed by running microcomputer programs *ABNSTL*, *BBNSTL*, *ABLNSTL*, *BBLNSTL*, and *BLISTL* can exhibit marked variability under continual replication of the experiment test program. Accordingly, we make the following recommendations:

> *Recommendation one*: Instead of reporting the *A*-basis or *B*-basis statistical tolerance limit computed by running microcomputer program *ABNSTL*, *BBNSTL*, *ABLNSTL*, *BBLNSTL*, or *BLISTL*,

Table 8.5 Proportions of the Conceptual Two-Parameter $\log_e$–Normal Distribution Mean that Delimit the Central $p\%$ of the Empirical Sampling Distribution for A-Basis Statistical Tolerance Limits Based on 30,000 Replicate $\log_e$–Normal Pseudorandom Data Sets, Each of Size n_{db}[a]

n_{db}	(stddev)/(mean) = 0.05			(stddev)/(mean) = 0.1			(stddev)/(mean) = 0.2			(stddev)/(mean) = 0.4		
	90%	95%	99%	90%	95%	99%	90%	95%	99%	90%	95%	99%
6	0.68–0.89	0.66–0.91	0.62–0.94	0.47–0.79	0.44–0.82	0.39–0.88	0.22–0.62	0.19–0.67	0.15–0.76	0.05–0.38	0.04–0.44	0.03–0.58
8	0.73–0.89	0.72–0.90	0.68–0.93	0.53–0.79	0.51–0.81	0.47–0.86	0.29–0.62	0.26–0.66	0.22–0.73	0.08–0.38	0.07–0.43	0.05–0.53
10	0.76–0.89	0.75–0.90	0.77–0.92	0.57–0.79	0.55–0.81	0.51–0.85	0.33–0.62	0.31–0.65	0.27–0.71	0.11–0.38	0.10–0.42	0.08–0.50
12	0.77–0.89	0.76–0.90	0.74–0.92	0.60–0.79	0.58–0.81	0.55–0.84	0.36–0.62	0.34–0.65	0.30–0.70	0.13–0.38	0.12–0.42	0.09–0.49
16	0.80–0.89	0.79–0.90	0.77–0.91	0.63–0.79	0.62–0.80	0.59–0.83	0.40–0.62	0.38–0.64	0.35–0.69	0.16–0.38	0.15–0.41	0.12–0.46
24	**0.82–0.89**	0.81–0.90	0.80–0.91	0.67–0.79	0.66–0.80	0.63–0.82	0.44–0.62	0.43–0.64	0.40–0.67	0.20–0.38	0.19–0.40	0.16–0.45
32	**0.83–0.89**	**0.82–0.89**	0.81–0.91	0.69–0.79	0.68–0.80	0.66–0.82	0.47–0.62	0.46–0.64	0.43–0.62	0.22–0.38	0.21–0.40	0.19–0.44

[a]Bold entries denote that $p\%$ of replicate A-basis statistical tolerance limits lie within $\pm5\%$ of the estimated mean of their empirical sampling distribution.

Table 8.6 Proportions of the Conceptual Two-Parameter $\log_e$–Normal Distribution Mean that Delimit the Central $p\%$ of the Empirical Sampling Distribution for B-Basis Statistical Tolerance Limits Based on 30,000 Replicate $\log_e$–Normal Pseudorandom Data Sets, Each of Size n_{dv} [a]

n_{dv}	(stddev)/(mean) = 0.05			(stddev)/(mean) = 0.1			(stddev)/(mean) = 0.2			(stddev)/(mean) = 0.4		
	90%	95%	99%	90%	95%	99%	90%	95%	99%	90%	95%	99%
6	0.79–0.94	0.78–0.95	0.75–0.78	0.50–0.87	0.47–0.91	0.41–0.96	0.25–0.76	0.22–0.82	0.18–0.92	0.07–0.57	0.07–0.66	0.03–0.82
8	0.83–0.94	0.82–0.95	0.79–0.97	0.68–0.88	0.66–0.89	0.63–0.93	0.46–0.76	0.44–0.79	0.40–0.86	0.22–0.57	0.20–0.62	0.16–0.71
10	**0.84–0.94**	0.84–0.95	0.82–0.96	0.71–0.88	0.70–0.89	0.67–0.92	0.51–0.76	0.49–0.79	0.45–0.84	0.25–0.57	0.23–0.61	0.20–0.69
12	**0.86–0.94**	0.85–0.94	0.83–0.96	0.73–0.88	0.72–0.89	0.69–0.92	0.53–0.76	0.51–0.78	0.48–0.83	0.28–0.57	0.26–0.60	0.23–0.68
16	**0.87–0.94**	**0.86–0.94**	0.85–0.95	0.76–0.88	0.74–0.89	0.72–0.91	0.57–0.76	0.55–0.78	0.52–0.82	0.32–0.57	0.30–0.59	0.27–0.66
24	**0.89–0.94**	**0.88–0.94**	**0.87–0.95**	0.78–0.88	0.77–0.89	0.76–0.90	0.61–0.76	0.60–0.78	0.60–0.81	0.37–0.57	0.35–0.59	0.33–0.64
32	**0.89–0.94**	**0.89–0.94**	**0.88–0.95**	**0.80–0.88**	0.79–0.88	0.78–0.90	0.63–0.76	0.62–0.77	0.60–0.80	0.40–0.57	0.38–0.59	0.35–0.63

[a]Bold entries denote that $p\%$ of replicate B-basis statistical tolerance limits lie within $\pm5\%$ of the estimated mean of their empirical sampling distribution.

469

Table 8.7 Proportions of the Conceptual Two-Parameter Weibull Distribution Mean that Delimit the Central $p\%$ of the Empirical Sampling Distribution for A-Basis Statistical Tolerance Limits Based on 30,000 Replicate Weibull Pseudorandom Data Sets, Each of Size n_{dv}[a]

n_{dv}	(stddev)/(mean) = 0.05			(stddev)/(mean) = 0.1			(stddev)/(mean) = 0.2			(stddev)/(mean) = 0.4		
	90%	95%	99%	90%	95%	99%	90%	95%	99%	90%	95%	99%
6	0.52–0.85	0.48–0.88	0.42–0.92	0.26–0.71	0.23–0.76	0.17–0.84	0.06–0.49	0.04–0.56	0.02–0.67	0.00–0.20	0.00–0.26	0.00–0.42
8	0.59–0.85	0.57–0.87	0.51–0.91	0.34–0.72	0.31–0.75	0.26–0.82	0.10–0.49	0.08–0.54	0.06–0.65	0.01–0.20	0.00–0.25	0.00–0.37
10	0.63–0.85	0.61–0.87	0.57–0.90	0.39–0.71	0.36–0.74	0.31–0.80	0.14–0.49	0.12–0.53	0.08–0.62	0.01–0.20	0.01–0.24	0.00–0.34
12	0.66–0.85	0.64–0.87	0.60–0.90	0.43–0.72	0.40–0.74	0.35–0.79	0.16–0.49	0.14–0.53	0.11–0.61	0.02–0.20	0.02–0.24	0.01–0.33
16	0.70–0.85	0.68–0.86	0.65–0.89	0.47–0.71	0.45–0.73	0.41–0.78	0.21–0.49	0.19–0.52	0.15–0.59	0.03–0.20	0.03–0.23	0.02–0.30

[a] Bold entries denote that $p\%$ of replicate A-basis statistical tolerance limits lie within $\pm5\%$ of the estimated mean of their empirical sampling distribution. Note that there are no bold entries in this table.

Table 8.8 Proportions of the Conceptual Two-Parameter Weibull Distribution Mean that Delimit the Central $p\%$ of the Empirical Sampling Distribution for B-Basis Statistical Tolerance Limits Based on 30,000 Replicate Weibull Pseudorandom Data Sets, Each of Size n_{dv}[a]

n_{dv}	(stddev)/(mean) = 0.05			(stddev)/(mean) = 0.1			(stddev)/(mean) = 0.2			(stddev)/(mean) = 0.4		
	90%	95%	99%	90%	95%	99%	90%	95%	99%	90%	95%	99%
6	0.72–0.93	0.70–0.95	0.64–0.98	0.51–0.86	0.47–0.89	0.41–0.95	0.24–0.73	0.21–0.78	0.15–0.87	0.04–0.49	0.03–0.56	0.02–0.72
8	0.77–0.93	0.75–0.95	0.71–0.97	0.58–0.87	0.55–0.89	0.50–0.93	0.31–0.73	0.28–0.78	0.22–0.85	0.08–0.49	0.06–0.55	0.04–0.68
10	0.80–0.93	0.78–0.94	0.75–0.96	0.62–0.87	0.60–0.89	0.55–0.93	0.37–0.73	0.34–0.77	0.28–0.84	0.11–0.49	0.09–0.54	0.06–0.65
12	0.82–0.93	0.80–0.94	0.77–0.96	0.66–0.87	0.63–0.88	0.59–0.92	0.40–0.73	0.38–0.76	0.33–0.82	0.14–0.49	0.12–0.53	0.09–0.63
16	0.84–0.93	0.83–0.94	0.80–0.96	0.69–0.87	0.67–0.88	0.64–0.91	0.45–0.73	0.43–0.76	0.39–0.81	0.17–0.49	0.15–0.53	0.12–0.60

[a]Bold entries denote that $p\%$ of replicate B-basis statistical tolerance limits lie within $\pm 5\%$ of the estimated mean of their sampling distribution. Note that there are no bold entries in this table.

report the value that is the mean of the pragmatic sampling distribution computed by running microcomputer program *SNABSTL, SLNBBST, SNABSTL, SLNBBST, SWABSTL,* or *SWBBSTL* with its pragmatic bias-corrected coefficient of variation (Supplemental Topic 8.D) as input data. This methodology increases the precision of the reported *A*-basis and *B*-basis statistical tolerance limits, viz., it reduces the range of the sampling distribution percentiles reported in Tables 8.3 through 8.8.

Recommendation Two: Because the value for the coefficient of variation is so useful in terms of perspective and test planning it is good statistical practice to compute and compile (record) values for the estimated mean, standard deviation, and coefficient of variation for all experiment test programs that are commonly conducted. Otherwise the actual value for the coefficient of variation will most likely be markedly underestimated in the statistical planning of life (endurance) experiment test programs.

8.A.4. Classical *Distribution-Free* (Nonparametric) *A*-Basis and *B*-Basis Statistical Tolerance Limits

The following *A*-basis and *B*-basis statistical tolerance limits find only occasional application in mechanical reliability, but are nevertheless included in this supplemental topic for completeness.

Consider a conceptual statistical distribution with a continuous metric whose analytical PDF expression is *unspecified* (unknown). Distribution-free (nonparametric) *A*-basis and *B*-basis statistical tolerance limits are computed using the classical distribution-free (nonparametric) lower (one-sided) statistical tolerance limit expression:

$$\text{probability}[X(1) < (1 - pp^*)] = \left[1 - (pp^*)^{n_{dv}}\right]$$
$$= \text{actual value for the statistical confidence probability} = scp^*$$

in which $X(1)$ is the smallest of the n_{dv} replicate datum values, pp^* is the population proportion of specific interest, either 0.99 or 0.90, that lies above $X(1)$ with the actual value for the statistical confidence probability $scp = scp^*$, and n_{dv} is the number of replicate datum values generated in a quantitative CRD experiment test program. Note that, because classical distribution-free (nonparametric) statistical tolerance limits depend only on the smallest of the n_{dv} replicate datum values, the experiment test program can involve *Type I* censored (or even arbitrarily suspended) tests.

The problem with distribution-free (nonparametric) lower (one-sided) statistical tolerance limits is that n_{dv} must be at least equal to approximately 300 and 30, respectively, to attain A-basis and B-basis statistical tolerance limits. Otherwise, the actual values for the statistical confidence probability scp^* are less than 0.95. Clearly, these minimum values for n_{dv} are too large to allow distribution-free (nonparametric) A-basis and B-basis statistical tolerance limits to have wide practical application. In fact, the minimum number of replicate datum values n_{dv} is so large that a compilation of material behavior datum values pertaining to several nominally identical experiment test programs is typically required even to establish a B-basis distribution-free (nonparametric) statistical tolerance limit. (Although a strict quantitative interpretation of a statistical tolerance limit based on a compilation of material behavior datum values pertaining to several nominally identical experiment test programs may be obscure, such compilations are preferable to replicate datum values generated in a single experiment test program. Statistical tolerance limits based on compilations of datum values virtually eliminate the chance that all such values pertain to atypical batches of experimental units or atypical test conduct procedures.)

Exercise Set 5

These exercises are intended to verify the text assertion that extremely large experiment test programs are required to establish A-basis and B-basis statistical tolerance limits using the classical distribution-free (nonparametric) lower (one-sided) statistical tolerance limit expression.

1. Compute the respective minimum experiment test program size to establish A-basis and B-basis statistical tolerance limits using the classical distribution-free (nonparametric) lower (one-sided) statistical tolerance limit expression.
2. Compute the minimum experiment test program size required to establish the classical distribution-free (nonparametric) 99% lower (one-sided) statistical tolerance limit that allegedly bounds the 10th and 01th percentiles of the unspecified conceptual statistical distribution of specific interest.
3. Based on the results of Exercises 1 and 2, do you see why the $\log_e$–normal and Weibull life (endurance) distributions are so widely presumed when computing A-basis and B-basis statistical tolerance limits? Does this "necessity" improve the credibility of either distribution?

8.A.5. Modified Distribution-Free (Nonparametric) *B*-Basis Lower (One-Sided) Statistical Tolerance Limit

Suppose we assert that (a) the PDF for the unspecified continuous conceptual life (endurance) distribution of specific interest is equal to zero when the life (endurance) metric is equal to zero and that (b) the slope (derivative) of this PDF is strictly increasing in the interval from zero to the metric value pertaining to p^*. Then, given this PDF behavior, we can assert that

$$\text{probability}\left[\frac{X(1)}{c} < (1-pp^*)\right] = 1 - \left\{\left[1 - (1-pp^*)c^2\right]^{n_{dv}}\right\}$$

$$= \text{statistical confidence probability} = scp$$

provided that $c > 1$ and that $(1-pp^*)(c^2) < (p^*)$. This modified expression is useful in establishing *B*-basis statistical tolerance limits when n_{dv} is slightly less than the minimum value required to attain the required scp value for the classical distribution-free (nonparametric) *B*-basis statistical tolerance limit. Its credibility depends on the minimum value for p^* that appears to be reasonable to presume for the unspecified continuous life (endurance) PDF. For purposes of perspective, the slope of the PDF is strictly increasing for p^* values up to approximately 0.16 for a conceptual (two-parameter) normal distribution, up to approximately 0.21 for a conceptual (two-parameter) logistic distribution, up to approximately 0.32 for a conceptual (two-parameter) smallest-extreme-value distribution (skewed to the left), and up to approximately 0.07 for a conceptual (two-parameter) largest-extreme-value distribution (skewed to the right). The p^* value for the conceptual two-parameter Weibull distribution depends on the actual value for *cdp2*: its PDF is equal to zero when the life (endurance) metric is equal to zero for all values of *cdp2* greater than one, and its slope is strictly increasing for p^* between approximately 0.07 and 0.26 when the actual value for the *cdp2* is between 2.5 and 10.0.

Microcomputer program *MDFBBSTL* (modified distribution-free *B*-basis statistical tolerance limit) computes the modified distribution-free (nonparametric) *B*-basis tolerance limit for replicate life (endurance) datum values that are presumed to have been randomly selected from an unspecified conceptual statistical distribution with a continuous metric. As mentioned above, this program has application (only) for estimating distribution-free (nonparametric) *B*-basis tolerance limits in situations where the experiment test program size is slightly smaller than the minimum size required to establish the classical distribution-free (nonparametric) *B*-basis statistical tolerance limit.

```
C> COPY MDFDATA DATA

    1 file(s) copied

C> MDFBBSTL
```

The modified distribution-free (nonparametric) *B*-basis statistical tolerance limit is equal to 0.104074D + 00.

8.A.6. Perspective and Closure

It is extremely unlikely that a sufficient number of "replicate" datum values can ever be compiled to establish a credible distribution-free (nonparametric) *A*-basis statistical tolerance limit. Moreover, although it is possible that a sufficient number of "replicate" datum values can be generated in a mechanical reliability test program to establish a credible distribution-free (nonparametric) *B*-basis statistical tolerance limit, it will not be a common occurrence. Accordingly, almost all statistical tolerance limits in mechanical reliability analyses are based on the presumption that the experiment test program datum values are (were) randomly selected from a conceptual (two-parameter) normal, a conceptual two-parameter $\log_e$–normal, or a conceptual two-parameter Weibull distribution. It is, therefore, very important in terms of perspective to understand that these conceptual statistical distributions are merely analytical models and do not (and never will) model any mechanical mode of failure exactly. Thus, it is never rational to attempt to "improve" on *A*-basis or *B*-basis statistical tolerance limits by computing a lower 100(*scp*)% (one-sided) statistical confidence limit that allegedly bounds the $(pf)^{th}$ percentile of the presumed conceptual distribution by selecting *scp* to be greater than 0.95 or selecting *pf* to be less than 0.01. In fact, even selecting *scp* to be equal to 0.95 while simultaneously selecting *pf* to be equal to 0.01 (as for an *A*-basis statistical tolerance limit) typically stretches the bounds of common sense.

> *Reminder*: Since all components (experimental units) are manufactured (produced) in batches, the critical issue in rationally interpreting any quantitative statistical estimate is whether the given batch of components (experimental units) is actually representative of the population of specific interest.

8.B. SUPPLEMENTAL TOPIC: MAXIMUM LIKELIHOOD ANALYSIS FOR OUTCOME OF A LIFE (ENDURANCE) EXPERIMENT TEST PROGRAM WITH *TYPE I* CENSORING, PRESUMING log$_e$-NORMAL DATUM VALUES

Recall that given replicate life (endurance) datum values that are presumed to have been randomly selected from a conceptual two-parameter Weibull distribution, the natural logarithms of these datum values can be presumed to have been randomly selected from the corresponding conceptual (two-parameter) smallest-extreme-value distribution with a logarithmic metric. Similarly, given life (endurance) datum values that are presumed to have been randomly selected from a conceptual two-parameter log$_e$–normal distribution, the natural logarithms of these datum values can be presumed to have been randomly selected from the corresponding conceptual (two-parameter) normal distribution with a logarithmic metric. Accordingly, we now perform a ML analysis for log$_e$–normal life (endurance) datum values by taking the natural logarithms of the respective datum values and presuming that these transformed datum values are normally distributed. The resulting point and interval estimates of specific interest are then restated in the original metric by taking their exponentials (antilogarithms).

When the life (endurance) CRD quantitative experiment test program includes *Type I* censored tests, the likelihood can be expressed as

$$\text{likelihood} = \frac{(n_f + n_s)!}{n_f \cdot !n_s!} \cdot \prod_{i=1}^{n_f} f[\log_e(fnc_i)] \cdot \prod_{j=1}^{n_s} \{1 - F[\log_e(snc^*)]\}$$

in which n_f denotes the number of failed items and n_s denotes the number of *Type I* censored tests. The conventional parameterization for the conceptual (two-parameter) normal distribution is

$$f[\log_e(fnc_i)] = \frac{1}{\sqrt{2\pi} \cdot csp} \cdot \exp\left\{-\frac{1}{2} \cdot \left[\frac{\log_e(fnc_i) - clp}{csp}\right]^2\right\}$$

in which, as mentioned above, the metric is $\log_e(fnc)$, and

$$F[\log_e(snc_j)] = \frac{1}{\sqrt{2\pi} \cdot csp} \cdot \int_{-\infty}^{\log_e(snc_j)} \exp\left[-\frac{1}{2} \cdot \left(\frac{u - clp}{csp}\right)^2\right] du$$

in which u is the dummy variable of integration. Recall, however, that there are four alternative parameterizations that can be used to specify a conceptual (two-parameter) normal distribution.

It is analytically convenient to re-express the likelihood in terms of the standardized conceptual normal distribution variate y, which for this conventional parameterization is

$$y = \frac{\log_e(fnc) - clp}{csp}$$

The likelihood expression can then be re-expressed as

$$\text{likelihood} = \frac{(n_f + n_s)!}{n_f \cdot !n_s!} \cdot \prod_{i=1}^{n_f} \frac{1}{csp} \cdot f(y_i) \cdot \prod_{j=1}^{n_s} [1 - F(y^*)]$$

Thus, ignoring the nonessential terms, the $\log_e$(likelihood) expression becomes

$$\log_e(\text{likelihood}) = -n_f \cdot \log_e(csp) - \frac{1}{2} \cdot \sum_{i=1}^{n_f} y_i^2 + \sum_{j=1}^{n_s} \log_e[1 - F(y^*)]$$

In turn, the maximum likelihood estimates of the actual values for the clp and the csp are obtained by simultaneously solving the two (nonlinear) equations that are generated by setting the partial derivatives of this $\log_e$(likelihood) expression with respect to clp and csp equal to zero, viz.,

$$\frac{\partial \log_e(\text{likelihood})}{\partial clp} = -\frac{1}{2} \cdot \sum_{i=1}^{n_f} 2 \cdot y_i \cdot \frac{\partial y_i}{\partial clp} + \sum_{j=1}^{n_s} \frac{1}{1 - F(y^*)} \cdot \frac{\partial[1 - F(y^*)]}{\partial clp}$$

and

$$\frac{\partial \log_e(\text{likelihood})}{\partial csp} = -\frac{n_f}{csp} - \frac{1}{2} \cdot \sum_{i=1}^{n_f} 2 \cdot y_i \cdot \frac{\partial y_i}{\partial csp} + \sum_{j=1}^{n_s} \frac{1}{1 - F(y^*)} \cdot \frac{\partial[1 - F(y^*)]}{\partial csp}$$

Then, substituting

$$\frac{\partial y}{\partial clp} = -\frac{1}{csp}$$

$$\frac{\partial[1 - F(y)]}{\partial clp} = \frac{\partial[1 - F(y)]}{\partial y} \cdot \frac{\partial y}{\partial clp} = -f(y) \cdot \left(-\frac{1}{csp}\right) = \frac{f(y)}{csp}$$

$$\frac{\partial y}{\partial csp} = [\log_e(fnc_i) - clp] \cdot \left(-\frac{1}{csp^2}\right) = \frac{-y}{csp}$$

$$\frac{\partial[1 - F(y)]}{\partial csp} = \frac{\partial[1 - F(y)]}{\partial y} \cdot \frac{\partial y}{\partial csp} = -f(y) \cdot \left(-\frac{y}{csp}\right) = \frac{y \cdot f(y)}{csp}$$

into the respective partial derivative expressions generates the following simultaneous (nonlinear) equations:

$$\frac{\partial \log_e(\text{likelihood})}{\partial clp} = \frac{1}{csp} \cdot \left[\sum_{i=1}^{n_f} y_i + \sum_{j=1}^{n_s} \frac{f(y^*)}{1 - F(y^*)} \right] = 0$$

and

$$\frac{\partial \log_e(\text{likelihood})}{\partial csp} = \frac{1}{csp} \cdot \left[-n_f + \sum_{i=1}^{n_f} y_i^2 + \sum_{j=1}^{n_s} \frac{y^* \cdot f(y^*)}{1 - F(y^*)} \right] = 0$$

in which the respective numerical values of y, $f(y)$, and $F(y)$ each depend on the corresponding values of est(clp), est(csp), fnc_i, and snc^*. The Newton–Raphson (N–R) method of numerical solution for est(clp) and est(csp) is very effective when reasonably accurate initial guestimates are used to begin this iterative methodology. Recall that it requires analytical expressions for the respective second partial derivatives, viz.,

$$\frac{\partial^2 \log_e(\text{likelihood})}{\partial clp^2} = -\frac{1}{csp^2} \cdot \left\{ n_f - \sum_{j=1}^{n_s} \frac{y^* \cdot f(y^*)}{1 - F(y^*)} + \sum_{j=1}^{n_s} \frac{f(y^*)^2}{[1 - F(y^*)]^2} \right\}$$

$$\frac{\partial^2 \log_e(\text{likelihood})}{\partial clp\, \partial csp} = -\frac{1}{csp^2} \cdot \left\{ 2 \sum_{i=1}^{nf} y_i - \sum_{j=1}^{n_s} \frac{y^{*2} \cdot f(y^*)}{[1 - F(y^*)]} \right.$$

$$\left. + \sum_{j=1}^{n_s} \frac{y^* \cdot f(y^*)^2}{[1 - F(y^*)]^2} + \sum_{j=1}^{n_s} \frac{f(y^*)}{1 - F(y^*)} \right\}$$

$$\frac{\partial^2 \log_e(\text{likelihood})}{\partial csp^2} = -\frac{1}{csp^2} \cdot \left\{ -n_f + 3 \cdot \sum_{i=1}^{n_f} y_i^2 - \sum_{j=1}^{n_s} \frac{y^{*3} \cdot f(y^*)}{[1 - F(y^*)]} \right.$$

$$\left. + \sum_{j=1}^{n_s} \frac{y^* \cdot 2f(y^*)^2}{[1 - F(y^*)]^2} + 2 \cdot \sum_{j=1}^{n_s} \frac{y^* \cdot f(y^*)}{1 - F(y^*)} \right\}$$

that are evaluated numerically by substituting est(clp) for the clp and est(csp) for the csp in the respective expressions for y, $f(y)$, and $F(y)$. The negatives of the resulting values for these second partial derivatives form the respective elements of a 2 by 2 array, whose inverse is the symmetrical estimated asymptotic covariance matrix, viz.,

$|\text{est}\{\text{covar}[\text{est}(clp), \text{est}(csp)]\}|$

$$= \begin{vmatrix} \text{est}\{\text{var}[\text{est}(clp)]\} & \text{est}\{\text{covar}[\text{est}(clp), \text{est}(csp)]\} \\ \text{est}\{\text{covar}[\text{est}(clp), \text{est}(csp)]\} & \text{est}\{\text{var}[\text{est}(csp)]\} \end{vmatrix}$$

This estimated covariance matrix establishes the values needed to compute classical lower $100(scp)\%$ (one-sided) asymptotic statistical confidence limits that allegedly bound the actual value for the $fnc(pf)$ of specific interest using the propagation of variability methodology (Supplemental Topic 7.A.). The associated pragmatic bias-corrected lower $100(scp)\%$ (one-sided) statistical confidence limits are computed by running microcomputer programs *LNPBCPV* and *LNPBCLR*. (Supplemental Topics 8.D and 8.E).

> *Remark*: If the life (endurance) experiment test program includes arbitrarily suspended tests, then the n_s (equal) snc^*'s are merely replaced by the respective (different) snc_j's in microcomputer file *WBLDATA*. Recall, however, that the fundamental statistical concept of a continually replicated experiment test program is obscure when arbitrarily suspended tests occur.

8.B.1. Numerical Examples

The following numerical examples for the conceptual two-parameter $\log_e$–normal distribution are obtained by running microcomputer programs *LN1A*, *LN2A*, *LN3A*, and *LN4A* with the fatigue life datum values found in microcomputer file *WBLDATA*. These numerical examples correspond directly to the numerical examples that were obtained by running microcomputer programs *LSEV1A*, *LSEV2A*, *LSEV3A*, and *LSEV4A*. The $LN(J)A$ series of microcomputer programs confirm that when the respective lower $100(scp)\%$ (one-sided) asymptotic statistical confidence limits that allegedly bound the actual $fnc(pf)$ value of specific interest are computed using propagation of variance methodology, these confidence limits are independent of the parameterization used to state the CDF.

8.C. SUPPLEMENTAL TOPIC: MAXIMUM LIKELIHOOD ANALYSIS FOR OUTCOME OF A STRENGTH (RESISTANCE) EXPERIMENT TEST PROGRAM WITH TWO OR MORE STRESS (STIMULUS) LEVELS

We begin this section by considering the ML analysis of the outcome of a generic strength (resistance) test. We then simplify this ML analysis to

```
C> COPY WBLDATA DATA

    1 files(s) copied

C> LN1A
```

Given the Standardized Conceptual Normal Distribution Variate
$y = csp \cdot [\log_e(fnc) - clp]$

$\text{est}(clp) = 0.5956793964\text{D} + 01$
$\text{est}(csp) = 0.4171442213\text{D} + 01$
$\text{est}\{\text{var}[\text{est}(clp)]\} = 0.9987394347\text{D} - 02$
$\text{est}\{\text{var}[\text{est}(csp)]\} = 0.1903271765\text{D} + 01$
$\text{est}\{\text{covar}[\text{est}(clp),\text{est}(csp)]\} = -0.1346739280\text{D} - 01$
$\text{est}(conceptual\ correlation\ coefficient) = -0.9768030920\text{D} - 01$

fnc	est(*y*)	est(*pf*)
277.000	−1.3881578	0.0825445
310.000	−0.9186420	0.1791414
374.000	−0.1357311	0.4460169
402.000	0.1654316	0.5656978
456.000	0.6912032	0.7552811

snc	est(*y*)	est(*pf*)
500.000	1.0754568	0.8589149

Classical Lower 95% (One-Sided) Asymptotic Statistical Confidence
Limits that Allegedly Bound the Actual Value for *fnc*(01)

158.933 – Computed Using the Propagation of Variability Expression
for est[var(est{log$_e$[*fnc*(01)]})]
112.714 – Computed Using the Propagation of Variability Expression
for est{var[est(*y* given *fnc* = *closascl*)]}

```
C> COPY WBLDATA DATA

    1 files(s) copied

C> LN2A
```

Given the Standardized Conceptual Normal Distribution Variate
$y = -clp + csp \cdot \log_e fnc)$

est(clp) = 0.2484842180D + 02
est(csp) = 0.4171442213D + 01
est{var[est(clp)]} = 0.6703904559D + 02
est{var[est(csp)]} = 0.1903271765D + 01
est{covar[est(clp),est(csp)]} = −0.1128121931D + 02
est($conceptual\ correlation\ coefficient$) = −0.9987153578D + 00

fnc	est(y)	est(pf)
277.000	−1.3881578	0.0825445
310.000	−0.9186420	0.1791414
374.000	−0.1357311	0.4460169
402.000	0.1654316	0.5656978
456.000	0.6912032	0.7552811

snc	est(y)	est(pf)
500.000	1.0754568	0.8589149

Classical Lower 95% (One-Sided) Asymptotic Statistical Confidence
Limits that Allegedly Bound the Actual Value for fnc(01)

158.933 – Computed Using the Propagation of Variability Expression
for est[var(est{$\log_e$[fnc(01)]})]
112.714 – Computed Using the Propagation of Variability Expression
for est{var[est(y given $fnc = closascl$)]}

```
C> COPY WBLDATA DATA

    1 files(s) copied

C> LN3A
```

Given the Standardized Conceptual Normal Distribution Variate
$y = [\log_e(fnc) - clp]/csp$

$\text{est}(clp) = 0.5956793964\text{D}+01$
$\text{est}(csp) = 0.2397252434\text{D}+00$
$\text{est}\{\text{var}[\text{est}(clp)]\} = 0.99787394347\text{D}-02$
$\text{est}\{\text{var}[\text{est}(csp)]\} = 0.6285732256\text{D}-02$
$\text{est}\{\text{covar}[\text{est}(clp),\text{est}(csp)]\} = 0.7739467195\text{D}-03$
$\text{est}(conceptual\ correlation\ coefficient) = 0.9768030920\text{D}-01$

fnc	est(y)	est(pf)
277.000	−1.3881578	0.0825445
310.000	−0.9186420	0.1791414
374.000	−0.1357311	0.4460169
402.000	0.1654316	0.5656978
456.000	0.6912032	0.7552811

snc	est(y)	est(pf)
500.000	1.0754568	0.8589149

Classical Lower 95% (One-Sided) Asymptotic Statistical Confidence
Limits that Allegedly Bound the Actual Value for $fnc(01)$

158.933 – Computed Using the Propagation of Variability Expression
for $\text{est}[\text{var}(\text{est}\{\log_e[fnc(01)]\})]$
112.714 – Computed Using the Propagation of Variability Expression
for $\text{est}\{\text{var}[\text{est}(y\ \text{given}\ fnc = closascl)]\}$

```
C> COPY WBLDATA DATA

    1 files(s) copied

C> LN4A
```

Given the Standardized Conceptual Normal Distribution Variate
$y = \{[\log_e(fnc)]/csp\} - clp$

$\text{est}(clp) = 0.2484842180D + 02$
$\text{est}(csp) = 0.2397252434D + 00$
$\text{est}\{\text{var}[\text{est}(clp)]\} = 0.6703904559D + 02$
$\text{est}\{\text{var}[\text{est}(csp)]\} = 0.6285732256D - 02$
$\text{est}\{\text{covar}[\text{est}(clp),\text{est}(csp)]\} = -0.6483112811D + 02$
$\text{est}(conceptual\ correlation\ coefficient) = -0.9987153578D + 00$

fnc	est(*y*)	est(*pf*)
277.000	−1.3881578	0.0825445
310.000	−0.9186420	0.1791414
374.000	−0.1357311	0.4460169
402.000	0.1654316	0.5656978
456.000	0.6912032	0.7552811

snc	est(*y*)	est(*pf*)
500.000	1.0754568	0.8589149

Classical Lower 95% (One-Sided) Asymptotic Statistical Confidence
Limits that Allegedly Bound the Actual Value for *fnc*(01)

158.933 − Computed Using the Propagation of Variability Expression
for est[var(est{log$_e$[*fnc*(01)]})]
112.714 − Computed Using the Propagation of Variability Expression
for est{var[est(*y* given *fnc* = *closascl*)]}

pertain to the outcome of a strength (resistance) test conducted using the up-and-down strategy. The former analysis presumes a conceptual two-parameter strength (resistance) distribution, whereas the latter presumes a conceptual one-parameter strength (resistance) distribution, viz., presumes that the two-parameter distribution *csp* is known.

8.C.1. Conceptual Two-Parameter Strength (Resistance) Distributions

Consider a strength (resistance) experiment test program that is conducted to estimate the actual values for the *clp* and the *csp* of a conceptual two-parameter strength (resistance) distribution. First, presume that a series of independent strength (resistance) tests has been conducted at a single stress (stimulus) level and each test item was sequentially tested until failure occurred or until a predetermined test duration $d*$ was endured without failure. If so, recall that each of the respective tests can be statistically viewed as a binomial trial and that the probability of failure, *pf*, before duration $d*$ at this stress (stimulus) level can be estimated. Next, presume that analogous strength (resistance) tests have been conducted at two or more stress (stimulus) levels. Let the number of these stress (stimulus) levels be denoted n_{sl} and let the ith stress (stimulus) level be denoted s_i, where $i = 1$ to n_{sl}. Then, presume that n_{it_i} items were tested at s_i and that n_{if_i} of these test items failed prior to enduring duration $d*$. Accordingly, ignoring nonessential (factorial) terms, the likelihood expression for the resulting strength (resistance) experiment test program can be written as

$$\text{likelihood} = \prod_{i=1}^{n_{sl}} pf_i^{n_{if_i}} \cdot (1 - pf_i)^{\left(n_{it_i} - n_{if_i}\right)}$$

in which pf_i is shorthand notation for *pf* given $s = s_i$ and $pf = F(y)$, where $y = clp + csp \cdot s$ and $F(y)$ is the conceptual (two-parameter) strength (resistance) CDF. The corresponding $\log_e$(likelihood) expression is

$$\log_e(\text{likelihood}) = \sum_{i=1}^{n_{sl}} n_{if_i} \cdot \log_e(pf_i) + \sum_{i=1}^{n_{sl}} \left(n_{it_i} - n_{if_i}\right) \cdot \log_e(1 - pf_i)$$

In turn, est(*clp*) and est(*csp*) are computed by simultaneously solving the partial derivative equations:

$$\frac{\partial \log_e(\text{likelihood})}{\partial clp} = \sum_{i=1}^{n_{sl}} \frac{n_{if_i}}{pf_i} \cdot \frac{\partial pf_i}{\partial clp} + \sum_{i=1}^{n_{sl}} \left(\frac{n_{it_i} - n_{if_i}}{1 - pf_i}\right) \cdot \frac{\partial(1 - pf_i)}{\partial clp} = 0$$

and

$$\frac{\partial \log_e(\text{likelihood})}{\partial csp} = \sum_{i=1}^{n_{sl}} \frac{n_{if_i}}{pf_i} \cdot \frac{\partial pf_i}{\partial csp} + \sum_{i=1}^{n_{sl}} \left(\frac{n_{it_i} - n_{if_i}}{1 - pf_i}\right) \cdot \frac{\partial(1 - pf_i)}{\partial csp} = 0$$

However, these two nonlinear simultaneous partial derivative equations can be written more concisely as

$$\frac{\partial \log_e(\text{likelihood})}{\partial clp} = \sum_{i=1}^{n_{sl}} \left\{ \frac{n_{st_i} \cdot \left[\left(n_{if_i}/n_{it_i}\right) - pf_i\right]}{pf_i \cdot (1 - pf_i)} \right\} \cdot \frac{\partial pf_i}{\partial clp} = 0$$

and

$$\frac{\partial \log_e(\text{likelihood})}{\partial csp} = \sum_{i=1}^{n_{sl}} \left\{ \frac{n_{st_i} \cdot \left[\left(n_{if_i}/n_{it_i}\right) - pf_i\right]}{pf_i \cdot (1 - pf_i)} \right\} \cdot \frac{\partial pf_i}{\partial csp} = 0$$

in which the respective (n_{if_i}/n_{it_i})'s are the respective ML estimates of the actual values for the corresponding conceptual pf_i's (Section 8.11). Moreover, it is analytically convenient to restate these two nonlinear partial derivative equations in terms of the standardized CDF metric y. Recall that $y = clp + csp \cdot s$. Thus,

$$\frac{\partial y}{\partial clp} = 1$$

$$\frac{\partial y}{\partial csp} = s$$

$$\frac{\partial pf}{\partial clp} = \frac{\partial pf}{\partial y} \cdot \frac{\partial y}{\partial clp}$$

$$\frac{\partial pf}{\partial csp} = \frac{\partial pf}{\partial y} \cdot \frac{\partial y}{\partial csp}$$

Substituting these expressions into the two nonlinear partial derivative equations stated in terms of the *clp* and the *csp* gives the restated expressions:

$$\frac{\partial \log_e(\text{likelihood})}{\partial clp} = \sum_{i=1}^{n_{sl}} \left\{ \frac{n_{st_i} \cdot \left[\left(n_{if_i}/n_{it_i}\right) - pf_i\right]}{pf_i \cdot (1 - pf_i)} \right\} \cdot \frac{\partial pf_i}{\partial y_i} = 0$$

and

$$\frac{\partial \log_e(\text{likelihood})}{\partial csp} = \sum_{i=1}^{n_{sl}} \left\{ \frac{n_{st_i} \cdot \left[\left(n_{if_i}/n_{it_i}\right) - pf_i\right]}{pf_i \cdot (1 - pf_i)} \right\} \cdot s_i \cdot \frac{\partial pf_i}{\partial y_i} = 0$$

In turn, the corresponding second partial derivative expressions are

$$
\frac{\partial^2 \log_e(\text{likelihood})}{\partial clp^2} = -\sum_{i=1}^{n_{sl}} \left[\frac{n_{it_i}}{pf_i \cdot (1 - pf_i)} \right] \cdot \left(\frac{\partial pf_i}{\partial y_i} \right)^2
$$

$$
+ \sum_{i=1}^{n_{sl}} \left\{ \frac{n_{it_i} \cdot \left[\left(n_{if_i}/n_{it_i} \right) - pf_i \right]}{pf_i \cdot (1 - pf_i)} \right\} \cdot \frac{\partial^2 pf_i}{\partial y_i^2}
$$

$$
- \sum_{i=1}^{n_{sl}} \left\{ \frac{n_{it_i} \cdot \left[\left(n_{if_i}/n_{it_i} \right) - pf_i \right] \cdot [1 - (2pf_i)]}{[pf_i \cdot (1 - pf_i)]^2} \right\} \cdot \left(\frac{\partial pf_i}{\partial y_i} \right)^2
$$

and

$$
\frac{\partial^2 \log_e(\text{likelihood})}{\partial csp^2} = -\sum_{i=1}^{n_{sl}} \left[\frac{n_{it_i}}{pf_i \cdot (1 - pf_i)} \right] \cdot s_i^2 \cdot \left(\frac{\partial pf_i}{\partial y_i} \right)^2
$$

$$
+ \sum_{i=1}^{n_{sl}} \left\{ \frac{n_{it_i} \cdot \left[\left(n_{if_i}/n_{it_i} \right) - pf_i \right]}{pf_i \cdot (1 - pf_i)} \right\} \cdot s_i^2 \cdot \frac{\partial^2 pf_i}{\partial y_i^2}
$$

$$
- \sum_{i=1}^{n_{sl}} \left\{ \frac{n_{it_i} \cdot \left[\left(n_{if_i}/n_{it_i} \right) - pf_i \right][1 - (2pf_i)]}{[pf_i \cdot (1 - pf_i)]^2} \right\} \cdot s_i^2 \cdot \left(\frac{\partial pf_i}{\partial y_i} \right)^2
$$

The associated mixed second partial derivative expression is

$$
\frac{\partial^2 \log_e(\text{likelihood})}{\partial clp \, \partial csp} = -\sum_{i=1}^{n_{sl}} \left[\frac{n_{it_i}}{pf_i \cdot (1 - pf_i)} \right] \cdot s_i \cdot \left(\frac{\partial pf_i}{\partial y_i} \right)^2
$$

$$
+ \sum_{i=1}^{n_{sl}} \left\{ \frac{n_{it_i} \cdot \left[\left(n_{if_i}/n_{it_i} \right) - pf_i \right]}{pf_i \cdot (1 - pf_i)} \right\} \cdot s_i \cdot \frac{\partial^2 pf_i}{\partial y_i^2}
$$

$$
- \sum_{i=1}^{n_{sl}} \left\{ \frac{n_{it_i} \cdot \left[\left(n_{if_i}/n_{it_i} \right) - pf_i \right] \cdot [1 - (2pf_i)]}{[pf_i \cdot (1 - pf_i)]^2} \right\} \cdot s_i \cdot \left(\frac{\partial pf_i}{\partial y_i} \right)^2
$$

These second derivative expressions can (should) be used in the iterative Newton–Raphson (N–R) method for estimating the actual values for the *clp* and the *csp*. However, these expressions are traditionally simplified by ignoring summations with $[(n_{if_i}/n_{it_i}) - pf_i]$ terms. The resulting simplified (approximate) second partial derivative expressions are sufficiently accurate that the convergence of the N–R estimation method is not compromised (but one or two more iterations are generally required to obtain the same accuracy as obtained when using the exact second partial derivative expressions). This simplification (approximation) was introduced to expedite hand-calculations, but it continues to be universally used—perhaps because it still has a strong conceptual advantage when the simplified second partial derivative expressions are re-expressed in terms of statistical weights, viz., when

$$\frac{\partial^2 \log_e(\text{likelihood})}{\partial clp^2} = -\sum_{i=1}^{n_{sl}} \left[\frac{n_{it_i}}{pf_i \cdot (1 - pf_i)} \right] \cdot \left(\frac{\partial pf_i}{\partial y_i} \right)^2$$

$$\frac{\partial^2 \log_e(\text{likelihood})}{\partial csp^2} = -\sum_{i=1}^{n_{sl}} \left[\frac{n_{it_i}}{pf_i \cdot (1 - pf_i)} \right] \cdot s_i^2 \cdot \left(\frac{\partial pf_i}{\partial y_i} \right)^2$$

and

$$\frac{\partial^2 \log_e(\text{likelihood})}{\partial clp \, \partial csp} = -\sum_{i=1}^{n_{sl}} \left[\frac{n_{it_i}}{pf_i \cdot (1 - pf_i)} \right] \cdot s_i \cdot \left(\frac{\partial pf_i}{\partial y_i} \right)^2$$

are rewritten as

$$\frac{\partial^2 \log_e(\text{likelihood})}{\partial clp^2} = -\sum_{i=1}^{n_{sl}} sw_i$$

$$\frac{\partial^2 \log_e(\text{likelihood})}{\partial csp^2} = -\sum_{i=1}^{n_{sl}} sw_i \cdot s_i^2$$

$$\frac{\partial^2 \log_e(\text{likelihood})}{\partial clp \, \partial csp} = -\sum_{i=1}^{n_{sl}} sw_i \cdot s_i$$

in which, as explained below,

$$sw_i = \left[\frac{n_{it_i}}{pf_i(1 - pf_i)} \right] \cdot \left(\frac{\partial pf_i}{\partial y_i} \right)^2$$

Note that these statistical weights must be estimated in ML analysis.

Recall that the actual value for a statistical weight is equal to the inverse of the actual value for the variance of the corresponding conceptual sampling distribution. In this ML analysis, the relevant statistical weights sw_i's are the inverses of the actual values for the variances of the asymptotic normally distributed sampling distributions for the respective [est(y_i)]'s. Accordingly, the generic expression for these statistical weights is developed as follows. Presume that tentative values for est(clp) and est(csp) have been guestimated. The corresponding values for the respective [est(y_i)]'s can then be computed. In turn, when the [est(y_i)]'s have been computed, then the corresponding respective [est(pf_i)]'s can be computed. Next, recall that the variances of the conceptual binomial sampling distributions for each respective est(pf_i) is equal to the $[pf_i \cdot (1 - pf_i)]/n_{it_i}$. Recall also that the conceptual binomial distribution can be asymptotically approximated by a conceptual (two-parameter) normal distribution that has identical actual values for its mean and variance. Thus, we now assert that the asymptotic conceptual sampling distributions for the respective [est(pf_i)]'s are appropriately scaled conceptual (two-parameter) normal distributions. In turn, we assert that (a) the asymptotic conceptual sampling distributions for the corresponding [est(y_i)]'s are also appropriately scaled conceptual (two-parameter) normal distributions, and (b) the actual values for their variances are related by the propagation of the variability expression:

$$\text{var}[\text{est}(y_i)] = \left(\frac{\partial y_i}{\partial pf_i}\right)^2 \cdot \text{var}[\text{est}(pf_i)]$$

in which the inverse of $\partial y_i/\partial pf_i$, the ordinate of the PDF of the underlying (presently unspecified) conceptual (two-parameter) strength distribution, is theoretically evaluated at pf_i. However, since pf_i is unknown, this partial derivative expression must be evaluated at est(pf_i).

The iterative N–R estimation method is conveniently employed to solve numerically for est(clp) and est(csp). These estimates are in turn used to compute the respective [est(y_i)]'s, [est(pf_i)]'s, {est[var(pf_i)]}'s, {est[-var(y_i)]}'s, and [est(sw_i)]'s. This iterative analysis begins with guestimated values for est(clp) and est(csp) and continues until successive N–R corrections are less than some preselected very small numerical value, say 10^{-12}. The associated estimated asymptotic covariance matrix is the inverse of the 2 by 2 array comprised of the negatives of the second partial and mixed derivatives of the $\log_e$(likelihood) expression with respect to clp and csp. (These derivative expressions are given above.) The numerical values for elements of this array are obtained by substituting est(clp) for clp and est(csp) for csp in these partial derivative expressions. Accordingly, the associated estimated asymptotic covariance matrix is stated in its inverse form as

$$|\text{est}\{\text{covar}[\text{est}(clp), \text{est}(csp)]\}| = \begin{vmatrix} \sum_{i=1}^{n_{sl}} \text{est}(sw_i) & \sum_{i=1}^{n_{sl}} [\text{est}(sw_i)] \cdot s_i \\ \sum_{i=1}^{n_{sl}} [\text{est}(sw_i)] \cdot s_i & \sum_{i=1}^{n_{sl}} [\text{est}(sw_i)] \cdot s_i^2 \end{vmatrix}^{-1}$$

Evaluating this inverse gives

$|\text{est}\{\text{covar}[\text{est}(clp), \text{est}(csp)]\}|$

$$= (1/\text{determinant}) \begin{vmatrix} \sum_{i=1}^{n_{sl}} [\text{est}(sw_i)] \cdot s_i^2 & -\sum_{i=1}^{n_{sl}} [\text{est}(sw_i)] \cdot s_i \\ -\sum_{i=1}^{n_{sl}} [\text{est}(sw_i)] \cdot s_i & \sum_{i=1}^{n_{sl}} [\text{est}(sw_i)] \end{vmatrix}$$

in which

$$\text{determinant} = \left[\sum_{i=1}^{n_{sl}} \text{est}(sw_i) \right] \cdot \left\{ \sum_{i=1}^{n_{sl}} [\text{est}(sw_i)] \cdot s_i^2 \right\} - \left\{ \sum_{i=1}^{n_{sl}} [\text{est}(sw_i)] \cdot s_i \right\}^2$$

Note that this estimated asymptotic covariance matrix is identical to the exact covariance matrix developed for simple linear weighted regression (Supplemental Topic 7.B), except that exact statistical weights are replaced by estimated statistical weights. In fact, this ML analysis generates the same numerical estimates as an *iterative* linear weighted regression analysis in which the unknown statistical weights are first guestimated and then successively corrected such that the corrected statistical weights used in the $(j+1)^{th}$ iteration were computed during the jth iteration.

Pearson's central $\chi^2_{n_{sl}-2}$ conceptual sampling distribution can be used in a statistical test of the adequacy of the presumed conceptual (two-parameter) strength (resistance) distribution. The associated asymptotic test statistic is

$$\text{"residual" } \chi^2_{n_{sl}-2;p} = \sum_{i=1}^{n_{sl}} \left\{ \frac{n_{it_i}}{\text{est}(pf_i) \cdot [1 - \text{est}(pf_i)]} \right\} \cdot \left[(n_{if_i}/n_{it_i}) - \text{est}(pf_i) \right]^2$$

and $(1-p)$ is the rejection probability for the null hypothesis that the presumed conceptual (two-parameter) distribution is correct. Unfortunately, the statistical power of the this test is quite poor.

Remark: Little (unpublished) compared the respective values of the "residual" $\chi^2_{n_{sl}-2}$ values for approximately 30 sets of fatigue limit (endurance limit) experiment test program data sets in which the normal, logistic, smallest-extreme-value, and largest-extreme-value conceptual (two-parameter) strength (resistance) distributions were presumed in ML analysis, each with both a linear and logarithmic alternating stress metric. This study indicated that the "residual" $\chi^2_{n_{sl}-2}$ value was almost always smallest for either the logistic or the normal distribution with a linear alternating stress metric.

Exercise Set 6

These exercises are intended to verify relevant expressions for the first and second partial derivatives of *pf* with respect to *y*, given the standardized form of the presumed conceptual strength (resistance) distribution.

1. Verify that, for the standardized conceptual normal distribution, $(\partial pf/\partial y)$ is equal to *z*, the ordinate of the normal PDF evaluated at *y*. In turn, verify that the second partial derivative of *pf* with respect to *y* is equal to $(-y \cdot z)$.
2. Verify that, for the standardized conceptual logistic distribution, $(\partial pf/\partial y)$ is equal to $[pf \cdot (1 - pf)]$. In turn, verify that the second partial derivative of *pf* with respect to *y* is equal to $\{[pf \cdot (1 - pf)] \cdot (1 - 2pf)\}$.
3. Verify that, for the standardized conceptual smallest-extreme-value distribution, $(\partial pf/\partial y)$ is equal to $\{-(1 - pf) \cdot [\log_e(1 - pf)]\}$. In turn, verify that the second partial derivative of *pf* with respect to *y* is equal to $\{-(1 - pf) \cdot [\log_e(1 - pf)] \cdot [1 + \log_e(1 - pf)]\}$.
4. Verify that, for the standardized conceptual largest-extreme-value distribution, $(\partial pf/\partial y)$ is equal to $\{-pf \cdot [\log_e(pf)]\}$. In turn, verify that the second partial derivative of *pf* with respect to *y* is equal to $\{pf \cdot [\log_e(pf)] \cdot [1 + \log_e(pf)]\}$.

8.C.1.1. Example Microcomputer Programs

Although the standardized form of the presumed conceptual strength (resistance) distribution can be expressed using the same four parameterizations previously presented for life (endurance) data, viz.,

(1) $\quad y(pf) = csp \cdot [s(pf) - clp]$

(2) $\quad y(pf) = clp + csp \cdot s(pf)$

(3) $\quad y(pf) = \dfrac{s(pf) - clp}{csp}$

(4) $\quad y(pf) = \dfrac{s(pf)}{cps} - clp$

only parameterization (2) is traditionally used in ML analysis. Microcomputer programs *N2AS50*, *L2AS50*, *SEV2AS50*, *LEV2AS50*, *N2ALCL*, *L2ALCL*, *SEV2ALCL*, and *LEV2ALCL* (pages 494–499) respectively pertain to conceptual (two-parameter) normal, logistic, smallest-extreme-value, and largest-extreme-value strength (resistance) distributions. The first set of microcomputer programs have outputs that are analogous to the outputs of microcomputer programs *N1A*, *L1A*, *SEV1A*, and *LEV1A* (pages 438–439). The second set of microcomputer programs have outputs that are analogous to the outputs of microcomputer programs *WEIBULL* (page 422) and *LSEV1A*, *LSEV2A*, *LSEV31A*, and *LSEV4A* (pages 424–427). Note that for the datum values that appear in microcomputer file *ASDATA* the respective estimated median strengths (resistances) are identical for practical purposes. On the other hand, the respective classical lower $100(scp)\%$ (one-sided) asymptotic statistical confidence limits can differ markedly for small values of *pf*, depending on (a) which conceptual strength (resistance) distribution is presumed in ML analysis, and (b) whether the propagation of the variability expression used to compute this asymptotic statistical confidence limit pertains to est(var{est[$s(pf)$]} or to est{var[est(y given $s = closastl$)]}.

> *Remark*: A *scp*-based value for Student's central t variate is traditionally used to compute the lower $100(scp)\%$ (one-sided) statistical confidence limit that allegedly bounds the actual value for $s(50)$.

8.C.1.2. Statistical Weights

Figure 8.8 illustrates how the statistical weight per item depends on the probability of failure *pf*. Note that, given a symmetrical strength (resistance) distribution, tests that are conducted with the *pf* close to zero or one generate very small statistical weights per test item relative to tests conducted with the *pf* near 0.5. Obviously, such tests should be avoided (except when required by the optimal variance strategy discussed below). Note also that, given a nonsymmetrical strength (resistance) distribution, the direction of skewness must be known to state the *pf* value that generates the most effective statistical weight per item. Although Figure 8.8 provides valuable perspective and enhances intuition, the proper use of statistical weights lies in adopting a minimum variance strategy that allocates test items to stimulus levels such that a statistically more precise ML estimate of the actual value for the quantity of specific interest is obtained. As previously mentioned, minimum variance strategies are easily programmed such that, at any given stage of the experiment test program, the microcomputer program will calculate the statistically most effective stimulus level for testing the next test item. Suppose, for example, the objective is to maximize the numerical value for the lower $100(scp)\%$ (one-sided) asymptotic statistical confidence limit that allegedly bounds the actual value for $s(pf)$. Suppose further that this lower limit is computed by minimizing the estimated variance of the asymptotic conceptual (normally distributed) sampling distribution that consists of all possible replicate realization values for est(y) given the *s* value of specific interest. Then, the most effective stimulus level for testing

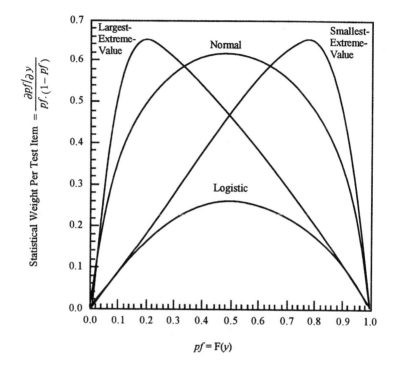

Figure 8.8 Plot of the statistical weight *sw* per test item versus *pf* for the standardized conceptual (two-parameter) normal, logistic, smallest-extreme-value and largest-extreme-value strength (resistance) distributions.

the next item is computed by running microcomputer program *OTPNLCLY* (optimal two-parameter normal distribution lower confidence limit, version Y). Analogous microcomputer program *OTPNLCLS* also establishes the most effective stimulus level for testing the next item, but it minimizes the estimated variance of the corresponding asymptotic conceptual (normally distributed) sampling distribution that consists of all possible replicate realization values for est(*s*) given the *y* value of specific interest. The example datum values for these two programs appear in microcomputer file *OSADATA*.

Exercise Set 7

These exercises are intended to verify the numerical values of the statistical weight per test item that are plotted in Figure 8.8 versus the actual probability of failure *pf* for that test item.

1. Starting with the expression:

$$sw = \frac{(\partial pf / \partial y)^2}{pf(1 - pf)}$$

state the appropriate analytical expression for the statistical weight sw per test item when the partial derivative of pf with respect to y pertains to the standardized conceptual (a) normal distribution, (b) logistic distribution, (c) smallest-extreme-value distribution, and (d) largest-extreme-value distribution.

2. Evaluate the statistical weight per test item expressions in Exercise 1 for $pf = 0.01$, 0.05, 0.10, 0.25, 0.50, 0.75, 0.90, 0.95, and 0.99 to verify the numerical values plotted in Figure 8.8.

8.C.2. Conceptual One-Parameter Strength (Resistance) Distributions

When small sample up-and-down experiment test programs are used to estimate the actual value for the median of a conceptual one-parameter strength (resistance) distribution, the actual value for the csp of this conceptual strength distribution must be presumed to be known. Then, ML estimation merely requires solving the single nonlinear partial derivative equation:

$$\frac{\partial \log_e(\text{likelihood})}{\partial clp} = \sum_{i=1}^{n_{sl}} \left\{ \frac{n_{st_i}\left[\left(n_{if\,i}/n_{it_i}\right) - pf_i\right]}{pf_i(1 - pf_i)} \right\} \frac{\partial pf_i}{\partial y_i} = 0$$

for est(clp). In turn, the actual value of the median of the presumed conceptual one-parameter strength (resistance) distribution, denoted $s(50)$, is estimated using the expression:

$$y(50) = \text{est}(clp) + csp \cdot \text{est}[s(50)]$$

Note that $y(50)$ is equal to zero for the symmetrical conceptual normal and logistic strength (resistance) distributions, then $\text{est}[s(50)] = \{-[\text{est}(clp)]/csp\}$. (Otherwise, see Exercise Set 8.)

The actual value for the variance of the normally distributed asymptotic sampling distribution that consists of all possible replicate realization values for est(clp) is conventionally estimated using the expression:

$$\text{est}\{\text{var}[\text{est}(clp)]\} = \frac{1}{\displaystyle\sum_{i=1}^{n_{sl}} \text{est}(sw_i)}$$

The corresponding actual value for the variance of the normally distributed asymptotic sampling distribution that consists of all possible replicate realization values for est[$s(50)$] is conventionally estimated using the expression:

$$\text{est}\left(\text{var}\left\{\text{est}[s(50)]\right\}\right) = \frac{1}{csp^2 \displaystyle\sum_{i=1}^{n_{sl}} \text{est}(sw_i)}$$

Since the csp is presumed to be known, the lower $100(scp)\%$ (one-sided) asymptotic statistical confidence band that allegedly bounds the actual CDF is parallel to the estimated CDF and passes through the corresponding lower $100(scp)\%$ (one-sided) asymptotic statistical confidence limit that allegedly bounds the actual value for $s(50)$.

Recall that microcomputer program *N1A* pertains to a conceptual one-parameter normal strength (resistance) distribution with its presumed known csp (its standard deviation) as input in microcomputer file *UADDATA*. It can be run at any time during the conduct of a small sample up-and-down strength (resistance) experiment test program to establish the most effective stimulus level amplitude for the next test specimen provided that sufficient data have previously been accumulated in the experiment test program to generate a ML estimate of the actual value for $s(50)$. (Recall also

---

```
C> COPY ASDATA DATA

    1 files(s) copied

C> N2AS50
```

Presuming a Conceptual Two-Parameter Normal Distribution

$$\text{est}[s(50)] = 83.3$$

Lower 95% (One-Sided) Asymptotic Statistical Confidence Limit that Allegedly Bounds the Actual Value for $s(50)$

76.1 – Computed using the Propagation of Variability Expression for est{var[est(y given $s = closascl$)]} and Student's central t(1,5;0.95) = 2.0150

```
C> COPY ASDATA DATA

    1 files(s) copied

C> L2AS50
```

Presuming a Conceptual Two-Parameter Logistic Distribution

$$\text{est}[s(50)] = 83.7$$

Lower 95% (One-Sided) Asymptotic Statistical Confidence Limit that Allegedly Bounds the Actual Value for $s(50)$

75.7 – Computed using the Propagation of Variability Expression for est{var[est(y given $s = closascl$)]} and Student's central $t(1,5;0.95) = 2.0150$

```
C> COPY ASDATA DATA

    1 files(s) copied

C> SEV2AS50
```

Presuming a Conceptual One-Parameter Smallest-Extreme-Value Distribution

$$\text{est}[s(50)] = 85.0$$

Lower 95% (One-Sided) Asymptotic Statistical Confidence Limit that Allegedly Bounds the Actual Value for $s(50)$

76.4 – Computed using the Propagation of Variability Expression for est{var[est(y given $s = closascl$)]} and Student's central $t(1,5;0.95) = 2.0150$

```
C> COPY ASDATA DATA

    1 files(s) copied

C> LEV2AS50
```

Presuming a Conceptual Two-Parameter Largest-Extreme-Value Distribution

$$\text{est}[s(50)] = 81.7$$

Lower 95% (One-Sided) Asymptotic Statistical Confidence Limit that Allegedly Bounds the Actual Value for $s(50)$

75.2 – Computed using the Propagation of Variability Expression for $\text{est}\{\text{var}[\text{est}(y \text{ given } s = closascl)]\}$ and Student's central $t(1,5;0.95) = 2.0150$

```
C> COPY ASDATA DATA

    1 files(s) copied

C> N2ALCL
```

Given Standardized Conceptual Normal Distribution Variate $y(pf) = clp + csp \cdot s(pf)$

$\text{est}(clp) = 0.7438137674\text{D}+01$
$\text{est}(csp) = 0.8926409259\text{D}-01$
$\text{est}\{\text{var}[\text{est}(clp)]\} = 0.4315338215\text{D}+01$
$\text{est}\{\text{var}[\text{est}(csp)]\} = 0.5899274299\text{D}-03$
$\text{est}\{\text{covar}[\text{est}(clp),\text{est}(csp)]\} = -0.5012866877\text{D}-01$
$\text{est}(conceptual\ correlation\ coefficient) = -0.9935265033\text{D}+00$

s	$\text{est}(y)$	$\text{est}(pf)$
100.000	1.4882716	0.9316604
95.000	1.0419511	0.8512828
90.000	0.5956307	0.7242890
85.000	0.1493102	0.5593456

```
80.000  −0.2970103  0.3832293
70.000  −1.1896512  0.1170918
65.000  −1.6359717  0.0509228
```

Classical Lower 95% (One-Sided) Asymptotic Statistical Confidence Limits that Allegedly Bound the Actual Value for $s(01)$

44.124 – Computed Using the Propagation of Variability Expression for est(var{est[$s(01)$]})

34.074 – Computed Using the Propagation of Variability Expression for est{var[est(y given $s = closascl$)]}

C> COPY ASDATA DATA

    1 files(s) copied

C> L2ALCL

Given Standardized Conceptual Logistic Distribution Variate $y(pf) = clp + csp \times s(pf)$

est(clp) = $-0.1316853685\mathrm{D}+02$
est(csp) = $0.1572425344\mathrm{D}-00$
est{var[est(clp)]} = $0.1719712841\mathrm{D}+02$
est{var[est(csp)]} = $0.2331274510\mathrm{D}-02$
est{covar[est(clp),est(csp)]} = $-0.19923002537\mathrm{D}+00$
est($conceptual\ correlation\ coefficient$) = $-0.9950172579\mathrm{D}+00$

s	est(y)	est(pf)
100.000	2.5557166	0.9279566
95.000	1.7965039	0.8543960
90.000	0.9832912	0.7277608
85.000	0.1970786	0.5491108
80.000	−0.5891341	0.356336
70.000	−2.1615594	0.1032560
65.000	−2.9477721	0.0498419

Classical Lower 95% (One-Sided) Asymptotic Statistical Confidence Limits that Allegedly Bound the Actual Value for $s(01)$

38.311 – Computed Using the Propagation of Variability Expression
for est(var{est[s(01)]})
22.363 – Computed Using the Propagation of Variability Expression
for est{var[est(y given $s = closascl$)]}

```
C> COPY ASDATA DATA

    1 files(s) copied

C> SEV2ALCL
```

Given Standardized Conceptual Smallest-Extreme-Value Distribution
Variate $y(pf) = clp + csp \times s(pf)$

est(clp) = $-0.1003518158D + 02$
est(csp) = $0.1137870314D + 00$
est{var[est(clp)]} = $0.8451907194D + 01$
est{var[est(csp)]} = $0.1057426055D - 02$
est{covar[est(clp),est(csp)]} = $-0.9417710941D - 01$
est(*conceptual correlation coefficient*) = $-0.9961918698D + 00$

s	est(y)	est(pf)
100.000	1.3435216	0.9783449
95.000	0.7745864	0.8857875
90.000	0.2056512	0.7072173
85.000	−0.3632839	0.5011196
80.000	−0.9322191	0.3254295
70.000	−2.0700894	0.1185390
65.000	−2.6390245	0.0689394

Classical Lower 95% (One-Sided) Asymptotic Statistical Confidence
Limits that Allegedly Bound the Actual Value for s(01)

28.010 – Computed Using the Propagation of Variability Expression
for est(var{est[s(01)]})
10.789 – Computed Using the Propagation of Variability Expression
for est{var[est(y given $s = closascl$)]}

```
C> COPY ASDATA DATA

    1 files(s) copied

C> LEV2ALCL
```

Given Standardized Conceptual Largest-Extreme-Value Distribution Variate $y(pf) = clp + csp \times s(pf)$

est(clp) = $-0.6761135294D+01$
est(csp) = $0.8725876010D-01$
est{var[est(clp)]} = $0.3662399509D+01$
est{var[est(csp)]} = $0.5512958654D-03$
est{covar[est(clp),est(csp)]} = $-0.4531384182D-01$
est(*conceptual correlation coefficient*) = $-0.9906437076D+00$

s	est(y)	est(pf)
100.000	1.9647407	0.8691911
95.000	1.5284469	0.8050328
90.000	1.0921531	0.7149853
85.000	0.6558593	0.5951178
80.000	0.2195655	0.4480423
70.000	-0.6530221	0.1464109
65.000	-1.0893159	0.0511885

Classical Lower 95% (One-Sided) Asymptotic Statistical Confidence Limits that Allegedly Bound the Actual Value for $s(01)$

49.975 – Computed Using the Propagation of Variability Expression for est(var{est[$s(01)$]})
42.814 – Computed Using the Propagation of Variability Expression for est{var[est(y given $s = closascl$)]}

```
C> TYPE OSADATA

10
3000 1 1
2750 1 1
2500 1 0
2750 1 1
2500 1 0
2750 1 0
2710 1 1
2655 1 0
2700 1 0
2740 1 0
01          CDF Percentile of Specific Interest (Integer Value)
```

These datum values are identical to the datum values that appear in microcomputer file *UADDATA* except as indicated.

```
C> COPY OSADATA DATA

    1 file(s) copied

C> OTPNLCLY
```

Test the next item with its stimulus level *s* equal to 2639.7

```
C> COPY OSADATA DATA

    1 file(s) copied

C> OTPNLCLS
```

Test the next item with its stimulus level *s* equal to 2647.1

that for the conceptual one-parameter logistic, smallest-extreme-value, and largest-extreme values distributions, the *csp* is not equal to the standard deviation.)

8.D. SUPPLEMENTAL TOPIC: GENERATING PSEUDORANDOM DATUM VALUES AND COMPUTING PRAGMATIC STATISTICAL BIAS CORRECTIONS FOR ML-BASED ESTIMATES

Recall that microcomputer simulations can be used to examine statistical behaviors empirically, both from a qualitative (understanding) and quantitative (estimation) perspective. These simulations require the generation of appropriate pseudorandom datum values that are allegedly randomly selected from the conceptual two-parameter statistical distribution of specific interest. Given such pseudorandom datum values, simulation-based empirical and pragmatic (defined later) sampling distributions for any ML-based estimate of specific interest can be developed. The statistical bias and variability of this ML-based estimate can then be examined. In turn, statistical bias corrections can be made and the precision of the resulting empirical (pragmatic) bias-corrected estimate can be assessed.

8.D.1. Generating Pseudorandom Datum Values

Suppose that we wish to generate pseudorandom datum values from the conceptual two-parameter Weibull distribution that is expressed as

$$F(x) = 1 - \exp^{-\left(\frac{x}{cdp1}\right)^{cdp2}}$$

First, presume that the actual values for *cdp1* and *cdp2* are known. Then, when a series of n_{dv} uniform pseudorandom numbers, zero to one, are successively substituted for $F(x)$, the resulting values for x can be viewed as having been randomly selected from a conceptual two-parameter Weibull distribution. Next, presume that the actual values for the mean and variance are known. If so, the corresponding values for *cdp1* and *cdp2* can be computed by numerically solving the mean and variance expressions given below and then the pseudorandom datum values of specific interest can be generated.

The following analytical expressions for the actual values of the mean and variance for the conceptual two-parameter Weibull distribution involve the gamma function (Abramowitz and Stegun, 1964), viz.,

$$\text{mean}\,(X) = cdp1 \cdot \text{gamma}\left[1 + \left(\frac{1}{cdp2}\right)\right]$$

$$\text{var}(X) = cdp1^2 \cdot \left(\text{gamma}\left[1 + \left(\frac{2}{cdp2}\right)\right] - \left\{\text{gamma}\left[1 + \left(\frac{1}{cdp2}\right)\right]\right\}^2\right)$$

Also,

$$\text{median}\,(X) = cdp1 \cdot log_e(2.0)^{\frac{1}{cdp2}}$$

and

$$\text{mode}\,(X) = cdp1 \cdot \left(\frac{cdp2 - 1}{cdp2}\right)^{\frac{1}{cdp2}}$$

The numerical complexity of the Weibull distribution is due to its logarithmic metric. Recall that the statistically equivalent conceptual (two-parameter) smallest-extreme-value distribution with its metric equal to $log_e(x)$ rather than x is much more convenient both numerically and geometrically.

This pseudorandom number generation methodology pertains to all continuous conceptual two-parameter statistical distributions with explicit CDF expressions and known conceptual distribution parameter values. The conceptual (two-parameter) normal distribution involves an integral in its CDF expression and thus presents an exception. However, the polar method (Knuth, 1969) provides a simple, exact procedure to generate replicate pseudorandom datum values from a conceptual (two-parameter) normal distribution when the actual values for its mean and variance are known.

Exercise Set 8

These exercises provide experience in generating pseudorandom datum values from the various conceptual two-parameter statistical distributions of specific interest in mechanical reliability analyses. It is also intended to provide perspective regarding the difficulty of correctly identifying the actual conceptual distribution when two or more alternative conceptual distributions must be considered.

1. Given the following information pertaining to the conceptual (two-parameter) smallest-extreme-value distribution written as

$$F(x) = 1 - \exp^{-\exp\left(\frac{x - clp}{csp}\right)}$$

$$f(x) = \left(\frac{1}{csp}\right) \exp^{\left(\frac{x - clp}{csp}\right) - \exp\left(\frac{x - clp}{csp}\right)}$$

$$\text{mean}(X) = clp - \text{Euler's constant} \cdot csp$$

(where Eurler's constant

$$= 0.5\,772\,156\,649\,015\,328)$$

$$\text{var}(X) = \frac{\pi^2}{6} \cdot csp^2$$

$$\text{median}(X) = clp + \log_e\left[-\log_e(0.5)\right] \cdot csp$$

$$\text{mode}(X) = clp$$

run microcomputer program *SEV* to generate six sets of 10 replicate pseudorandom datum values from this conceptual distribution when the actual values for both its mean and variance are equal to 100. Then, plot these six sets of 10 replicate pseudorandom datum values on both normal and smallest-extreme-value probability papers using $y(pp)_i$ values that correspond to the empirical plotting positions $p(pp)_i = (i - 0.5)/n_{dv}$.

2. Given the following information pertaining to the conceptual (two-parameter) largest-extreme-value distribution written as

$$F(x) = \exp^{-\exp^{-\left(\frac{x - clp}{csp}\right)}}$$

$$f(x) = \left(\frac{1}{csp}\right) \cdot \exp^{-\left(\frac{x - clp}{csp}\right) - \exp^{-\left(\frac{x - clp}{csp}\right)}}$$

$$\text{mean}(X) = clp + \text{Euler's constant} \cdot csp$$

(where Euler's constant $= 0.5\,772\,156\,649\,015\,328)$

$$\text{var}(X) = \frac{\pi^2}{6} \cdot csp^2$$

$$\text{median}(X) = clp - \log_e\left[-\log_e(0.5)\right] \cdot csp$$

$$\text{mode}(X) = clp$$

run microcomputer program *LEV* to generate six sets of 10 replicate pseudorandom datum values from this conceptual distribution, given that the actual values for both its mean and variance
are equal to 100. Then, plot these six sets of 10 replicate pseudorandom datum values on both normal and largest-extreme-value
probability paper using $y(pp)_i$ values that correspond to the
empirical plotting positions $p(pp)_i = (i - 0.5)/n_{dv}$.

3. Given the following information pertaining to the conceptual
(two-parameter) logistic distribution written as

$$F(x) = \left[1 + \exp^{-\left(\frac{x - clp}{csp}\right)} \right]^{-1}$$

$$f(x) = \text{(Exercise: derive this expression)}$$
$$\text{mean}(X) = clp$$
$$\text{var}(X) = \frac{\pi^2}{3} \cdot csp^2$$
$$\text{median}(X) = clp$$
$$\text{mode}(X) = clp$$

run microcomputer program *LOG* to generate six sets of 10 replicate pseudorandom datum value from this conceptual distribution, given that the actual values for both its mean and variance
are equal to 100. Then, plot these six sets of 10 replicate pseudorandom datum values on both normal and logistic probability
paper using $y(pp)_i$ values that correspond to the empirical plotting positions $p(pp)_i = (i - 0.5)/n_{dv}$.

4. This exercise provides practical perspective regarding the results
of Exercises 1 through 3. First run microcomputer program *NOR*
to generate six sets of 10 replicate pseudorandom datum values
from a conceptual (two-parameter) normal distribution and plot
these six sets of 10 datum values on normal probability paper.
Identify the actual conceptual distribution *only* on the back side
of the resulting 48 sheets of normal probability paper. Then,
thoroughly mix these six plots with the 18 plots on normal probability paper for Exercises 1 through 3. In turn, sort through the
resulting 24 plots and attempt to identify by visual inspection the
six sets of normally distributed datum values. State the number of
plots correctly identified and compare this number to its expected

value given the null hypothesis that it is impossible to discern among the respective datum plots.

5. First verify the following information pertaining to the conceptual two-parameter $\log_e$–normal distribution written as

$$F(x) = \frac{1}{\sqrt{2\pi}\,cdp2} \int_0^x \frac{1}{u} \exp\left\{-\frac{1}{2}\left[\frac{\log_e(u) - \log_e(cdp1)}{cdp2}\right]^2\right\} du$$

where $x > 0$.

$$f(x) = \frac{1}{\sqrt{2\pi}\,cdp2}\frac{1}{x} \exp^{-\frac{1}{2}\left[\frac{\log_e(x) - \log_e(cdp1)}{cdp2}\right]^2}$$

$$\text{mean}(X) = cdp1 \exp^{\left(\frac{1}{2}cdp2^2\right)}$$

$$\text{var}(X) = cdp1^2 \exp^{\left(cdp2^2\right)} \cdot \left(\exp^{cdp2^2} - 1\right)$$

$$\text{median}(X) = cdp1$$

$$\text{mode}(X) = cdp1 \exp^{\left(-cdp2^2\right)}$$

Then, generate six sets of 10 replicate pseudorandom datum values from this conceptual distribution, given that the actual values for both its mean and variance are equal to 100. In turn, plot these six sets of datum values on both $\log_e$–normal and Weibull probability papers using the $y(pp)_i$ values that corresponds to the empirical plotting positions $p(pp)_i = (i - 0.5)/n_{dv}$.

6. Given the conceptual two-parameter Weibull distribution CDF written as

$$F(x) = 1 - \exp^{-\left(\frac{x}{cdp1}\right)^{cdp2}}$$

Generate six sets of 10 pseudorandom datum values from this conceptual distribution, given that the actual values for both its mean and variance are equal to 100. Plot these six sets of datum values on both Weibull and $\log_e$–normal probability papers using the $y(pp)_i$ that corresponds to the empirical plotting positions $p(pp)_i = (i - 0.5)/n_{dv}$.

7. This exercise provides practical perspective regarding the results of Exercises 5 and 6. Identify the actual conceptual distribution *only* on the back side of the 12 sheets of $\log_e$–normal probability paper

and the 12 sheets of Weibull probability paper. Then, thoroughly mix each set of 12 $\log_e$–normal and Weibull plots. In turn, sort through each of these two sets of 12 plots and attempt to identify by visual inspection the six sets of datum values that are actually plotted on the correct probability paper. State the number of plots correctly identified and compare this number to its expected value given the null hypothesis that it is impossible to discern between $\log_e$–normal and Weibull pseudorandom datum values.

8.D.2. Computing Empirical Statistical Bias Corrections for ML Estimates of the Actual Values for the Coefficient of Variation

Recall that although the ML estimates of the actual values for the parameters of a given conceptual statistical distribution have known asymptotic (normally distributed) conceptual sampling distributions, their conceptual sampling distributions are unknown for experiment test programs of practical sizes. Accordingly, the associated conceptual sampling distribution for the ML estimate of the actual value for the coefficient of variation is unknown. Thus, the actual value for the statistical bias for the ML estimated coefficient of variation is unknown. Nevertheless, this statistical bias can be estimated as the metric distance that the simulation-based empirical sampling distribution for the ML estimated coefficient of variation must be translated to have its mean located at the actual value for the coefficient of variation. Microcomputer programs *EBCNCOV*, *EBCLNCOV*, and *EBCWCOV* compute simulation-based empirical statistical bias corrections for the conceptual (two-parameter) normal distribution, the conceptual two-parameter $\log_e$–normal distribution, and the conceptual two-parameter Weibull distribution, respectively.

8.D.3. Computing Empirical Statistical Bias Corrections for ML Analyses Pertaining to Life (Endurance) Experiment Test Programs with No Censoring

Suppose that the life (endurance) datum values of specific interest are presumed to have been randomly selected from a given conceptual two-parameter statistical distribution with known parameters. If so, we could generate n_{sim} replicate pseudorandom data sets of size n_{dv} from this conceptual statistical distribution. In turn, we could compute the respective n_{sim} ML estimated values for each conceptual distribution parameter, construct the associated empirical sampling distribution, and compute the desired empirical statistical bias correction as the difference between the mean of this empirical sampling distribution and the known conceptual distribution parameter value. (Recall that the accuracy of this empirical bias correction improves statistically as n_{sim} increases.) Next, suppose that our objective is to compute A-basis statistical tolerance limits. If so, we could generate another n_{sim} replicate pseudorandom data sets of size n_{dv} from this conceptual statistical distribution. In turn, we could compute the respective n_{sim} ML-based est[$fnc(01)$] values, construct the associated empirical sampling distribution, and translate this empirical sampling distribution a metric distance such that its 95th percentile coincides with the known (computed) value for the $fnc(01)$. Then, given any ML-based est[$fnc(01)$] value that is randomly selected from this translated empirical sampling distribution, the probability is 0.95 that it will be smaller than the known (computed) value for the $fnc(01)$. Microcomputer programs *ABNOR*, *BBNOR*, *ABLNOR*, *BBLNOR*, *ABW*, and *BBW* compute the respective metric translation distances and state the resulting A-basis and B-basis statistical tolerance limits in terms of their proportion of the underlying known (computed) conceptual two-parameter statistical distribution mean (see Tables 8.9–8.11). The associated proportions presented in these tables in parentheses pertain to the means of the sampling distributions for the exact A-basis and B-basis statistical tolerance limits that were computed by running microcomputer programs *SABNSTL*, *SBBNSTL*, *SABLNSTL*, *SBBLBSTL*, *SWABSTL*, and *SWBBSTL*. Note that the former proportions, although consistently larger, are remarkably accurate for most of the cases tabulated. Note, however, that the disparity increases as the coefficient of variation increases and n_{dv} decreases.

Clearly, when the disparity between the corresponding entries in Tables 8.9–8.11 is negligible for practical purposes, ML-based A-basis and B-basis statistical tolerance limits for uncensored replicate datum values are equivalent for practical purposes to corresponding exact (unbiased) A-basis and B-basis statistical tolerance limits. Thus, we assert that, for the associated values of n_{dv} and the coefficient of variation, the respective ML-based A-basis and B-basis statistical tolerance limits are almost exact (unbiased) for experiment test programs with no censoring.

8.D.4. Computing Empirical Statistical Bias Corrections for ML Analyses of Life (Endurance) Experiment Test Programs with Only a Limited Amount of *Type I* Censoring

Suppose that the life (endurance) datum values of specific interest are presumed to have been randomly selected from a given conceptual two-parameter statistical distribution with known parameters. Suppose further that the value of the life (endurance) metric pertaining to *Type I* censoring

Table 8.9 Proportions of the Conceptual (Two-Parameter) Normal Distribution Mean that Empirically Delimit the ML-estimated A-Basis and B-Basis Statistical Tolerance Limits, Based on 30,000 Replicate Normal Pseudorandom Data Sets, Each of Size n_{dv}

n_{dv}	(stddev)/(mean) = 0.0125		(stddev)/(mean) = 0.025		(stddev)/(mean) = 0.05		(stddev)/(mean) = 0.1	
	A-basis	B-basis	A-basis	B-basis	A-basis	B-basis	A-basis	B-basis
6	0.95 (0.94)[a]	0.97 (0.97)	0.90 (0.88)	0.94 (0.93)	0.81 (0.76)	0.88 (0.86)	0.62 (0.52)	0.77 (0.71)
8	0.96 (0.95)	0.97 (0.97)	0.91 (0.89)	0.96 (0.94)	0.82 (0.79)	0.89 (0.88)	0.64 (0.58)	0.78 (0.75)
10	0.96 (0.95)	0.97 (0.97)	0.91 (0.90)	0.95 (0.94)	0.83 (0.81)	0.90 (0.89)	0.65 (0.61)	0.79 (0.77)
12	0.96 (0.95)	0.98 (0.97)	0.92 (0.91)	0.95 (0.95)	0.83 (0.82)	0.90 (0.89)	0.66 (0.63)	0.80 (0.78)
16	0.96 (0.96)	0.98 (0.98)	0.92 (0.91)	0.95 (0.95)	0.84 (0.83)	0.91 (0.90)	0.68 (0.66)	0.81 (0.80)
24	0.96 (0.96)	0.98 (0.98)	0.92 (0.92)	0.96 (0.95)	0.85 (0.84)	0.91 (0.91)	0.70 (0.69)	0.82 (0.82)
32	0.96 (0.96)	0.98 (0.98)	0.93 (0.92)	0.96 (0.96)	0.85 (0.85)	0.92 (0.91)	0.71 (0.70)	0.83 (0.83)

[a]The values in parentheses are the corresponding exact values.

Table 8.10 Proportions of the Conceptual Two-Parameter $\log_e$–Normal Distribution Mean that Empirically Delimit the ML-estimated A-Basis and B-Basis Statistical Tolerance Limits, Based on 30,000 Replicate Normal Pseudorandom Data Sets, Each of Size n_{dv}

n_{dv}	(stddev)/(mean) = 0.05		(stddev)/(mean) = 0.1		(stddev)/(mean) = 0.2		(stddev)/(mean) = 0.4	
	A-basis	B-basis	A-basis	B-basis	A-basis	B-basis	A-basis	B-basis
6	0.82 (0.79)[a]	0.89 (0.87)	0.68 (0.62)	0.79 (0.75)	0.46 (0.40)	0.62 (0.57)	0.21 (0.17)	0.38 (0.33)
8	0.83 (0.81)	0.90 (0.88)	0.69 (0.66)	0.80 (0.78)	0.48 (0.44)	0.64 (0.61)	0.23 (0.20)	0.40 (0.37)
10	0.84 (0.82)	0.90 (0.89)	0.70 (0.68)	0.81 (0.79)	0.49 (0.46)	0.65 (0.63)	0.24 (0.22)	0.42 (0.40)
12	0.84 (0.83)	0.90 (0.90)	0.71 (0.69)	0.82 (0.80)	0.50 (0.48)	0.66 (0.64)	0.25 (0.24)	0.43 (0.41)
16	0.85 (0.84)	0.91 (0.90)	0.72 (0.71)	0.82 (0.82)	0.52 (0.50)	0.68 (0.66)	0.27 (0.26)	0.45 (0.44)
24	0.86 (0.85)	0.91 (0.91)	0.74 (0.73)	0.83 (0.83)	0.54 (0.53)	0.69 (0.68)	0.29 (0.28)	0.47 (0.46)
32	0.86 (0.86)	0.92 (0.92)	0.74 (0.74)	0.84 (0.84)	0.55 (0.54)	0.70 (0.70)	0.30 (0.29)	0.48 (0.48)

[a]The values in parentheses are the corresponding exact values.

Table 8.11 Proportions of the Conceptual Two-Parameter Weibull Distribution Mean that Empirically Delimit the ML-estimated A-Basis and B-Basis Statistical Tolerance Limits, Based on 30,000 Replicate Normal Pseudorandom Data Sets, Each of Size n_{dv}

n_{dv}	(stddev)/(mean) = 0.05		(stddev)/(mean) = 0.1		(stddev)/(mean) = 0.2		(stddev)/(mean) = 0.4	
	A-basis	B-basis	A-basis	B-basis	A-basis	B-basis	A-basis	B-basis
6	0.76 (0.69)[a]	0.88 (0.84)	0.56 (0.48)	0.76 (0.69)	0.30 (0.23)	0.56 (0.47)	0.07 (0.06)	0.27 (0.21)
8	0.77 (0.73)	0.88 (0.86)	0.58 (0.52)	0.77 (0.73)	0.32 (0.27)	0.58 (0.52)	0.08 (0.07)	0.29 (0.25)
10	0.78 (0.74)	0.89 (0.87)	0.60 (0.55)	0.78 (0.75)	0.33 (0.29)	0.59 (0.55)	0.09 (0.08)	0.31 (0.27)
12	0.78 (0.76)	0.89 (0.88)	0.61 (0.57)	0.79 (0.76)	0.34 (0.31)	0.60 (0.57)	0.10 (0.09)	0.32 (0.29)
16	0.79 (0.78)	0.90 (0.89)	0.62 (0.59)	0.80 (0.78)	0.36 (0.34)	0.62 (0.59)	0.11 (0.10)	0.34 (0.32)

[a]The values in parentheses are the corresponding exact values.

is known so that we can compute the exact value for pf at which *Type I* censoring occurs. If so, we could generate n_{sim} replicate sets of n_{dv} pseudorandom datum values from this conceptual two-parameter statistical distribution, and censor the pseudorandom datum values that exceed the life (endurance) metric pertaining to *Type I* censoring. Then, in theory, empirical statistical bias corrections for life (endurance) experiment test programs with *Type I* censoring would be computed exactly as empirical statistical bias corrections are computed for experiment test programs with uncensored data. However, in practice, especially for small experiment test programs, a simulation-based replicate data set will eventually be generated that has an excessive number of suspended tests. Then, even a microcomputer program that employs a very robust ML estimation algorithm will fail to estimate the actual values for the conceptual distribution parameters.

8.D.5. Computing Pragmatic Bias-Corrected Coefficients of Variation

Microcomputer programs *PBCNCOV*, *PBCLNCOV*, and *PBCWCOV* are analogous to microcomputer programs *EBCNCOV EBCLNCOV*, and *EBCWCOV* except that the former pertain to the experiment test program that was actually conducted. Accordingly, the ML estimate of the actual value for the coefficient of variation replaces the actual value for the coefficient of variation in these microcomputer programs. As discussed later, when the experiment test program that was actually conducted includes *Type I* censored life (endurance) datum values, microcomputer programs *PBCLNCOV* and *PBCWCOV* compute *median* pragmatic bias-corrected coefficients of variation.

8.D.6. Computing *Ad Hoc* Pragmatic Bias-Corrected A-Basis and B-Basis Statistical Tolerance Limits

Suppose that the disparity between the corresponding entries in Tables 8.9–8.11 is not regarded as being negligible for practical purposes. If so, then we cannot assert that ML estimates of A-basis or B-basis statistical tolerance limits are almost exact (unbiased) for the associated values of n_{dv} and the coefficient of variation. Thus, various alternative statistics must be examined to find one that generates almost unbiased *ad hoc* pragmatic bias-corrected A-basis and B-basis statistical tolerance limits for these values of n_{dv} and the coefficient of variation—as judged by comparing the resulting *ad hoc* pragmatic bias-corrected A-basis and B-basis statistical tolerance limits to corresponding exact (unbiased) A-basis and B-basis statistical tolerance limits.

In turn, given a satisfactory alternative statistic, we assert that its *ad hoc* pragmatic bias-corrected *A*-basis and *B*-basis statistical tolerance limits are almost exact (unbiased) under continual replication for experiment test programs.

Recall that, for experiment test programs of practical sizes without censoring, microcomputer programs *ABLNSTL* and *BBLNSTL* respectively compute exact (unbiased) *A*-basis and *B*-basis statistical tolerance limits for replicate $\log_e$–normal datum values. Similarly, microcomputer program *BLISTL* respectively computes exact (unbiased) *A*-basis and *B*-basis statistical tolerance limits for replicate Weibull datum values. These exact (unbiased) *A*-basis and *B*-basis statistical tolerance limit values were used in screening various alternative statistics relative to the agreement between their resulting *ad hoc* pragmatic bias-corrected *A*-basis and *B*-basis statistical tolerance limits and the corresponding exact (unbiased) reference *A*-basis and *B*-basis statistical tolerance limits. The two *ad hoc* pragmatic bias-corrected *A*-basis and *B*-basis statistical tolerance limits that exhibited the closest overall agreement are presented in Sections 8.D.7, 8.D.8, and 8.E.

Fortunately, we do not have to compute each ML-based est[$fnc(pf)$] that is used to generate its pragmatic sampling distribution when only an *ad hoc* pragmatic bias-corrected *A*-basis or *B*-basis statistical tolerance limit is of specific interest. Classical ML-based lower $100(scp)\%$ (one-sided) asymptotic statistical confidence limits that allegedly bound the actual value for $fnc(pf)$ are always smaller for pseudorandom data sets with several *Type I* censored tests than for pseudorandom data sets with few if any *Type I* censored tests. Thus, pseudorandom data sets with several *Type I* censored tests need only be counted. In turn, when the respective classical ML-based lower $100(scp)\%$ (one-sided) asymptotic statistical confidence limits for pseudorandom data sets with few if any *Type I* censored tests are computed and ordered from smallest to largest, the actual ranks for these ordered values are easily computed.

8.D.7. Computing *Ad Hoc* Pragmatic Bias-Corrected A-Basis and *B*-basis Statistical Tolerance Limits, Given Replicate $\log_e$–Normal Life (Endurance) Datum Values with Only a Limited Amount of *Type I* Censoring

Microcomputer program *LNPBCPV* can be used to compute *ad hoc* pragmatic bias-corrected *A*-basis and *B*-basis statistical tolerance limits, given replicate $\log_e$–normal datum values with only a limited amount of *Type I* censoring. This microcomputer program generates a simulation-based pragmatic sampling distribution that consists of n_{rep} "replicate" realizations for the classical lower $100(scp)\%$ (one-sided) asymptotic statistical confidence limit that allegedly bounds the actual value for $fnc(pf)$, given a conceptual two-parameter $\log_e$–normal life (endurance) distribution. The respective n_{rep} realizations are computed using the propagation of variability expression for est$\{$var$[$est$(y$ given $fnc = closascl)]\}$. The $100(scp)^{th}$ percentile of this pragmatic sampling distribution is then translated so that $100(scp)\%$ of the 30,000 "replicate" pseudorandom data sets generate est$[fnc(pf)]$ values that fall below the ML-based est$[fnc(pf)]$ value pertaining to the experiment test program that was actually conducted.

Ad hoc pragmatic bias-corrected A-basis and B-basis statistical tolerance limits ($scp = 0.95$ and $pf = 0.01$ and 0.10, respectively) are compared to their corresponding exact (unbiased) reference A-basis and B-basis statistical tolerance limits in Tables 8.12–8.17. The respective entries in these tables are the ratios of the *ad hoc* pragmatic bias-corrected A-basis and B-basis statistical tolerance limits computed by running microcomputer programs *LNPBCPV* and *LNPBCLR* (presented in Supplemental Topic 8.E) to corresponding exact (unbiased) reference A-basis and B-basis statistical tolerance limits computed by running microcomputer programs *ABLNSTL* and *BBLNSTL*. Tables 8.12–8.17 also include, for comparative purposes, analogous ratios pertaining to corresponding A-basis and B-basis statistical tolerance limits computed by running microcomputer program *LN1A*.

8.D.7.1. Discussion

Ad hoc pragmatic bias-corrected A-basis and B-basis statistical tolerance limits computed by running microcomputer program *LNPBCPV* or *LNPBCLR* are remarkably accurate as judged by their agreement with exact (unbiased) A-basis and B-basis statistical tolerance limits computed by running microcomputer programs *ABLNSTL* and *BBLBSTL*. Microcomputer program *LNPBCLR* is more accurate than microcomputer program *LNPBCPV* for n_{dv} less than 10–12. However, it takes several hours to run on a microcomputer with a 450MHz microprocessor. Program *LNPBCPV* is just as accurate as program *LNPBCLR* for n_{dv} greater than 12–16 and its run-time is much shorter. Overall, the accuracy is so remarkable that we assert (conjecture) that *ad hoc* pragmatic bias-corrected A-basis and B-basis statistical tolerance limits are almost exact (unbiased) *even* for small experiment test programs with only a limited amount of *Type I* censoring (say 20–25% maximum).

```
C> COPY AWBLDATA DATA

    1 files(s) copied

C> LNPBCPV
```

Given the Standardized Conceptual Normal Distribution Variate
$y = csp[\log_e(fnc) - clp]$

est(clp) = 0.5956793964D + 01
est(csp) = 0.4171442213D + 01
est{var[est(clp)]} = 0.9987394347D − 02
est{var[est(csp)]} = 0.1903271765D + 01
est{covar[est(clp),est(csp)]} = −0.1346739280D − 01
est(*conceptual correlation coefficient*) = −0.9768030920D − 01

fnc	est(y)	est(pf)
277.000	−1.3881578	0.0825445
310.000	−0.9186420	0.1791414
374.000	−0.1357311	0.4460169
402.000	0.1654316	0.5656978
456.000	0.6912032	0.7552811

snc	est(y)	est(pf)
500.000	1.0754568	0.8589149

Ad Hoc Pragmatic Bias-Corrected Lower 95% (One-Sided) Statistical
Confidence (Tolerance) Limit that Allegedly Bounds the Actual Value
for fnc(01)

97.070

Based on the 95th Percentile of the Sampling Distribution Comprised
of 30,000 "Replicate" Realizations for the Classical
Lower 95% (One-Sided) Asymptotic Statistical Confidence Limit that
Allegedly Bounds the Actual Value for fnc(01),
Each Realization Computed Using the Propagation of Variability
Expression for est{var[est(y given $fnc = closascl$)]}

Table 8.12 Ratios of *Ad Hoc* Pragmatic Bias-Corrected *A*-Basis Statistical Confidence (Tolerance) Limits Computed by Running Microcomputer Programs *LNPBCPV* and *LNPBCLR* to the Corresponding Exact (Unbiased) *A*-Basis Statistical Tolerance Limits Computed by Running Microcomputer Program *ABLNSTL*—for 12 Sets of 6 Uncensored Pseudorandom Datum Values—and the Corresponding Ratios Pertaining to the Smaller More Accurate of the Two Lower (One-Sided) Asymptotic Statistical Confidence Limits Computed by Running Microcomputer *LN1A* (Which does not Include a Statistical Bias Correction)

(stddev)/(mean) = 0.1			(stddev)/(mean) = 0.2			(stddev)/(mean) = 0.4		
LN1A	LNPBCPV	LNPBCLR	LN1A	LNPBCPV	LNPBCLR	LN1A	LNPBCPV	LNPBCLR
1.058	1.027	**1.004**	1.210	1.052	**1.016**	1.336	0.943	**0.963**
1.104	1.037	**1.003**	1.169	1.045	**1.005**	1.330	0.966	**0.989**
1.120	1.037	**0.999**	1.164	1.035	**0.994**	1.128	1.038	**1.000**
1.061	1.027	**1.003**	1.198	1.036	**0.998**	1.485	0.694	**0.939**
1.062	1.028	**1.003**	1.124	1.040	**1.003**	1.534	0.538	**0.873**
1.049	1.023	**1.004**	1.139	1.043	**1.004**	1.510	0.640	**0.909**
1.089	1.032	**1.002**	1.233	1.036	**0.991**	1.155	1.043	**1.005**
1.120	1.039	**1.002**	1.120	1.038	**0.999**	1.351	0.954	**1.003**
1.052	1.020	**1.003**	1.077	1.033	**1.004**	1.357	0.916	**0.964**
1.086	1.037	**1.004**	1.118	1.041	**1.005**	1.507	0.652	**0.930**
1.117	1.042	**1.007**	1.196	1.038	**1.001**	1.185	1.048	**0.995**
1.057	1.039	**1.005**	1.145	1.037	**0.998**	1.470	0.730	**0.938**

Notes: (a) Bold entries are the most accurate. (b) The 12 exact (unbiased) *A*-basis statistical tolerance limits computed by running program *ABLNSTL* ranged from 0.492 to 0.770 for (stddev)/(mean) = 0.1, from 0.263 to 0.723 for (stddev)/(mean) = 0.2, and from 0.066 to 0.446 for (stddev)/(mean) = 0.4.

Table 8.13 Ratios of *Ad Hoc* Pragmatic Bias-Corrected *B*-Basis Statistical Confidence (Tolerance) Limits Computed by Running Microcomputer Programs *LNPBCPV* and *LNPBCLR* to the Corresponding Exact (Unbiased) *B*-Basis Statistical Tolerance Limits Computed by Running Microcomputer Program *BBLNSTL*—for 12 Sets of 6 Uncensored Pseudorandom Datum Values—and the Corresponding Ratios Pertaining to the Smaller More Accurate of the Two Lower (One-Sided) Asymptotic Statistical Confidence Limits Computed by Running Microcomputer *LN1A* (Which does not Include a Statistical Bias Correction)

(stddev)/(mean) = 0.1			(stddev)/(mean) = 0.2			(stddev)/(mean) = 0.4		
LN1A	*LNPBCPV*	*LNPBCLR*	*LN1A*	*LNPBCPV*	*LNPBCLR*	*LN1A*	*LNPBCPV*	*LNPBCLR*
1.034	1.015	**0.998**	1.118	1.044	**1.003**	1.186	1.017	**0.977**
1.060	1.023	**0.995**	1.096	1.032	**0.995**	1.183	1.031	**0.992**
1.069	1.025	**0.993**	1.094	1.029	**0.989**	1.073	1.026	**0.993**
1.035	1.016	**0.998**	1.112	1.033	**0.992**	1.262	**0.975**	0.972
1.036	1.016	**0.998**	1.071	1.026	**0.995**	1.286	0.939	**0.957**
1.029	1.014	**0.999**	1.080	1.029	**0.994**	1.274	0.966	**0.977**
1.048	1.020	**0.997**	1.131	1.040	**0.997**	1.088	1.031	**0.994**
1.069	1.024	**0.991**	1.069	1.025	**0.992**	1.194	1.033	**0.998**
1.023	1.011	**0.999**	1.045	1.020	**0.998**	1.197	**1.018**	0.977
1.049	1.020	**0.996**	1.068	1.026	**0.996**	1.273	**0.970**	0.968
1.067	1.026	**0.995**	1.111	1.035	**0.992**	1.105	1.039	**0.997**
1.057	1.023	**0.997**	1.083	1.027	**0.991**	1.255	**0.977**	0.968

Notes: (a) Bold entries are the most accurate. (b) The 12 exact (unbiased) *B*-basis statistical tolerance limits computed by running program *BBLNSTL* ranged from 0.648 to 0.848 for (stddev)/(mean) = 0.1; from 0.439 to 0.866 for (stddev)/(mean) = 0.2, and from 0.173 to 0.633 for (stddev)/(mean) = 0.4.

Table 8.14 Ratios of *Ad Hoc* Pragmatic Bias-Corrected *A*-Basis Statistical Confidence (Tolerance) Limits Computed by Running Microcomputer Programs *LNPBCPV* and *LNPBCLR* to the Corresponding Exact (Unbiased) *A*-Basis Statistical Tolerance Limits Computed by Running Microcomputer Program *ABLNSTL*—for 12 Sets of 8 Uncensored Pseudorandom Datum Values—and the Corresponding Ratios Pertaining to the Smaller More Accurate of the Two Lower (One-Sided) Asymptotic Statistical Confidence Limits Computed by Running Microcomputer *LN1A* (Which does not Include a Statistical Bias Correction)

(stddev)/(mean) = 0.1			(stddev)/(mean) = 0.2			(stddev)/(mean) = 0.4		
LN1A	LNPBCPV	LNPBCLR	LN1A	LNPBCPV	LNPBCLR	LN1A	LNPBCPV	LNPBCLR
1.045	1.014	**0.997**	1.079	1.023	**0.997**	1.192	0.981	**0.988**
1.033	1.012	**0.998**	1.122	1.015	**0.998**	1.221	0.927	**1.017**
1.053	1.019	**0.996**	1.095	1.015	**0.998**	1.115	1.015	**0.993**
1.048	1.017	**0.998**	1.104	1.019	**0.997**	1.261	0.889	**0.997**
1.035	1.013	**0.999**	1.077	1.022	**0.997**	1.195	0.975	**1.012**
1.074	1.020	**0.995**	1.082	1.019	**1.000**	1.152	**1.002**	1.010
1.044	1.015	**0.999**	1.082	1.025	**0.998**	1.203	0.974	**1.000**
1.052	1.015	**0.997**	1.110	1.012	**1.003**	1.153	1.006	**1.003**
1.047	1.018	**1.000**	1.092	1.018	**0.999**	1.241	0.916	**1.032**
1.052	1.018	**0.998**	1.093	1.022	**0.997**	1.239	0.947	**1.019**
1.040	1.016	**0.996**	1.090	1.019	**0.999**	1.123	1.014	**1.000**
1.061	1.022	**1.001**	1.113	1.017	**0.994**	1.252	0.915	**1.018**

Notes:(a) Bold entries are the most accurate. (b) The 12 exact (unbiased) *A*-basis statistical tolerance limits computed by running program *ABLNSTL* ranged from 0.539 to 0.731 for (stddev)/(mean) = 0.1, from 0.377 to 0.812 for (stddev)/(mean) = 0.2, and from 0.125 to 0.373 for (stddev)/(mean) = 0.4.

Table 8.15 Ratios of *Ad Hoc* Pragmatic Bias-Corrected *A*-Basis Statistical Confidence (Tolerance) Limits Computed by Running Microcomputer Programs *LNPBCPV* and *LNPBCLR* to the Corresponding Exact (Unbiased) *B*-Basis Statistical Tolerance Limits Computed by Running Microcomputer Program *ABLNSTL*—for 12 Sets of 8 Uncensored Pseudorandom Datum Values—and the Corresponding Ratios Pertaining to the Smaller More Accurate of the Two Lower (One-Sided) Asymptotic Statistical Confidence Limits Computed by Running Microcomputer *LN1A* (Which does not Include a Statistical Bias Correction)

(stddev)/(mean) = 0.1			(stddev)/(mean) = 0.2			(stddev)/(mean) = 0.4		
LN1A	*LNPBCPV*	*LNPBCLR*	*LN1A*	*LNPBCPV*	*LNPBCLR*	*LN1A*	*LNPBCPV*	*LNPBCLR*
1.026	1.009	**0.996**	1.046	1.023	**0.996**	1.109	1.009	**0.993**
1.019	1.008	**0.998**	1.070	1.013	**0.995**	1.126	0.998	**1.000**
1.031	1.012	**0.995**	1.055	1.013	**0.995**	1.067	1.014	**0.997**
1.028	1.010	**0.997**	1.060	1.015	**0.996**	1.147	0.983	**0.994**
1.020	1.008	**0.994**	1.045	1.015	**0.994**	1.097	1.013	**0.999**
1.043	1.014	**0.994**	1.048	1.014	**0.996**	1.087	1.013	**0.997**
1.026	1.009	**0.997**	1.048	1.015	**0.996**	1.116	1.007	**0.993**
1.030	1.010	**0.996**	1.063	1.014	**0.998**	1.088	1.016	**0.998**
1.028	1.012	**0.998**	1.054	1.012	**0.996**	1.136	**0.998**	1.017
1.030	1.011	**0.996**	1.054	1.015	**0.994**	1.135	**0.999**	1.005
1.024	1.009	**0.996**	1.052	1.014	**0.997**	1.071	1.014	**0.996**
1.036	1.014	**0.998**	1.065	1.014	**0.993**	1.142	**0.996**	1.013

Notes: (a) Bold entries are the most accurate. (b) The 12 exact (unbiased) *B*-basis statistical tolerance limits computed by running program *BBLNSTL* ranged from 0.692 to 0.832 for (stddev)/(mean) = 0.1, from 0.533 to 0.744 for (stddev)/(mean) = 0.2, and from 0.267 to 0.560 for (stddev)/(mean) = 0.4.

Table 8.16 Ratios of *Ad Hoc* Pragmatic Bias-Corrected *A*-Basis Statistical Confidence (Tolerance) Limits Computed by Running Microcomputer Programs *LNPBCPV* and *LNPBCLR* to the Corresponding Exact (Unbiased) *A*-Basis Statistical Tolerance Limits Computed by Running Microcomputer Program *ABLNSTL*—for 12 Sets of Uncensored Pseudorandom Datum Values Whose Coefficient of Variation is Equal to 0.4—and the Corresponding Ratios Pertaining to the Smaller More Accurate of the Two Lower (One-Sided) Asymptotic Statistical Confidence Limits Computed by Running Microcomputer *LN1A* (Which does not Include a Statistical Bias Correction)

| | $n_{dv} = 10$ | | | $n_{dv} = 16$ | | | $n_{dv} = 32$ | |
LN1A	LNPBCPV	LNPBCLR	LN1A	LNPBCPV	LNPBCLR	LN1A	LNPBCPV	LNPBCLR
1.196	0.902	**1.022**	1.073	0.988	**1.005**	1.027	1.000	—
1.170	**0.969**	1.038	1.073	**0.995**	1.016	1.028	0.996	—
1.129	**0.997**	1.013	1.070	0.987	**1.001**	1.027	0.997	—
1.084	1.018	**1.007**	1.081	0.982	**1.013**	1.029	0.995	—
1.257	0.811	**1.100**	1.058	**1.000**	1.002	1.031	0.996	—
1.137	**0.988**	1.014	1.060	0.990	**0.996**	1.026	1.001	—
1.117	**1.002**	1.005	1.079	**0.987**	1.015	1.029	0.999	—
1.126	0.992	**1.005**	1.067	**0.997**	1.009	1.032	0.993	—
1.182	0.931	**1.023**	1.074	**0.994**	1.014	1.028	0.997	—
1.135	0.984	**1.010**	1.042	**1.001**	0.994	1.031	1.001	—
1.184	0.924	**1.019**	1.080	**1.005**	1.023	1.030	0.993	—
1.112	0.995	**0.997**	1.045	1.003	**1.000**	1.033	1.002	—

Notes: (a) Bold entries are the most accurate. (b) The 12 exact (unbiased) *A*-basis statistical tolerance limits computed by running program *ABLNSTL* ranged from 0.073 to 0.415 for $n_{dv} = 10$, from 0.175 to 0.464 for $n_{dv} = 16$, and from 0.246 to 0.341 for $n_{dv} = 32$.

Table 8.17 Ratios of *Ad Hoc* Pragmatic Bias-Corrected *A*-Basis Statistical Confidence (Tolerance) Limits Computed by Running Microcomputer Programs *LNPBCPV* and *LNPBCLR* to the Corresponding Exact (Unbiased) *B*-Basis Statistical Tolerance Limits Computed by Running Microcomputer Program *ABLNSTL*—for 12 Sets of Uncensored Pseudorandom Datum Values Whose Coefficient of Variation is Equal to 0.4—and the Corresponding Ratios Pertaining to the Smaller More Accurate of the Two Lower (One-Sided) Asymptotic Statistical Confidence Limits Computed by Running Microcomputer *LN1A* (Which does not Include a Statistical Bias Correction)

$n_{dv} = 10$			$n_{dv} = 16$			$n_{dv} = 32$		
LN1A	*LNPBCPV*	*LNPBCLR*	*LN1A*	*LNPBCPV*	*LNPBCLR*	*LN1A*	*LNPBCPV*	*LNPBCLR*
1.112	0.979	**1.002**	1.043	0.998	**0.999**	1.016	1.001	—
1.102	**1.004**	1.012	1.043	1.003	**1.001**	1.016	1.000	—
1.077	1.006	**0.998**	1.041	**0.998**	0.997	1.016	0.999	—
1.049	1.013	**1.000**	1.047	0.998	**1.000**	1.017	1.001	—
1.145	**0.974**	1.042	1.034	1.003	**0.999**	1.018	1.000	—
1.079	1.010	**1.002**	1.035	**0.997**	0.992	1.015	1.002	—
1.067	1.010	**0.998**	1.046	0.997	**1.000**	1.017	1.001	—
1.073	1.005	**0.997**	1.039	1.004	**1.001**	1.019	0.997	—
1.104	0.992	**1.003**	1.044	0.997	**1.000**	1.017	1.001	—
1.078	**1.004**	0.996	1.025	**1.002**	0.994	1.018	0.999	—
1.105	**0.994**	1.006	1.047	**1.005**	1.007	1.018	0.998	—
1.065	1.007	0.994	1.026	**1.003**	**0.997**	1.020	1.003	—

Notes: (a) Bold entries are the most accurate. (b) The 12 exact (unbiased) *B*-basis statistical tolerance limits computed by running program *BBLNSTL* ranged from 0.231 to 0.604 for $n_{dv} = 10$, from 0.325 to 0.651 for $n_{dv} = 16$, and from 0.432 to 0.555 for $n_{dv} = 32$.

8.D.8. Computing *Ad Hoc* Pragmatic Bias-Corrected *A*-Basis and *B*-Basis Statistical Tolerance Limits, Given Replicate Weibull Life (Endurance) Datum Values with Only a Limited Amount of *Type I* Censoring

Microcomputer programs *WPBCPV* and *WPBCLR* are analogous to microcomputer programs *LNPBCPV* and *LNPBCLR*. Similarly, Tables 8.18–8.21 are analogous to Tables 8.12–8.17. Given uncensored replicate Weibull datum values, these tables compare *ad hoc* pragmatic bias-corrected *A*-basis and *B*-basis statistical tolerance limits computed by running microcomputer programs *WPBCPV* and *WPBCLR* to corresponding exact (unbiased) reference *A*-basis and *B*-basis statistical tolerance limits computed by running microcomputer program *BLISTL*. Tables 8.18–8.23 also include, for comparative purposes, analogous ratios pertaining to the corresponding *A*-basis and *B*-basis statistical tolerance limits computed by running microcomputer program *WEIBULL*.

8.D.8.1. Discussion

Note that, in contrast to the *A*-basis and *B*-basis statistical tolerance limits computed by running microcomputer program *LN1A*, the same limits computed by running microcomputer program *WEIBULL* are sometimes smaller than corresponding exact (unbiased) *A*-basis and *B*-basis statistical tolerance limits computed by running microcomputer program *BLISTL*. If so, then *ad hoc* pragmatic statistical bias corrections should not be employed. Otherwise, considering the marked variability of exact (unbiased) *A*-basis and *B*-basis statistical tolerance limits computed by running microcomputer program *BLISTL*, microcomputer programs *WPBCPV* and *WPBCLR* compute *A*-basis and (especially) *B*-basis statistical tolerance limits that are also remarkably accurate when the coefficient of variation is equal to 0.2 or less. Thus, we assert (conjecture) that, when the coefficient of variation is equal to 0.2 or less, *ad hoc* pragmatic bias-corrected *A*-basis and *B*-basis statistical tolerance limits computed by running microcomputer programs *WPBCPV* and *WPBCLR* are almost exact (unbiased) *even* for small experiment test programs with only a limited amount of *Type I* censoring (say 20–25% maximum).

Microcomputer program *WPBCLR* is slightly more accurate than microcomputer program *WPBCPV* for n_{dv} less than 10–12. However, it takes several hours to run on a microcomputer with a 450 MHz microprocessor. Microcomputer program *WPBCPV* is just as accurate as microcomputer program *WPBCLR* for n_{dv} greater than 12–16 and its run-time is considerably shorter.

C> COPY AWBLDATA DATA

   1 files(s) copied

C> WPBCPV

Given $F(fnc) = 1 - \exp^{-\left(\frac{fnc}{cdp1}\right)^{cdp2}}$

est($cdp1$) = 0.4289595976D + 03
est($cdp2$) = 0.4731943406D + 01
est{var[est($cdp1$)]} = 0.1666932095D + 04
est{var[est($cdp2$)]} = 0.3013638227D + 01
est{covar[est($cdp1$),est($cdp2$)]} = 0.8393830857D + 01
est($conceptual\ correlation\ coefficient$) = 0.1184283562D + 00

fnc	est(y)	est(pf)
277.000	−2.0694929	0.1186053
310.000	−1.5368900	0.1934980
374.000	−0.6487823	0.4070717
402.000	−0.3071535	0.5207523
456.000	0.2892640	0.7369587

snc	est(y)	est(pf)
500.000	0.7251484	0.8731865

Ad Hoc Pragmatic Bias-Corrected Lower 95% (One-Sided) Statistical
Confidence (Tolerance) Limit that Allegedly Bounds the Actual Value
for $fnc(01)$

35.138

Based on the 95th Percentile of the Sampling Distribution Comprised
of 30,000 "Replicate" Realizations for the Classical
Lower 95% (One-Sided) Asymptotic Statistical Confidence Limit that
Allegedly Bounds the Actual Value for $fnc(01)$,
Each Realization Computed Using the Propagation of Variability
Expression for est{var[est(y given $fnc = closascl$)]}

Table 8.18 Ratios of *Ad Hoc* Pragmatic Bias-Corrected *A*-Basis Statistical Confidence (Tolerance) Limits Computed by Running Microcomputer Programs *WPBCPV* and *WPBCLR* to Corresponding *A*-Basis Statistical Tolerance Limits Computed by Running Microcomputer Program *BLISTL*—for 12 Sets of 6 Uncensored Pseudorandom Datum Values—and the Corresponding Ratios Pertaining to the Smaller More Accurate of the Two Lower (One-Sided) Asymptotic Statistical Confidence Limits Computed by Running Microcomputer *WEIBULL* (Which does not Include a Statistical Bias Correction)

(stddev)/(mean) = 0.1			(stddev)/(mean) = 0.2			(stddev)/(mean) = 0.4		
WEIBULL	*WPBCPV*	*WPBCLR*	*WEIBULL*	*WPBCPV*	*WPBCLR*	*WEIBULL*	*WPBCPV*	*WPBCLR*
1.143	1.100	**1.063**	1.212	1.128	**1.059**	1.896	Value < 0	Value < 0
1.033	1.011	**1.009**	1.093	0.730	0.894	1.393	1.061	1.091
1.135	1.078	**1.055**	1.558	1.152	**1.115**	2.368	Value < 0	0.895
1.165	1.102	**1.064**	1.087	*1.000*	1.011	0.960	0.513	0.806
1.152	1.117	**1.067**	1.252	1.143	**1.092**	1.306	0.134	0.685
1.147	1.092	**1.025**	1.101	0.849	**0.969**	1.550	1.186	1.146
1.011	0.988	**0.995**	1.428	0.807	**1.016**	0.754	Value < 0	0.533
1.012	0.938	0.968	**0.982**	0.849	0.918	1.667	Value < 0	0.298
1.133	1.091	**1.039**	1.010	*0.985*	**0.993**	1.286	1.177	1.105
0.933	0.867	0.896	1.185	*1.029*	1.046	1.452	0.920	1.049
0.965	0.950	0.964	1.113	*0.996*	1.018	1.503	0.725	1.035
1.090	1.067	**1.040**	1.043	1.026	**1.011**	1.110	0.886	0.956

Notes: (a) Bold entries are the most accurate. (b) Italic entries connote a reversal of the anticipated order for the respective values. (c) The 12 exact (unbiased) *A*-basis statistical tolerance limits computed by running program *BLISTL* ranged from 0.318 to 0.721 for (stddev)/(mean) = 0.1, from 0.051 to 0.648 for (stddev)/(mean) = 0.2, and from 0.003 to 0.331 for (stddev)/ (mean) = 0.4.

Table 8.19 Ratios of *Ad Hoc* Pragmatic Bias-Corrected *B*-Basis Statistical Confidence (Tolerance) Limits Computed by Running Microcomputer Programs *WPBCPV* and *WPBCLR* to Corresponding *B*-Basis Statistical Tolerance Limits Computed by Running Microcomputer Program *BLISTL*—for 12 Sets of 6 Uncensored Pseudorandom Datum Values—and the Corresponding Ratios Pertaining to the Smaller More Accurate of the Two Lower (One-Sided) Asymptotic Statistical Confidence Limits Computed by Running Microcomputer *WEIBULL* (Which does not Include a Statistical Bias Correction)

(stddev)/(mean) = 0.1			(stddev)/(mean) = 0.2			(stddev)/(mean) = 0.4		
WEIBULL	WPBCPV	WPBCLR	WEIBULL	WPBCPV	WPBCLR	WEIBULL	WPBCPV	WPBCLR
1.057	1.043	**1.023**	1.080	1.060	**1.024**	1.260	*0.830*	**1.071**
1.008	1.003	**0.999**	*1.010*	0.959	0.983	1.135	1.083	**1.048**
1.048	1.036	**1.020**	1.193	1.140	**1.070**	1.408	*1.220*	1.229
1.060	1.045	**1.024**	1.025	1.009	**1.002**	**0.946**	0.885	0.931
1.059	1.051	**1.026**	1.093	1.072	**1.040**	1.089	0.970	**1.008**
1.054	1.042	**1.009**	1.018	*0.982*	**0.996**	1.191	1.142	**1.081**
0.998	0.993	0.993	1.140	*1.054*	1.057	*0.827*	0.714	*0.842*
0.990	0.972	0.980	**0.967**	0.937	0.954	1.194	*0.977*	1.114
1.049	1.039	**1.011**	**0.995**	0.989	0.991	1.106	1.085	**1.047**
0.951	0.934	0.944	1.061	1.035	**1.020**	1.152	1.089	**1.068**
0.977	0.973	0.979	1.032	1.007	**1.002**	1.162	1.090	1.062
1.035	1.028	**1.015**	1.014	1.009	**1.002**	1.021	0.978	**0.987**

Notes: (a) Bold entries are the most accurate. (b) Italic entries connote a reversal of the anticipated order for the respective values. (c) The 12 exact (unbiased) *B*-basis statistical tolerance limits computed by running program *BLISTL* ranged from 0.561 to 0.875 for (stddev)/(mean) = 0.1, from 0.229 to 0.812 for (stddev)/(mean) = 0.2, and from 0.053 to 0.706 for (stddev)/(mean) = 0.4.

Table 8.20 Ratios of *Ad Hoc* Pragmatic Bias-Corrected *A*-Basis Statistical Confidence (Tolerance) Limits Computed by Running Microcomputer Programs *WPBCPV* and *WPBCLR* to Corresponding *A*-Basis Statistical Tolerance Limits Computed by Running Microcomputer Program *BLISTL*—for 12 Sets of 8 Uncensored Pseudorandom Datum Values—and the Corresponding Ratios Pertaining to the Smaller More Accurate of the Two Lower (One-Sided) Asymptotic Statistical Confidence Limits Computed by Running Microcomputer *WEIBULL* (Which does not Include a Statistical Bias Correction)

(stddev)/(mean) = 0.1			(stddev)/(mean) = 0.2			(stddev)/(mean) = 0.4		
WEIBULL	*WPBCPV*	*WPBCLR*	*WEIBULL*	*WPBCPV*	*WPBCLR*	*WEIBULL*	*WPBCPV*	*WPBCLR*
1.026	1.007	**0.999**	1.049	0.997	**0.998**	1.190	0.788	**1.097**
1.066	1.013	**0.991**	1.186	1.050	**1.043**	1.341	0.859	1.241
1.034	1.010	**0.994**	1.112	1.056	**1.048**	1.372	0.834	1.181
0.986	0.959	0.958	1.090	1.065	**1.038**	1.195	*0.994*	1.140
0.982	0.960	0.962	1.162	1.075	**1.061**	1.735	0.461	1.804
0.996	0.985	0.982	1.054	1.015	**0.999**	1.262	0.748	1.208
1.055	1.013	**1.006**	1.059	1.008	**1.001**	1.217	*1.058*	1.089
1.074	1.039	**0.999**	1.083	1.054	**1.029**	0.825	0.600	0.834
0.971	0.955	0.957	0.990	0.919	0.945	1.469	0.723	1.345
1.043	1.023	**1.004**	1.123	1.075	**1.053**	1.099	*0.975*	1.028
1.030	**0.992**	0.991	1.250	1.054	**1.052**	1.327	0.863	1.245
1.035	1.021	**1.007**	1.100	**1.040**	1.060	1.219	0.815	1.208

Notes: (a) Bold entries are the most accurate. (b) Italic entries comote a reversal of the anticipated order for the respective values. (c) The 12 exact (unbiased) *A*-basis statistical tolerance limits computed by running program *BLISTL* ranged from 0.356 to 0.767 for (stddev)/(mean) = 0.1, from 0.110 to 0.523 for (stddev)/(mean) = 0.2, and from 0.010 to 0.174 for (stddev)/(mean) = 0.4.

Table 8.21   Ratios of *Ad Hoc* Pragmatic Bias-Corrected *B*-Basis Statistical Confidence (Tolerance) Limits Computed by Running Microcomputer Programs *WPBCPV* and *WPBCLR* to Corresponding *B*-Basis Statistical Tolerance Limits Computed by Running Microcomputer Program *BLISTL*—for 12 Sets of 8 Uncensored Pseudorandom Datum Values—and the Corresponding Ratios Pertaining to the Smaller More Accurate of the Two Lower (One-Sided) Asymptotic Statistical Confidence Limits Computed by Running Microcomputer *WEIBULL* (Which does not Include a Statistical Bias Correction)

(stddev)/(mean) = 0.1			(stddev)/(mean) = 0.2			(stddev)/(mean) = 0.4		
WEIBULL	*WPBCPV*	*WPBCLR*	*WEIBULL*	*WPBCPV*	*WPBCLR*	*WEIBULL*	*WPBCPV*	*WPBCLR*
1.009	**1.003**	**0.997**	1.013	**1.001**	0.991	1.057	**1.005**	0.940
1.025	1.012	**0.994**	1.071	1.045	*1.014*	1.122	*1.059*	1.085
1.011	**1.004**	0.993	1.043	1.032	**1.018**	1.136	*1.047*	1.059
0.988	0.980	0.976	1.037	1.030	**1.015**	1.083	*1.040*	1.043
0.988	0.981	0.978	1.064	1.044	**1.024**	1.258	*1.141*	1.225
0.995	0.992	0.989	1.019	1.008	**0.995**	1.090	*1.025*	1.052
1.019	1.008	**0.998**	1.020	1.009	**0.997**	1.083	1.052	**1.032**
1.027	1.018	**0.995**	1.033	1.026	**1.011**	**0.899**	0.860	*0.899*
0.983	0.979	0.977	**0.986**	0.971	0.970	1.171	*1.079*	1.110
1.016	1.010	**1.000**	1.049	1.038	**1.022**	1.033	1.010	**1.006**
1.007	**0.997**	0.991	1.096	1.056	**1.018**	1.138	*1.051*	1.075
1.014	1.010	**1.002**	1.051	1.030	**1.008**	1.071	*1.024*	1.062

Notes: (a) Bold entries are the most accurate. (b) Italic entries connote a reversal of the anticipated order for the respective values. (c) The 12 exact (unbiased) *B*-basis statistical tolerance limits computed by running program *BLISTL* ranged from 0.580 to 0.901 for (stddev)/(mean) = 0.1, from 0.323 to 0.758 for (stddev)/(mean) = 0.2, and from 0.099 to 0.475 for (stddev)/(mean) = 0.4.

Table 8.22 Ratios of *Ad Hoc* Pragmatic Bias-Corrected *A*-Basis Statistical Confidence (Tolerance) Limits Computed by Running Microcomputer Programs *WPBCPV* and *WPBCLR* to Corresponding *A*-Basis Statistical Tolerance Limits Computed by Running Microcomputer Program *BLISTL*—for 12 Sets of 10 Uncensored Pseudorandom Datum Values—and the Corresponding Ratios Pertaining to the Smaller More Accurate of the Two Lower (One-Sided) Asymptotic Statistical Confidence Limits Computed by Running Microcomputer *WEIBULL* (Which does not Include a Statistical Bias Correction)

(stddev)/(mean) = 0.1			(stddev)/(mean) = 0.2			= 0.4		
WEIBULL	*WPBCPV*	*WPBCLR*	*WEIBULL*	*WPBCPV*	*WPBCLR*	*WEIBULL*	*WPBCPV*	*WPBCLR*
0.996	0.953	0.960	1.081	*1.018*	1.021	1.164	**0.910**	1.141
1.034	1.022	**1.009**	1.070	1.021	**1.007**	1.402	*1.034*	*1.470*
1.050	1.023	**0.998**	1.086	1.060	**1.037**	1.225	*1.101*	1.124
1.084	1.054	**1.034**	1.075	1.039	**1.029**	1.112	*1.017*	1.065
1.044	1.026	**1.012**	1.077	1.025	**0.999**	1.191	0.958	1.176
0.999	0.985	0.978	1.026	**0.995**	0.991	1.106	0.767	1.101
1.059	1.040	**1.022**	1.079	*0.998*	1.018	1.211	0.193	**1.588**
1.049	1.030	**1.004**	1.060	1.018	**1.007**	1.185	0.889	1.148
1.010	0.979	0.979	1.024	0.960	**0.982**	1.209	0.555	**1.514**
1.034	1.014	**1.010**	**0.957**	0.917	0.929	1.208	0.857	*1.251*
0.984	0.970	0.965	1.077	1.028	**1.020**	1.166	*1.072*	1.101
1.018	**0.993**	0.989	1.084	1.044	**1.035**	1.224	**1.018**	1.111

Notes: (a) Bold entries are the most accurate. (b) Italic entries connote a reversal of the anticipated order for the respective values. (c) The 12 exact (unbiased) *A*-basis statistical tolerance limits computed by running program *BLISTL* ranged from 0.379 to 0.765 for (stddev)/(mean) = 0.1, from 0.226 to 0.478 for (stddev)/(mean) = 0.2, and from 0.011 to 0.199 for (stddev)/(mean) = 0.4.

Table 8.23 Ratios of *Ad Hoc* Pragmatic Bias-Corrected *B*-Basis Statistical Confidence (Tolerance) Limits Computed by Running Microcomputer Programs *WPBCPV* and *WPBCLR* to Corresponding *B*-Basis Statistical Tolerance Limits Computed by Running Microcomputer Program *BLISTL*—for 12 Sets of 10 Uncensored Pseudorandom Datum Values—and the Corresponding Ratios Pertaining to the Smaller More Accurate of the Two Lower (One-Sided) Asymptotic Statistical Confidence Limits Computed by Running Microcomputer *WEIBULL* (Which does not Include a Statistical Bias Correction)

(stddev)/(mean) = 0.1			(stddev)/(mean) = 0.2			(stddev)/(mean) = 0.4		
WEIBULL	*WPBCPV*	*WPBCLR*	*WEIBULL*	*WPBCPV*	*WPBCLR*	*WEIBULL*	*WPBCPV*	*WPBCLR*
0.992	0.979	0.974	1.032	1.017	**1.004**	1.056	*1.013*	1.027
1.014	1.010	**1.003**	1.027	1.015	**1.001**	1.153	*1.105*	1.125
1.019	1.011	**0.996**	1.035	1.028	**1.013**	1.090	1.062	**1.051**
1.035	1.027	**1.014**	1.030	1.021	**1.010**	1.042	1.025	**1.018**
1.018	1.014	**1.004**	1.028	1.015	**0.994**	1.073	*1.031*	1.045
0.997	0.992	0.987	1.007	**1.000**	0.992	1.030	0.975	**1.012**
1.024	1.019	**1.008**	1.028	1.010	**1.000**	1.070	***0.942***	1.098
1.019	1.014	**1.000**	1.021	1.012	**0.997**	1.066	***1.018***	1.040
1.001	0.993	0.987	*1.002*	0.989	0.982	1.067	***0.996***	1.013
1.014	1.011	1.003	**0.974**	0.961	0.958	1.073	*1.015*	1.076
0.989	0.985	0.981	1.029	1.018	**1.004**	1.066	1.046	**1.034**
1.003	*0.998*	0.991	**1.033**	1.023	**1.011**	1.087	1.052	**1.041**

Notes: (a) Bold entries are the most accurate. (b) Italic entries connote a reversal of the anticipated order for the respective values. (c) The 12 exact (unbiased) *B*-basis statistical tolerance limits computed by running program *BLISTL* ranged from 0.619 to 0.917 for (stddev)/(mean) = 0.1, from 0.482 to 0.746 for (stddev)/(mean) = 0.2, and from 0.108 to 0.481 for (stddev)/(mean) = 0.4.

Table 8.24 Ratios of *Ad Hoc* Pragmatic Bias-Corrected *A*-Basis Statistical Confidence (Tolerance) Limits Computed by Running Microcomputer Programs *WPBCPV* and *WPBCLR* to Corresponding *A*-Basis Statistical Tolerance Limits Computed by Running Microcomputer Program *BLISTL*—for 12 Sets of 16 Uncensored Pseudorandom Datum Values—and the Corresponding Ratios Pertaining to the Smaller More Accurate of the Two Lower (One-Sided) Asymptotic Statistical Confidence Limits Computed by Running Microcomputer *WEIBULL* (Which does not Include a Statistical Bias Correction)

(stddev)/(mean) = 0.1			(stddev)/(mean) = 0.2			(stddev)/(mean) = 0.4		
WEIBULL	WPBCPV	WPBCLR	WEIBULL	WPBCPV	WPBCLR	WEIBULL	WPBCPV	WPBCLR
1.023	1.011	**1.003**	1.038	1.021	**1.014**	1.109	**0.991**	1.139
1.024	1.012	**1.001**	1.031	**1.000**	1.009	1.018	0.883	**1.060**
0.997	0.989	0.984	1.045	0.995	1.010	1.113	**0.992**	1.079
1.029	1.017	**1.004**	**0.999**	0.972	0.971	1.102	**0.963**	1.184
1.029	1.009	**0.999**	*1.005*	0.952	0.984	1.155	**1.050**	1.236
0.999	0.993	0.988	1.065	1.042	**1.035**	*1.018*	0.939	1.034
1.033	1.024	**1.014**	1.033	1.006	**1.002**	1.074	**0.977**	1.111
0.997	0.988	0.982	1.045	1.022	**1.021**	*1.017*	0.927	1.040
1.020	1.008	**0.996**	1.026	0.995	**1.000**	*1.040*	0.921	1.147
1.010	**1.002**	0.995	1.038	1.016	**1.009**	*1.068*	0.911	1.205
1.017	**1.003**	0.994	1.026	0.992	**0.998**	1.079	**0.949**	1.103
1.029	1.020	**1.010**	**0.967**	0.938	0.943	**0.905**	0.758	0.951

Notes: (a) Bold entries are the most accurate. (b) Italic entries connote a reversal of the anticipated order for the respective values. (c) The 12 exact (unbiased) *A*-basis statistical tolerance limits computed by running program *BLISTL* ranged from 0.487 to 0.740 for (stddev)/(mean) = 0.1, from 0.212 to 0.463 for (stddev)/(mean) = 0.2, and from 0.044 to 0.137 for (stddev)/(mean) = 0.4.

Table 8.25 Ratios of *Ad Hoc* Pragmatic Bias-Corrected *B*-Basis Statistical Confidence (Tolerance) Limits Computed by Running Microcomputer Programs *WPBCPV* and *WPBCLR* to Corresponding *B*-Basis Statistical Tolerance Limits Computed by Running Microcomputer Program *BLISTL*—for 12 Sets of 16 Uncensored Pseudorandom Datum Values—and the Corresponding Ratios Pertaining to the Smaller More Accurate of the Two Lower (One-Sided) Asymptotic Statistical Confidence Limits Computed by Running Microcomputer *WEIBULL* (Which does not Include a Statistical Bias Correction)

(stddev)/(mean) = 0.1			(stddev)/(mean) = 0.2			(stddev)/(mean) = 0.4		
WEIBULL	*WPBCPV*	*WPBCLR*	*WEIBULL*	*WPBCPV*	*WPBCLR*	*WEIBULL*	*WPBCPV*	*WPBCLR*
1.009	1.005	**0.999**	1.014	1.009	**1.002**	1.047	**1.023**	1.038
1.009	1.005	**0.999**	1.009	**0.999**	0.994	**0.997**	0.967	0.987
0.996	0.996	0.990	1.018	1.007	**0.997**	1.039	**1.016**	1.017
1.011	1.008	**1.000**	**0.995**	0.988	0.981	1.033	**1.008**	1.035
1.011	**1.004**	0.995	**0.994**	0.982	0.979	*1.057*	1.066	1.062
0.998	0.996	0.993	1.026	1.020	**1.010**	**0.999**	0.982	0.992
1.013	1.010	**1.004**	1.013	1.005	**0.996**	1.026	**1.008**	1.020
0.997	0.994	0.990	1.017	1.011	**1.002**	**0.997**	0.978	0.989
1.006	**1.002**	0.995	1.008	**1.000**	0.992	*1.004*	0.990	1.021
1.005	**1.000**	0.994	1.014	1.008	**1.000**	1.017	**0.987**	1.028
1.006	**1.002**	0.996	1.096	**1.009**	0.990	1.032	**1.007**	1.020
1.017	1.009	**1.003**	**0.980**	0.972	0.965	**0.950**	0.916	0.941

Notes: (a) Bold entries are the most accurate. (b) Italic entries connote a reversal of the anticipated order for the respective values. (c) The 12 exact (unbiased) *B*-basis statistical tolerance limits computed by running program *BLISTL* ranged from 0.704 to 0.872 for (stddev)/(mean) = 0.1, from 0.453 to 0.713 for (stddev)/(mean) = 0.2, and from 0.211 to 0.404 for (stddev)/(mean) = 0.4.

Remark One: Given the same value for the coefficient of variation, exact (unbiased) *B*-basis statistical tolerance limits are typically smaller for the Weibull distribution than for the $\log_e$–normal distribution. Moreover, the ratio of exact (unbiased) *A*-basis statistical tolerance limits to their corresponding exact (unbiased) *B*-basis statistical tolerance limits is typically smaller for the Weibull distribution than for the $\log_e$–normal distribution. Accordingly, the denominators for the ratios in Tables 8.18 through 8.25 are typically smaller than the denominators for the ratios in Tables 8.12 through 8.17. Thus, even when the respective actual values for the statistical biases are equal, the ratios in Tables 8.18 through 8.25 typically deviate further from 1.000 than the ratios in Tables 8.12 through 8.18. Unfortunately, these deviations increase markedly as the coefficient of variation increases. Thus, the maximum value for the coefficient of variation is more critical for the Weibull distribution than for the $\log_e$–normal distribution.

Remark Two: Microcomputer programs *WPBCPV* and *WPBCLR* are much more likely to fail to compute an *ad hoc* pragmatic bias-corrected *A*-basis or *B*-basis statistical tolerance limit than are microcomputer programs *LNPBCPV* and *LNPBCLR*. However, a change in the input seed numbers will usually overcome this problem.

8.D.9. Computing Ad Hoc Pragmatic Statistical Bias Corrections for ML Analyses Pertaining to Conceptual s_a–$\log_e[fnc(pf)]$ Models with a Homoscedastic Fatigue Strength Distribution

As mentioned in Section 8.11, ML analyses for conceptual s_a–$\log_e[fnc(pf)]$ models with a homoscedastic fatigue strength distribution are conditional on the observed fatigue lives (endurances). Thus, to obtain a pragmatic set of "replicate" experiment test program datum values, we must generate a new pseudorandom alternating stress amplitude datum value at each respective observed fatigue life (instead of generating a new pseudorandom fatigue life datum value at each respective alternating stress amplitude). The associated pseudorandom alternating stress datum values for *Type I* censoring pertain either to a fatigue failure or to a suspended test, depending on whether the associated uniform pseudorandom number is less than or greater than the estimated CDF at the *Type I* censoring number of fatigue cycles. When a quadratic (parabolic) conceptual s_a–$\log_e[fnc(pf)]$ model is presumed in ML analysis, some of the "replicate" estimated s_a–$\log_e[fnc(50)]$ curves can exhibit a reversed curvature. If so, then there are three options:

(a) assert that no statistically credible pragmatic sampling distribution can be generated; or (b) include the associated "replicate" values for the statistical estimate of specific interest in its pragmatic sampling distribution; or (c) exclude the associated "replicate" values for the statistical estimate of specific interest from its pragmatic sampling distribution.

Microcomputer programs *SAFNCM11*, *SAFNCM12*, and *SAFNCM13* pertain to a linear (straight line) conceptual s_a–$\log_e[fnc(pf)]$ model that is based on the standardized normal distribution variate $y = [s_a - clp0 - clp1 \cdot \log_e(fnc)]/csp$. These programs are illustrated using the s_a–$\log_e(fnc)$ datum values that appear in microcomputer file *SAFNCDTA*. Each program first computes the associated ML estimates for the parameters of the linear (straight line) conceptual s_a–$\log_e[fnc(pf)]$ model. Each program then constructs n_{rep} "replicate" sets of *Type I* censored pseudoran-

---

```
C> COPY SAFNCDBC DATA

    1 file(s) copied

C> SAFNCM11
```

Presumed Linear s_a–$\log_e[fnc(pf)]$ Model:

standardized normal distribution variate $y = (s_a - clp)/csp$, where
$$clp = clp0 + clp1\,[\log_e(fnc)]$$

fnc	s_a	est[$s_{fs}(50)$]
56430	320	312.0
99000	300	298.9
183140	280	284.5
479490	260	262.0
909810	240	247.0
3632590	220	214.6
4917990	200	207.5
19186790	180	175.7

snc	s_a	est[$s_{fs}(50)$]
32250000	160	163.6

Ad Hoc Pragmatic Bias-Corrected Estimate of the Actual Value for $s_{fs}(50)$ at 25000000 Cycles

168.3

C> SAFNCM12

Presumed Linear s_a–$\log_e[fnc(pf)]$ Model:

standardized normal distribution variate $y = (s_a - clp)/csp$, where
$$clp = clp0 + clp1\,[\log_e(fnc)]$$

Ad Hoc Pragmatic Bias-Corrected Lower 95% (One-Sided) Statistical
Confidence Limit that Allegedly Bounds
the Actual Value for $s_{fs}(50)$ at 25000000 Cycles

161.9

Computed Using the Propagation of Variability Expression for
est{var[est(y given $s_{fs} = closascl$)]}

C> SAFNCM13

Presumed Linear s_a–$\log_e[fnc(pf)]$ Model:

standardized normal distribution variate $y = (s_a - clp)/csp$, where
$$clp = clp0 + clp1\,[\log_e(fnc)]$$

Ad Hoc Pragmatic Bias-Corrected Lower 95% (One-Sided) Statistical
Confidence Limit that Allegedly Bounds the Actual Value for $s_{fs}(10)$ at
25000000 Cycles

151.7

Computed Using the Propagation of Variability Expression for
est{var[est(y given $s_{fs} = closascl$)]}

dom s_a–$\log_e(fnc)$ datum values and in turn computes the pragmatic bias-corrected value of specific interest. Program *SAFNCM11* computes the pragmatic bias-corrected estimate of the actual value for $s_{fs}(50)$ at $fnc = fnc^*$. Program *SAFNCM12* computes the pragmatic bias-corrected lower $100(scp)\%$ (one-sided) statistical confidence limit that allegedly bounds the actual value for $s_{fs}(50)$ at $fnc = fnc^*$, whereas program *SAFNCM13* computes the pragmatic bias-corrected lower $100(scp)\%$ (one-sided) statistical confidence limit that allegedly bounds the actual value for $s_{fs}(p)$ at $fnc = fnc^*$. The latter two programs employ propagation of variability expressions for est{var[est(y given $s_{fs} = closascl$)]} in their respective calculations.

The analogous quadratic (parabolic) conceptual s_a–$\log_e[fnc(pf)]$ model is based on the standardized conceptual normal distribution variate $y = \{s_a - clp0 - clp1 \log_e(fnc) - clp2 [\log_e(fnc)]^2\}/csp$. As mentioned above, the pragmatic sampling distribution that consists of n_{sim} "replicate" estimated quadratic (parabolic) conceptual s_a–$\log_e[fnc(pf)]$ models can include estimated models that exhibit a reversed curvature, viz., est($clp2$) is negative. These estimated models are included in the pragmatic sampling distributions employed in microcomputer programs *SAFNCM31*,

---

```
C> COPY SAFNCDBC DATA

    1 file(s) copied

C> SAFNCM31
```

Presumed Quadratic s_a–$\log_e[fnc(pf)]$ Model:

standardized normal distribution variate $y = (s_a - clp)/csp$, where
$$clp = clp0 + clp1 [\log_e(fnc)] + clp2 [\log_e(fnc)]^2$$

fnc	s_a	est[$s_{fs}(50)$]
56430	320	317.6
99000	300	300.9
183140	280	283.5
479490	260	258.0
909810	240	242.3
3632590	220	211.6
4917990	200	205.5
19186790	180	180.7

snc	s_a	est[$s_{fs}(50)$]
32250000	160	172.4

Ad Hoc Pragmatic Bias-Corrected Estimate of the Actual Value for $s_{fs}(50)$ at 25000000. Cycles

176.3

Note: 71 of the estimated "replicate" quadratic (parabolic) conceptual s_a–$\log_e[fnc(50)]$ models exhibited a reversed curvature. The associated est[$s_{fs}(50)$] values at $fnc = 25000000$ cycles were included in computing the pragmatic bias correction.

SAFNCM32, and *SAFNCM33* to compute their respective pragmatic bias-corrected estimates and limits. In contrast, these estimated models are excluded from the pragmatic sampling distributions employed in microcomputer programs *SAFNCM34*, *SAFNCM35*, and *SAFNCM36* to compute their respective pragmatic bias-corrected estimates and limits. These two sets of programs compute pragmatic bias-corrected estimates and limits that are identical for practical purposes when the proportion of estimated models that exhibit a reversed curvature is relatively small.

8.D.9.1. Discussion

The likelihood ratio test should be used to decide whether a linear or a quadratic s_a–$\log_e[fnc(pf)]$ model is statistically appropriate. The null hypothesis is that a linear (straight line) conceptual s_a–$\log_e[fnc(pf)]$ model is correct, whereas the alternative hypothesis is that the quadratic (parabolic) conceptual s_a–$\log_e[fnc(pf)]$ model is correct. Then,

$$\text{Bartlett's LR test statistic} = -2 \cdot \log_e\left(\frac{\text{est(ML)}_{n_{cp}=3}}{\text{est(ML)}_{n_{cp}=4}}\right)$$

$$= 2 \cdot \left\{ \log_e\left[\text{est(ML)}_{n_{cp}=4}\right] - \log_e\left[\text{est(ML)}_{n_{cp}=3}\right]\right\} = \chi^2_{n_{sdf}=4-3=1}$$

which, for the example datum values in microcomputer file *SAFNCDBC* gives

$$2 \cdot (-22.4748) - 2 \cdot (-25.2905) = 5.6314$$

Thus, the null hypothesis rejection probability is equal to 0.0176. Accordingly, we reject the null hypothesis and adopt the quadratic (parabolic) conceptual s_a–$\log_e[fnc(pf)]$ model.

The precision of the pragmatic bias-corrected values computed by running microcomputer programs *SAFNCM(XX)* can be examined by additionally outputting the $(p^*/2)$th and $[1 - (p^*/2)]$th percentiles of their associated pragmatic sampling distributions. A decision can then be made regarding the adequacy of this precision and thus the credibility of the pragmatic bias-corrected value of specific interest.

8.D.10. Computing *Ad Hoc* Pragmatic Statistical Bias Corrections for ML Analyses Pertaining to Strength (Resistance) Experiment Test Programs

Suppose that it is presumed that a certain conceptual (two-parameter) statistical distribution describes the probability of failure for the stress ampli-

C> SAFNCM32

Presumed Quadratic s_a–$\log_e[fnc(pf)]$ Model:

standardized normal distribution variate $y = (s_a - clp)/csp$, where
$$clp = clp0 + clp1\,[\log_e(fnc)] + clp2\,[\log_e(fnc)]^2$$

Ad Hoc Pragmatic Bias-Corrected Lower 95% (One-Sided) Statistical
Confidence Limit that Allegedly Bounds
the Actual Value for $s_{fs}(50)$ at 25000000 Cycles

167.0

Computed Using the Propagation of Variability Expression for
$\mathrm{est}\{\mathrm{var}[\mathrm{est}(y$ given $s_{fs} = closascl)]\}$

Note: 71 of the estimated "replicate" quadratic (parabolic) conceptual
s_a–$\log_e[fnc(50)]$ models exhibited a reversed curvature. The associated
lower 95% (one-sided) asymptotic statistical confidence limits that
allegedly bound the actual value for $s_{fs}(50)$ at 25000000 cycles were
included in computing the pragmatic bias correction.

C> SAFNCM33

Presumed Quadratic s_a–$\log_e[fnc(pf)]$ Model:

standardized normal distribution variate $y = (s_a - clp)/csp$, where
$$clp = clp0 + clp1\,[\log_e(fnc)] + clp2\,[\log_e(fnc)]^2$$

Ad Hoc Pragmatic Bias-Corrected Lower 95% (One-Sided) Statistical
Confidence Limit that Allegedly Bounds the Actual Value for $s_{fs}(10)$ at
25000000 Cycles

159.0

Computed Using the Propagation of Variability Expression for
$\mathrm{est}\{\mathrm{var}[\mathrm{est}(y$ given $s_{fs} = closascl)]\}$

Note: 71 of the estimated "replicate" quadratic (parabolic) conceptual
s_a–$\log_e[fnc(10)]$ models exhibited a reversed curvature. The associated
lower 95% (one-sided) asymptotic statistical confidence limits that
allegedly bound the actual value for $s_{fs}(10)$ at 25000000 cycles were
included in computing the pragmatic bias correction.

C> SAFNCM34

Presumed Quadratic s_a–$\log_e[fnc(pf)]$ Model:

standardized normal distribution variate $y = (s_a - clp)/csp$, where
$clp = clp0 + clp1\,[\log_e(fnc)] + clp2\,[\log_e(fnc)]^2$

fnc	s_a	est[$s_{fs}(50)$]
56430	320	317.6
99000	300	300.9
183140	280	283.5
479490	260	258.0
909810	240	242.3
3632590	220	211.6
4917990	200	205.5
19186790	180	180.7

snc	s_a	est[$s_{fs}(50)$]
32250000	160	172.4

Ad Hoc Pragmatic Bias-Corrected Estimate of the Actual Value for
$s_{fs}(50)$ at 25000000 Cycles

176.3

Note: 71 of the estimated "replicate" quadratic (parabolic) conceptual
s_a–$\log_e[fnc(50)]$ models exhibited a reversed curvature. The associated
est[$s_{fs}(50)$] values at 25000000 cycles were not included in computing
the pragmatic bias correction.

tudes (stimulus levels) employed in the experiment test program of specific
interest. If so, then we can use the ML-based estimates of the *cdp*'s and
uniform pseudorandom numbers to decide whether each respective specimen in each "replicate" experiment test program fails (or not). Accordingly,
for any set of stress amplitudes (stimulus levels) and their associated numbers of replicate tests of specific interest, we can generate n_{sim} "replicate"
pseudorandom strength (resistance) data sets, with each of these sets differ-
ing only by the (randomly selected) numbers of specimens failed at the

---

C> SAFNCM35

Presumed Quadratic s_a–$\log_e[fnc(pf)]$ Model:

standardized normal distribution variate $y = (s_a - clp)/csp$, where
$$clp = clp0 + clp1\,[\log_e(fnc)] + clp2\,[\log_e(fnc)]^2$$

Ad Hoc Pragmatic Bias-Corrected Lower 95% (One-Sided) Statistical Confidence Limit that Allegedly Bounds the Actual Value for $s_{fs}(50)$ at 25000000 Cycles

167.0

Computed Using the Propagation of Variability Expression for
$$\text{est}\{\text{var}[\text{est}(y \text{ given } s_{fs} = closascl)]\}$$

Note: 71 of the estimated "replicate" quadratic (parabolic) conceptual s_a–$\log_e[fnc(50)]$ models exhibited a reversed curvature. The associated lower 95% (one-sided) asymptotic statistical confidence limits that allegedly bound the actual value for $s_{fs}(50)$ at 25000000 cycles were not included in computing the pragmatic bias correction.

---

respective stress amplitudes (stimulus levels). Thus, in theory, we can generate the pragmatic sampling distribution for any ML-based estimate of specific interest. Then, in turn, we can compute either its pragmatic statistical bias correction or the associated pragmatic bias-corrected *A*-basis or *B*-basis values. However, in practice, the probability of generating a "replicate" data set that either has no rational explanation or no valid ML solution increases as n_{sim} increases. Thus, the computation of a pragmatic statistical bias correction for a small strength (resistance) experiment test program may be impossible.

We demonstrated in Supplemental Topic 8.C that the ML estimates of the actual values for the $[s_{fs}(50)]$'s pertaining to the four alternative conceptual two-parameter strength (resistance) distributions considered herein are identical for practical purposes. On the other hand, the respective estimated classical lower $100(scp)\%$ (one-sided) asymptotic confidence limits that allegedly bound the actual value of the $[s_{fs}(p)]$'s pertaining to these four conceptual strength (resistance) distributions differ quite markedly, even for extremely large strength (resistance) experiment test programs. In fact,

C> SAFNCM36

Presumed Quadratic s_a–$\log_e[fnc(pf)]$ Model:

standardized normal distribution variate $y = (s_a - clp)/csp$, where
$$clp = clp0 + clp1\,[\log_e(fnc)] + clp2\,[\log_e(fnc)]^2$$

Ad Hoc Pragmatic Bias-Corrected Lower 95% (One-Sided) Statistical
Confidence Limit that Allegedly Bounds the Actual Value for $s_{fs}(10)$ at
25000000 Cycles

159.0

Computed Using the Propagation of Variability Expression for
$\text{est}\{\text{var}[\text{est}(y \text{ given } s_{fs} = closascl)]\}$

Note: 71 of the estimated "replicate" quadratic (parabolic) conceptual
s_a–$\log_e[fnc(10)]$ models exhibited a reversed curvature. The associated
lower 95% (one-sided) asymptotic statistical confidence limits that
allegedly bound the actual value for $s_{fs}(10)$ at 25000000 cycles were
not included in computing the pragmatic bias correction.

these differences are so large that pragmatic statistical bias corrections are
almost always negligible in comparison. Thus, the computation of pragmatic statistical bias corrections for strength (resistance) data is not illustrated herein.

8.D.11. Perspective and Closure

Simulation is by far the most intuitive way to examine the variability that is
intrinsic in the outcome of any quantitative statistical analysis. Ideally, the
empirical sampling distribution that consists of n_{nep} replicate realization
values for its associated statistic can be readily generated. If so, the metric
values pertaining to certain of its percentiles can be established and in turn
related to the required (desired) precision for the quantitative estimate of
specific interest. If not, then a pragmatic sampling distribution is an acceptable alternative to this empirical sampling distribution. At the very least,
the following methodology should be adopted:

Use the estimated values for the parameters of the presumed conceptual statistical model to generate at least 12 sets of "replicate"
experiment test program datum values. Then perform the 12 asso-

ciated quantitative statistical analyses and tabulate and/or plot the respective outcomes appropriately. In turn, compare the extreme outcomes of these 12 statistical analyses relative to the precision required (desired) for the quantitative outcome of the given statistical analysis. Finally, reassess your quantitative outcome in the light of this simulation-based information.

As a very crude rule of thumb, the range of the (extreme) outcomes for 12 "replicate" statistical analyses includes (bounds) approximately 90% of all possible replicate outcomes for a given quantitative statistical analysis.

> *Remember*: An incorrect experimental result is worse than no result *because it can be very misleading*, but even a correct experimental result can be very misleading if it lacks a dependable index to its precision.

8.E. SUPPLEMENTAL TOPIC: CLASSICAL AND LIKELIHOOD RATIO LOWER 100(*SCP*)% (ONE-SIDED) ASYMPTOTIC STATISTICAL CONFIDENCE BANDS AND LIMITS

8.E.1. Classical and LR Lower 100(*scp*)% (One-Sided) Asymptotic Statistical Confidence Bands for Both Life (Endurance) and Strength (Resistance) Data

Recall that the 100(*scp*)% joint confidence region for the actual values of the *clp0* and the *clp1* was used in simple linear regression to compute hyperbolic 100(*scp*)% (two-sided) statistical confidence bands that allegedly include mean(*APRCRHNDRDV*'s) for all values of the *ivv* employed in the regression experiment test program. Analogously, classical and LR 100(*scp*)% (two-sided) asymptotic statistical confidence bands that allegedly bound the actual values for all percentiles of the presumed conceptual two-parameter distribution can be computed using the 100(*scp*)% joint asymptotic statistical confidence region for the *clp* and the *csp*. In turn, these classical and LR 100(*scp*)% (two-sided) asymptotic statistical confidence bands can be reinterpreted as classical and LR lower 100[(1 + *scp*)/2])% (one-sided) asymptotic statistical confidence (tolerance limits) that pertain to the pth percentile of the presumed conceptual two-parameter distribution.

Recall that the elliptical boundary of the 100(*scp*)% joint statistical confidence region in simple linear regression is established by the *scp*-based selected value for Snedecor's central F test statistic with 2 and $n_{rdv}-2$ statistical degrees of freedom (Supplemental Topic 7.C). However, in ML analysis, the elliptical 100(*scp*)% joint statistical confidence region that allegedly includes the actual values for the *clp* and the *csp* is asymptotic. Accordingly, the number of denominator statistical degrees of freedom for the analogous Snedecor's central F test statistic is infinite. Thus, the analogous ML-based test statistic is Pearson's central χ^2 test statistic with 2 statistical degrees of freedom. (Recall Figure 5.7). Accordingly, the joint $100(scp)\%$ asymptotic statistical confidence region that allegedly includes the actual values for the *clp* and the *csp* is established by the quadratic expression:

$$
\begin{aligned}
\chi^2_{2;scp} = &-\frac{\partial^2 \log_e(\text{likelihood})}{\partial clp^2} \cdot [\text{est}(clp) - clp_{\text{critical}}]^2 \\
&-\frac{\partial^2 \log_e(\text{likelihood})}{\partial csp^2} \cdot [\text{est}(csp) - csp_{\text{critical}}]^2 \\
&- 2 \cdot \frac{\partial^2 \log_e(\text{likelihood})}{\partial clp\, \partial csp} \cdot [\text{est}(clp) - clp_{\text{critical}}] \cdot [\text{est}(csp) - csp_{\text{critical}}]
\end{aligned}
$$

in which the partial derivatives must be evaluated at est(*clp*) and est(*csp*). This joint $100(scp)\%$ asymptotic statistical confidence region is elliptical regardless of the parameterization of the presumed conceptual two-parameter distribution.

The joint $100(scp)\%$ asymptotic statistical confidence region generated by the LR method is not elliptical. Its boundary is defined by the trace on the $\log_e$(likelihood) surface at which the magnitude of the corresponding $\log_e$(likelihood) is less than $\log_e$(maximized likelihood) by $\left(\frac{1}{2} \cdot \chi^2_{2;scp}\right)$. The shape of this trace depends on the parameterization selected for the presumed conceptual two-parameter distribution. Nevertheless, the associated LR $100(scp)\%$ (two-sided) asymptotic statistical confidence bands do not depend on this parameterization.

All joint $100(scp)\%$ asymptotic statistical confidence regions can be depicted on *clp,csp* co-ordinates by plotting numerous points that lie on their respective boundaries. These boundary points pertain to coupled values of clp_{critical} and csp_{critical} for which the null hypothesis that these hypothetical values for the *clp* and the *csp* are correct will just be rejected when the acceptable probability of committing a *Type I* error is equal to $(1 - scp)$. In our microcomputer programs 3600 coupled boundary point values for clp_{critical} and csp_{critical} are computed, one boundary point value for each one-tenth of a degree of counterclockwise rotation around the (estimated) center of the $100(scp)\%$ joint asymptotic statistical confidence region. These 3600 coupled values of clp_{critical} and csp_{critical} are used to compute 3600 corresponding hypothetical CDF's. Then, for each *y* value or metric value of specific interest, the respective minimum and maximum [est(metric value given the *y* of specific interest)]'s or [est(*y* value given the metric value of specific interest)]'s associated with these 3600 hypothetical CDF's establish reasonably accurate numerical values for the 100(*scp*)% (two-sided) asymptotic statistical confidence bands that allegedly include the actual CDF. When *y* pertains to the value of *pf* of specific interest, the respective minimums of the 3600 [est(metric value given the *y* of specific interest)'s are classical and LR lower 100[(1 + *scp*)/2]% (one-sided) asymptotic statistical confidence (tolerance) limits.

The respective maximums and minimums of the 3600 computed hypothetical [est(metric value given the *y* of specific interest)]'s do not correspond one-to-one to the associated maximums and minimums of the 3600 [est(*y* value given the metric value of specific interest)]'s, or vice versa, for any of the four parameterizations considered herein. Moreover, only the two linear parameterizations, viz., (2) and (3) on this page, have their elliptical joint 100(*scp*)% asymptotic statistical confidence region bisected by the straight line that connects the hypothetical [*clp*,*csp*] points associated with the respective maximums and minimums of the 3600 computed values of the [est(metric value given the *y* of specific interest)'s and [est(*y* value given the metric value of specific interest)]'s. Accordingly, only linear parameterizations (2) and (3) generate hyperbolic 100(*scp*)% (two-sided) asymptotic statistical confidence bands such that the probability that the actual CDF lies (in part) above the upper band exactly equals the probability that the actual CDF lies (in part) below the lower band, viz., (1 − *scp*)/2. This probability behavior is only asymptotically approached for parameterizations (1) and (4), and for LR-based bands.

8.E.2. Life (Endurance) Data

Recall that we presume that life (endurance) can be modeled using either a conceptual (two-parameter) smallest-extreme-value distribution or a conceptual (two-parameter) normal distribution with a logarithmic metric and that one of the following parameterizations is of specific interest:

(1) $y = csp \cdot [\log_e(fnc) - clp]$

(2) $y = clp + csp \cdot [\log_e(fnc)]$

(3) $y = \dfrac{\log_e(fnc) - clp}{csp}$

(4) $y = \dfrac{\log_e(fnc)}{cps} - clp$

The *LSEV(J)B* and *LN(J)B* series of microcomputer programs compute classical and LR lower 100(*scp*)% (one-sided) asymptotic statistical confidence bands, where *scp* is the input value that appears in microcomputer file *WBLDATA*. These programs also compute classical and LR lower $100(scp)\%$ (one-sided) asymptotic statistical confidence (tolerance) limits that allegedly bound $100(p)\%$ of the presumed conceptual two-parameter $\log_e$–normal and Weibull life (endurance) distributions, where p is the probability complement to the CDF percentile of specific interest. The eight respective *A*-basis asymptotic statistical confidence (tolerance) limits are presented in Table 8.26 for comparative purposes. Note that these asymptotic statistical confidence (tolerance) limits can differ markedly for small experiment test programs, especially when suspended tests occur. However, the differences among the respective asymptotic statistical tolerance limits decrease as n_{dv} increases. This behavior is evident in Table 8.27, where it is also evident that these differences are markedly decreased by including *ad hoc* pragmatic statistical bias corrections.

8.E.3. Strength (Resistance) Data

Recall that for strength (resistance) data the four analogous parameterizations are:

(1) $y = csp \cdot [s(pf) - clp]$

(2) $y = clp + csp \cdot s(pf)$

(3) $y = \dfrac{s(pf) - clp}{csp}$

(4) $y = \dfrac{s(pf)}{cps} - clp$

These four parameterizations generate results akin to those presented in Table 8.26 and 8.27. However, recall that ML strength (resistance) analyses traditionally employ only parameterization (2). Accordingly, we employ only parameterization (2) in microcomputer programs *N2B*, *L2B*, *SEV2B*, and *LEV2B* that are respectively based on the presumption of a conceptual (two-parameter) normal, logistic, smallest-extreme-value, and largest-extreme-value distribution to compute classical and LR lower $100(scp)\%$ (one-sided) asymptotic statistical confidence bands that allegedly bound the actual CDF. Note that, although our example strength experiment test program is much larger than is practical in most mechanical reliability applications, there are marked differences among the respective classical and LR lower $100(scp)\%$ (one-sided) asymptotic statistical confidence bands. In particular, note that points on these bands can even take on nonsensical negative values when *pf* is sufficiently small.

Table 8.26 Respective Classical and LR Lower 95% (One-Sided) Asymptotic Statistical Tolerance Limits that Allegedly Bound 99% of All Possible Replicate Datum Values that Comprise the Presumed Conceptual Two-Parameter Weibull and $\log_e$–Normal (Endurance) Distributions[a]

	CDF parameterization			
	1	2	3	4
Conceptual Two-Parameter Weibull distribution				
Lower (one-sided) asymptotic statistical confidence limit computed by running the $LSEV(J)A$ series of microcomputer programs (for comparison)		34.584		
Computed using the joint asymptotic confidence region	4.331	3.901	71.811	1.679
Computed using the likelihood ratio method		32.254		
Conceptual Two-Parameter $\log_e$-Normal Distribution				
Lower (one-sided) asymptotic statistical confidence limit computed by running the $LN(J)A$ series of microcomputer programs (for comparison)		112.714		
Computed using the joint asymptotic confidence region	57.493	57.357	143.704	7.625
Computed using the likelihood ratio method		92.828		

[a]Computed by running the $LSEV(J)B$ and $LN(J)B$ series of microcomputer programs with the input datum values that appear in microcomputer file *WBLDATA*.

Table 8.27 Respective Classical and LR Lower 95% (One-Sided) Asymptotic Statistical Tolerance Limits that Allegedly Bound 99% of All Possible Replicate Datum Values that Comprise the Presumed Conceptual Two-Parameter $\log_e$–Normal Life (Endurance) Distribution[a]

		CDF parameterization			
		1	2	3	4
Conceptual Two-Parameter $\log_e$-Normal Distribution	Computed by running microcomputer program *ABLNSTL* (for reference)		0.11434		
	Lower (one-sided) asymptotic statistical confidence limit computed by running the $LN(J)A$ series of microcomputer programs (for comparison)		0.114709		
	Computed using the associated *ad hoc* pragmatic bias correction based on 30,000 replicate pseudorandom data sets (microcomputer program *LNPBCPV*)		0.114406		
	Computed using the joint asymptotic statistical confidence region	0.113663	0.113447	0.114679	0.116010
	Computed using the associated *ad hoc* pragmatic bias correction based on 30,000 replicate pseudorandom data sets	0.114164	0.114103	0.114083	0.114712
	Computed using the likelihood ratio method		0.113723		
	Computed using the associated *ad hoc* pragmatic bias correction based on 30,000 replicate pseudorandom data sets (microcomputer program *LNPBCLR*)		0.114134		

[a]Computed by running the $LN(J)B$ series of microcomputer programs with the input datum values that appear in microcomputer file *ASTLDATA*.

```
C> LSEV1B
```

Given the Standardized Conceptual Smallest-Extreme-Value Distribution Variate $y = csp \cdot [\log_e(fnc) - clp]$

Points on the Lower 95% (One-Sided) Asymptotic Statistical Confidence Bands that Allegedly Bound the Actual CDF Computed Using 3600 Points on the Boundary of the Elliptical Joint Asymptotic Statistical Confidence Region for the Actual Values of the clp and the csp (with 2 Statistical Degrees of Freedom)

Lower Boundary Point =	at fnc =
53.438	277.000
90.623	310.000
217.892	374.000
302.833	402.000
371.903	456.000
	at snc =
403.022	500.000
	at $fnc(01)$
Lower Boundary Point =	est(fnc) =
4.331	162.262

Computed Using 3600 Points on the Boundary of the Corresponding Region Established by the Likelihood Ratio Method

Lower Boundary Point =	at fnc =
126.062	277.000
165.957	310.000
255.244	374.000
296.154	402.000
369.401	456.000
	at snc =
417.964	500.000
	at $fnc(01)$
Lower Boundary Point =	est(fnc) =
32.254	162.262

```
C> LSEV2B
```

Given the Standardized Conceptual Smallest-Extreme-Value Distribution Variate $y = clp + csp \cdot \log_e(fnc)$

Points on the Lower 95% (One-Sided) Asymptotic Statistical Confidence Bands that Allegedly Bound the Actual CDF Computed Using 3600 Points on the Boundary of the Elliptical Joint Asymptotic Statistical Confidence Region for the Actual Values of the *clp* and the *csp* (with 2 Statistical Degrees of Freedom)

Lower Boundary Point =	at *fnc* =
46.425	277.000
77.873	310.000
176.313	374.000
234.472	402.000
342.820	456.000
	at *snc* =
403.004	500.000
	at *fnc*(01)

Lower Boundary Point =	est(*fnc*) =
3.901	162.262

Computed Using 3600 Points on the Boundary of the Corresponding Region Established by the Likelihood Ratio Method

Lower Boundary Point =	at *fnc* =
126.062	277.000
165.957	310.000
255.244	374.000
296.154	402.000
369.401	456.000
	at *snc* =
417.964	500.000
	at *fnc*(01)

Lower Boundary Point =	est(*fnc*) =
32.254	162.262

```
C> COPY WBLDATA DATA

    1 files(s) copied

C> LSEV3B
```

Given Standardized Conceptual Smallest-Extreme-Value Distribution
Variate $y = [\log_e(fnc) - clp]/csp$

Points on the Lower 95% (One-Sided) Asymptotic Statistical
Confidence Bands that Allegedly Bound the Actual CDF
Computed Using 3600 Points on the Boundary of the Elliptical Joint
Asymptotic Statistical Confidence Region for the Actual Values of the
clp and the csp (with 2 Statistical Degrees of Freedom)

Lower Boundary Point =	at fnc =
181.888	277.000
219.406	310.000
293.594	374.000
323.796	402.000
371.768	456.000
	at snc =
399.424	500.000
	at $fnc(01)$
Lower Boundary Point =	est(fnc) =
71.811	162.262

Computed Using 3600 Points on the Boundary of the Corresponding
Region Established by the Likelihood Ratio Method

Lower Boundary Point =	at fnc =
126.062	277.000
165.957	310.000
255.244	374.000
296.154	402.000
369.401	456.000
	at snc =
417.964	500.000
	at $fnc(01)$
Lower Boundary Point =	est(fnc) =
32.254	162.262

```
C> COPY WBLDATA DATA

    1 files(s) copied

C> LSEV4B
```

Given the Standardized Conceptual Normal Distribution Variate
$y = \{[\log_e(fnc)]/csp\} - clp$

$\text{est}(clp) = 0.2868202543\text{D} + 02$
$\text{est}(csp) = 0.2113296619\text{D} + 00$
$\text{est}\{\text{var}[\text{est}(clp)]\} = 0.1120467627\text{D} + 03$
$\text{est}\{\text{var}[\text{est}(csp)]\} = 0.6010809287\text{D} - 02$
$\text{est}\{\text{covar}[\text{est}(clp),\text{est}(csp)]\} = -0.8199326557\text{D} + 00$
$\text{est}(\textit{conceptual correlation coefficient}) = -0.9991071169\text{D} + 00$

fnc	$\text{est}(y)$	$\text{est}(pf)$
277.000	−2.0694929	0.1186053
310.000	−1.5368900	0.1934980
374.000	−0.6487823	0.4070717
402.000	−0.3071535	0.5207523
456.000	0.2892640	0.7369587

snc	$\text{est}(y)$	$\text{est}(pf)$
500.000	0.7251484	0.8731865

Points on the Lower 95% (One-Sided) Asymptotic Statistical
Confidence Bands that Allegedly Bound the Actual CDF
Computed Using 3600 Points on the Boundary of the Elliptical Joint
Asymptotic Statistical Confidence Region for the Actual Values of the
clp and the csp (with 2 Statistical Degrees of Freedom)

Lower Boundary Point =	at fnc =
4.365	277.000
5.338	310.000
7.465	374.000
8.493	402.000
10.229	456.000
	at snc =
10.432	500.000

at *fnc*(01)

Lower Boundary Point = est(*fnc*) =

1.679   162.262

Computed Using 3600 Points on the Boundary of the Corresponding
Region Established by the Likelihood Ratio Method

Lower Boundary Point =	at *fnc* =
126.062	277.000
165.957	310.000
255.244	374.000
296.154	402.000
369.401	456.000
	at *snc* =
417.964	500.000
	at *fnc*(01)
Lower Boundary Point =	est(*fnc*) =
32.254	162.262

The example outputs for programs *N2B*, *L2B*, *SEV2B*, and *LEV2B*
are typical of most ML analyses in that LR statistical confidence intervals,
limits, and bands are intermediate to the extremes of the associated classical
statistical confidence intervals, limits, and bands. This relationship is examined further in the next section.

8.E.4. Classical and LR Lower 100(*scp*)% (One-Sided) Asymptotic Statistical Confidence Limits for Both Life (Endurance) and Strength (Resistance) Data

Classical and LR lower 100(*scp*)% (one-sided) asymptotic statistical confidence limits that allegedly bound the presumed conceptual two-parameter
distribution metric value, given any $y(pf)$ value of specific interest, can also
be computed using the numerical computation procedure outlined above.
However, these limits are established by a *pf* based on Pearson's central χ^2
conceptual sampling distribution with only one statistical degree of
freedom.

```
C> COPY WBLDATA DATA

    1 files(s) copied

C> LN1B
```

Given the Standardized Conceptual Normal Distribution Variate
$y = csp \cdot [\log_e(fnc) - clp]$

Points on the Lower 95% (One-Sided) Asymptotic Statistical
Confidence Bands that Allegedly Bound the Actual CDF
Computed Using 3600 Points on the Boundary of the Elliptical Joint
Asymptotic Statistical Confidence Region for the Actual Values of the
clp and the *csp* (with 2 Statistical Degrees of Freedom)

Lower Boundary Point =	at *fnc* =
124.351	277.000
182.454	310.000
302.115	374.000
323.179	402.000
358.474	456.000
	at *snc* =
384.429	500.000
	at *fnc*(01)
Lower Boundary Point =	est(*fnc*) =
57.493	221.209

Computed Using 3600 Points on the Boundary of the Corresponding
Region Established by the Likelihood Ratio Method

Lower Boundary Point =	at *fnc* =
155.627	277.000
198.266	310.000
282.348	374.000
315.504	402.000
368.378	456.000
	at *snc* =
401.906	500.000
	at *fnc*(01)
Lower Boundary Point =	est(*fnc*) =
92.828	221.209

```
C> COPY WBLDATA DATA

    1 files(s) copied

C> LN2B
```

Given the Standardized Conceptual Normal Distribution Variate
$y = clp + csp \cdot \log_e(fnc)$

est(clp) = 0.2484842180D + 02
est(csp) = 0.4171442213D + 01
est{var[est(clp)]} = 0.6703904559D + 02
est{var[est(csp)]} = 0.1903271765D + 01
est{covar[est(clp),est(csp)]} = −0.1128121931D + 02
est(*conceptual correlation coefficient*) = −0.9987153578D + 00

fnc	est(y)	est(pf)
277.000	−1.3881578	0.0825445
310.000	−0.9186420	0.1791414
374.000	−0.1357311	0.4460169
402.000	0.1654316	0.5656978
456.000	0.6912032	0.7552811

snc	est(y)	est(pf)
500.000	1.0754568	0.8589149

Points on the Lower 95% (One-Sided) Asymptotic Statistical
Confidence Bands that Allegedly Bound the Actual CDF
Computed Using 3600 Points on the Boundary of the Elliptical Joint
Asymptotic Statistical Confidence Region for the Actual Values of the
clp and the csp (with 2 Statistical Degrees of Freedom)

Lower Boundary Point =	at fnc =
120.103	277.000
169.937	310.000
275.414	374.000
313.548	402.000
368.208	456.000
	at snc =
401.704	500.000
	at fnc(01)

Lower Boundary Point =	est(*fnc*) =
57.357	221.209

Computed Using 3600 Points on the Boundary of the Corresponding
Region
Established by the Likelihood Ratio Method

Lower Boundary Point =	at *fnc* =
155.627	277.000
198.266	310.000
282.348	374.000
315.504	402.000
368.378	456.000
	at *snc* =
401.906	500.000
	at *fnc*(01)
Lower Boundary Point =	est(*fnc*) =
92.828	221.209

```
C> COPY WBLDATA DATA

    1 files(s) copied

C> LN3B
```

Given the Standardized Conceptual Normal Distribution Variate
$y = [\log_e(fnc) - clp]/csp$

$est(clp) = 0.5959062671D + 01$
$est(csp) = 0.2412998143D + 00$
$est\{var[est(clp)]\} = 0.9921960558D - 02$
$est\{var[est(csp)]\} = 0.6305038312D - 02$
$est\{covar[est(clp),est(csp)]\} = 0.6974504622D - 03$
$est(conceptual\ correlation\ coefficient) = 0.8818013378D - 01$

fnc	est(*y*)	est(*pf*)
277.000	−1.3881578	0.0825445
310.000	−0.9186420	0.1791414
374.000	−0.1357311	0.4460169

402.000	0.1654316	0.5656978
456.000	0.6912032	0.7552811
snc	est(y)	est(pf)
500.000	1.0754568	0.8589149

Points on the Lower 95% (One-Sided) Asymptotic Statistical
Confidence Bands that Allegedly Bound the Actual CDF
Computed Using 3600 Points on the Boundary of the Elliptical Joint
Asymptotic Statistical Confidence Region for the Actual Values of the
clp and the *csp* (with 2 Statistical Degrees of Freedom)

Lower Boundary Point =	at *fnc* =
204.569	277.000
240.769	310.000
302.115	374.000
322.932	402.000
353.552	456.000
	at *snc* =
372.193	500.000
	at *fnc*(01)
Lower Boundary Point =	est(*fnc*) =
143.704	221.209

Computed Using 3600 Points on the Boundary of the Corresponding
Region
Established by the Likelihood Ratio Method

Lower Boundary Point =	at *fnc* =
155.627	277.000
198.266	310.000
282.348	374.000
315.504	402.000
368.378	456.000
	at *snc* = —
401.906	500.000
	at *fnc*(01)
Lower Boundary Point =	est(*fnc*) =
92.828	221.209

```
C> COPY WBLDATA DATA

    1 files(s) copied

C> LN4B
```

Given the Standardized Conceptual Normal Distribution Variate
$y = \{[\log_e(fnc)]/csp\} - clp$

$\text{est}(clp) = 0.2484842180D + 02$
$\text{est}(csp) = 0.2397252434D + 00$
$\text{est}\{\text{var}[\text{est}(clp)]\} = 0.6703904559D + 02$
$\text{est}\{\text{var}[\text{est}(csp)]\} = 0.6285732256D - 02$
$\text{est}\{\text{covar}[\text{est}(clp),\text{est}(csp)]\} = -0.6483112811D + 02$
$\text{est}(conceptual\ correlation\ coefficient) = -0.9987153578D + 00$

fnc	$\text{est}(y)$	$\text{est}(pf)$
277.000	−1.3881578	0.0825445
310.000	−0.9186420	0.1791414
374.000	−0.1357311	0.4460169
402.000	0.1654316	0.5656978
456.000	0.6912032	0.7552811

snc	$\text{est}(y)$	$\text{est}(pf)$
500.000	1.0754568	0.8589149

Points on the Lower 95% (One-Sided) Asymptotic Statistical
Confidence Bands that Allegedly Bound the Actual CDF
Computed Using 3600 Points on the Boundary of the Elliptical Joint
Asymptotic Statistical Confidence Region for the Actual Values of the
clp and the csp (with 2 Statistical Degrees of Freedom)

Lower Boundary Point =	at fnc =
11.200	277.000
13.577	310.000
18.703	374.000
19.390	402.000
20.114	456.000
	at $fnc(01)$
	at snc =
20.660	500.000

Lower Boundary Point = est(*fnc*) =
7.625 221.209

Computed Using 3600 Points on the Boundary of the Corresponding
Region
Established by the Likelihood Ratio Method

Lower Boundary Point =	at *fnc* =
155.627	277.000
198.266	310.000
282.348	374.000
315.504	402.000
368.378	456.000
	at *snc* =
401.906	500.000
	at *fnc*(01)
Lower Boundary Point =	est(*fnc*) =
92.828	221.209

8.E.5. Life (Endurance) Data

Microcomputer programs *LSEV1C, LSEV2C, LSEV3C, LSEV4C, LN1C, LN2C, LN3C,* and *LN4C* respectively involve only a minor modification of microcomputer programs *LSEV1B LSEV2B, LSEV3B, LSEV4B, LN1B, LN2B, LN3B,* and *LN4B*. These modified programs compute classical and LR lower $100(scp)\%$ (one-sided) asymptotic statistical confidence limits, where *scp* is the input value that appears in microcomputer file *WBLDATA*. The eight respective *A*-base asymptotic statistical confidence (tolerance) limits are presented in Table 8.28 for comparative purposes. Note again that the LR lower $100(scp)\%$ (one-sided) asymptotic statistical confidence limit does not depend on the parameterization of the CDF.

The respective classical lower $100(scp)\%$ (one-sided) asymptotic statistical confidence limits for linear parameterizations (2) and (3) are theoretically identical to those computed using propagation of variability expressions. The numerical discrepancy in Table 8.28 for parameterization (2) is due to the very large value of the estimated conceptual correlation coefficient for est(*clp*) and est(*csp*).

```
C> COPY AWBLDATA DATA

    1 files(s) copied

C> WPBCLR
```

Given $F(fnc) = 1 - \exp^{-\left(\frac{fnc}{cdp1}\right)^{cdp2}}$

$est(cdp1) = 0.4289595976D + 03$
$est(cdp2) = 0.4731943406D + 01$
$est\{var[est(cdp1)]\} = 0.1666932095D + 04$
$est\{var[est(cdp2)]\} = 0.3013638227D + 01$
$est\{covar[est(cdp1),est(cdp2)]\} = 0.8393830857D + 01$
$est(conceptual\ correlation\ coefficient) = 0.1184283562D + 00$

fnc	$est(y)$	$est(pf)$
277.000	−2.0694929	0.1186053
310.000	−1.5368900	0.1934980
374.000	−0.6487823	0.4070717
402.000	−0.3071535	0.5207523
456.000	0.2892640	0.7369587

snc	$est(y)$	$est(pf)$
500.000	0.7251484	0.8731865

Ad Hoc Pragmatic Bias-Corrected Lower 95% (One-Sided)
Asymptotic Statistical Confidence (Tolerance)
Limit that Allegedly Bounds the Actual Value for $fnc(01)$

38.579

Based on the 95th Percentile of the Sampling Distribution Comprised
of 30,000 "Replicate" Realizations for the Classical
Lower 95% (One-Sided) Asymptotic Statistical Confidence Limit that
Allegedly Bounds the Actual Value for $fnc(01)$,
Each Realization Computed Using 360 Points on the Boundary of the
Likelihood Ratio Asymptotic Statistical
Confidence Region that Allegedly Includes the Actual Values for the
clp and the *csp* (with 2 Statistical Degrees of Freedom).

```
C> COPY AWBLDATA DATA

    1 files(s) copied

C> LNPBCLR
```

Given Standardized Conceptual Normal Distribution Variate
$y = csp \cdot [\log_e(fnc) - clp]$

$\text{est}(clp) = 0.5956793964\text{D} + 01$
$\text{est}(csp) = 0.4171442213\text{D} + 01$
$\text{est}\{\text{var}[\text{est}(clp)]\} = 0.9987394347\text{D} - 02$
$\text{est}\{\text{var}[\text{est}(csp)]\} = 0.1903271765\text{D} + 01$
$\text{est}\{\text{covar}[\text{est}(clp), \text{est}(csp)]\} = -0.1346739280\text{D} - 01$
$\text{est}(conceptual\ correlation\ coefficient) = -0.9768030920\text{D} - 01$

fnc	est(*y*)	est(*pf*)
277.000	−1.3881578	0.0825445
310.000	−0.9186420	0.1791414
374.000	−0.1357311	0.4460169
402.000	0.1654316	0.5656978
456.000	0.6912032	0.7552811

snc	est(*y*)	est(*pf*)
500.000	1.0754568	0.8589149

Ad Hoc Pragmatic Bias-Corrected Lower 95% (One-Sided)
Asymptotic Statistical Confidence (Tolerance)
Limit that Allegedly Bounds the Actual Value for *fnc*(01)

95.817

Based on the 95th Percentile of the Sampling Distribution Comprised
of 30,000 "Replicate" Realizations for the Classical
Lower 95% (One-Sided) Asymptotic Statistical Confidence Limit that
Allegedly Bounds the Actual Value for *fnc*(01).
Each Realization Computed Using 360 Points on the Boundary of the
Likelihood Ratio Asymptotic Statistical
Confidence Region that Allegedly Includes the Actual Values for the
clp and the *csp* (with 2 Statistical Degrees of Freedom).

8.E.6. Strength (Resistance) Data

Microcomputer programs *N2C*, *L2C*, *SEV2C*, and *LEV2C* involve only a minor modification of microcomputer programs *N2B*, *L2B*, *SEV2B*, and *LEV2B*, viz., employing Pearson's central χ^2 conceptual sampling distribution with only one statistical degree of freedom rather than two. These modified programs compute classical and LR lower 100(*scp*)% (one-sided) asymptotic statistical confidence limits that allegedly bound the metric pertaining to the pth percentile of the presumed conceptual (two-parameter) strength (resistance) distribution, where *scp* and *p* are input values that are specified in microcomputer file *ASDATA*. Note that the respective classical lower (one-sided) asymptotic statistical confidence limits are identical to those computed using the propagation of variability expression for est{-var[est(*y* given *s* = *closascl*)]} in microcomputer programs *N2ALCL*, *L2ALCL*, *SEV2ALCL*, and *LEV2ALCL*. [Recall that linear parameterization (2) is employed in these microcomputer programs.]

8.E.7. Summary

Ad hoc pragmatic bias-corrected LR-based lower 100(*scp*)% (one-sided) statistical confidence and tolerance limits (a) are statistically equivalent, for practical purposes, to corresponding exact limits for uncensored life (endurance) experiment test programs, when the coefficients of valuation is less than about 0.2, and (b) have no peer for life (endurance) experiment test programs with only a limited amount of *Type I* censoring. However, it is not practical to compute *ad hoc* pragmatic bias-corrected lower 100(*scp*)% (one-sided) statistical confidence and tolerance limits for strength (endurance) experiment test programs. Nevertheless, asymptotic LR-based limits are clearly preferable to classical ML-based limits.

8.F. SUPPLEMENTAL TOPIC: TESTING STATISTICAL ADEQUACY OF PRESUMED CONCEPTUAL MODEL USING THE LR METHOD

Recall that it is always good statistical practice to test the statistical adequacy of the presumed conceptual model. Most often the focus is on the statistical adequacy of the deterministic (so-called physical) component of the conceptual model. However, the presumption of homoscedasticity typically requires examination because experimental datum values typically exhibit variabilities that depend on the magnitude of the respective experiment test program outcomes, viz., small datum values typically exhibit small variabilities whereas large datum values typically exhibit large variabilities. The LR test is remarkably suited to testing the statistical adequacy of the presumed

Table 8.28 Respective Classical and LR Lower 95% (One-Sided) Asymptotic Statistical Confidence Limits that Allegedly Bound the Actual Value for the Metric Pertaining to the 01th Percentile of the Presumed Life (Endurance) Distribution, Computed by Running the $LN(J)A$, $LN(J)C$, $LSEV(J)A$, and $LSEV(J)C$ Series of Microcomputer Programs with the Input Datum Values that Appear in Microcomputer File *WBLDATA*

Distribution	Method	CDF parameterization			
		1	2	3	4
conceptual (two-parameter) Weibull distribution	Computed using the propagation of variability expression for est[var(est{log$_e$[fnc(01)]})]			86.868	
	Computed using the propagation of variability expression for est{var[est(y given fnc = closascl)]}		**34.584**		
	Computed using the joint asymptotic confidence region	36.167	**34.585**	86.868	9.598
	Computed using the likelihood ratio method		58.511		
conceptual (two-parameter) log$_e$-normal distribution	Computed using the propagation of variability expression for est[var(est{log$_e$[fnc(01)]})]			*158.933*	
	Computed using the propagation of variability expression for est{var[est(y given fnc = closascl)]}		**112.714**		
	Computed using the joint asymptotic confidence region	114.499	**112.735**	*158.933*	28.526
	Computed using the likelihood ratio method		128.175		

Note: The 112.735 value should be exactly equal to 112.174 and the 34.585 value should be exactly equal to 34.584. However, the est(*conceptual correlation coefficient*) for this parameterization is so large that the elliptical joint asymptotic confidence region is very long and slender. Thus, given equal spacing of the 3600 boundary points in terms of their angular orientation, too few of these points lie at the extreme ends of the joint confidence region to generate exact agreement with the corresponding propagation of variability estimate.

C> N2B

Given Standardized Conceptual Normal Distribution Variate
$y(pf) = clp + csp \cdot s(pf)$

Points on the Lower 95% (One-Sided) Asymptotic Statistical
Confidence Bands that Allegedly Bound the Actual CDF
Computed Using 3600 Points on the Boundary of the Elliptical Joint
Asymptotic Statistical Confidence Region for the Actual Values of the
clp and the *csp* (with 2 Statistical Degrees of Freedom)

Lower Boundary Point =	at s =
92.738	100.000
88.885	95.000
84.314	90.000
78.025	85.000
69.165	80.000
47.258	75.000
35.640	65.000

at $s(01)$

Lower Boundary Point =	est(s) =
17.407	57.226

Computed Using 3600 Points on the Boundary of the Corresponding
Region Established by the Likelihood Ratio Method

Lower Boundary Point =	at s =
92.302	100.000
88.560	95.000
84.262	90.000
78.768	85.000
71.308	80.000
52.421	75.000
42.275	65.000

at $s(01)$

Lower Boundary Point =	est(s) =
26.305	57.226

C> L2B

Given Standardized Conceptual Logistic Distribution Variate
$y(pf) = clp + csp \cdot s(pf)$

Points on the Lower 95% (One-Sided) Asymptotic Statistical
Confidence Bands that Allegedly Bound the Actual CDF
Computed Using 3600 Points on the Boundary of the Elliptical Joint
Asymptotic Statistical Confidence Region for the Actual Values of the
clp and the csp (with 2 Statistical Degrees of Freedom)

Lower Boundary Point =	at s =
92.637	100.000
88.915	95.000
84.306	90.000
77.126	85.000
65.976	80.000
38.630	75.000
24.317	65.000
	at $s(01)$
Lower Boundary Point =	est(s) =
−6.021	54.524

Computed Using 3600 Points on the Boundary of the Corresponding
Region Established by the Likelihood Ratio Method

Lower Boundary Point =	at s =
91.941	100.000
88.420	95.000
84.208	90.000
78.558	85.000
70.519	80.000
50.101	75.000
39.234	65.000
	at $s(01)$
Lower Boundary Point =	est(s) =
16.101	54.524

```
C> SEV2B
```

Given Standardized Conceptual Smallest-Extreme-Value Distribution
Variate $y(pf) = clp + csp \cdot s(pf)$

Points on the Lower 95% (One-Sided) Asymptotic Statistical
Confidence Bands that Allegedly Bound the Actual CDF
Computed Using 3600 Points on the Boundary of the Elliptical Joint
Asymptotic Statistical Confidence Region for the Actual Values of the
clp and the csp (with 2 Statistical Degrees of Freedom)

Lower Boundary Point =	at s =
94.256	100.000
90.171	95.000
84.443	90.000
75.300	85.000
63.767	80.000
38.817	75.000
26.079	65.000

at $s(01)$

Lower Boundary Point =	est(s) =
-18.174	47.765

Computed Using 3600 Points on the Boundary of the Corresponding
Region Established by the Likelihood Ratio Method

Lower Boundary Point =	at s =
94.122	100.000
90.254	95.000
85.168	90.000
77.969	85.000
68.828	80.000
48.310	75.000
37.705	65.000

at $s(01)$

Lower Boundary Point =	est(s) =
0.707	47.765

```
C> LEV2B
```

Given Standardized Conceptual Largest-Extreme-Value Distribution
Variate $y(pf) = clp + csp \cdot s(pf)$

Points on the Lower 95% (One-Sided) Asymptotic Statistical
Confidence Bands that Allegedly Bound the Actual CDF
Computed Using 3600 Points on the Boundary of the Elliptical Joint
Asymptotic Statistical Confidence Region for the Actual Values of the
clp and the csp (with 2 Statistical Degrees of Freedom)

Lower Boundary Point =	at s =
90.718	100.000
87.111	95.000
83.186	90.000
78.562	85.000
72.397	80.000
53.545	75.000
42.328	65.000

	at $s(01)$
Lower Boundary Point =	est(s) =
30.678	59.982

Computed Using 3600 Points on the Boundary of the Corresponding
Region Established by the Likelihood Ratio Method

Lower Boundary Point =	at s =
89.751	100.000
86.229	95.000
82.488	90.000
78.227	85.000
72.612	80.000
55.836	75.000
46.098	65.000

	at $s(01)$
Lower Boundary Point =	est(s) =
36.025	59.982

```
C> LSEV1C
```

Given the Standardized Conceptual Smallest-Extreme-Value Distribution Variate $y = csp \cdot [\log_e(fnc) - clp]$

Classical Lower 95% (One-Sided) Asymptotic Statistical Confidence Limits that Allegedly Bound the Actual Value for $fnc(01)$

36.167 – Computed Using 3600 Points on the Boundary of the Elliptical Joint Asymptotic Statistical Confidence Region for the Actual Values of the clp and the csp (with 1 Statistical Degree of Freedom)

58.511 – Computed Using 3600 Points on the Boundary of the Corresponding Region Established by the Likelihood Ratio Method

```
C> LSEV2C
```

Given the Standardized Conceptual Smallest-Extreme-Value Distribution Variate $y = clp + csp \cdot \log_e(fnc)$

Classical Lower 95% (One-Sided) Asymptotic Statistical Confidence Limits that Allegedly Bound the Actual Value for $fnc(01)$

34.585 – Computed Using 3600 Points on the Boundary of the Elliptical Joint Asymptotic Statistical Confidence Region for the Actual Values of the clp and the csp (with 1 Statistical Degree of Freedom)

58.511 – Computed Using 3600 Points on the Boundary of the Corresponding Region Established by the Likelihood Ratio Method

---

```
C> LSEV3C
```

Given Standardized Conceptual Smallest-Extreme-Value Distribution
Variate $y = [\log_e(fnc) - clp]/csp$

Classical Lower 95% (One-Sided) Asymptotic Statistical Confidence
Limits that Allegedly Bound the Actual Value for $fnc(01)$

86.868 – Computed Using 3600 Points on the Boundary of the
Elliptical Joint Asymptotic Statistical Confidence
Region for the Actual Values of the clp and the csp (with 1 Statistical
Degree of Freedom)

58.511 – Computed Using 3600 Points on the Boundary of the
Corresponding Region Established by the Likelihood Ratio Method

---

```
C> LSEV4C
```

Given the Standardized Conceptual Normal Distribution Variate
$y = \{[\log_e(fnc)]/csp\} - clp$

Classical Lower 95% (One-Sided) Asymptotic Statistical Confidence
Limits that Allegedly Bound the Actual Value for $fnc(01)$

9.598 – Computed Using 3600 Points on the Boundary of the
Elliptical Joint Asymptotic Statistical Confidence
Region for the Actual Values of the clp and the csp (with 1 Statistical
Degree of Freedom)

58.511 – Computed Using 3600 Points on the Boundary of the
Corresponding Region Established by the Likelihood Ratio Method

C> LN1C

Given the Standardized Conceptual Normal Distribution Variate
$$y = csp \cdot [\log_e(fnc) - clp]$$

Classical Lower 95% (One-Sided) Asymptotic Statistical Confidence Limits that Allegedly Bound the Actual Value for $fnc(01)$

114.499 – Computed Using 3600 Points on the Boundary of the Elliptical Joint Asymptotic Statistical Confidence Region for the Actual Values of the clp and the csp (with 1 Statistical Degree of Freedom)

128.175 – Computed Using 3600 Points on the Boundary of the Corresponding Region Established by the Likelihood Ratio Method

---

C> LN2C

Given the Standardized Conceptual Normal Distribution Variate
$$y = clp + csp \log_e(fnc)$$

Classical Lower 95% (One-Sided) Asymptotic Statistical Confidence Limits that Allegedly Bound the Actual Value for $fnc(01)$

112.735 – Computed Using 3600 Points on the Boundary of the Elliptical Joint Asymptotic Statistical Confidence Region for the Actual Values of the clp and the csp (with 1 Statistical Degree of Freedom)

128.175 – Computed Using 3600 Points on the Boundary of the Corresponding Region Established by the Likelihood Ratio Method

C> LN3C

Given the Standardized Conceptual Normal Distribution Variate
$y = [\log_e(fnc) - clp]/csp$

Classical Lower 95% (One-Sided) Asymptotic Statistical Confidence
Limits that Allegedly Bound the Actual Value for $fnc(01)$

158.933 – Computed Using 3600 Points on the Boundary of the
Elliptical Joint Asymptotic Statistical Confidence
Region for the Actual Values of the clp and the csp (with 1 Statistical
Degree of Freedom)

128.175 – Computed Using 3600 Points on the Boundary of the
Corresponding Region Established by the Likelihood Ratio Method

C> LN4C

Given the Standardized Conceptual Normal Distribution Variate
$y = [\log_e(fnc)]/csp - clp$

Classical Lower 95% (One-Sided) Asymptotic Statistical Confidence
Limits that Allegedly Bound the Actual Value for $fnc(01)$

28.526 – Computed Using 3600 Points on the Boundary of the
Elliptical Joint Asymptotic Statistical Confidence
Region for the Actual Values of the clp and the csp (with 1 Statistical
Degree of Freedom)

128.175 – Computed Using 3600 Points on the Boundary of the
Corresponding Region Established by the Likelihood Ratio Method

C> N2C

Given the Standardized Conceptual Normal Distribution Variate
$y(pf) = clp + csp \cdot s(pf)$

Lower 95% (One-Sided) Asymptotic Statistical Confidence Limits that
Allegedly Bound the Actual Value for $s(01)$

34.074 – Computed Using 3600 Points on the Boundary of the
Elliptical Joint Asymptotic Statistical Confidence
Region for the Actual Values of the clp and the csp (with 1 Statistical
Degree of Freedom)

37.678 – Computed Using 3600 Points on the Boundary of the
Corresponding Region Established by the Likelihood Ratio Method

C> L2C

Given Standardized Conceptual Logistic Distribution Variate
$y(pf) = clp + csp \cdot s(pf)$

Lower 95% (One-Sided) Asymptotic Statistical Confidence Limits that
Allegedly Bound the Actual Value for $s(01)$

22.363 – Computed Using 3600 Points on the Boundary of the
Elliptical Joint Asymptotic Statistical Confidence
Region for the Actual Values of the clp and the csp (with 1 Statistical
Degree of Freedom)

30.027 – Computed Using 3600 Points on the Boundary of the
Corresponding Region Established by the Likelihood Ratio Method

```
C> SEV2C
```

Given Standardized Conceptual Smallest-Extreme-Value Distribution
Variate $y(pf) = clp + csp \cdot s(pf)$

Lower 95% (One-Sided) Asymptotic Statistical Confidence Limits that
Allegedly Bound the Actual Value for $s(01)$

10.789 – Computed Using 3600 Points on the Boundary of the
Elliptical Joint Asymptotic Statistical Confidence
Region for the Actual Values of the clp and the csp (with 1 Statistical
Degree of Freedom)

17.593 – Computed Using 3600 Points on the Boundary of the
Corresponding Region Established by the Likelihood Ratio Method

```
C> LEV2C
```

Given Standardized Conceptual Largest-Extreme-Value Distribution
Variate $y(pf) = clp + csp \cdot s(pf)$

Lower 95% (One-Sided) Asymptotic Statistical Confidence Limits that
Allegedly Bound the Actual Value for $s(01)$

42.814 – Computed Using 3600 Points on the Boundary of the
Elliptical Joint Asymptotic Statistical Confidence
Region for the Actual Values of the clp and the csp (with 1 Statistical
Degree of Freedom)

45.238 – Computed Using 3600 Points on the Boundary of the
Corresponding Region Established by the Likelihood Ratio Method

conceptual model. We illustrate this LR methodology using a presumed conceptual quadratic normal regression model for stopping distance datum values given by Ezekiel and Fox (1963).

Consider the stopping distance datum values, (*sddv*'s), listed in Table 8.29, for an unspecified automobile given its initial speed value *isv* at the instant that the brakes are applied.

Table 8.29 Stopping Distance Datum
Values (*sddv*'s) Given the Associated Initial
Speed Value (*isv*) at the Instant the Brakes
of an Unidentified Automobile Are Applied

Initial speed (mph)	Stopping distance (feet)
4	4
5	2,8,8,4
7	6,7
8	9,8,13,11
9	5,13,5
10	8,17,14
12	11,21,19
13	18,27,15
14	14,16
15	16
16	19,14,34
17	29,22
18	47,29,34
19	30
20	48
21	55,39,42
22	35
24	56
25	33,59,48,56
26	39,41
27	78,57
28	64,84
29	68,54
30	60,101,67
31	77
35	85,107
36	79
39	138
40	110,134

Visual examination of Figure 8.9 suffices to indicate that a quadratic conceptual regression model is appropriate, viz.,

$$CRHNDSDDV_{i,j}\text{'s} = \text{mean}_i(APRCRHNDSDDV\text{'s}) + CRHNDSDEE_{i,j}\text{'s}$$
$$= clp1 \cdot isv_i + clp2 \cdot isv_i^2 + CRHNDSDEE_{i,j}\text{'s}$$

The primary issue of interest in examining this model is whether the H in $CRHNDSDDV$'s connotes homoscedasticity or heteroscedasticity. Accordingly, we now test the null hypothesis H_n: the standardized conceptual normal distribution homoscedastic variate $y_{i,j}$ is equal to $(sddv_{i,j} \quad clp1$ $isv_i + clp2\ isv_i^2)/csp$ versus the alternative hypothesis H_a: the standardized conceptual normal distribution heteroscedastic variate $y_{i,j}$ is equal to $(sddv_{i,j} - clp1 \cdot isv_i - clp2 \cdot isv_i^2)/(csp0 + csp1 \cdot isv_i)$. Microcomputer programs $QNRMHOSD$ (quadratic normal regression statistical model with a homoscedastic standard deviation) and $QNRHESD$ (quadratic normal regression statistical model with a heteroscedastic standard deviation) compute the respective est($clpj$'s) and est($cspj$'s) as well as the associated data-based values for the respective maximum estimated $\log_e$(likelihoods) given H_n and given H_a.

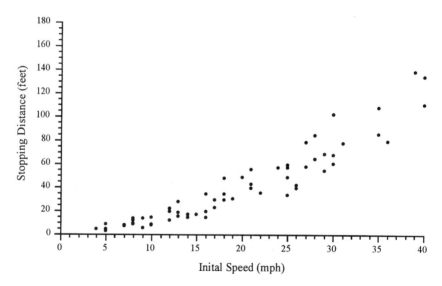

Figure 8.9 Stopping distance datum values ($sddv_{i,j}$'s) plotted versus the corresponding initial speed values (isv_i's).

Recall that the LR test statistic is equal to $(2 \cdot \{\text{est}[\log_e(\text{likelihood})$ given $H_a]\} - 2 \cdot \{\text{est}[\log_e(\text{likelihood})$ given $H_n]\})$ and that its respective realization values asymptotically generate Pearson's central $\chi^2_{4-3=1}$ conceptual sampling distribution under continual replication of the experiment test program. Thus, the data-based value for the LR test statistic of specific interest for our stopping distance example is

$$[2 \cdot (-155.38656) - 2 \cdot (-174.40727)] = 38.04 = \chi^2_{1;p}$$

and the associated asymptotic null hypothesis rejection probability is equal to 0.0000. Clearly, the null hypothesis of a homoscedastic standard deviation for our stopping distance datum values must rationally be rejected. Accordingly, we conclude statistically that the appropriate quadratic conceptual stopping distance statistical model has a heteroscedastic standard deviation that increases with increasing initial speeds. Thus, the width of the classical $100(scp)\%$ (two-sided) asymptotic statistical confidence interval that allegedly includes [mean($APRCRHNDSDDV$'s) given $isv_i = isv_i*$] increases with increasing initial speeds.

> *Remark*: Recall that, because the mean and median have the same value for a conceptual normal distribution, [mean ($APRCRH NDSDDV$'s) given $isv_i = isv_i*$] can also be written as $sd(50)$ given isv_isv*.

Observe that, given H_a and our example stopping distance datum values, est($csp0$) is negative. Thus, we now consider the following revised alternative hypothesis RH_a: the conceptual normal distribution heteroscedastic standard deviation is equal to $csp1 \cdot isv_i$. Microcomputer $RQNRM$ (revised quadratic normal regression statistical model) computes the respective est($clpj$'s) and est($cspj$'s) as well as the associated data-based value for the maximum estimated $\log_e$(likelihood) given RH_a.

Now let RH_a be considered the null hypothesis relative to H_a. The resulting data-based value for Bartlett's likelihood ratio test statistic generates an asymptotic null hypothesis rejection probability equal to 0.6843, viz.,

$$[2 \cdot (-155.38656) - 2 \cdot (-155.46923)] = 0.16534 = \chi^2_{1;p}$$

Accordingly, the conceptual normal distribution heteroscedastic standard deviation $csp1 \cdot isv_i$ cannot rationally be rejected in favor of the (physically noncredible) conceptual normal distribution heteroscedastic standard deviation $(csp0 + csp1 \cdot isv_i)$. We thus adopt the RH_a conceptual statistical stopping distance statistical model.

```
C> COPY DATAHOM DATA

C> QNRMHOSD
```

Presumed Conceptual Normal Distribution Mean $= clp1 \cdot isv_i + clp2 \cdot isv_i^2$:

Presumed Conceptual Normal Distribution Homoscedastic Standard Deviation $= csp$

$\text{est}(clp1) = 0.555256\text{D} + 00$
$\text{est}(clp2) = 0.626918\text{D} - 01$
$\text{est}(csp) = 0.966363\text{D} + 01$

Estimated Maximum $\log_e(\text{Likelihood})] = -0.17440727\text{D} + 03$

```
C> COPY DATAHEM DATA

C> QNRMHESD
```

Presumed Conceptual Normal Distribution Mean $= clp1 \cdot isv_i + clp2 \cdot isv_i^2$:

Presumed Conceptual Normal Distribution Heteroscedastic Standard Deviation $= csp0 + csp1 \cdot isv_i$

$\text{est}(clp1) = 0.655211\text{D} + 00$
$\text{est}(clp2) = 0.587758\text{D} - 01$
$\text{est}(csp0) = 0.300930\text{D} + 00$
$\text{est}(csp1) = 0.464066\text{D} + 00$

Estimated Maximum $\log_e(\text{Likelihood})] = -0.15538656\text{D} + 03$

```
C> COPY DATARHEM DATA

C> RQNRM
```

Presumed Conceptual Normal Distribution Mean $= clp1 \cdot isv_i + clp2 \cdot isv_i^2$:

Presumed Conceptual Normal Distribution Heteroscedastic Standard Deviation $= csp1 \cdot isv_i$

$\text{est}(clp1) = 0.646591\text{D}+00$
$\text{est}(clp2) = 0.591359\text{D}-01$
$\text{est}(csp1) = 0.441880\text{D}+00$

Estimated Maximum $\log_e(\text{Likelihood})] = -0.155469\text{D}+03$

8.F.1. Discussion

When the experiment test program datum values are presumed to be normally distributed and homoscedastic and when there are no suspended tests, then both covar[est($clp1$),est(csp)] and covar[est($clp2$),est(csp)] are equal to zero. Thus, given our example stopping distance datum values, the estimated asymptotic covariance matrix pertaining to H_n is:

$$\begin{vmatrix} 0.03821396 & -0.00126131 & 0.00000000 \\ -0.00126131 & 0.00004550 & 0.00000000 \\ 0.00000000 & 0.00000000 & 0.74115673 \end{vmatrix}$$

On the other hand, when the experiment test program datum values are presumed to be normally distributed and heteroscedastic, then these covariances are not necessarily equal to zero. For example, given our example stopping distance datum values, the estimated asymptotic covariance matrix pertaining to H_a is:

$$\begin{vmatrix} 0.01428673 & -0.00060385 & -0.01554849 & 0.00119109 \\ -0.00060385 & 0.00003262 & 0.00065202 & -0.00004995 \\ -0.01554849 & 0.00065202 & 0.50325707 & -0.03960520 \\ 0.00119109 & -0.00004995 & -0.03960520 & 0.00465823 \end{vmatrix}$$

8.F.2. Extension

Suppose we wish to assert that, based on the RH_a conceptual statistical stopping distance statistical model, we have approximately 95% statistical confidence that 99% of all possible replicate stopping distances will actually be less than some specific (calculated) stopping distance value given the $isv = isv^*$ of specific interest. First, we write

$$
\begin{aligned}
\text{est}[sd(99) \text{ given } isv = isv^*] &= \text{est}[\text{mean}(APRCRHNDSDDV\text{'s}) \\
&\qquad \text{given } isv = isv^*] \\
&\qquad + 2.3263 \cdot \text{est}(csp1)(isv^*) \\
&= \text{est}(clp1) \cdot (isv^*) + \text{est}(clp2) \cdot (isv^*)^2 \\
&\qquad + 2.3263 \cdot \text{est}(csp1) \cdot (isv^*)
\end{aligned}
$$

and then we use the propagation of variability methodology to generate the following asymptotic expression for the est(var{est[sd(99) given $isv = isv^*$]}):

$$
\begin{aligned}
\text{est}(\text{var}\{\text{est}[sd(99) \text{ given } isv = isv^*]\}) &= (isv^*)^2 \cdot \text{est}\{\text{var}[\text{est}(clp1)]\} \\
&\quad + (isv^*)^4 \cdot \text{est}\{\text{var}[\text{est}(clp2)]\} \\
&\quad + 2 \cdot (isv^*)^3 \cdot \text{est}\{covar[\text{est}(clp1), \\
&\qquad \text{est}(clp2)]\} \\
&\quad + (2.3263)^2 \cdot (isv^*)^2 \cdot \text{est}\{\text{var}[\text{est}(csp1)]\}
\end{aligned}
$$

In turn, we obtain numerical values for est{var[est($clp1$)]}, est{var[est($clp2$)]}, est{covar[est($clp1$),est($clp2$)]}, and est{var[est($csp1$)]} from the following estimated asymptotic covariance matrix:

$$
\begin{vmatrix}
0.01470763 & -0.00061199 & 0.00000000 \\
-0.00061199 & 0.00003226 & 0.00000000 \\
0.00000000 & 0.00000000 & 0.00154966
\end{vmatrix}
$$

The upper 95% (one-sided) asymptotic statistical confidence limit that allegedly bounds the actual value for the sd(99) is depicted in Figure 8.10. However, the plotted curve is extremely deceptive. This statistical confidence limit actually pertains only to the single initial speed value isv^* of specific interest. Moreover, the credibility of this upper 95% (one-sided) asymptotic statistical confidence limit is dubious at low speeds, say below 12 to 15 miles per hour, because the associated 90% (two-sided) asymptotic statistical confidence interval improperly includes negative values for the actual value of the stopping distance.

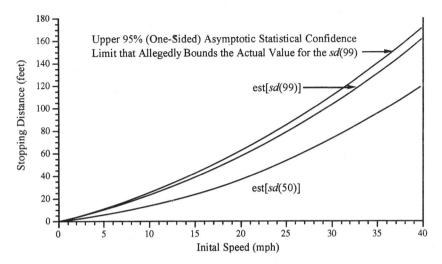

Figure 8.10 Estimated quadratic conceptual stopping distance statistical model defined by RH_a. At 40 mph, the ML estimate of the actual value for the $sd(50)$ is 120.48 ft, and the upper 95% (one-sided) asymptotic statistical confidence limit that allegedly bounds the actual value for the $sd(99)$ is 172.15 ft. Accordingly, we can say with approximately 95% statistical confidence that 99% of all replicate stopping distance datum values will actually be less than or equal to 172.15 ft.

Exercise Set 9

These exercises are intended to support the numerical results pertaining to H_n, H_a, and RH_a to provide perspective regarding the likelihood ratio test for the adequacy of a conceptual statistical model.

1. Given H_n, state the appropriate first and second derivative expressions of the $\log_e$(likelihood) with respect to the conceptual parameters and verify analytically that the respective asymptotic covariances of the estimated conceptual homoscedastic standard deviation and the estimated conceptual location parameters are equal to zero.

2. Given H_a, state the appropriate first and second derivative expressions of the $\log_e$(likelihood) with respect to the conceptual parameters and verify analytically that the respective asymptotic covariances of the estimated conceptual heteroscedastic standard deviation and the estimated conceptual location parameters are not equal to zero.

3. Given RH_a, state the appropriate first and second derivative expressions of the $\log_e$(likelihood) with respect to the conceptual parameters and verify analytically that the actual values for the respective asymptotic covariances of the estimated heteroscedastic standard deviation and the estimated conceptual location parameters are equal to zero.

Exercise Set 10

These exercises are intended to provide perspective regarding ML analysis by reconsidering the simple linear regression conceptual statistical model.

1. (a) Given the following the conceptual simple linear regression statistical model:

$$CRHNDRDV_i\text{'s} = \text{mean}_i(APRCRHNDRDV\text{'s}) + CRHNDREE_i\text{'s}$$
$$= \quad clp0 + clp1 \cdot ivv_i \quad\quad + CRHNDREE_i\text{'s}$$

first write the standardized conceptual normal distribution variate as $y_i = (clp0 + clp1 \cdot ivv_i)/csp$ and then state the joint PDF for n_{rdv} regression datum values. Next, (b) restate this joint PDF expression as likelihood and analytically establish expressions for the ML estimates of the actual values for the parameters in the conceptual simple linear regression model. In turn, (c) compare these ML estimation expressions to the corresponding least-squares estimation expressions. Then, (d) suggest an *ad hoc* multiplicative statistical bias correction factor for the ML estimate of the actual value for the *csp*. Finally, (e) do you recommend using a *scp*-based value for Student's central $t_{1,n_{dsdf}=n_{rdv}-n_{clp}}$ variate instead of the asymptotically correct *scp*-based value for the standardized normal y variate to compute the upper $100(scp)\%$ (one-sided) asymptotic statistical confidence limit for the actual value of mean($APRCRHNDRDV$'s) give $ivv = ivv^*$? Discuss.

2. This exercise extends the ML analysis begun in Exercise 1. (a) Derive analytical expressions for the respective elements of the estimated asymptotic covariance matrix. Then, (b) compare these expressions to the corresponding variance and covariance expressions pertaining to simple linear regression and comment appropriately.

References

Abramowitz M, Stegun IA. Handbook of Mathematical Functions. National Bureau of Standards Applied Mathematics Series No. 55. Washington, DC: US Government Printing Office, 1964.

Box GEP, Hunter WG, Hunter JS. Statistics for Experimenters. New York: John Wiley, 1978.

Draper NR, Smith H. Applied Regression Analysis. New York: John Wiley, 1966.

Dumonceaux R, Antle CE. Discrimination between the log-normal and Weibull distributions. Technometrics 15: 923–926, 1973.

Ezekiel M, Fox KA. Methods of Correlation and Regression Analysis, Linear and Curvilinear. 3rd ed. New York: John Wiley, 1963.

Hahn GJ, Shapiro SS. Statistical Models in Engineering. New York: John Wiley, 1967.

Kastenbaum MA, Hoel DG, Bowman KO. Sample size requirements: randomized block designs. Biometrika 57: 573–577, 1970.

Knuth DE. The Art of Computer Programming: Vol. II, Seminumerical Methods. Reading, MA: Addison-Wesley, 1969.

Little RE. The up-and-down method for small samples with two specimens "in series." J Am Stat Assoc 70: 846–851, 1975.

Little RE. Tables for Estimating Median Fatigue Limits. Special Technical Publication 731. Philadelphia: American Society for Testing and Materials.

Little RE. Optimal stress amplitude selection in estimating median fatigue limits using small samples. ASTM, J Test Eval, 18: 115–122, 1990.

Little RE. Effect of specimen thickness on the long-life fatigue performance of a randomly-oriented continuous-strand glass-mat-reinforced polypropylene composite. ASTM, J Test Eval 25: 491–496, 1997.

Little RE, Jebe EH. Statistical Design of Fatigue Experiments. London: Applied Science Publishers, 1975.

Little RE, Thomas JJ. Up-and-down methodology applied to statistically planned experiments. J Test Eval 21: 14–20, 1993.

Mann NR, Fertig KW. Table for obtaining Weibull confidence bounds and tolerance bounds based on best linear invariant estimates of parameters of the extreme-value distribution. Technometrics 15: 87–101, 1973.

Mantel N. Calculation of scores for a Wilcoxon generalization applicable to data subject to arbitrary right censorship. Am Statistician 35: 244–247, 1981.

Michael JR. The stabilized probability plot. Biometrika 70: 11–17, 1983.

Natrella MG. Experimental Statistics. National Bureau of Standards Handbook 91. Washington, Dc: US Government Printing Office, 1963.

Noreen EW. Computer-Intensive Methods for Testing Hypotheses: An Introduction. New York: John Wiley, 1989.

Parzen E. Modern Probability Theory and its Applications. New York: John Wiley, 1960.

Snedecor GW, Cochran WG. Statistical Methods. 6th ed. Ames, IA: Iowa State University Press, 1967.

Tukey JW. Sunset salvo. Am Statistician 40: 72–76, 1986.

Wichmann BA, Hill ID. Algorithm AS183: an efficient and portable pseudorandom number generator. Appl Stat 31: 188–190, 1981.

Index

Alternative hypotheses, 2
  omnibus, 231
  omnibus composite, 245
  simple (one-sided), 88
  specific simple (one-sided), 245
  specific composite (two-sided), 245
ANOVA, 231–307
  ctesc's, 232
  enumeration-based, 288
  equally replicated CRD experiment
    test program, 246
  fixed effects, 231
  normality, 278
  randomization-based, 288
  snedecor's central F statistic, 232
  summary and perspective, 275
  unreplicated RCBD experiment test
    program, 247
  unreplicated SPD experiment test
    program, 262
Arithmetic averages, 19
Asymptotic sampling distributions, 124

Asymptotically unexcelled, 410

Batches, 33
Batch-to-batch effects, 16, 305
between(MS), 231
Bias (*see also* Statistical bias), 12
Binomial distribution, 132
Block design experiment test programs,
  57
Blocking (*see also* Planned grouping),
  9, 10, 63–68
Blocks (*see also* Time blocks), 12, 28
Bogey, 74
  analytical, 387
  experimental, 387

Classical:
  analysis of variance, 231
  distribution-free (non-parametric) *A*-
    basis, *B*-basis statistical
    tolerance limits, 472

581